Power Systems

Federico Milano

Power System Modelling and Scripting

Springer

Dr. Federico Milano
ETSII, University of Castilla - La Mancha
13071, Ciudad Real
Spain
E-mail: Federico.Milano@uclm.es

ISSN 1612-1287 e-ISSN 1860-4676
ISBN 978-3-642-13668-9 e-ISBN 978-3-642-13669-6
DOI 10.1007/978-3-642-13669-6
Springer London Dordrecht Heidelberg New York

British Library Cataloguing in Publication Data
A catalogue record for this book is available from the British Library

Library of Congress Control Number: 2010928724

Cover Design: deblik, Berlin, Germany

Printed on acid-free paper

Springer is part of Springer Science+Business Media (www.springer.com)

To Yolanda and Alessandro

Τί δῆτα, ὦ ξένε, εἴδωλον ἄν φαῖμεν
εἶναι πλήν γε τὸ πρὸς τἀληχινόν
ἀφωμοιωμένον ἕτερον τοιοῦτον;

Plato, *Sophist*, 365-361 B.C.

2.1	We make ourselves pictures of facts.
2.12	The picture is a model of reality.
2.225	There is no picture which is a priori true.

Ludwig Wittgenstein, *Tractatus Logico-Philosophicus*, 1922 A.D.

Preface

History the Book

The first draft of these notes was born in the winter of 2002. At that time, I was a visiting scholar at the University of Waterloo. Originally, those notes were not intended as a book, but as a quick reference for not forgetting the models I was implementing for my research. After eight years, I am with Universidad de Castilla-La Mancha. During these years, the notes have been growing up little by little, ceaselessly. During the summer of 2009, I have reorganized the notes in the present book.

Justification of the Title

Power system modelling and scripting is a quite general and ambitious title. Of course, to embrace all existing aspects of power system modelling would lead to an encyclopedia. Thus, the book focuses on a subset of power system models based on the following assumptions: (i) devices are modelled as a set of nonlinear differential algebraic equations, (ii) all alternate-current devices are operating in three-phase balanced fundamental frequency, and (iii) the time frame of the dynamics of interest ranges from tenths to tens of seconds. These assumptions basically restrict the analysis to transient stability phenomena and generator controls. The modelling step is not self-sufficient. Mathematical models have to be translated into computer programming code in order to be analyzed, understood and "experienced". It is an object of the book to provide a general framework for a power system analysis software tool and hints for filling up this framework with versatile programming code.

Objectives of the Book

This book is for all students and researchers that are looking for a quick reference on power system models or need some guidelines for starting the challenging adventure of writing their own code. Thus, the objectives of this book are twofold.

The primary objective is to provide a selection of the most used device models ranging from static models for power flow, continuation power flow and optimal power flow analyses to as complete as possible dynamic electromechanical models for small-signal stability analysis and time domain simulations. This selection includes classical devices (e.g., synchronous machines) as well as non-conventional distributed energy resources (e.g., wind turbines), static voltage dependent loads as well as emerging energy storage devices. While describing each device, no matter if it is a well-known PV bus or a very specific pitch angle control for wind turbines, the focus is on the model hypotheses and on the implications of adopted simplifications.

The second objective is to provide a guide for organizing and translating mathematical models into computer programming code. The purpose is that the reader understands that there is always a gap between printed equations and software applications running on computers. Fortunately, this gap is not so huge and the book attempts to provide the methodological approach to fill it.

Choice of the Programming Language

When dealing with programming issues, one has to face and answer a tricky question: which is the most adequate computer language for tackling power system analysis? Then, after deciding on the language, one already knows that in a decade that language will be inevitably obsolete and a newer, easier, classier language will be available. To avoid a quick obsolescence, the goal of the book is not to provide code, but rather to teach how to design, organize and eventually write it. Programming issues will be always the same, at least as far as power systems will be the way they are. Thus, the adopted language is not so important.

At the end of a careful one-year-long study, I finally opted for the Python programming language. This language is well documented on the Internet, is elegant and neat, is fully based on classes and provides efficient libraries for solving linear algebra, handling sparse matrices and producing publication quality figures. Last but not least, the Python interpreter is free and open source. These characteristics do not guarantee that Python will last forever, but make it very appropriate for educational purposes.

Organization of the Book

The material included in this book is organized in a somewhat unorthodox way. Since the purpose is to concentrate on modelling, main power system analysis tools and basic programming concepts are introduced before describing the devices. The book is organized in five parts, as follows.

Part I contains introductory concepts. Chapter 1 provides the motivation of the book, some *philosophical* foundations of the art of modelling physical systems and defines the general mathematical model used for describing the behavior of power systems. Chapter 2 introduces the structure and the features of a software package for power system analysis while Chapter 3 discusses on the concept of *scripting* applied to power system analysis. The latter chapter also attempts to provide general guidelines for *thinking* power systems analysis in terms of computer programming. I hope that the results can be useful for Ph.D. students that, at the very end, will be the only readers of this book that have time to implement their own software applications.

Part II introduces basic tools for power system analysis. The viewpoint used for describing these tools is as general as possible. Chapter 4 describes the power flow analysis, Chapter 5 the continuation power flow, Chapter 6 the optimal power flow, Chapter 7 the small signal stability analysis, and Chapter 8 the time domain integration. Each topic is huge and, thus, only a very reduced selection of methods and algorithms is presented. The object is to provide a starting point for further investigations as well as a basement on top of which the following part dedicated to device modelling can be built.

Part III is the barycentric and most extended part of the book. It embraces the most important families of power system devices in an as systematic and exhaustive way as possible. Chapter 9 provides an introduction to the basic mathematical aspects of a generic electrical device. Following Chapters from 10 to 20 describe static power flow devices, transmission lines, static and regulating transformers, optimal power flow models, faults, protections, measurement devices, non-conforming static and dynamic loads, synchronous and induction machines, primary frequency and voltage regulators and power system stabilizers, dc devices, ac-dc devices, FACTS devices, and wind turbines and other distributed energy resources.

Part IV discusses spare topics that are relevant for power system analysis but are seldom included in power system books. Chapter 21 introduces the variegated world of data formats and discusses the challenges for creating a common model for exchanging power system data. Chapter 22 discusses the advantages of the Unix-style command line approach versus graphical user interfaces. Chapter 22 also describes plotting utilities aimed to power system visualization ranging from conventional plots to advanced 2D and 3D temperature maps. Chapter 23 describes some relevant educational aspects of free and open source power system software packages.

Finally Part V contains supporting material in form of appendices. Appendix A provides a minimal introduction to the Python non-standard scientific libraries used in the book. The aim of Appendix A is to make the book as self-contained as possible. Appendix B defines Python structures and classes that are used in the examples of the book. Appendix C discusses control diagrams and hard limit models. Finally, Appendix D provides the power system data used in the example of previous chapters whereas Appendix E describes the software requirements for working with the book as well as some useful links related to power system analysis.

Style of the Book

The style used in the book is somewhat unconventional with respect to traditional references about power system analysis. The will of merging together two worlds, namely power system modelling and computer programming for computational science, leads to the necessity of using a hybrid style that is unusual for both worlds. The major risk is perhaps to end up writing a software manual. To avoid that, I have tried to be as rigorous as possible and to make the examples based on computer code a supporting material rather than an essential part of the book, so that readers that despise computer code can skip it. I have also tried to apply the lesson of the Venikov's "Theory of Similarity and Simulation" [325]: whenever possible, I have included analogies and similarities taken from any mathematical and scientific field.

The material is organized in several parts, each part in several chapters and each chapter in several sections and subsections. This fragmentation can remind Seneca's style *arena sine calce* (i.e., sand without concrete) and is a kind of deformation due to the habit of object-oriented programming. However, this style is also dictated by the hope that in this way each topic can be easily found and fixed in mind.

For those interested in very technicalities, to write this book, I used LaTeX 3 with some useful packages such as PSfrag for the fine adjusting of figures and the IEEE style for formatting the bibliography data base. Python 2.6.2 was used as main environment while modules CVXOPT 1.1.2 and NumPy 1.3 were used for linear algebra, sparse matrix and eigenvalue analysis. Matplotlib 0.99 was used for generating simulation plots and Xfig 3.2.5 for drawing all other figures.

Acknowledgments

There is a beautiful Italian word that defines someone able to teach such that he changes someone else life and makes it irremediably better. This word is *maestro*. I have been lucky enough to have good ones: my grandfather Cesare, my father Guido and my mother Silvana, Profs. Bruno Delfino, Gio Battista Denegri and Marco Invernizzi from Università degli Studi di Genova, Prof. Claudio Cañizares from University of Waterloo and Prof. Antonio Conejo from Universidad de Castilla-La Mancha.

Concluding Remark

While completing this preface, I realize that much material has been left out of the book. However, I hope that what is included will be enough to transmit to the reader my passion for power system modelling and scripting. The book will accomplish its ultimate object if the next time the reader looks at some differential algebraic equations defining a power system device, he or she will be seized by a vague intellectual pleasure and a subtle ardent curiosity.

Waterloo, Genova, Ciudad Real 2002-2010

Contents

Part V: Appendices

List of Figures

List of Tables

List of Examples

List of Scripts

Notation

To set up a complete list of symbols for a book of this kind is a complex task. The variety of models and physical quantities used in Part III leads to a huge list of symbols. On the other hand, Parts I and II present general mathematical concepts that require a careful notation consistency. Hence, two notation approaches are used throughout the book.

The first notation concerns common quantities and general mathematical functions and variables. This notation is common to the whole book and is followed as rigorously as possible.

The second notation concerns local parameters that are needed to describe each device model included in Part III. These parameters are defined the first time they appear or gathered in tables. The organization of the book in several chapters helps avoid confusion and allows keeping each device model or group of models well separated from the others.

General Notation Rules

General notation rules are as follows.

1. Scalar functions, variables and parameters expressed in pu are in lower case Latin fonts. For example: x, v, p.
2. Angles and angular speeds are given in lower case Greek fonts. For example: θ, δ, α.
3. Upper case Latin fonts indicate scalar variables and parameters expressed in absolute values. For example: S_n [MVA], T_1 [s].
4. Function and variable vectors are in lower case, bold face fonts. For example: $\boldsymbol{f}$, $\boldsymbol{g}$, $\boldsymbol{x}$.
5. Jacobian matrices are in lower case, bold face fonts with a sub-index that indicates the variables with respect to which derivatives are computed. For example: $\boldsymbol{g_y} = \nabla_{\boldsymbol{y}}^T \boldsymbol{g}$.
6. Hessian matrices are indicated as Jacobian ones but with two sub-indexes. For example: $\boldsymbol{g_{yy}} = \nabla_{\boldsymbol{yy}}^T \boldsymbol{g}$.

7. Other matrices are in upper case, bold face fonts. For example: $\boldsymbol{A}$, $\boldsymbol{I}_n$, $\boldsymbol{W}$.
8. Sets are indicated by calligraphic fonts. For example: $\mathcal{I}$, $\mathcal{N}$.
9. Iterations are indicated with a superscript in brackets. For example: $\boldsymbol{x}^{(i)}$, $\boldsymbol{y}^{(i+1)}$, $\boldsymbol{v}^{(0)}$. The latter indicates the initial guess.
10. The subscript 0 indicates the equilibrium point or the initial value. For example: $\boldsymbol{x}_0 = \boldsymbol{x}(t_0)$.
11. A bar on top of a symbol indicates a phasor. For example: $\bar{v} = ve^{j\theta}$.
12. In case of complex quantities, a superscript asterisk indicates the conjugate. For example: $\bar{v}^* = ve^{-j\theta}$. In case of real quantities, a superscript asterisk indicates the optimal value. For example: p_w^*.
13. A superscript T indicates transpose. For example: $\boldsymbol{A}^T$.
14. Verbatim fonts are used to indicate Python scripts. For example: `list`, `class`. The symbol >>> indicates the Python interactive command line prompt.

Frequent Symbols

The following is only a selection of most common quantities used in the book.

Functions and Equations

$\boldsymbol{f}$ differential equations.
$\boldsymbol{g}$ algebraic equations.
$\boldsymbol{h}$ inequality constraints.
$\mathscr{L}$ Lagrangian function.
$\hat{\boldsymbol{q}}^{(i)}$ vector used for implicit numerical methods.
$\boldsymbol{r}$ reduced gradient.
s differential operator in block diagrams ($s = \frac{d}{dt}$).
ϱ continuation equation.
φ objective function.
$\boldsymbol{\varphi}$ general vector of differential equations.
$\boldsymbol{\psi}$ homotopy map.

Variables and Parameters

a_m converter modulating amplitude.
$\mathscr{E}$ energy.
k_G distributed slack bus variable.
ℓ transmission line length.
m transformer tap ratio.
p_h, $\boldsymbol{p}$ active powers.
q_h, $\boldsymbol{q}$ reactive powers.
$\boldsymbol{s}$ slack variables.
t time.
$\boldsymbol{u}$ discrete variables.

v_h, $\boldsymbol{v}$ bus voltage magnitudes.
$\boldsymbol{w}$ left eigenvector.
$\boldsymbol{x}$ state variables.
$\dot{\boldsymbol{x}}$ first time derivatives of state variables.
$\boldsymbol{y}$ algebraic variables.
$\boldsymbol{z}$ compound vector of variables (e.g., $\boldsymbol{z} = [\boldsymbol{x}^T, \boldsymbol{y}^T]^T$).
α firing angle.
δ_j, $\boldsymbol{\delta}$ generator rotor angles.
$\boldsymbol{\eta}$ controllable variables or parameters.
θ_h, $\boldsymbol{\theta}$ bus voltage angles.
Θ temperature.
λ eigenvalue.
μ loading level (continuation parameter).
$\hat{\mu}$ barrier parameter for the interior point method.
$\boldsymbol{\mu}$ independent variables or uncontrollable parameters.
$\boldsymbol{\nu}$ right eigenvector.
$\boldsymbol{\xi}$ general vector of state variables.
$\boldsymbol{\pi}$ dual variables associated with inequality constraints.
$\boldsymbol{\rho}$ dual variables associated with equality constraints.
$\boldsymbol{\tau}$ tangent vector.
τ_e electrical torque.
τ_m mechanical torque.
ϕ transformer phase shift.
ω_j, $\boldsymbol{\omega}$ generator rotor speeds.

Matrices

$\boldsymbol{A}$ generic constant element matrix.
$\boldsymbol{A}_C$ complete system Jacobian matrix.
$\boldsymbol{A}_c^{(i)}$ matrix defined for implicit numerical methods.
$\boldsymbol{A}_S$ state matrix.
$\boldsymbol{B}$ admittance matrix used in dc power flow.
$\boldsymbol{B}'$ and $\boldsymbol{B}''$ admittance matrices used in fast decoupled power flow.
$\boldsymbol{C}$ incidence matrix.
$\boldsymbol{f}_{\boldsymbol{x}}$ Jacobian matrix $\nabla_{\boldsymbol{x}}^T \boldsymbol{f}$.
$\boldsymbol{f}_{\boldsymbol{y}}$ Jacobian matrix $\nabla_{\boldsymbol{y}}^T \boldsymbol{f}$.
$\boldsymbol{G}$ nodal conductance matrix.
$\boldsymbol{g}_{\boldsymbol{x}}$ Jacobian matrix $\nabla_{\boldsymbol{x}}^T \boldsymbol{g}$.
$\boldsymbol{g}_{\boldsymbol{y}}$ Jacobian matrix $\nabla_{\boldsymbol{y}}^T \boldsymbol{g}$.
$\boldsymbol{N}$ matrix of right eigenvectors.
$\boldsymbol{T}$ connectivity matrix.
$\boldsymbol{W}$ matrix of left eigenvectors.
$\bar{\boldsymbol{Y}}$ admittance matrix.
$\boldsymbol{\Lambda}$ diagonal matrix of eigenvalues.
$\boldsymbol{\Sigma}$ diagonal matrix of singular values.

Constants

A surface area.
B, b_{h0}, b_{k0}, b_{hk} susceptances.
c_p specific heat.
D rotor damping.
f_n nominal frequency in Hz.
G, g_{h0}, g_{k0}, g_{hk} conductances.
H, H_m, H_t inertia constants.
K, K_0, K_i, K_p gains.
ℓ_t total transmission line length.
m_g mass.
R, r_{hk}, r_T resistances.
S_b base power.
T, T_a, T_1 time constants.
V_b base voltage.
X, x_{hk}, x_T reactances.
$\bar{z}$, $\bar{z}_T$ impedances.
$\bar{y}_{hk}$, $\bar{y}_T$ admittances.
Ω_b nominal synchronous speed in rad/s.
ω_s synchronous speed in pu.

Numbers

n_b number of ac buses.
n_c number of series connections or branches.
n_g number of algebraic functions $\boldsymbol{g}$.
n_G number of generators.
n_u number of discrete variables $\boldsymbol{u}$.
n_x number of state variables $\boldsymbol{x}$.
n_y number of algebraic variables $\boldsymbol{y}$.
n_z number of variables $\boldsymbol{z}$.
n_η number of controllable variables $\boldsymbol{\eta}$.
n_μ number of uncontrollable variables $\boldsymbol{\mu}$.
n_ξ number of general state variables $\boldsymbol{\xi}$.

Sets

$\mathcal{B}$ set of ac buses.
$\mathcal{C}$ set of series connections or branches.
$\mathcal{D}$ set of demands.
$\mathcal{G}$ set of generators.
$\mathcal{N}$ set of dc nodes.
$\mathcal{R}$ set of generator power reserves.

$\mathcal{S}$ set of supplies.
$\mathcal{T}$ set of times.
Ω set of all devices.
Ω_h set of devices connectes to bus h.

Device Model Notation

In Part III, the notation is aimed to maintain light expressions. The use of indexes is avoided wherever is possible. Device equations imply the device index i in all variables and parameters. Thus, v_h and θ_h indicate the bus voltage magnitude and phase angle, respectively, while p_h and q_h indicate the active and power injections, respectively. For series devices connected to two buses, the indexes h and k are used.

Bases for Per Unit Values

Throughout the whole book, the bases used for ac values are:

$S_b^{\rm ac}$ three-phase power in MVA.
$V_b^{\rm ac}$ phase-to-phase voltage in kV.

Thus, the current base $I_b^{\rm ac}$ and the impedance base $Z_b^{\rm ac}$ are:

$$I_b^{\rm ac} = \frac{S_b^{\rm ac}}{\sqrt{3} \cdot V_b^{\rm ac}}$$

$$Z_b^{\rm ac} = \frac{V_b^{\rm ac}}{\sqrt{3} \cdot I_b^{\rm ac}} = \frac{(V_b^{\rm ac})^2}{S_b^{\rm ac}}$$

For dc values, the following bases are used:

$S_b^{\rm dc}$ power in MW.
$V_b^{\rm dc}$ voltage in kV.

Thus, the current base $I_b^{\rm dc}$ and the resistance base $R_b^{\rm dc}$ are obtained as follows:

$$I_b^{\rm dc} = \frac{S_b^{\rm dc}}{V_b^{\rm dc}}$$

$$R_b^{\rm dc} = \frac{V_b^{\rm dc}}{I_b^{\rm dc}} = \frac{(V_b^{\rm dc})^2}{S_b^{\rm dc}}$$

For systems where there are both ac and dc devices, it is assumed that $S_b^{\rm dc} = S_b^{\rm ac}$.

Part I
Introduction

Chapter 1
Power System Modelling

This chapter introduces basic modelling concepts that are used throughout the book. Section 1.1 defines a power system and provides most relevant references related to power system analysis. Section 1.2 states the philosophical background of the book and general motivations. Section 1.3 presents programmatic assumptions and the proposed methodological approach for power system modelling. Finally, Section 1.4 defines the general equations that are used for modelling power systems.

1.1 Background

In essence, an electrical power systems is a set of interacting devices that transform primary energy sources, e.g., heat, into electricity and then transform electricity into another form of energy, e.g., the mechanical one. The electricity is transmitted at various voltage levels, through an adequate series of transformers, transmission lines and cables, from generators to loads. It is noteworthy to observe that electricity is never used as is. Therefore, the expression "electric power system" is somewhat incomplete, since a power system is essentially an energy conversion system. For this reason, the expression "power system" is used in this book, without specifying the form of the energy involved.

Energy is what engineering is all about. Thus, it is not surprising that interconnected power systems are the largest and most complex systems ever built by man. For example, Figure 1.1 depicts a simplified scheme of the UCTE high-voltage transmission system [318]. Fortunately, one does not need a real-world interconnected network to approach the study of power systems. Simple benchmark systems are often enough to understand the basic functioning as well as advanced topics of power systems. This remarkable property is exploited in the book to simplify as much as possible the examples.

However, even the simplest example requires some background. In the remainder, it is assumed that basic power system concepts are known by the reader. The level required for fully taking advantage of this book is that

F. Milano: Power System Modelling and Scripting, Power Systems, pp. 3–17.
springerlink.com

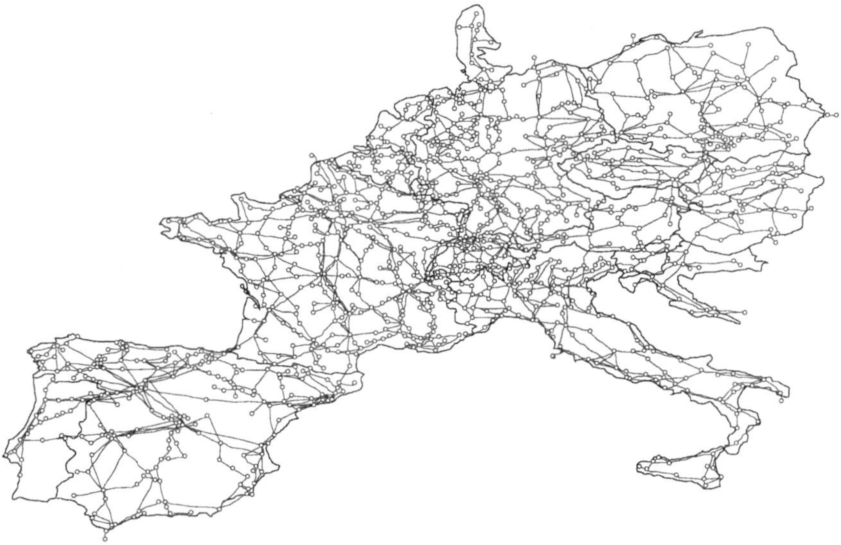

Fig. 1.1 UCTE interconnected system

provided by basic undergraduate courses on electrical machines and power systems. Moreover, several excellent books in the literature provide the fundamentals of power system operation, analysis, control and stability. Traditional references are [10, 52, 114, 160, 163, 179, 269, 294, 330, 355]. References that cover advanced topics while maintaining a comprehensive approach are [110, 147, 184, 263].

1.2 Motivations

As a student, I suffered the deep hiatus between the aforementioned references, which provide theoretical tools, and the commercial computer-based implementations of power system models, which are almost impenetrable black boxes. There is often a huge gap between the equations typeset on a book page and the code that represents those equations in a form suitable for a computer. Unfortunately, it is not easy to find books that tries to cover this gap. This is quite surprising since nowadays no one is really doing any calculation by hand, at least for power system analysis. Thus, given that a book like Venikov's "Transient Processes in Electrical Power Systems" is an evidence of what a pencil and a long Russian winter can inspire to a brilliant mind [330], it is time to systematically provide side by side to mathematical models the implementation of such models in some adequate computer language. Some examples of books that have attempted to follow such approach are [291] and [227] and, more recently [14], [17] and [2].

As a professor, I suffer, even more than when I was a student myself, when observing students accepting acritically the results provided by some software package. The idea that a computer only executes code and that the code has been written by some error-prone human being is too often beyond students' imagination. Students also very often ignore the old programmer saying *garbage in, garbage out*, which means that if the input data (provided by the student) are inconsistent, the computer cannot do the miracle and return correct results.

Although based on the well established background of the references cited in Section 1.1, the purpose of this book is not to concentrate on some specific detail of power systems. Rather, the main goal is to provide, through a variety of examples, very general tools for approaching a systematic study of power systems. These tools are both methodological (*modelling*), structural (*architecture*) and practical (*scripting*). The ultimate object is to help the reader develop the ability of approaching power system analysis in a both critical and constructive way. This chapter is dedicated to modelling issues, while architecture and scripting matters are tackled in Chapters 2 and 3, respectively.

1.3 Modelling Physical Systems

In the very first class of electrical circuits that I attended at the University of Genoa, the professor drew on the blackboard a scheme similar to the one depicted in Figure 1.2. The scheme shows the logical steps that an engineer has to follow to study a physical system. The first step is to define the model, which requires hypotheses and simplifications. Then, one has to formalize the model into equations. Finally, the equations have to be solved either in a closed form or numerically.

In this book, the "physical system" is a power system. In the past, analogical Transient Network Analyzers (TNAs) were the only simulation tools available for research and education in power engineering [10]. However, the advent of digital analysis has led to a more convenient way of performing simulations through digital computers [156]. Thus, in the book, it is assumed that the *model* is a simplified representation of the physical system suitable for being expressed in terms of mathematical equations and translated into computer programing code.

The scheme of Figure 1.2 is actually an old concept by Plato. In the "Sophist", Plato defines the science (e.g., the real knowledge) as τήν μετ' ἐπιστήμης ἰστοριχήν τινα μίμησιν (e.g., an imitation technique based on the experience). Thus, the engineer that formulates mathematical models corresponds to the philosopher. Actually, one of the goals of this book is to show that modelling does not consist simply in writing a set of equations but implies a "philosophical" view of physical phenomena.

A delicate point that has to be emphasized is that both simplifications and hypotheses are often driven by the numerical solution methods that are available for solving the given system equations. Even more critical is the importance of simplifications if one looks for an analytical explicit solution. Indeed, it is a pity that students do not solve anything by hand anymore.

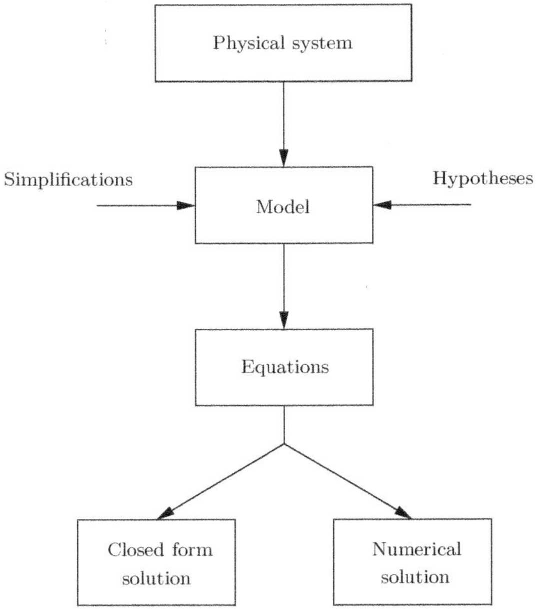

Fig. 1.2 General approach for studying a physical system

Good ideas often comes out just to avoid a repetitive work. In this regard, a computer programming saying states that very good programmers are often very lazy men.

In conclusion, the simple feed-forward scheme shown in Figure 1.2 has to be modified as in Figure 1.3. For the sake of realism, the analytical solution block has been removed from Figure 1.3. The key point of Figure 1.3 is that between the equations and the numerical solution blocks there is a key step, namely the translation of the equations in a suitable code using available algorithms. As any translation, programming is not a trivial step and cannot be neglected.

Example 1.1 Optimal Placement of Capacitor Banks

In order to describe with a qualitative example the modelling issues discussed above, consider the problem of finding the optimal bus positions of a given number n_c of capacitor banks that minimizes power system losses. The positions of the capacitor banks can be conveniently described by a vector $\boldsymbol{u}$ of order n_b of binary variables, where n_b is the number of system buses. The Boolean variable associated to each bus is 1 if the bank is installed at that bus, 0 otherwise. Since system losses are defined by the set of power flow equations that are intrinsically nonlinear, the resulting optimization problem belongs to the class of mixed integer nonlinear programming (MINLP) problems.

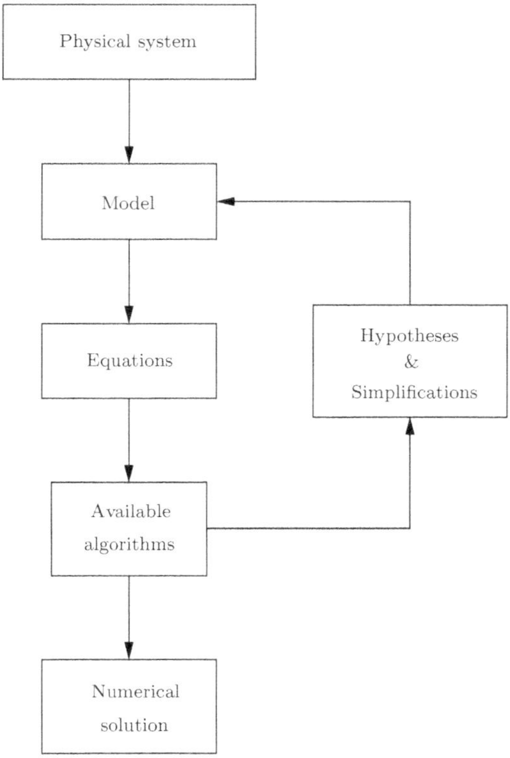

Fig. 1.3 Modified general approach for studying a physical system

This MINLP problem can certainly be solved, it is just a matter of time. In fact, an evident solution is to try all possible permutations $n_b!/(n_b - n_c)!$ of the vector $\boldsymbol{u}$. Unfortunately, even for small values of n_b and n_c, the number of permutations is huge and the required simulation time of trying all permutations is intractable. Actually, MINLP problems cannot be solved using a well-assessed and commonly accepted algorithm similar to, for example, the LU factorization. Thus, formulating the problem as a MINLP does not conclude the modelling phase. Further modelling effort is needed to make the problem solvable.

A common strategy is to adopt adequate simplifications or decomposition techniques to transform a MINLP either in a mixed integer linear programming (MILP) problem, thus eliminating nonlinearity, or in a nonlinear programming (NLP), thus eliminating integer variables.

In order to describe the need of the feedback included in Figure 1.3, assume, only for the sake of example, that $\boldsymbol{u}$ can be modelled as a vector of continuous variables through the following constraint:

$$u_i(u_i - 1) = 0, \quad \forall i \in 1, \ldots, n_b. \tag{1.1}$$

Equation (1.1) may lead to think that the original MINLP problem can be promptly converted into a nonlinear programming (NLP) one. Unfortunately equation (1.1) is a highly nonlinear constraint that generally leads to so severe convexity issues that it is practically impossible to find the global optimum. Thus this is an example of modelling inconsistency driven by a poor knowledge of NLP solvers.

Of course, in the literature, one can find a variety of methods for solving the proposed MINLP problem. In particular, it can be approached using a Benders' decomposition [204]. Other techniques based on artificial intelligence can be efficacious as well [169]. All these techniques attempt to reduce as much and as "smartly" as possible the number of permutations to be taken into account.

The first conclusion is that there is generally no easy shortcut for avoiding the complexity due to a combinatorial problem. The second conclusion, which is more relevant for the scope of this book, is that without an adequate knowledge of algorithms and numerical methods, one may set up an inconsistent model. Since engineers are often interested more in the solution than in the model itself, an unsolvable model is useless.

Example 1.2 Flyball Governor Model

As another qualitative example, consider the problem of the classical James Watt's centrifugal control (see Figure 1.4[1]). This control system allows controlling the speed of an engine by regulating the amount of fluid flow. The centrifugal control has been widely used for steam turbines of electric power generators for maintaining synchronism (e.g., flyball and isochronous governors). The main principle is as follows: a set of flyballs (or fly weights), which rotate driven by the engine to be controlled, is connected to levers whose position actuates on the steam valve (throttle). An increase in the engine speed results in a centrifugal force increase. The latter raises the weights and thereby reduces the steam flow into the engine. Finally, the steam flow decrease causes a reduction in the engine speed.

There is always a delay between the time at which the engine speed varies and the time at which the steam flow is reduced by the centrifugal control [182]. Actually, any steam turbine governor control intrinsically suffers of time delays. As a matter of fact, a thermal plant consists of several heat transfer apparatus connected together by pipes. An adequate modelling of transport delays introduced by these pipes is desirable to calculate system transient behavior. If the fluid velocity through the pipe is constant, the

[1] Picture obtained from Science Fair Project Encyclopedia, available at www.all-science-fair-projects.com

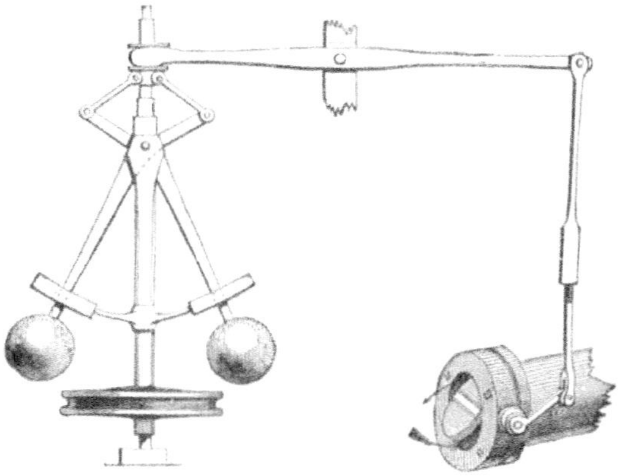

Fig. 1.4 Flyball governor

transfer function of the piping lag is also constant. However, if the velocity of the fluid is a function of time, the delays become time dependent.

An accurate modelling of variable time delays complicates considerably the analysis since it is not possible to define a standard transfer function of the system. As in Example 1.1, the main issue is the mathematical complexity inherent variable time delay (or *retarded*) systems. These belong to the class of *functional differential equations* (FDEs) that are infinite dimensional, as opposed to the finite dimension of ordinary differential equations (ODEs). In order to explain the complexities that raises when studying retarded systems, it suffices to say that, for FDEs, there is an infinite number of trajectories that passes for a given state $\boldsymbol{x}(t)$. As a matter of fact, in FDEs, the state is a function of a past time-interval $[t - \tau(t), t]$, where $\tau > 0$ is the time delay, which is possibly a function of time. Another difficulty is that it is not possible to define a state matrix for FDEs as it is usual practice for ODEs. For this reason research is in progress to assess the effects of time delays on system stability [258].

Regarding the steam turbine regulator, the common solution, that avoids time-delay issues, is to model the control as an ODE. As a matter of fact, time delays are often neglected and, along with other simplifications, the resulting differential equations used for modelling turbine governors are linear [10]. Further discussion on turbine governors is given in Chapter 16.

Example 1.3 Inductor Model

Any basic circuit theory course introduces the concept of inductance, whose well-known constitutive differential equation is

$$v = L\frac{di}{dt} \tag{1.2}$$

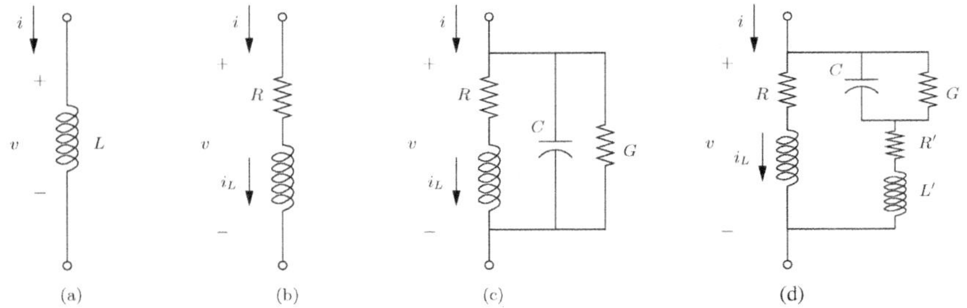

Fig. 1.5 Various detail degree models of a inductor winding: (a) ideal inductance; (b) fundamental frequency model; (c) second-order model; and (d) third-order model. The schemes only show electrical quantities

where L is a constant parameter called inductance and di/dt is the first time derivative of the current flowing in the device (see Figure 1.5.a).

However, ideal inductances do *not* exist. In other words, there is no physical device whose terminal voltage can be exactly described by the differential equation (1.2). The physical device whose behavior is close to an ideal inductance is the inductor. Inductors are made by a coil wound on an iron core. A winding is characterized by a resistance R, hence the model of a real device have to take into account Ohmic losses. Moreover, the magnetic properties of the iron are not ideal, hence the relation between the total magnetic flux ϕ and the electrical current i is not linear. A more realistic model of the inductive winding is as follows (see Figure 1.5.b):

$$v = \frac{d\phi}{dt} + Ri(\phi) \tag{1.3}$$

This model can be adequate in case of constant temperature and low frequency. Otherwise, one has to take into account the dependence of the resistance R on the temperature Θ and the capacitive effects among coils and between the coils an the ground. Such effects can be modeled using introducing a lumped parasitic (or hearting) capacity C in parallel with the inductance (see Figure 1.5.c). One can also include a conductance G in parallel with the capacitance to take into account the skin resistance of the winding. The resulting model is as follows:

$$\begin{aligned} v &= \frac{d\phi}{dt} + R(\Theta)i_L(\phi) \\ i &= C\frac{dv}{dt} + i_L(\phi) + Gv \\ Ri_L^2 &= m_g c_p \frac{d\Theta}{dt} + kA(\Theta - \Theta_a) \end{aligned} \tag{1.4}$$

where i_L is the current circulating in the winding, i the total current that flows in the inductor, m_g is the winding mass, c_p is the equivalent specific heat of the winding, k the convective heat transfer coefficient, A the winding active area and Θ_a the ambient temperature. Both the electrical and the thermodynamical models can be further complicated. For example, Figure 1.5.d shows a third-order model of an inductor [168].

The most precise electrical model that can be formulated for the inductor is based on Maxwell's and constitutive equations:

$$\begin{aligned} \nabla \times \vec{\boldsymbol{h}} = \vec{\boldsymbol{j}} + \frac{d\vec{\boldsymbol{d}}}{dt}, \quad \nabla \times \vec{\boldsymbol{e}} = -\frac{d\vec{\boldsymbol{b}}}{dt} \\ \nabla \cdot \vec{\boldsymbol{b}} = 0, \quad \nabla \cdot \vec{\boldsymbol{d}} = \varrho \\ \vec{\boldsymbol{b}} = \mu \vec{\boldsymbol{h}}, \quad \vec{\boldsymbol{j}} = \varsigma \vec{\boldsymbol{e}}, \vec{\boldsymbol{d}} = \varepsilon \vec{\boldsymbol{e}} \end{aligned} \tag{1.5}$$

where $\vec{\boldsymbol{h}}$ is the magnetic field, $\vec{\boldsymbol{b}}$ is the magnetic flux density, $\vec{\boldsymbol{e}}$ is the electric field, $\vec{\boldsymbol{d}}$ is the electric flux density, $\vec{\boldsymbol{j}}$ is the electric current density, ϱ is the charge density, μ is the magnetic permeability, ς is the electric conductivity, and ε is the electric permittivity. Equations (1.5) have to be particularized with adequate boundary conditions and, in general, have to be solved numerically using, for example, a finite-element three-dimensional approach [82].

Even without entering into the mathematical details of modelling the inductor as a continuum, one can easily anticipate that the computational burden of modelling an entire power system using such approach as well as the amount of information that should be processed is intractable. Too much information can result in no information. Even though in the literature there are some proposals of modelling a power system as a continuum [274, 308], the continuum approach remains so far only an academic, though intriguing, exercise.

The main conclusion that can be drawn by this example is that the upper boundary to the details that can be taken into account in a mathematical model is driven by the computational limits of available computers and by the ability of processing and analyzing results. As it is discussed in Part II, in the past, such limits have led to some approximations that are nowadays considered standards *de facto* but that are not really binding anymore.

1.4 Hybrid Dynamical Model

One should gather from the discussions and the examples provided in the previous sections that a power system consists of a set of devices that can be described by continuous dynamics as well as discrete events. Synchronous machines can be adequately modelled using a set of nonlinear ordinary differential equations (ODEs) as long as lumped parameters are used. Most controllers can also be modelled using ODEs as long as time-dependent delays are approximated with lags or simply neglected. Discrete events or, which is the same, discrete variables are also an essential part of power systems.

For example, the tap position of voltage regulating transformers and breaker statuses are discrete variables.

If one assumes that the status of discrete variables is known or can be *reasonably* guessed, then, for each discrete variable change, the system *jumps* from one set of continuous equations to another. In order to clarify this point, it suffices to think of the consequence of opening a breaker of a transmission line. Before opening the breaker the line is connected and the power flow through the line itself can be defined using well-known continuous equations. After opening the breaker, the line is disconnected and thus its equations have to be removed from the model.[2] Except for the fact that before and after the breaker intervention, power flow equations are different, there is no need of modelling the status of the breaker as a discrete variable. Of course, a completely different issue would be to find the best transmission line arrangement that minimize a certain aspect of the power system, say, for example, losses. In this case the problem is again combinatorial and breaker statuses have to be used as Boolean variables. However, while simulating the power system behavior, the status of the lines (as well as of most devices) is known and one has not have to deal with a MINLP problem.

The most general model that can be used for representing power systems is as follows:

$$\dot{\boldsymbol{\xi}} = \boldsymbol{\varphi}(\boldsymbol{\xi}, \boldsymbol{u}, t) \tag{1.6}$$

where $\boldsymbol{\xi}$ ($\boldsymbol{\xi} \in \mathbb{R}^{n_\xi}$) is the vector of state variables, $\boldsymbol{u}$ ($\boldsymbol{u} \in \mathbb{R}^{n_u}$) the vector of discrete variables, t ($t \in \mathbb{R}^+$) the time, and $\boldsymbol{\varphi}$ ($\boldsymbol{\varphi} : \mathbb{R}^{n_\xi} \times \mathbb{R}^{n_u} \times \mathbb{R}^+ \mapsto \mathbb{R}^{n_\xi}$) the vector of differential equations. As previously discussed, $\boldsymbol{u}$ have generally not to be determined but are fixed to a given value and then changed to enforce system transients. For example, opening a breaker can be required for clearing a fault and can be driven by an over-current protection. If one want to avoid using the vector $\boldsymbol{u}$, (1.6) can be rewritten as follows:

$$\dot{\boldsymbol{\xi}} = \begin{cases} \boldsymbol{\varphi}^-(\boldsymbol{\xi}, t) & \text{if} \quad t < t_i \\ \boldsymbol{\varphi}^+(\boldsymbol{\xi}, t) & \text{if} \quad t \geq t_i \end{cases} \tag{1.7}$$

where t_i is the breaker intervention time. Clearly, if more than one discrete variable changes value, the if-then rule is more complex. In conclusion, whenever is possible to transform a discrete variable u as an if-then condition, the discrete variable can be eliminated from the equation at the cost of splitting $\boldsymbol{\varphi}$ as in (1.7).

It has to be noted that, in (1.7), both $\boldsymbol{\varphi}^-$ and $\boldsymbol{\varphi}^+$ are continuous and smooth (i.e., differentiable) functions. This fact is important since one can use any numerical method for nonlinear equations (e.g., Newton's method) and standard stability analysis (e.g., eigenvalue analysis) when studying (1.7).

---

[2] The case of under load tap changers is slightly more complex and is discussed in Subsection 11.2.2 of Chapter 11.

Equations (1.6) as well as (1.7) imply that any variable of the system is a *state*. This assumption is in agreement with the Leibniz's axiom: *natura non facit saltus* (i.e., nature does not jump). This is certainly an excellent approximation for macroscopic dynamical systems. However, when dealing with power system analysis the time constants of different phenomena can span a wide time range. Figure 1.6 depicts the time scales of a variety of typical phenomena and controls that are of interest in power system analysis. Since the time scales span from micro-seconds to several hours, taking into account the dynamics of each phenomena in an unique model would results in an intractable system.

The common solution to this issue is to divide the vector of general state variables $\boldsymbol{\xi}$ into three sub-vectors: (i) state variables $\boldsymbol{\xi}_s$ characterized by slow dynamics (i.e., *big* time constants), (ii) state variables $\boldsymbol{\xi}_i$ whose dynamics are of interest, and (iii) state variables $\boldsymbol{\xi}_f$ characterized by fast dynamics (i.e., *small* time constants). In mathematical terms, one has:

$$\begin{aligned} \dot{\boldsymbol{\xi}}_s &= \boldsymbol{\varphi}_s(\boldsymbol{\xi}_s, \boldsymbol{\xi}_i, \boldsymbol{\xi}_f, \boldsymbol{u}, t) \\ \dot{\boldsymbol{\xi}}_i &= \boldsymbol{\varphi}_i(\boldsymbol{\xi}_s, \boldsymbol{\xi}_i, \boldsymbol{\xi}_f, \boldsymbol{u}, t) \\ \dot{\boldsymbol{\xi}}_f &= \boldsymbol{\varphi}_f(\boldsymbol{\xi}_s, \boldsymbol{\xi}_i, \boldsymbol{\xi}_f, \boldsymbol{u}, t) \end{aligned} \tag{1.8}$$

where $\boldsymbol{\xi} = [\boldsymbol{\xi}_s^T, \boldsymbol{\xi}_i^T, \boldsymbol{\xi}_f^T]^T$. In (1.8), the time evolution of $\boldsymbol{\xi}_s$ can be considered so slow that their variations can be neglected (e.g., $\boldsymbol{\xi}_s$ are *frozen* at a given value). On the other hand, the dynamics of $\boldsymbol{\xi}_f$ can be considered so fast that their variations can be considered instantaneous.

The consequences of the approximation described above are twofold: $\boldsymbol{\xi}_s$ are constant and can be considered controllable parameters; and the state variables $\boldsymbol{\xi}_f$ can be considered algebraic variables, i.e., variables whose variations are instantaneous. Thus, algebraic variables can *facere saltus* (i.e., jump) from one value to another and can be discontinuous.

The resulting model is thereby a set of nonlinear differential algebraic equations (DAEs) with discrete variables, as follows:

$$\begin{aligned} \dot{\boldsymbol{x}} &= \boldsymbol{f}(\boldsymbol{x}, \boldsymbol{y}, \boldsymbol{\eta}, \boldsymbol{u}, t) \\ \boldsymbol{0} &= \boldsymbol{g}(\boldsymbol{x}, \boldsymbol{y}, \boldsymbol{\eta}, \boldsymbol{u}, t) \end{aligned} \tag{1.9}$$

where $\boldsymbol{x}$ ($\boldsymbol{x} \in \mathbb{R}^{n_x}$) indicates the vector state variables, $\boldsymbol{y}$ ($\boldsymbol{y} \in \mathbb{R}^{n_y}$) are the algebraic variables, $\boldsymbol{\eta}$ ($\boldsymbol{\eta} \in \mathbb{R}^{n_\eta}$) are the controllable parameters, $\boldsymbol{f}$ ($\boldsymbol{\varphi} : \mathbb{R}^{n_x} \times \mathbb{R}^{n_y} \times \mathbb{R}^{n_\eta} \times \mathbb{R}^{n_u} \times \mathbb{R}^+ \mapsto \mathbb{R}^{n_x}$) are the differential equations, and $\boldsymbol{g}$ ($\boldsymbol{\varphi} : \mathbb{R}^{n_x} \times \mathbb{R}^{n_y} \times \mathbb{R}^{n_\eta} \times \mathbb{R}^{n_u} \times \mathbb{R}^+ \mapsto \mathbb{R}^{n_y}$). Comparing (1.9) with (1.8), the following correspondences hold:

$$\boldsymbol{\eta} \equiv \boldsymbol{\xi}_s, \quad \boldsymbol{x} \equiv \boldsymbol{\xi}_i, \quad \boldsymbol{y} \equiv \boldsymbol{\xi}_f, \quad \boldsymbol{f} \equiv \boldsymbol{\varphi}_i, \quad \boldsymbol{g} \equiv \boldsymbol{\varphi}_f \tag{1.10}$$

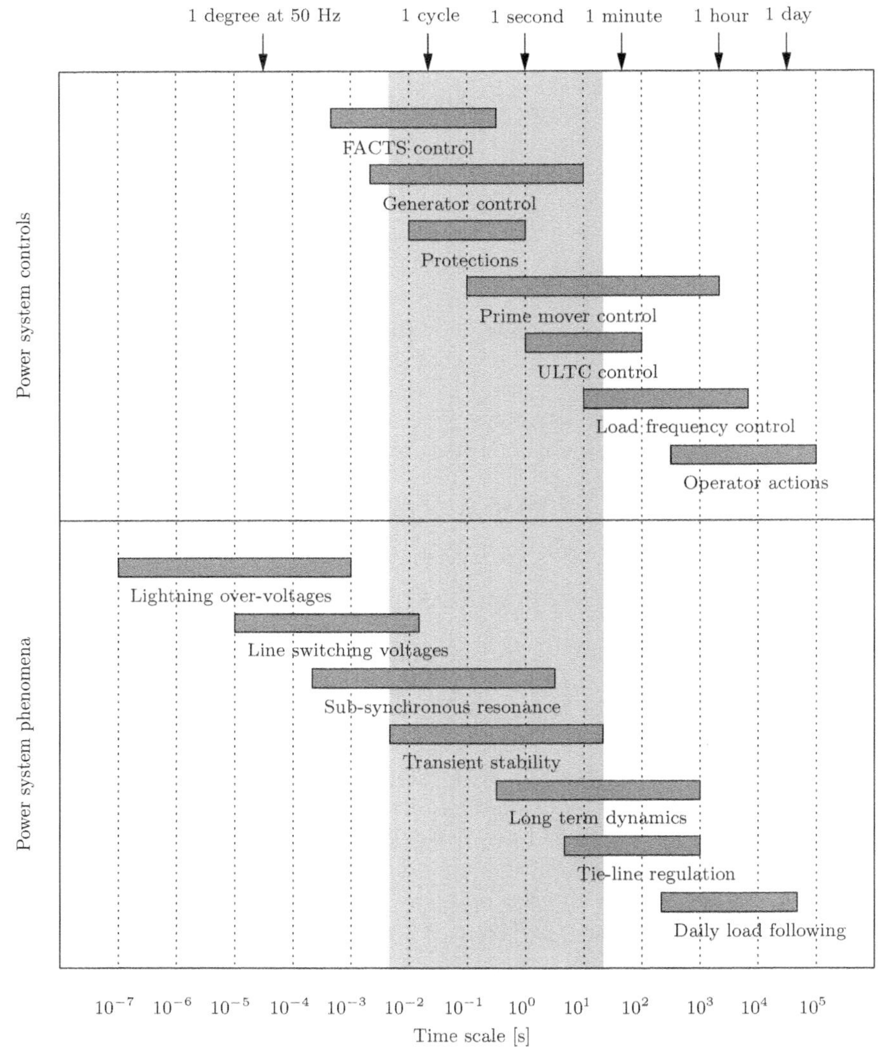

Fig. 1.6 Time scales of relevant power system dynamics. The gray strip indicates the time frame object of this book

In (1.9), the time t is an independent variable as opposed to *dependent* variables $\boldsymbol{x}$ and $\boldsymbol{y}$. More in general, independent variables (or parameters) can be a vector, say $\boldsymbol{\mu}$ ($\boldsymbol{\mu} \in \mathbb{R}^{n_\mu}$), hence (1.9) becomes:

$$\begin{aligned} \dot{\boldsymbol{x}} &= \boldsymbol{f}(\boldsymbol{x}, \boldsymbol{y}, \boldsymbol{\eta}, \boldsymbol{u}, \boldsymbol{\mu}) \\ \boldsymbol{0} &= \boldsymbol{g}(\boldsymbol{x}, \boldsymbol{y}, \boldsymbol{\eta}, \boldsymbol{u}, \boldsymbol{\mu}) \end{aligned} \tag{1.11}$$

For example, in voltage stability analysis, load power consumptions are independent variables. In most applications considered in this book, only a scalar independent variable is considered. For example, the loading level μ for continuation power flow analysis (see Chapter 5) and the time t for time domain integration (see Chapter 8).

It is worth noting the difference between controllable parameters $\boldsymbol{\eta}$ and device parameters such as transmission line series resistances or synchronous generator sub-transient short-circuit time constants. The latter cannot vary unless the device itself is changed. On the contrary, $\boldsymbol{\eta}$ are constant because their dynamic is slow with respect to the considered time scale. However, $\boldsymbol{\eta}$ can vary according to a given strategy or control. For example, $\boldsymbol{\eta}$ may represent the vector of dispatched generator active powers in optimal power flow analysis (see Chapter 6).

Analogously to (1.7), equations (1.11) can be split into a collection of subsystems if discrete variables $\boldsymbol{u}$ are substituted for if-then rules. Thus, (1.11) can be conveniently rewritten as a collection of continuous DAEs, one per each discrete variable change. Such a system is also known as *hybrid automaton* or *hybrid dynamical system*. An in-depth description and formalization of hybrid systems for power system analysis can be found in [131] and other works by the same author.

The DAE system (1.11) can be defined for any time scale. The terms *slow* and *fast* do not mean anything unless the reference time scale is defined. In this book, the time scale of interest concerns transient stability analysis, i.e., from 0.01 s to 10 s or, which is the same, from 0.1 Hz to 100 Hz. Subsynchronous resonance and some long term dynamics (such as under load tap changer controls) overlap transient stability dynamics and are thus also taken into account in this book (see Figure 1.6). In some specific cases, also faster dynamics, such as synchronous machine magnetic fluxes are considered, mostly to perform comparisons with transient models.

Example 1.4 Transient Behavior of State and Algebraic Variables

Figure 1.7 illustrates the difference between state and algebraic variable transient behavior. The figure shows the time evolution of two variables of the IEEE 14-bus system, namely the phase angle θ of the voltage at bus 5, and the rotor angle δ of synchronous machines at bus 3. The simulation shows the transient following line 2-4 outage occurring at $t = 1.0$ s. As explained above, the line disconnection can be modelled as a switch of a discrete variable. After the occurrence of the line outage, the bus voltage phase angle trajectory shows a discontinuity, while the synchronous machine rotor angle, as expected, is always continuous. Discontinuities only appear if a discrete variable changes its value. Hence, algebraic variable trajectories before and after discrete variable switches are continuous.

In Figure 1.7, the transition of the bus phase angle θ from t^- to t^+ is indicated by a dotted line vertical line. However, for simplicity, it is common

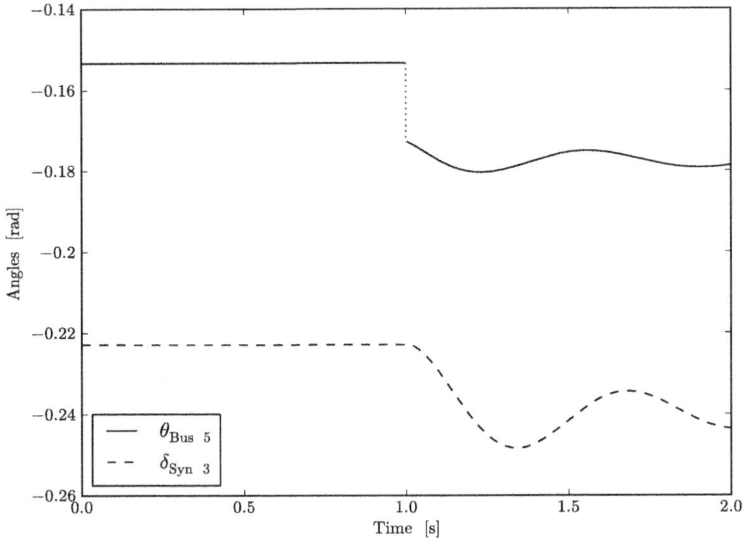

Fig. 1.7 Time evolution of state and algebraic variables

practice (also used in this book hereinafter) to plot pre- and post-disturbance values of algebraic variables as they were connected by a continuous trajectory. Thus, the reader has to be aware that straight vertical lines in the trajectories of as algebraic variables indicate a discontinuity or, in other words, a *jump* from t^- to t^+.

Example 1.5 Reactor Transient Stability Model

Let us consider again the inductor model discussed in Example 1.3 and define a model suitable for transient stability studies. Neglecting the ideal model (1.3) which is of no engineering interest and assuming lumped parameters, the most adequate starting point is the model (1.4).

Reactors (or *chokes*) used in industrial applications for short circuit protections or filters have an iron core and copper windings. Given that the copper and iron specific heat are $c_{p,\text{Cu}} = 0.385$ and $c_{p,\text{Fe}} = 0.450$ kJ/kgK, respectively, we can assume a mean specific heat of about $\hat{c}_p = 0.4$ kJ/KgK. The density of the copper and the iron are $\rho_{\text{Cu}} = 8\,920$ and $\rho_{\text{Fe}} = 7\,870$ kg/m^3, respectively, we can assume a mean reactor density of about $\hat{\rho} = 8\,500$ kg/m^3. If the reactor volume is Vol $= 0.5$ m^3, an active area of $A = 4.0$ m^2 and a convective heat transfer coefficient of about $k_c = 1.0$ kW/m^{2}K, then the thermal time constant can estimated as:

$$T_\Theta = \frac{\hat{\rho} \cdot \mathrm{Vol} \cdot \hat{c}_p}{k_c A} \approx \frac{8\,500 \cdot 0.5 \cdot 0.4}{4.0 \cdot 1.0} = 425 \text{ s}$$

The thermal time constant is thus well beyond the transient stability time scale (e.g., 10 s) and we can consider that the reactor temperature Θ is *frozen* at a given value during transient stability simulations. Therefore, the reactor resistance value R is constant.

On the other hand, the parasitic capacity C and conductance G are very small. If the capacity is about 0.1 nF and the conductance is about 1.0 μS, then the time constant associated to parasitic effects is:

$$T_p = \frac{C}{G} \approx \frac{10^{-10}}{10^{-6}} = 10^{-4} \text{ s}$$

that is much smaller than the limit of transient stability time scale (e.g., 0.01 s). Hence, the voltage at the reactor terminal can be considered an algebraic variable. Moreover, since both the conductance G and the capacity C are comparatively small, the total current can be approximated as $i \approx i_L$.

Neglecting magnetic flux saturation and assuming that the reactance of the reactor at the fundamental frequency of $f = 50$ Hz is, say, $X_L = \omega L = 5.0$ Ω and its series resistance is $R = 2.0$ Ω, then the time constant associated with the inductive coil is:

$$T_i = \frac{L}{R} = \frac{X_L}{\omega R} \approx \frac{5.0}{2 \cdot \pi \cdot 50 \cdot 2.0} = 7.96 \cdot 10^{-3} \text{ s}$$

where $\omega = 2\pi f$ is the fundamental angular frequency of the system. Hence, the time constant T_i is sufficiently small to allow considering also i_L an algebraic variable. In conclusion, the reactor model for transient stability studies, at 50 Hz, is as follows:

$$\bar{V} = (R + jX_L)\bar{I} \tag{1.12}$$

where $\bar{V}$ and $\bar{I}$ are the phasor voltage and phasor current, respectively.

In case the reactor coil is made of a superconductor as in superconducting magnetic energy storage (SMES) devices, then the resistance R is some orders of magnitude lower than for standard copper coils. Hence, the time constant T_i can fall into the range of transient stability time scale. This justifies considering the SMES current as a state variable. Further insights on the SMES model are given in Example 18.3 of Chapters 18.

Chapter 2
Power System Architecture

The main concept that is developed in the chapter is that any complex project can be conveniently handled if correctly planned and structured. With this aim, Section 2.1 discusses the fragmentation of software packages. An example is also given in this section. Section 2.2 describes the main components that compose a general-purpose software package, namely *classes* and *procedures*, while Section 2.3 introduces the concept of modularity. A simple example on how organizing a modular software package is also provided in Section 2.3. Section 2.3 also discusses the modularity of power system structure. Finally, Section 2.4 applies the concepts previously discussed and proposes the structure of a general power system analysis tool.

2.1 Structure of Software Projects

The Cantorian triadic bar, also known as Cantor's set, is built as follows. From a straight segment, one removes the central third. Then, from each remaining lateral segments, one removes the central thirds. And so on. The results is depicted in Figure 2.1. A software project, no matter how complex it can appear at a first glance, is very similar in its internal structure to the Cantor's set. A software project is not a monolithic straight line, i.e., is not a unique, long function, in the same way a book is not a unique breathtaking sentence.

Any software project can be divided into a variety of modules and each module in a variety of tasks, etc., until each operation is comparatively simple, single-purpose and easy to solve. For example, task fragmentation is the philosophy on which Unix and Linux are based. These operating systems are a collections of a huge number of small projects, each one solving a very specific task and independent from the others. Like an ant colony, each application (e.g., each ant) accomplishes a very basic function and only knows how to solve that function. However, the sum of each program is not simply a collection of functions but the whole operating system (e.g., the colony).

F. Milano: Power System Modelling and Scripting, Power Systems, pp. 19–30.
springerlink.com

Fig. 2.1 Cantorian triadic bar with horizontal section dimension $D = \ln 2/\ln 3 = 0.6309$

Clearly, some applications are more complicated than the average. For example the kernel of the operating system is a monolithic application. A current Linux kernel 2.6 with common modules has a size more than 8.5 millions of lines of code. However, the kernel is an exceptional application. And even the kernel is internally divided into several modules, each one taking care of a specific issue.

The fact that a complex and general purpose operating system such as Unix, can be reduced into a set of generally small programs and scripts, is common to any software project. The key point is to find an efficient, yet simple, mechanism to make all programs interact smoothly together. Ants have solved the issue basing their complex communication system on fer-ormons. The Unix solution is the assumption that everything is a file or a file-like object. Thus, applications communicate writing and reading files. Folders, Internet sockets, process forks and other fancy stuff commonly used in informatics are treated as files.

An objective of this book is to formalize fundamental concepts for creating an application oriented to power system analysis. The conceptual issues that have to be solved are two: (i) to divide the project into a series of small functions, and (ii) to find a convenient *glue* that allows a good integration among all functions yet leaving each function as independent and stand-alone as possible. These issues are addressed in the remainder of this chapter.

Example 2.1 Unix Shell Script

To clarify the idea presented in the previous section, consider the following example. In the Unix (and Linux) environment, it is a common practice to create small shell scripts to execute a given series of commands. For the sake of example, the script used for generating the portable document format (pdf) file of this manuscript from source LaTeX files is as follows:

```
latex book.tex
makeindex -s index book.idx
bibtex book.aux
latex book.tex
latex book.tex
dvips -tletter -Ppdf -G0 -o book.ps book.dvi
ps2pdf book.ps book.pdf
```

The script above calls five executable programs that are supposed to be available in the system, namely `latex`, `makeindex`, `bibtex`, `dvips`, and `ps2pdf`. The first command compiles the `book.tex` source file (which is divided into a variety of files). The `latex` command, by itself, is not a monolithic application, but is based on the TEX compiler and on a myriad of specialized packages. Then the `makeindex` command create the analytical index, while `bibtex` parses the entries of the reference database for populating the book bibliography section. The following two `latex` commands consolidates the references and make a consistent document. The result of the `latex` commands is a dvi file which has to be transformed into a postscript (`dvips`) and finally into a pdf file (`ps2pdf`).

This lengthy description is aimed to show the fractal nature of Unix systems. A complex document such as this typeset manuscript is the results of a high fragmentation of the operations needed to obtain it. The pdf document is created using a variety of "macroscopic" applications (e.g., `latex` and `ps2pdf`). Then, each application internally calls a variety of other applications, packages, libraries, etc. (see Figure 2.2). If the function solved by each applications is "small" or "simple" enough, the function is executed, otherwise it is divided into simpler functions, and so on. The branching process ends up if a function is so simple that it cannot be conveniently further divided into other functions. Simplicity means that a specific function can be implemented and maintained by a single person or by a very reduced group of people. Moreover, simplicity generally also implies robustness. For example, the execution of the script provided in this example rarely fails despite the huge number of applications, functions and libraries that are directly or indirectly called.[1]

A failure occurs only if the source code (e.g., the `book.tex` file) contains errors, but this does not depend on the script nor on the application called by the script. Unfortunately, being able to recognize the errors due to the input data and those due to application bugs is not always straightforward as in this case.

2.2 Classes and Procedures

The discussion and the simple provided in the previous section suggests that any software project can be assessed and mastered if adequately divided into a tree of single-goal possibly simple tasks. This is the strategy that allowed Romans to govern their huge empire, namely *divide et impera* (e.g., "divide and conquer").

There are mainly two kinds of elementary pieces into which a software project can be divided: (i) *classes* which accomplishes a specific functionality (e.g., computing the current value of a function) and (ii) *procedures* that

[1] Actually, the TEX version 3.141592 program which is the kernel of the `latex` command, is said to be one of very few programs that are virtually bug free.

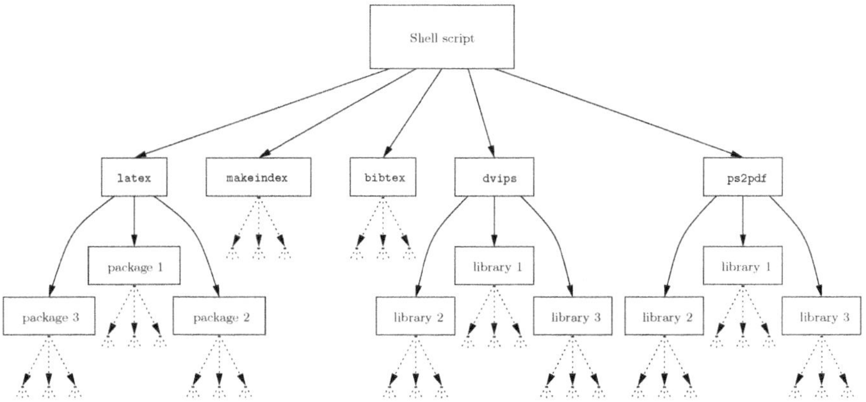

Fig. 2.2 Tree of applications called by a simple shell script

coordinate the activities of classes (e.g., the Newton's method for determining the solution of a set of nonlinear equations). Procedures typically provide *algorithms*, while classes typically provide specific functions or, as they are known in the slang of object oriented programming, *methods*.

Using a modern concept of computer science, a class can be assimilated to an *agent*. A rigorous definition of an agent is as follows: "A software routine that waits in the background and performs an action when a specified event occurs. For example, agents could transmit a summary file on the first day of the month or monitor incoming data and alert the user when a certain transaction has arrived." In this context, the details of agent theory are not relevant (e.g., the four key notions that distinguish agents from arbitrary programs: reaction to the environment, autonomy, goal-orientation and persistence) and agent-based programming techniques is not a concern.[2] However, the agent definition fits well with the structures proposed in the next sections of this chapter.

According to the analogy with computer agents, procedures can be identified with *agencies* or *authorities* that decide which (and if) agents have to be called and which action or method is appropriate. After being initialized, each class waits for an *event*, i.e., a call from a procedure. For example, the power flow analysis calls device algebraic equations and power flow Jacobians. In case one is interested in solving a time domain simulations, the numerical integration routine calls differential equations of all dynamic devices. Note that if a class does not take part to a certain kind of analysis, the procedure simply does not call it.

[2] The popularity of agent-based analysis is increasing in power system analysis, even though mostly oriented to the study and the modelling of electricity markets. The interested reader can find a comprehensive introduction to this topic in [347].

Even though class details have not been introduced yet, it should be easy to understand that classes have to be able to properly interact with procedures. This issue is discussed in the following section.

2.3 Modularity

An important aspect that is worth further discussion is how classes and procedures communicate. Each arrow in Figure 2.2 (see also following Figures 2.3 and 2.5) represents a *call* to a function or to a *method* of some class. Following each call, the method has to perform some action and, optionally, return some object/data in a given format.

It is not hard to see that if each class has its own custom methods and each method requires custom calls, the procedures have to take care of each class in a different way. Furthermore, any change in a class method would require a change in the procedures that call that class. Even for a reduced number of classes, the whole project would quickly become intractable and difficult to maintain.

The critical issue is how to define the way classes communicate with the rest of the world. The set of rules that define the communication grammar and syntax is often called *protocol* in computer data communication and telecommunication. For example, the server message block (SNB) or the systems network architecture (SNA) provide a shared access to files, printers, serial ports and other resources between nodes (e.g., computers) of a network. More sophisticated examples of communications protocols are human languages, like English or Italian. In the same vein, computer languages are the communication protocol between human beings and computers.

A protocol should be as flexible and as complete as possible to avoid raising exceptions. An example of completeness and flexibility are human languages: a modern language such as English allows expressing any concept. However, computer applications never require such level of completeness and flexibility. Furthermore, too much flexibility may lead to ambiguity, which has to be carefully avoided in any computer application. Fortunately, simplicity (and, thereby, consistency) and flexibility are not necessarily in opposition. Example 2.2 attempts to clarify this point.

The definition of a common communication protocol for classes of the same kind should be accompanied by a systematic programming approach, i.e., *modularity*. Generally speaking, modularity is the property of a software package to be organized in fixed though flexible sections whose structure recurs with little of no changes in several parts of the package itself. Modularity is strictly linked to object-oriented programming and helps one implement, reuse and maintain a robust code. Chapter 3 provides further discussion and examples of class-based (i.e., modular) scripting for power system analysis.

Properly understanding the concept of modularity is fundamental for mastering the architecture of power systems. In fact, although the number of devices that compose a real-world power system is huge, these devices can be divided into a few basic classes. Thus, even huge power systems such the one shown in Figure 1.1 of Chapter 1 can be assessed by defining a very reduced number of basic devices. In particular, the classes of devices that are of interest in this book are generators, transmission systems, transformers, loads, motors, regulators, power electronic converters and protections. Example 2.3 discusses the device taxonomy of the IEEE 14-bus system. This well-known benchmark system is extensively used throughout the book as main test board.

Example 2.2 Zero of a Scalar Function

To clarify the concepts introduced above, this example provides a qualitative description of a simple mathematical application containing modular classes and procedures.

The objective is to write a software application that finds the solution $f(x) = 0$ of a scalar function f that is defined by the user. Although f is not known, its general structure is fixed. Say that f is of the form:

$$f(x) = \sum_{i=1,\ldots,n} f_i(x) \tag{2.1}$$

There is no particular reason for imposing such structure, it is just for the sake of example. The important point is that the user can instantiate any number n of functions f_i. Such functions f_i are collected in a library of classes, each of which defining a specific function kind. In this example two classes are considered: the quadratic polynomial and the sine function. The class that defines polynomial calculates the following function:

$$f_i(x) = a_i x^2 + b_i x + c_i \tag{2.2}$$

while the sine class computes:

$$f_i(x) = A_i \sin(\omega_i x + \phi_i) \tag{2.3}$$

Any other function $f_i(x)$ can be defined just by adding a new class. The user has to define in a file a triplet (a_i, b_i, c_i) for each polynomial and a triplet (A_j, ω_j, ϕ_j) for each sine function. This can be conveniently accomplished using a text file with a given format.

Clearly, the polynomials and the sine functions could be merged together in a unique function. However, in this example we are not interested in efficiency or in reducing as much as possible the lines of code, but rather in providing an example of a modular application.

The user can also pass to the application some settings, such as the initial guess x_0, which algorithm has to be used for finding the solution $f(x) = 0$, the tolerance that has to be used for defining the convergence of the algorithm,

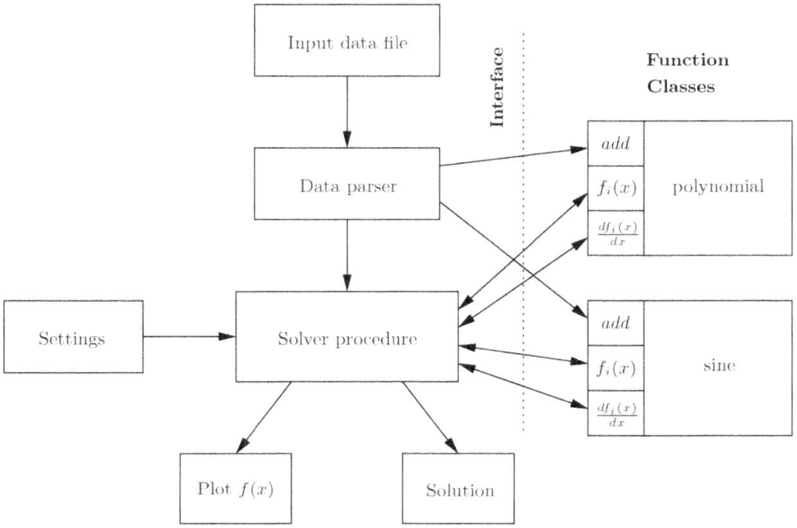

Fig. 2.3 Structure of a simple application that finds the zero of a general scalar function $f(x)$

the maximum number of iterations before stopping the algorithm in case of no convergence, etc.

After parsing the input data file, the application calls the main procedure for solving $f(x) = 0$. At the each iteration k, the procedure calls the classes to compute each $f_i(x^k)$, and the derivatives $df_i(x)/dx|_k$. If the method converges, the application returns to the user the solution x and, optionally, the final tolerance, the number of iterations, the plot of $f(x)$ in the neighborhood of the initial guess x_0, etc. Otherwise, the application displays an error message stating that the solver has not converged. The structure of this simple application is depicted in Figure 2.3.

The most important point that has to be retained is that any number of classes implementing a particular function can be added without changing the structure of the program or of the procedures outside the $f_i(x)$ classes.[3] However, in order to accomplish this modularity, each $f_i(x)$ class has to communicate using a standard protocol. In this simple example, the communication protocol simply consists in passing to the $f_i(x)$ classes the current value of the variable x^k and expecting that the classes return the values $f_i(x^k)$ and $df_i(x)/dx|_{x^k}$. It can also be helpful to use a common interface for simplifying the operations of handling several $f_i(x)$ classes. The interface can be a class itself or simply a collection of methods that organize the

[3] The only exception to this rule are data parsers that strictly depend on the functionality and features of the whole application. Adding a new $f_i(x)$ class likely affects also the parser, since the input data format changes.

initialization and the calls to the $f_i(x)$ classes. A possible implementation of this application is given in Script 3.1 of Chapter 3.

The example provided in this section is quite simple and has a limited practical use, however it is useful to understand the behavior of more complex tools, such as a power system software package. As a matter of fact, a power system package works in a very similar way as the tool described above. For example, classes that define dynamic power system devices may contain a method that evaluates differential equations. This method requires the current simulation time and the state variable vector as input and return first time derivatives of those state variables of its competence. The duty of the procedures that calls dynamic devices (e.g., the time integration algorithm) is to collect the time derivatives in an ordered vector. However, there is no need for the time integration algorithm to know the details of the implementation of each device as long as the device returns the first time derivatives. Thus, adding a new device does not require to modify existing code. Furthermore, except for the multi-dimensional variable vector, the analysis techniques described in Part II consists, at the very end, in finding the solution of a set of nonlinear equations in the form:

$$\mathbf{0} = \hat{\boldsymbol{\varphi}}(\hat{\boldsymbol{\xi}}) \tag{2.4}$$

where $\hat{\boldsymbol{\varphi}}$ and $\hat{\boldsymbol{\xi}}$ changes depending on the application. For example, in case of standard power flow analysis, $\hat{\boldsymbol{\varphi}}$ describes the power injections at network buses and $\hat{\boldsymbol{\xi}}$ are the voltage magnitudes and phases at load buses. In case of time domain analysis, (2.4) is solved for a sequence of given times and $\hat{\boldsymbol{\varphi}}$ is a set of equations that depends on the integration algorithm while $\hat{\boldsymbol{\xi}}$ are the state and algebraic variables of the DAE system. The same can be said for continuation and optimization methods. Thus the main issue that is addressed in Part II is the proper definition of the set of nonlinear equations $\hat{\boldsymbol{\varphi}}(\hat{\boldsymbol{\xi}})$.

Example 2.3 Structure of the IEEE 14-Bus System

In order to clarify the structure of a power system, Figure 2.4 depicts the IEEE 14-bus benchmark system that consists of two generators, three synchronous compensators, two two-winding and one three-winding transformers, fifteen transmission lines, eleven loads and one shunt capacitor. Not depicted in Figure 2.4, but implicitly included, are generator controllers, such as the primary voltage and frequency regulators, transmission line and transformer protections and breakers, etc. This simple network (as well as most benchmark systems) can be used to illustrate virtually all power system phenomena and, if properly modified, to show the behavior of any device model. This is a byproduct of the modularity of the power system structure.

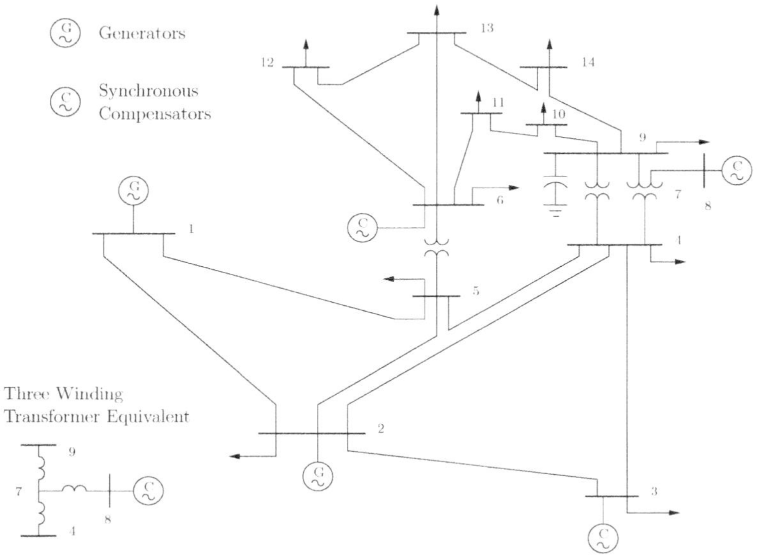

Fig. 2.4 IEEE 14-bus test system

2.4 Architecture of a Power System Software Tool

This section attempts to define a basic and as general as possible architecture for a power system analysis software package. The proposed structure is shown in Figure 2.5. The main parts that are shown in the figure are as follows.

1. *Parsing input data.* Input data can be defined as plain text files or through graphical tools, such as one-line network diagrams or graphical information systems (GIS). There is no particular need for using only one graphical library or GIS tool. The fact that proprietary software applications do not allow using graphical systems but the one embedded in the application itself is an example of traditional programming that, in this book, is deprecated. In the same vein, there is no reason for adopting a particular data format as long as suitable parsers for the input data are provided. A detailed discussion on data format issues is provided in Chapter 21.
2. *Initialization of power flow devices.* Given the input data, the initialization of power flow devices consists in creating an instance of all devices that are used in the power flow analysis and populating these instances with the data provided by the parser. In power flow analysis one has to define at least buses, transmission lines, static generators and loads. But, in principle, any device can be included in the power flow analysis. This point is further discussed in Chapter 4.

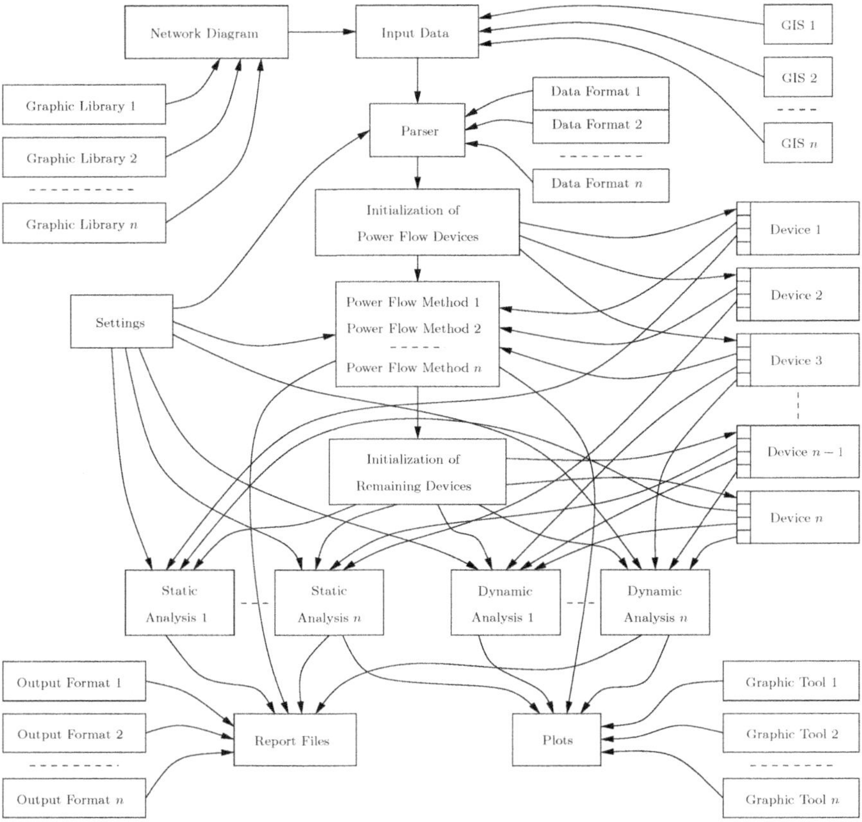

Fig. 2.5 Structure of a general purpose software suite for power system analysis

3. *Power flow analysis.* Since the focus of the book is on balanced systems at the fundamental frequency, the very kernel of the application is the power flow analysis. Of course, a similar structure can be extended or moved to other analysis, such as electro-magnetic transient (EMT) analysis. In this case, the core algorithm would be the EMT integration. However, the basic concepts that are beneath the proposed structure are independent of the specific analysis that is carried on.

   The power flow analysis is a general solver that looks for the zeros of a set of nonlinear equations but does not contain any information about the network or the devices included in the network. The solver is not necessarily unique. As a matter of fact, several algorithms have been proposed in the literature and are discussed in Chapter 4 (e.g., Newton's method, Seidel's method and fast decoupled).

4. *Initialization of remaining devices.* After completing the power flow analysis, it is common practice to initialize dynamic devices such as synchronous machines, primary voltage and frequency regulators, etc. The initialization consists in creating an instance of all required devices, assigning the data to the instances and computing the initial value of state and algebraic variables (see Subsection 9.1.1 of Chapter 7). At the end of the initializations, the power system model is at a steady-state equilibrium point that serves as initial condition for further analysis.
5. *Static or dynamic analyses.* Given a steady-state equilibrium point, several analysis can be performed. This book focuses on two static ones (namely, continuation power flow and optimal power flow) and two dynamical ones (namely, eigenvalue analysis and time domain simulation). Other relevant topics are harmonic analysis, fault analysis, state estimation, etc. However, the relevant point is that any algorithm can be included in the scheme of Figure 2.5 provided that the algorithm properly interacts with all available devices.
6. *Output storage and display.* The last step is to display the results in a convenient format. Report files, tables, plots and other visualization techniques help understand and interpret results. As for the input data and one-line diagram editors, the choice should not be limited to only one tool or output format. Further discussions on these topics are provided in Chapters 21 and 22.

From the observation of the scheme depicted in Figure 2.5, one can identify procedures and classes. It is easy to distinguish classes from procedures based on a pictorial difference: while classes are like terminals from which depart arrows, procedures are like hubs from and at which several arrows depart and arrive.

The procedures are: the input data parser; the initialization of power flow devices; the power flow analysis; the initialization of devices used after the power flow analysis; static and dynamic analyses; and output storage and display. The classes are: graphic libraries for defining network diagrams; graphical information systems; data format parser definitions; devices; and text and graphic output formatters. A special kind of class is the set of settings of the software package. Settings allow a fine tuning of the behavior of the application and of each procedure.

There are two kinds of procedures: the ones that implement mathematical analyses and algorithms (e.g., power flow analysis) and those that are mainly needed for organizing and putting in order the devices (e.g., input data parser). In the same vein, there are two kinds of classes: the ones that implements differential and algebraic equations of power system devices and those that are for non-engineering operations (e.g., data grammars and graphic libraries). In this book, the focus is only on algorithms (Part II) and algebraic differential equations (Part III). Part IV provides an overview of non-strictly-technical topics.

While procedures are totally application dependent and have thus to be written for the specific problem that has to be solved, classes can be in some cases imported from existing public-domain libraries. For example, GIS and graphical functions can be imported from external libraries. On the other hand, power system device classes or data parsers have to be implemented as they are highly application dependent. However, the key point of a fragmented (or fractal) structure, is the fact that device and parser classes can be collected in a library and, if one needs a new device or a new parser, only that specific device or parser has to be implemented.

Figure 2.5 also helps understand one of the main reasons why undergraduate and Ph.D. students are afraid of writing their own power system software package. Thinking, organizing, planning and implementing the whole project is, at a first glance, overwhelming. Furthermore, a great portion of the code is out of the scope of power system analysis and modelling. However, had the student only to implement a few device classes or an algorithm (procedure), the amount of code to implement would be relatively easy to master. Of course, the underlying assumption is that the main project to which the student adds his contributions has to be *open* in the sense discussed in Section 3.1. More details on educational aspects of power system software tools are given in Chapter 23.

Chapter 3
Power System Scripting

The topics of this chapter are threefold. The first topic is to find the most adequate scenario for a didactic and research-oriented programming environment (Section 3.1). With this aim, the concepts of *open* and *closed* software as well as the differences between traditional programming and scripting are introduced and discussed (Sections 3.2 and 3.3, respectively). The second topic is to describe the minimal features that a computer language should have to be suitable for power system analysis and simulation (Section 3.4). Comparisons among various modern programming languages commonly used in computational science is given. Finally, the chapter introduces the Python programming language and provides a complete simple example of modular application implemented in this language (Section 3.5).

The main thread of this chapter is that a computer language is not only a mean for translating mathematical formulæ into a computer-readable format. Rather, each computer language provides a particular "way of thinking". The choice of the computer language is thus more a philosophical issue than a technical one.

3.1 Open and Closed Programming

In a theoretical scenario, the steps to follow for studying the behavior of a power system are (i) to establish the model, (ii) to set up equations and (iii) to implement on a computer the resulting system using a suitable solution algorithm (see also Figure 1.3 of Chapter 1). This process is actually quite rare in practice, at least in what concerns standard analyses such as the solution of the power flow problem or time domain integrations. Students and, sad to say, even some researchers, often use a closed software package for solving the assigned problem. This scenario is illustrated in Figure 3.1.

In this context, the term *closed* refers to the lack of freedom to modify the source code of a certain software package. In this sense, commercial products

F. Milano: Power System Modelling and Scripting, Power Systems, pp. 31–58.
springerlink.com

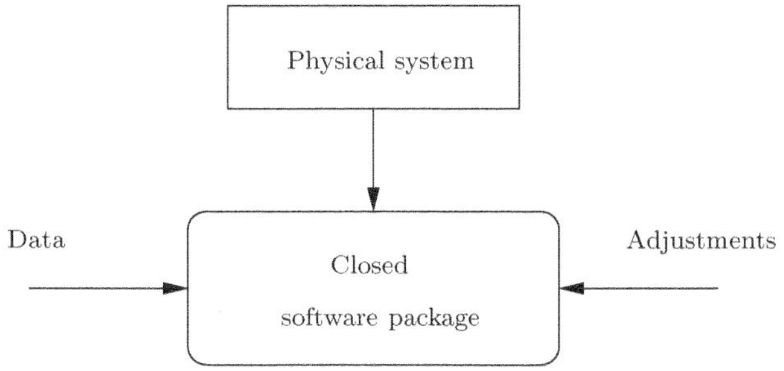

Fig. 3.1 Approach for studying a physical system based on a closed software package

are generally closed. However, also freely-distributed projects[1] can be closed at practical effects if the source code is not provided or, if provided, is too complicated to be mastered in a reasonable time. As it can be promptly observed, closed software packages embed and mask the most interesting parts, i.e., the modelling and computer implementation phases.

There are at least two important drawbacks in this approach. The first one is that the user has to accept the hypotheses and simplifications used by the authors of the software package. The other one is that the user often ignores the hypotheses and simplifications used by the authors of the software package. A byproduct drawback is also the absent or reduced possibility of modifying the equations and of replacing the algorithms used by the software package.

Clearly, the main advantage of using a closed software packages is to save time. For well-assessed and repetitive operations, such as most industry applications, it is also the correct approach. On the other hand, the educational weakness of closed software is evident. The user gives up the possibility of thinking in exchange for setting up input data and adjustments (e.g., software parameter settings). It has to be noted that setting up a set of data without having the control or the full knowledge of the model can lead to unpredictable results.

In any case, the approach illustrated in Figure 3.1 should be avoided as much as possible at least in the academic environment, since it promotes two dangerous habits: (i) to consider the implementation phase a non-relevant step of the study, and (ii) to accept acritically the model provided by the closed software application. The latter habit is in antithesis with the correct academic approach.

[1] The expression *open source* is widely used for defining a software application or library whose code is available to the final user. If the user is also free to copy, modify and re-utilize the code, provided that the resulting application remains open and free, the project pertain to the family of free and open source software (FOSS). Further details on this topic can be found in Chapter 23.

It has to be said that in most branches of engineering the approach depicted in Figure 3.1 is not so common as it is in power system analysis. This likely happens because, in power systems analysis, setting up the power flow problem and the transient analysis for a general system composed of an arbitrary number of buses, connections and devices implies the implementation of a considerable amount of code that accomplishes ancillary tasks (e.g., data parsing) and is not strictly related to power system analysis (see Section 2.4 of Chapter 2). Thus, using a closed software package is an easy shortcut that allows the students focusing on theoretical concepts more than on programming issues.

This book attempts to demonstrate that programming issues are not insuperable and actually help assimilate theoretical aspects. The proposed approach for studying a physical system is shown in Figure 3.2. In this figure, the *open* software package stands either for a self-made application or an available open source project that can be easily mastered and modified by the user. Of course, implementing the whole software package could result an overwhelming task for the average student. But, according to this approach, it is not necessary that the user implements the whole architecture, just a limited set of modifications, extensions, *add-ons* or *plug-ins*. If the extensions are worth, these can be used by others and the project grows up.

Actually, this approach is not new. It is the philosophical basis of the free and open source software community [292] and has proved to work well, at least for projects like Linux, Apache, LaTeX, Python, etc. Chapter 23 discusses these concepts and attempts to explain why the open source philosophy catches on with great difficulty in the power system community.

3.2 Scripting

In the previous section, the word *programming* is used for indicating the activity of writing computer code. The primary purpose of this section is to clarify the differences between two available approaches for computer programming, namely *traditional programming* and *scripting*. The second purpose of this section is to explain why the scripting approach is adopted in this book.

Traditional programming or *system programming* consists in the construction of a self-contained, stand-alone, typically complex and *monolithic* computer application.

Scripting or simply *programming* consists in the production of typically small *fractal* applications that get advantage of other existing packages to provide new functionalities.

The following remarks are useful to better explain the conceptual differences between traditional programming and scripting.

1. Languages that are suitable for traditional programming projects are typically C, C++, Java, C# and Scala. For engineering applications,

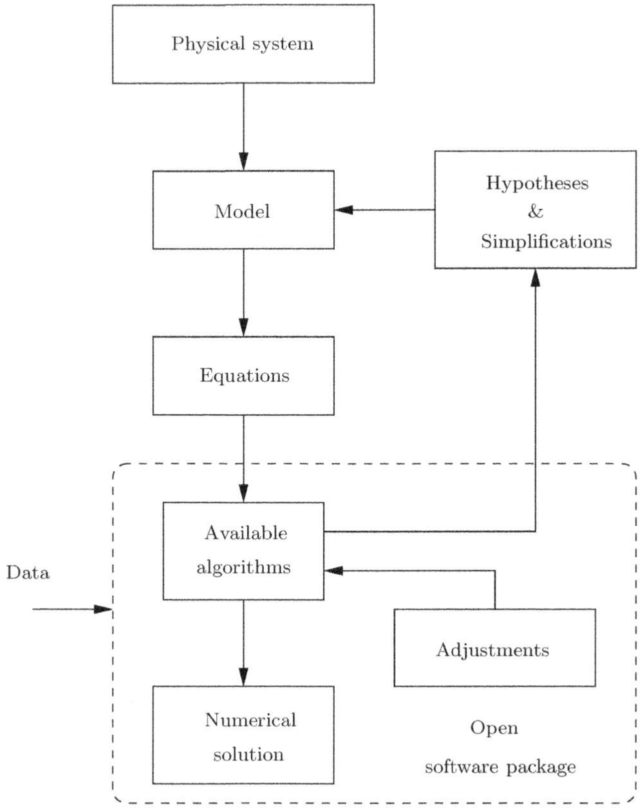

Fig. 3.2 Proposed approach for studying a physical system based on an open software package

FORTRAN (in any available versions, namely, 77, 90/95 and 2003) is also quite popular.

2. Languages commonly used for scripting are, for example, Perl, Python, Php, Tcl and Ruby. In the engineering world, high abstraction level languages such as R, Matlab, Mathematica and Octave are popular choices. The learning process of these languages is much faster than that of C, C++, Java and FORTRAN.
3. Traditional programming is oriented almost exclusively to the application and not to the re-usability of the code. However, one could do scripting also using C or FORTRAN. In the same vein, one could use Perl for a large project, although this would not likely be a good choice. The key point is that the difference between scripting and traditional programming does not concerns necessarily the computer language.

4. C, C++, Java and other languages used for traditional programming are *compiled*, while Perl, Python and other languages used for scripting are *interpreted*. An interpreted language facilitates the implementation phase and allows quickly prototyping new functions. On the other hand, compiled languages are able to provide faster applications than interpreted languages. Some nuances to the latter statement is provided in Example 3.1.
5. Traditional programming is general based on *low level* languages, while scripting prefers *high level* languages. Of course, the frontier that divides a low level from a high level language is somewhat subjective. However, we can assume that the language is *high level* for our purposes if we do not need to worry about the language idiosyncrasies and can concentrate on the kernel of our application.
6. Traditional programming generally leads to closed software applications, while scripting promotes open projects. In this context, the terms *open* and *closed* have the meanings discussed in the previous Section 3.1. Traditional programming would likely lead to a closed software application even if the source code were available. This is a result of the monolithic approach and low-level languages used in traditional programming.

In conclusion, the scripting language approach intrinsically promotes the reusability of the code and is suitable for quickly developing small applications and/or extensions of existing projects. These features are ideal for didactic purposes and fit well with the scheme shown in Figure 3.2.

3.3 Scripting Languages for Computational Science

Based on the discussion of the previous section, scripting appears an interesting option for any computer application. Actually, there are specific advantages that makes this programming approach suitable for computational science. The following is a list of remarks that are relevant to the material presented in this book but that can suite any other branch of engineering applications. For the interested reader, an excellent dissertation about scripting for computational science can be found in [164].

1. While system languages are adequate for big stand-alone applications that require time and a team of software developers, students and researchers are interested in a versatile tool for quickly developing their ideas. The scripting approach is able to provide such versatility and rapidity since most basic operations are already available in existing libraries. Moreover, scripting languages are characterized by a high-rate learning curve compared to system languages. This is due to the fact that scripting languages do not require worrying about type definition, memory allocation, etc. Furthermore the syntax of scripting languages is generally quite simple and clean.

2. As discussed in Chapter 2, scientific computing is not strictly limited to mathematical algorithms or *number crunching*. Focusing on power system analysis, it is important to have the possibility of a graphical interface for both drawing network one-line diagrams and visualizing results. Furthermore, a graphical user interface can be of great help in educational applications. Scripting languages simplify the process of building such graphical tools. Actually, most scripting languages are general purpose and are not explicitly oriented to scientific applications (such as FORTRAN). Thus, using scripting languages make generally easy or, at least, possible, any kind of task (e.g., web-based applications, multimedia, etc.)
3. Mathematical and scientific applications have been the original purpose of computers and are the noblest way of using such machines. No surprise that in the last decades very efficient libraries for linear algebra, matrix factorization, eigenvalue analysis, etc., have been developed and optimized with the help of thousands of scientists. BLAS, ATLAS, LAPACK, UMFPACK are good examples of such libraries. Most of these libraries are legacy code written in FORTRAN or in C. Using modern scripting languages do not mean to bury old code. On the contrary, the ability of scripting languages of gluing different applications allows creating very efficient interfaces and re-utilize legacy libraries. It is important to note that scripting language are generally a high-level layer between the programmer and a variety of low-level libraries.[2] Thus using a scripting languages allows taking advantage of all resources in a more efficient and seamless way than system languages.
4. Most scripting languages were born in a Unix or Unix-like environment. For example, Python, Perl, Php, and Ruby. This is not accidental but is largely due to the features of Unix and Unix-like operating systems that highly promote the development of custom applications. Most popular free and open source scripting languages provide a Windows version that thereby makes available the Unix power on this platform. A byproduct is that using popular scripting languages, the code is portable on various platforms with no or very limited modifications. Portability or *platform independence* can be of great importance in case a scientific project is carried out by different research groups or in case of popular projects that receive third-party contributions.

3.4 Computer Languages Suitable for Power System Analysis

Among the tens (or hundreds) of all available languages, only those that fit some basic requirements are eligible for power system analysis. These requirements are the availability of efficient and easy-to-use libraries for:

[2] The interpreter of the scripting language itself is generally written in some legacy system language such as C.

1. Basic mathematical functions (e.g., exponential, logarithm and trigonometric functions).
2. Complex numbers.
3. Multi-dimensional arrays (e.g., element by element operations and slicing).
4. Linear algebra (e.g., LU and Cholesky's factorization).
5. Sparse matrices.
6. Eigenvalue analysis of non-symmetrical matrices.

The reasons of the need for these libraries in power system analysis is quite evident. However, the discussions on mathematical tools given in Part II dissipate any possible doubt.

Strictly speaking, only basic mathematical functions are really necessary. All other routines can be implemented starting from there. However, this would require a really long time and results could hardly be as efficient as existing libraries. Graphical libraries, although certainly important for plotting results, have not been included in the basic requirements above. Plotting a curve is a post-processing operation, which can be done by means of external applications such as Gnuplot or Excel® once a plain ASCII file of data in tabular format is available. Thus, it suffices that the computer language is able to (efficiently) read and write files, which is a feature common to all languages.

One may argue that also the availability of algorithms for numerical integration of differential equations or optimization problem solvers can be useful. These algorithms perform properly if adjusted to the specific problem to be solved and is thereby a good idea to avoid using predefined functions unless these can be easily customized.

Computer languages that provide the features described above can be divided into four categories:

1. Legacy system languages (e.g., FORTRAN, C and C++).
2. Modern system languages (e.g., Delphi, Java, C# and Scala).
3. General purpose scripting languages with inclusion of adequate mathematical libraries (e.g., Python and Perl).
4. Scientific-oriented scripting languages (e.g., Matlab and Octave).

No computer language is perfect. Any choice has advantages and drawbacks. An assumption of this book is that the advantages of using general purpose scripting languages prevail on the advantages provided by other computer languages.

Sections 3.2 and 3.3 provide sufficient reasons for preferring scripting languages to system ones. Anyone that has programmed in FORTRAN or in C knows the suffering of building up a complex project. More difficult is to convince supporters of Java or other modern system languages. Java has become popular at the end of the last century as a substitute of C++. In my opinion, Java is not adequate for scientific applications that involve massive number crunching. This is a common drawback of most modern system languages that have been designed by and for computer programmers and not for engineers. I also think that languages that overuse punctuation marks (e.g., braces) are quite messy but, maybe, it is just a matter of tastes. Much more important for preferring scripting languages is the fact that Java and other modern system programming languages promote *closed* software tools and, thus, should be avoided in education and research.

More difficult is to justify why general purpose programming languages has to be preferred to scientific-oriented scripting languages. Actually, this may seem just a nonconformist attitude. Matlab along with the Simulink environment is a standard *de facto* in the scientific community for both education and research. Apart from Matlab, there are a variety of other scripting languages, for example Gauss, MathCAD, Mathematica, Maxima, Modelica, Octave, Q, R, SciLab and Yorick. All these languages are specifically oriented to scientific applications and abound with mathematical functions and libraries.

There are, however, a few drawbacks inherent to languages such as Matlab and other scientific scripting languages that should be carefully evaluated, as follows.

1. Scientific oriented languages are designed to make easy a specific mathematical task. For example, Matlab is systematically matrix oriented, thus any variable is by default a multi-dimensional, easy-to-reshape, complex array. This feature can be convenient in most cases, but one may sometimes simply need a scalar float or a short integer.
2. Most scientific languages do not support classes or provide a naïve class implementation. This fact makes difficult or impossible object-oriented programming.
3. While providing a good support for types useful for mathematical operations (e.g., multi-dimensional arrays), scientific scripting languages often lack or make difficult the use of some useful basic types such as tuples, lists or hashes.
4. Matlab is a proprietary software. Thus, the user has to rely on the solutions provided by the developers of Matlab. Of course, one could use some Matlab-like interpreters such as Octave or SciLab, but at the cost of using a slower interpreter, a limited programming language and a reduced set of libraries.

Despite the arguments above, the scientific community is clearly oriented to Matlab or to system languages. Table 3.1 depicts a rough comparison of a variety of open source software packages for power system analysis. Commercial packages are not considered since all of them are closed applications and, thus, are of no interest for the scope of this book. The features illustrated in Table 3.1 are standard power flow (PF), continuation power flow and/or voltage stability analysis (CPF), optimal power flow (OPF), eigenvalue or small signal stability analysis (EA), time domain simulation (TDS) for transient stability analysis, electro-magnetic transients (EMT), and some *æsthetic* features such as graphical user interface (GUI) and one-line diagram editor through computer aided design (CAD).

The tools shown in Table 3.1 are heterogeneous and are characterized by very different complexity levels. Thus, it is not fully fair to compare these projects. As a matter of fact, some packages are Ph.D. students' projects (e.g., MatDyn and Pylon) or niche research tools (e.g., AMES), while others are well-assessed projects with several years of experience (e.g., UWPFLOW). In any case, there is a clear preponderance of Matlab and system language-based tools. Furthermore, it is not by chance that the two industry-oriented packages, OpenDSS and InterPSS, are based on system languages and use Windows as operating system. In fact, these projects have been developed by companies used for producing proprietary software. On the other hand, Matlab is the preferred language in education and research environments.

3.5 Python Scripting Language

This section discusses the reasons for using Python as scripting language throughout this book. In the remainder of this chapter, it is assumed that the reader has some basic knowledge of this language. If this is not the case, among the wide variety of books and on-line documentations about Python, relevant references are [18, 32, 62, 164, 186]. The Python website (`www.python.org`) is also an excellent starting point for obtaining the latest Python version, learning about Python standard libraries and accessing a variety of spare material related to the Python language. Finally, Appendix A provides a quick reference of non-standard libraries used in this book.

Python is a safely, dynamically and strongly typed language. Hence polymorphism, meta-programming and introspection are easy to implement and to use. Concepts such as safe, dynamic and strong typing as well as meta-programming, polymorphism and introspection are relatively advanced (though intriguing) programming topics. While an in-depth discussion on types can be found in [239], intuitive definitions are as follows:

Safe typing: a programming language is considered "type-safe" if it does not allow operations or conversions that lead to erroneous or unpredictable results. The opposite of safe typing is *unsafe typing*.

Dynamic typing: a programming language is said to be dynamically typed if the majority of type checking is performed at run-time as opposed to at compile-time. Dynamic typing is easier to find in scripting languages (e.g., Python). The opposite of dynamic typing is *static typing*, which is typical of system programming languages (e.g., C). Dynamic and static typing can cohabit since a programming language can be dynamically typed in some aspects and statically typed in others (e.g., Matlab).

Strong typing: strongly typed programming languages prevent success for an operation on arguments which have the wrong type. The opposite of strong typing is *weak typing*.

Table 3.1 Open source packages for power system analysis

Package	Ref.	Language	PF	CPF	OPF	EA	TDS	EMT	GUI	CAD
AMES	[171]	Java	✓		✓					
EST	[338]	Matlab	✓			✓	✓			✓
InterPSS	[360]	Java	✓				✓		✓	✓
MatDyn	[64]	Matlab	✓				✓			
MatEMTP	[181]	Matlab					✓	✓	✓	✓
Matpower	[363]	Matlab	✓		✓					
ObjectStab	[166]	Modelica	✓			✓	✓		✓	✓
OpenDSS	[86]	Delphi	✓				✓		✓	✓
PAT	[271]	Matlab	✓			✓	✓			✓
PSAT	[195]	Matlab, Octave	✓	✓	✓	✓	✓		✓	✓
PST	[60]	Matlab	✓	✓		✓	✓			
Pylon	[172]	Python	✓		✓				✓	✓
UWPFLOW	[42]	C	✓	✓						
VST	[53]	Matlab	✓	✓		✓	✓		✓	

Meta-programming: meta-programming consists in writing computer programs that write or manipulate other programs (or themselves) as their data, or that do part of the work at compile-time that would otherwise be done at run-time.

Introspection: type introspection is the ability of some object-oriented programming languages to determine the type of an object at run-time.

Polymorphism: polymorphism is a programming language feature that allows values of different data types to be handled using a unique interface.

Other relevant features of Python are as follows.

1. Python is fully based on well-structured classes, which make easy creating, maintaining and reusing modular object-oriented code.
2. Libraries such as NumPy and CVXOPT provide a link to legacy libraries (e.g., BLAS, LAPACK, UMFPACK, etc.) for manipulating multidimensional arrays, linear algebra, eigenvalue analysis and sparse matrices.
3. Thanks to graphical libraries such as Matplotlib, the ability of producing publication quality 2D figures in Python is at least as powerful as in Matlab.
4. The huge variety of free third-party libraries available for Python, allows easily and quickly extending the features of an application well beyond the scope of the original project.
5. Python is free and open source. Hence Python promotes the implementation and distribution of open projects.
6. Python syntax is relatively simple, neat, compact and elegant. Hence Python is particularly adequate for education and illustrative examples.[3]

Although the features listed above can be likely found in other scripting languages and does not imply that Python is flawless, these are enough for developing the examples provided in this book.

Example 3.1 Python Performance

Before providing an example of Python script, it is worth checking the performance of Python in terms of power system analysis of a real-world power systems. If Python code were not fast enough, all arguments provided above would loose strength. However, computational efficiency is not the main issue in the context of this book. According to Moore's law, the number of transistors that can be placed inexpensively on an integrated circuit doubles approximately every two years. Thus, a software application that is comparatively slow today will be solved quickly in a few years. Nevertheless, most readers will likely argue that Python cannot be competitive with Matlab or with system programming languages. As a matter of fact, most power

[3] Some books use pseudo-code for illustrating implementation examples. The rationale behind the use of pseudo-code is the sake of generality. Actually pseudo-code is a sort of Esperanto of computer languages. However, Python syntax is almost as clear as pseudo-code and has the advantage of being a real scripting language.

system practitioners need a several thousand-bus system case study to believe the robustness and efficiency of novel techniques. Table 3.2 aims clearing any possible mistrust.

Table 3.2 Performance of open source packages for power system analysis

Application	Programming Language	Total time [s]	Jacob. matrix fact. time [s]
UWPFLOW	C	1.155	-
InterPSS	Java	15.97	-
Matpower	Matlab	1.464	0.0363
PSAT	Octave	5.221	0.0433
-	Python	1.420	0.0319

Table 3.2 shows the CPU times of different power system packages (i.e., UWPFLOW, InterPSS, Matpower, PSAT and a custom code written in Python) for solving the power flow analysis using a standard Newton's method for a 2746-bus 3514-line network. This case represents the Polish 400, 220 and 110 kV networks during winter 2003-04 evening peak conditions and is provided by the current Matpower release [363]. The processor used is a 2.4 GHz Intel Core 2 Duo with 2 GB of RAM. The software packages used in the comparison are: Java 1.6, Matlab 7.8, Python 2.6.3, Octave 3.0.1, UWPFLOW release 2006, InterPSS 1.4, Matpower 3.2 and PSAT 2.1.5. Simulations were solved using Windows XP® as operating system.

The comparison, although drawn using same environment conditions for all software packages, cannot be completely fair. For example, the time required to load the interpreter (e.g., the Java, Matlab, Octave or Python environments), and to convert the input data file, which vary considerably, are not taken into account. In case of InterPSS and PSAT, the time for creating the output file is not included. However, neglecting technicalities, relevant conclusions are:

1. Python can be competitive with Matlab and Octave.
2. System programming languages are not much faster than efficient scientific scripting languages. However, scripting languages do not perform directly heavy mathematical operations, but call efficient FORTRAN or C-based libraries. Thus, the main conclusion is that the interfaces that link scripting languages with external compiled libraries are quite efficient.
3. Array indexing in Octave is slower than in Matlab. However, the matrix factorization performance of these two languages is comparable.
4. The Java-based application is one order of magnitude slower than the other software tools considered in this comparison. This result is not surprising since Java was not born for number crunching.

5. The mean CPU time needed to factorize the power flow Jacobian matrix is slightly smaller for the Python implementation because the symbolic factorization provided by the module CVXOPT is used. Symbolic factorization allows pre-factorizing dense or sparse matrices based only on the non-zeros elements (see also Appendix A). Thus, if the number and the position of non-zero elements of the Jacobian matrix do not vary, the symbolic factorization is needed only once. The first full factorization required 0.0424 s.

Script 3.1 First Python Script

In Example 2.2 a general modular applications for finding the zero of a non-linear scalar function. This example provides a possible translation of the scheme shown in Figure 2.3 in the Python language. The goals of this example are to show how: (i) to implement each block of the scheme of Figure 2.3; and (ii) to organize the code in a modular and easily extensible way.

For the sake of simplicity, the whole project is contained in a unique file, say `zero.py`, and the algorithm used for finding the solution of $f(x) = 0$ is the Newton's method.

It is a good habit to begin a Python module with the declarations of the packages used by the module. In this case, the modules `optparse` and `sys` are needed for parsing the parameters passed to the module `zero.py` when it is called from the command line. Then, the modules `numpy` and `cvxopt` are needed for array operations and the module `matplotlib` for producing 2D plots. The script header is as follows:

```
from optparse import OptionParser
from cvxopt.base import matrix, mul, sin, cos
from numpy import linspace
import sys
import matplotlib.pyplot as pyplot
```

The second step consists in defining the common class for the functions $f_i(x)$. Defining a common base class is a good practice that is made possible by the ability of classes of inheriting methods and attributes from other classes. The base class is called *parent*, *ancestor*, *father* or *mother class* while the classes that inherit the method from the original class are called *child class* or *subclass*. Inheritance can be multiple, i.e., a subclass may have several parent classes. However, in this example, only simple inheritance is considered. The class `base` defines the common methods `add` for adding a function element, `setup` for initializing class parameters, and `list2matrix` for converting lists to double-precision arrays. Since the class `base` does not define a specific function $f_i(x)$, the methods `fcall` and `fxcall` for computing $f_i(x^k)$ and $df_i/dx|_k$ are void. The class `base` is as follows:

```
class base():

    """base class for functions"""

    def __init__(self):
```

```
        self.params = {}
        self.n = 0

    def setup(self):

        for key in self.params.keys():
            self.__dict__[key] = []

    def list2matrix(self):

        for key in self.params.keys():
            self.__dict__[key] = matrix(self.__dict__[key], (self.n, 1), 'd')

    def add(self, **kwargs):

        self.n += 1
        keys = self.params.keys()

        for key in keys:
            self.__dict__[key].append(self.params[key])

        for key, val in kwargs.iteritems():
            if not key in keys: continue
            self.__dict__[key][-1] = val

    def fcall(self, x):

        return 0

    def dfcall(self, x):

        return 0
```

The internal built-in dictionary[4] `__dict__` contains all attributes and method names of a class. Using `__dict__` makes possible to efficiently use meta-programming. For example, the method `setup` dynamically initializes at run-time the parameters of each specific function $f_i(x)$. These parameters are defined in the child classes `poly` and `sine`, i.e., the polynomial and sine functions. These are implemented as follows:

```
class poly(base):

    """polynomial class"""

    def __init__(self):

        base.__init__(self)
```

[4] *Dictionaries* are sometimes found in other languages as *associative memories*, *associative arrays* or *hashes*. Unlike sequences, which are indexed by a range of numbers, dictionaries are indexed by *keys*. Keys can be any immutable Python type, e.g., strings or numbers.

```
        self.params = {'a':0.0, 'b':0.0, 'c':0.0}
        self.setup()

    def fcall(self, x):

        fvec = self.c + x*(self.b + x*self.a)
        return sum(fvec)

    def dfcall(self, x):

        dfvec = self.b + 2.0*x*self.a
        return sum(dfvec)

class sine(base):

    """sine function class"""

    def __init__(self):

        base.__init__(self)
        self.params = {'A':0.0, 'omega':0.0, 'phi':0.0}
        self.setup()

    def fcall(self, x):

        fvec = mul(self.A, sin(self.omega*x + self.phi))
        return sum(fvec)

    def dfcall(self, x):

        dfvec = mul(mul(self.A, self.omega),
                    cos(self.omega*x + self.phi))
        return sum(dfvec)
```

As discussed above, the two classes `poly` and `sine` inherit the methods of the class `base`. Thus, to complete these classes it suffices to define function parameters and the methods `fcall` and `fxcall`. It is important to note that any other function $f_i(x)$ can be defined using the basic structure of the classes `poly` and `sine`. The functions `mul` and `cos` imported from the `cvxopt` package compute the element-by-element products and cosine, respectively, of double-precision arrays (see also Appendix A).

Figure 2.3 of Chapter 2 shows a dotted line for indicating an *interface* that takes care of the communication between the function $f_i(x)$ classes and the procedures. This interface is not strictly necessary but allows further generalizing the code and simplifies adding new $f_i(x)$ classes. In this example, the interface is implemented as the following class `function`.

```
class function():

    """interface for all specific function classes"""

    def __init__(self, flist):
```

```
        self.flist = flist
        for item in self.flist:
            self.__dict__[item] = eval(item + '()')

    def setup(self):

        for item in self.flist:
            if self.__dict__[item].n:
                self.__dict__[item].list2matrix()

    def fcall(self, x):

        f = 0
        for item in self.flist:
            if self.__dict__[item].n:
                f += self.__dict__[item].fcall(x)
        return f

    def dfcall(self, x):

        df = 0
        for item in self.flist:
            if self.__dict__[item].n:
                df += self.__dict__[item].dfcall(x)
        return df

flist = ['poly', 'sine']
Function = function(flist)
```

The constructor `__init__` of the class `function` requires as input the list (`flist`) of all defined function classes. In this case, the list is composed of only two items, but can be easily extended. The meta-dictionary `__dict__` and the meta-function `eval` allow using a general approach for handling all $f_i(x)$. By means of `__dict__` and `eval`, adding a new function $f_i(x)$ requires only including a new item in the function list `flist` apart from the definition of the new $f_i(x)$ class. But no procedure has to be modified. Thus, meta-programming and dynamical typing simplify the expansion of a project at the cost of an higher level of abstraction. The last line defines an instance `Function` of the class `function`. This instance is used by all procedures.

The kernel of the application is a routine `run` that collects all settings, calls the data parser, sets up the functions $f_i(x)$, launches the solver and, if required, executes the 2D plot.

```
def run(datafile, x0=0.0, plot=True, imax=20, tol=1e-5):

    """initialize function and run appropriate routines"""

    if not datafile:
        print '* Error: A data file must be defined!'
        print '* Type "dome -h" for help.'
        sys.exit(1)
```

```
    read(datafile)
    Function.setup()
    solve(x0, imax, tol)
    if plot: fplot(x0)
```

The settings are kept minimal for simplicity but one can easily include any adjustments in the form of input parameters. For example one could allow choosing among different solvers. In this case, available settings are the initial guess `x0`, the maximum number of iterations `imax`, the tolerance `tol` for the Newton's method, and a Boolean parameter `plot` that enforces the graphical output. Finally, observe the use of default values for the settings.

The parser of the data file is implemented in the routine `read`. The input data is a plain text ASCII file, as follows:

```
Poly   1.0   4.0  -12.0
Sine   3.0   5.0  -1.57079632679
Sine   5.0   3.0   0.0
```

The parser for this format is given below.

```
def read(datafile):

    """parse input data in plain text format"""

    fid = open(datafile, 'rt')

    for line in fid:

        data = line.split()
        if not len(data): continue
        if data[0] == 'Poly':
            Function.poly.add(a = float(data[1]),
                              b = float(data[2]),
                              c = float(data[3]))
        elif data[0] == 'Sine':
            Function.sine.add(A = float(data[1]),
                              omega = float(data[2]),
                              phi = float(data[3]))

    fid.close()
```

The parser directly calls the methods of the classes `poly` and `sine`. Thus, the parser has to know the parameters of these classes. Parsers are thus a *fragile* part of the application, since a change in the classes that implements functions $f_i(x)$ can make the parser inconsistent.

The solver `solve` implements a standard Newton's method. The solver only interacts with the instance `Function` of the interface class and does not know anything about the specific functions $f_i(x)$.

```
def solve(x0 = 0.0, imax = 20, tol = 1e-5):

    """simple Newton's method"""

    f = 1.0
    iteration = 0
    x = x0

    while abs(f) > tol:

        if iteration > imax: break
        f = Function.fcall(x)
        df = Function.dfcall(x)
        inc = f/df
        print 'Convergence error: %.8f' % inc
        x -= inc
        iteration += 1

    if iteration <= imax:
        print 'The solution is x = %.5f' % x
    else:
        print 'Reached maximum number of iterations'
```

The routine **fplot** takes care of graphical operations. In particular, **fplot** generates a 2D plot of the complete function $f(x)$ in a given interval around the initial guess x_0. This can be useful to double-check the solution obtained using the Newton's method. Most lines in the routine **fplot** are needed for defining the format of the 2D plot and are strictly dependent on the graphical module (Matplotlib in this case).

```
def fplot(x0):

    """plot f(x) in the neighborhood of the initial guess"""

    # build x and f vectors
    points = 200
    xmin = x0 - 5.0
    xmax = x0 + 5.0
    xvec = linspace(xmin, xmax, num = points, endpoint = True)
    fvec = matrix(0, (points, 1), 'd')
    for item, x in enumerate(xvec):
        fvec[item] = Function.fcall(x)

    # graphical commands
    fig = pyplot.figure()
    pyplot.hold(True)
    pyplot.plot(xvec, fvec, 'k')
    pyplot.axhline(linestyle = ':', color = 'k')
    pyplot.axvline(linestyle = ':', color = 'k')
    pyplot.xlabel('$x$')
    pyplot.ylabel('$f(x)$')
    pyplot.savefig('zeroplot.eps', format='eps')
    pyplot.show()
```

Finally, any script requires some command line interface to properly interact with the user. The following routine `main` accomplishes this task.

```
def main():

    """parse settings and launch solver"""

    parser = OptionParser(version=' ')
    parser.add_option('-x', '--x0', dest='x0',
                      default=0.0, help='Initial guess')
    parser.add_option('-p', '--plot', dest='plot',
                      action='store_true', default=False,
                      help='Plot f(x) around x0.')
    parser.add_option('-n', '--iterations', dest='imax',
                      help='Maximum number of iterations.',
                      default=20)
    parser.add_option('-t', '--tolerance', dest='tol',
                      help='Convergence tolerance.',
                      default=1e-5)

    options, args = parser.parse_args(sys.argv[1:])

    datafile = args[0]

    run(datafile,
        x0   = float(options.x0),
        plot = options.plot,
        imax = int(options.imax),
        tol  = float(options.tol))

# command line usage
if __name__ == "__main__": main()
```

The routine `main` uses the package `optparse` for providing a simple yet powerful parser of command line options. `optparse` is a good example of module that simplifies the effort of programming in Python.

Combining all code pieces depicted above in an unique file `zero.py` provides the complete Python script. Assuming that the input data are saved in the file `test.txt`, calling `zero.py` produces Figure 3.3 and the following output:

```
>>> python zero.py -p test.txt
Convergence error: -0.78947368
Convergence error: 0.16580881
Convergence error: -1.15519109
Convergence error: -0.13338833
Convergence error: -0.01723270
Convergence error: -0.00041200
Convergence error: -0.00000024
The solution is x = 1.92989
```

Although much more complex, a software package for power system analysis is very similar to this example. The main difference is the multi-dimensionality of the set of DAE. This implies that a particular care has to be devoted to index each variable so that each device operates on the correct element of the vector of DAE and the corresponding Jacobian matrices. The variable indexing issue is further discussed in Chapter 9.

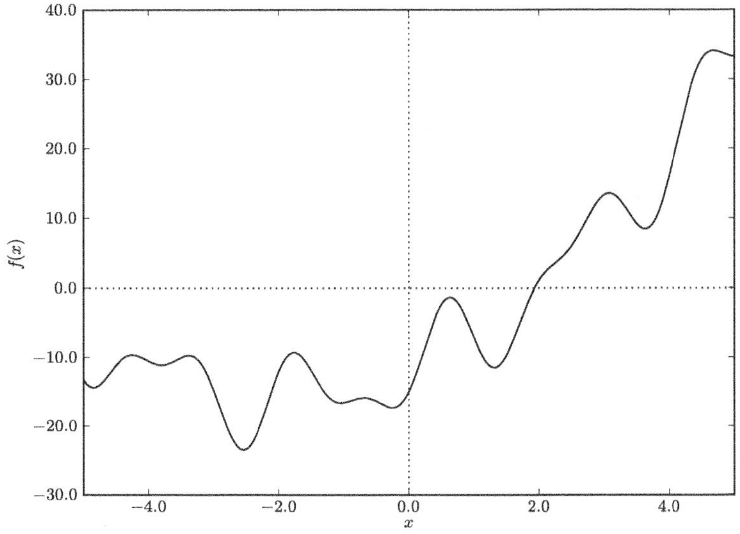

Fig. 3.3 Plot of the function around the initial guess point $x_0 = 0.0$

Script 3.2 Basis of a Power System Analysis Program

The main parts of a general script for power system analysis are the same as those described in the previous Script 3.1. Only, there are much more pieces that compose the puzzle. Listing all the code would take much more than the pages of this book. Thus, this example draws only a few outlines.

Following the logical order of the previous example and the synoptic scheme given in Figure 2.5 of Chapter 2, the following basic elements are needed:

1. External modules required by the program. This topic is discussed in Appendix E.
2. A structure whose attributes are all classes used by the program.
3. A main script that handles user inputs and options.
4. A general interface class that takes care of making routines to call device methods.
5. A general device class from which specific device classes inherit basic methods. This is the topic of Chapter 9.

6. A set of scripts that read input data.
7. A set of scripts that implement solvers. This is the topic of Part II.
8. A set of scripts that implement devices. This is the topic of Part III.
9. A set of scripts that process results (e.g., creating reports and drawing plots). This is the topic of Chapter 22.

Neglecting the issues that are treated in details in dedicated chapters and parts, only three topics remain to be discussed, namely the "system structure", the "main script" and the "interface class". The principal function of these three elements is to provide a common *environment* for all other parts of the software package. These are further described in the following items.

Structure `System`

The scheme of Figure 2.5 of Chapter 2 states that the number of classes and data types required by a complex program such as power system analysis software package is potentially huge. Each class and data type has at least one instance. Since most instances are shared by several methods and functions, it would be particularly lengthy to import instances one by one.

For example, if there are 100 device instances, one has to write 100 import statements in each solver. Furthermore, if a new device is added to the program, one has to revise all the program code an add the new component in the import list occurrences. This is clearly a weak and error-prone approach.

A more robust approach is to define a global structure which can be imported using simply one line of code. This structure is thus a "store" containing all instances of classes and data types required by the program. Thus, if a new class is added to this structure, all other parts of the program automatically inherit also the new class instance.

Assume that this global structure is called `system`. This name will be used hereinafter in all script examples. In Python, the easiest way to implement such structure is to define a module (i.e., a file), called for example `system.py`. A simple example of the contents of this module is given below.

```
# Settings
from settings.settings import settings
from settings.sssa import sssa
from settings.cpf import cpf
from settings.opf import opf

# Variables
from variables.dae import dae
from variables.device import device
from variables.varout import varout

# Devices
from devices.bus import bus
from devices.line import line
```

```
from devices.pv import pv, slack
from devices.pq import pq
from devices.shunt import shunt
from devices.fault import fault
from devices.breaker import breaker
from devices.zone import zone
from devices.synchronous import syn2, syn3, syn4, syn5a, syn5b
from devices.synchronous import syn5c, syn5d, syn6a, syn6b
from devices.avr import avr1, avr2, avr3
from devices.turbine import tg1, tg2
from devices.pss import pss1, pss2, pss3

# list of all active devices
device_list = ['Bus', 'Area', 'Region', 'System', 'Line',
               'Shunt', 'Breaker', 'Fault', 'PV', 'SW', 'Syn2',
               'Syn3', 'Syn4', 'Syn5a', 'Syn5b', 'Syn5c', 'Syn5d',
               'Syn6a', 'Syn6b', 'Avr1', 'Avr2', 'Avr3', 'Tg1',
               'Tg2', 'Pss1', 'Pss2', 'Pss3']

# settings
Settings = settings() # power flow and time domain settings
SSSA = sssa()         # eigenvalue analysis settings
CPF = cpf()           # continuation power flow settings
OPF = opf()           # optimal power flow settings

# variables
DAE = dae()
Varname = varname()
Varout = varout()
Device = device(device_list)

# D E V I C E S

# basic power flow devices
Bus = bus()
Line = line()
SW = slack()
PV = pv()
PQ = pq()
Shunt = shunt()
Area = zone('Area')
Region = zone('Region')
System = zone('System')

# switches
Fault = fault()
Breaker = breaker()

# synchronous machines
Syn2 = syn2()     # machine models
Syn3 = syn3()
Syn4 = syn4()
Syn5a = syn5a()
Syn5b = syn5b()
```

```
Syn5c = syn5c()
Syn5d = syn5d()
Syn6a = syn6a()
Syn6b = syn6b()

# synchronous machines controls
Avr1 = avr1() # automatic voltage regulators
Avr2 = avr2()
Avr3 = avr3()
Tg1 = tg1()   # turbine governors
Tg2 = tg2()
Pss1 = pss1() # power system stabilizers
Pss2 = pss2()
Pss3 = pss3()
```

The script is organized into two parts. In the first part, all modules are imported; in the second one, an instance of all classes is assigned to a variable with a meaningful name. Hence, any script importing the module `system` has access to any of these instances. For example `system.Settings` allows accessing methods and attributes of the instance of the class `settings` that, surprisingly enough, contains general settings for the system including power flow and numerical integration settings. The class `device` and its associated instance `Device` functions as the class `function` described in the Script 3.1 is further discussed at the end of this section. Other variable names should be self-explicative. The reader is invited to familiarize with the variable names of the code above because these will be used systematically throughout this book.

Another important point is the definition of a list of all devices implemented in the program (`device_list`). This list is used by the class `device` discussed below. Adding a new device is as simple as (i) importing the device module, (ii) adding a item to the list `device_list` and (iii) adding an instance of the new device. After these simple operations, the whole program treats the new device as any other pre-existing devices.

Main Script

The functions of the main script are:

1. Initializing all elements of the package.
2. Collecting the information about input data and custom options.
3. Checking the consistency of the options.
4. Calling the adequate parser for reading the input data.
5. Calling the power flow routine and checking the solution.
6. If required, initializing dynamic devices.
7. If required, calling a solver for further analysis (e.g. numerical integration, optimal power flow, etc.).

8. If required, calling the post-processing scripts for generating suitable output report and/or plots.
9. Terminating the execution and, if required, producing a log file.

No one of the operations above is *per se* particularly complicated. The main script is more a "secretary" that organizes the work of other routines. No "real" task should be solved in the main script rather those described above. This rule allows separating the *administrative* work for the *technical* one, which is carried out by specific solvers, parsers, device models, post-processing routines, etc.

The Python language is particularly well-suited for organizing the material in such tidy way. The main tool offered by Python is the possibility of wrapping all scripts (or *modules*) of the same kind within folders, that works as *meta-modules*. The duty of the the main script is thus only to launch the correct module for each task.

An example is probably easier to understand than several abstract explanations. Let us consider the task of choosing the correct parser for the input data. With this aim, suppose that the user can specify the format of the input data as a command line option and that this information is assigned to the variable `input_format`. Assume also that the set of all parsers is contained in the folder `filters`. Within this folder, there is a set of files (i.e., one per parser) and a special file `__init__.py` that is a kind of dedicated "secretary" for that module. The `__init__.py` script provides general methods for handling all defined parsers. Thus, the main script only needs to call the methods defined in the `__init__.py` script. For example, a code fragment that implements the concept above is as follows:

```
import filters

# parse data file using a suitable filter
if not input_format or input_format == 'all':
    input_format = filters.guess(datafile, path)
filters.read(input_format, datafile, path, addfile)
```

The code above takes into account the possibility that the user does not define any format for the input data. If this is the case, an heuristic procedure, namely `filters.guess()`, is called and tries to determine the input data format. Then, the main script call the method `filters.read()` (in the programming slang, this method is a *wrapper*) that tries to parse the input data files `datafile` and the optional additional file `addfile` located at the path `path` using the parser `input_format`. This code fragment remains valid regardless the kind of the data format and regardless the implementation of the `__init__.py` script. Changes are needed only if the input options of the method `filters.read` change.

The main difficulty when writing the main script is to find a general syntax for the modules to be called so that the arguments of wrapper methods are sufficiently general to be able to take into account any possible future modification of the modules. For example, in the case of the wrapper `filters.read()`, one should be sure that `input_format`, `datafile`, `path`, and `addfile` are enough information for any current and future parser. Since one cannot anticipate the future, this is clearly impossible. For example, if a specific data format requires two additional files, the argument list above is incomplete. Fortunately, thanks to the possibilities offered by modern scripting languages such as Python, it is sometime possible to anticipate the future or, in other words, to make that future changes do not affect the current structure of the program. In this case, the ability of *introspection* can be useful. If `addfile` is a list of strings rather than a simple string, one can define any number of additional data files. The only difference in the code is that the wrapper `filters.read()` has to check the type of `addfile`. This is easily done by:

```
if type(addfile) == list:
    some_action
else:
    some_other_action
```

The built-in Python function `type` is a simple way of implementing introspection.

Device Interface Class

The class `function` described in Script 3.1 is a quite simple solution to the problem of interfacing the solver with function types. The class `function` calls all functions defined in the list `flist`, no matter if calling these functions is required or not. If the number of elements is small, there is no real efficiency issue and the class `function` does not need to be optimized.

Unfortunately, for a power system software package, the number of possible devices can be huge. In EPRI Extended Transient-Midterm Stability Program manual [130], tens of AVRs, turbine governors and PSS devices are defined. Furthermore, scripting-based tools for power system analysis are the ideal platform on which building user-defined models.

The key point is that most of these models are likely never used together in the same network. Thus, there is a potential disproportion between the devices normally used and the ones that are actually defined in the package. For example, say that the defined devices are 100 and that those required in a simple benchmark system are 10. During a time domain analysis, device methods are called thousands of times. It is clearly inefficient to call all 100 device methods since 90% of them actually does nothing.

This problem is not new. For example, PSCAD [183] faces a similar issue for solving electro-magnetic transients. However, the solution is strongly dependent on the programming language used. In PSCAD, mathematical routines are written in FORTRAN which is a system language. The only solution in this case is to create for each case study a specific routine and compiling it with a FORTRAN compiler. Thus, in the case of PSCAD, the interface actually collects the information of the system topology and devices and writes the equations of the system in form of FORTRAN code. The results is certainly efficient since the program that is executed is specifically suited for the case study. However, one has to write and compile a certain amount of code before running the simulation.

Scripting languages allows implementing the approach discussed above, but also provide other interesting solutions. One of these solutions relies on meta-programming. The key point is that a scripting language is interpreted and not compiled. Thus, a script can be used for generating code fragments during its execution and then for executing that code "on the fly". This feature can be conveniently used for implementing the interface class.

For example, suppose that we want to solve a numerical integration for the IEEE 14-bus system. Assuming that an implicit solver is used, one has to call, at each iteration of each time step (see also Chapter 8 for mathematical details on implicit numerical methods) the differential and algebraic equations as well as Jacobian matrices of differential and algebraic equations.

The following code implements the concept of meta-programming previously discussed.

```
import system

class device:

    self.n = 0
    self.gcall = []
    self.gycall = []
    self.fcall = []
    self.fxcall = []

    def __init__(self, device_list):

        self.devices = device_list

    def setup(self):

        self.n = 0
        for item in self.devices:
            if system.__dict__[item].n:
                self.n += 1
                self.devices.append(item)
                properties = system.__dict__[item].properties
                for key in properties.keys():
                    self.__dict__[key].append(properties[key])

        string = '"""\n'
        for gcall, device in zip(self.gcall, self.devices):
            if gcall:  string += 'system.' + device + '.gcall(system.DAE)\n'
        string += '\n'
        for gycall, device in zip(self.gycall, self.devices):
            if gycall:  string += 'system.' + device + '.gycall(system.DAE)\n'
        string += '\n'
        for fcall, device in zip(self.fcall, self.devices):
```

```
            if fcall:  string += 'system.' + device + '.fcall(system.DAE)\n'
        string += '\n'
        for fxcall, device in zip(self.fxcall, self.devices):
            if fxcall:  string += 'system.' + device + '.fxcall(system.DAE)\n'
        string += '"""'
        self.call_int = compile(eval(string), '', 'exec')
```

The first part of the method `setup` scans all devices defined in the structure `system`. This operation is quite simple in Python thanks to the built-in dictionary `__dict__` that contains the names of all attributes of the structure. If a device is currently defined (i.e., if `system.__dict__[item].n` is a positive integer), then that device is added to the list `devices`. Furthermore, it assumed that each device has an attribute called `properties` that is a dictionary of the form:

```
self.properties = {'gcall':True, 'fcall':False,
                   'gycall':True, 'fxcall':False}
```

In other words, the dictionary `properties` specifies if a device has to be called when computing algebraic equations $\boldsymbol{g}$ (`gcall`), differential equations $\boldsymbol{f}$ (`fcall`), algebraic Jacobian matrix $\boldsymbol{g}_y$ (`gycall`) and remaining Jacobian matrices $\boldsymbol{f}_x$, $\boldsymbol{f}_y$ and $\boldsymbol{g}_x$ (`fxcall`).[5] The class `device` stores the information contained in the attribute `properties` of each device required in the current case study.

The second part of the method `setup` is a meta-code that creates another code and pre-compiles it using the built-in function `compile`. For example, for the IEEE 14-bus system, the variable `string` assumes the following value:

```
"""
system.Line.gcall(system.DAE)
system.PQ.gcall(system.DAE)
system.Shunt.gcall(system.DAE)
system.PV.gcall(system.DAE)
system.SW.gcall(system.DAE)
system.Syn5a.gcall(system.DAE)
system.Syn6a.gcall(system.DAE)
system.Avr1.gcall(system.DAE)

system.Line.gycall(system.DAE)
system.PQ.gycall(system.DAE)
system.Shunt.gycall(system.DAE)
system.PV.gycall(system.DAE)
system.SW.gycall(system.DAE)
system.Syn5a.gycall(system.DAE)
system.Syn6a.gycall(system.DAE)
system.Avr1.gycall(system.DAE)

system.Syn5a.fcall(system.DAE)
system.Syn6a.fcall(system.DAE)
system.Avr1.fcall(system.DAE)
```

[5] Chapter 9 provides further details on device methods and attributes.

```
system.Syn5a.fxcall(system.DAE)
system.Syn6a.fxcall(system.DAE)
system.Avr1.fxcall(system.DAE)
"""
```

All devices defined in the data file (see also Chapter 21) are called depending if it contains algebraic an/or differential equations. For example, the `Shunt` device is purely algebraic and is called only for computing $\boldsymbol{g}$ and $\boldsymbol{g}_{\boldsymbol{y}}$. Each device call accepts as argument the class `system.DAE` that is assumed to contain the value of the complete set of differential-algebraic equations as well as Jacobian matrices.

Part II
Power System Analysis

Chapter 4
Power Flow Analysis

This chapter describes the power flow[1] analysis from both analytic and algorithmic viewpoints. Section 4.1 introduces the power flow problem through a simple example and clarifies the differences between power flow and circuit analysis. Section 4.2 provides a taxonomy of the power flow problem, while Section 4.3 presents the standard power flow equations. Section 4.4 describes the most common algorithms used for solving this problem. These are the Gauss-Seidel's method, the Newton's method and its variants, the fast decoupled power flow and the dc power flow. A discussion about the single and distributed slack bus models and a comparative example are also included in Section 4.4. Section 4.5 provides a general mathematical framework for the power flow problem based on the continuous Newton's method. Finally, Section 4.6 summarizes the most relevant concepts provided in this chapter.

4.1 Background

A classical problem of circuit theory is to find all branch currents and all node voltages of an assigned circuit. Typical input data are generator voltages as well as the impedances of all branches. If all impedances are constant, the resulting set of equations that describe the circuit is linear.

For example, Figure 4.1 represents a single-phase steady-state ac system. Let us assume that the circuit shown in Figure 4.1 represents the single-phase equivalent of a symmetrical balanced three-phase transmission system where the branches between nodes 1, 2 and 3 are transmission lines, the impedance $\bar{z}_3$ is a load and the independent current sources $\bar{\imath}_1$ and $\bar{\imath}_2$ are generators. Assuming node 0 as the reference voltage, the current injections at nodes 1,

[1] In several papers and books, especially old ones, the power flow analysis is called *load flow* analysis. This notation should be avoided since, quoting Concordia and Tinney [335]: "Load does not flow, but power flows."

F. Milano: Power System Modelling and Scripting, Power Systems, pp. 61–101.
springerlink.com

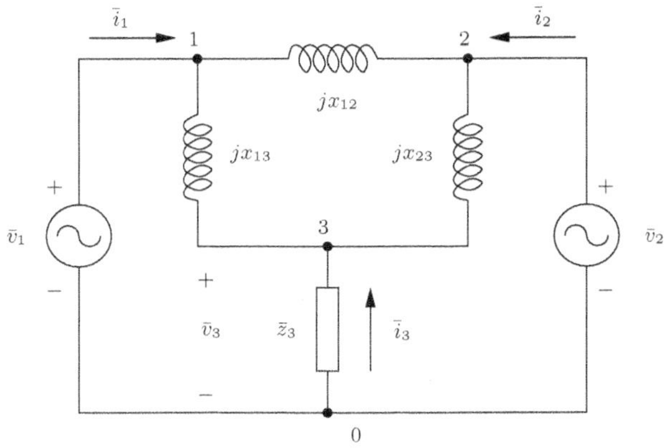

Fig. 4.1 Classical circuit problem

2 and 3 are obtained based on the well-known *branch current method*[2] as the solution of a simple set of linear equations:

$$\begin{aligned} 0 &= \frac{\bar{v}_1 - \bar{v}_2}{jx_{12}} + \frac{\bar{v}_1 - \bar{v}_3}{jx_{13}} - \bar{\imath}_1 \\ 0 &= \frac{\bar{v}_2 - \bar{v}_1}{jx_{12}} + \frac{\bar{v}_2 - \bar{v}_3}{jx_{23}} - \bar{\imath}_2 \\ 0 &= \frac{\bar{v}_3 - \bar{v}_1}{jx_{13}} + \frac{\bar{v}_3 - \bar{v}_2}{jx_{23}} - \bar{\imath}_3 \end{aligned} \tag{4.1}$$

where $\bar{\imath}_1$ and $\bar{\imath}_2$ are imposed by the generators and $\bar{\imath}_3$ depends on the voltage $\bar{v}_3$, as follows:

$$\bar{\imath}_3 = -\bar{v}_3/\bar{z}_3 \tag{4.2}$$

Rewriting (4.1) and (4.2) in vectorial form, one has:

$$\begin{bmatrix} \bar{\imath}_1 \\ \bar{\imath}_2 \\ 0 \end{bmatrix} = \left(\bar{\boldsymbol{Y}} + \boldsymbol{I}_3 \begin{bmatrix} 0 \\ 0 \\ 1/\bar{z}_3 \end{bmatrix} \right) \begin{bmatrix} \bar{v}_1 \\ \bar{v}_2 \\ \bar{v}_3 \end{bmatrix} = \bar{\boldsymbol{Y}}_{\rm tot} \bar{\boldsymbol{v}} \tag{4.3}$$

where $\boldsymbol{I}_3$ is a 3×3 identity matrix and $\bar{\boldsymbol{Y}}$ is the so-called admittance matrix:

$$\bar{\boldsymbol{Y}} = \begin{bmatrix} 1/jx_{12} + 1/jx_{13} & -1/jx_{12} & -1/jx_{13} \\ -1/jx_{12} & 1/jx_{12} + 1/jx_{23} & -1/jx_{23} \\ -1/jx_{13} & -1/jx_{23} & 1/jx_{13} + 1/jx_{23} \end{bmatrix} \tag{4.4}$$

[2] The *mesh* (or *loop*) *current method* is not used in this example because: (i) it can be used only for planar circuits, (ii) it is hard to implement in a computer code and, as a consequence of the previous issues, (iii) it has no relevant practical applications except for being a problem source for first-year students of electrical circuits.

More details about the admittance matrix are given in Section 11.1 of Chapter 11. For the moment, it suffices to say that this matrix can be easily built once the circuit topology is defined and that $-1/\bar{z}_3$ is taken apart from the admittance matrix since it does not belong to the transmission system. Since (4.3) is linear, the solution $\bar{\boldsymbol{v}}$ is unique and can be obtained, for example by means of an LU factorization of $\bar{\boldsymbol{Y}}_{\rm tot}$.[3]

The power flow problem is conceptually the same problem as solving a steady-state ac circuit as the one shown in Figure 4.1. The only, though substantial, difference is the set of input data. In power flow analysis, loads are expressed in terms of consumed active and reactive powers (*PQ load*) and generators are defined in terms of constant voltage magnitude and active power injection (*PV generator*). Finally, one generator is defined as a standard independent voltage source (i.e., as a constant voltage magnitude and phase angle) and is called *slack* or *swing generator*.

Defining one generator phase angle is needed for two reasons: (i) to fix an angle reference, and (ii) to balance system active losses. The first reason is mathematical: in ac systems, at least one phase angle has always to be assigned otherwise system equations are under-determined. For the same reason, in the branch current method discussed above, one node has to be chosen as the reference voltage level. The second reason can be explained based on a simple physical remark: one cannot fix all generator and load active powers since active power losses are not known *a priori*. Thus at least one active power has to be free to vary to account for transmission losses.[4]

In power flow analysis, it is also typical to represent the circuit as a one-line diagram as illustrated in Figure 4.2. The one-line diagram is topologically equivalent to the circuit of Figure 4.1 and implicitly assumes that generators and loads (and all shunt elements) are connected trough a common reference bus 0, which is called *ground* or *earth*.

For the system shown in Figure 4.2, one can write three complex equations (e.g., the expression of the complex power injection at each bus) or, as it is common practice to separate active and reactive power injections, six real equations. Assuming voltage polar coordinates (e.g., $\bar{v} = ve^{j\theta}$) and that the generator at bus 1 is the slack, generator at bus 2 is a PV and the load at bus 3 is a PQ, input data are: v_1, θ_1, p_2, v_2, p_3 and q_3. Thus, variables are p_1, q_1, q_2, θ_2, v_3 and θ_3.

The power flow problem is formulated in order to determine unknown voltage magnitudes and angles. Remaining unknowns, i.e., the power injections,

[3] In this simple example, (4.3) can be solved by hand. However practical systems have much more than three nodes and a numerical solution is thus the unique choice.

[4] In the example discussed in this section (see Figure 4.2), transmission lines have no resistance, thus the active power balance can be actually deduced before solving the power flow analysis. However, a loss-less transmission system is a mere approximation and is used in this example only for the sake of simplicity.

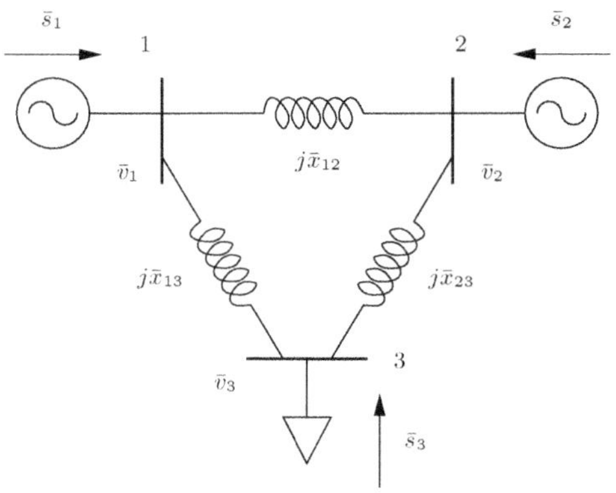

Fig. 4.2 Classical power flow problem

can be straightforwardly computed once all bus voltages are known. In conclusion, one has to write only the equations where active or reactive power injections are known. For example, according to the rule above, the power flow equations of the system depicted in Figure 4.5 are:

$$\begin{aligned}
0 &= \frac{v_2 v_1}{x_{12}} \sin(\theta_2 - \theta_1) - p_2 \\
0 &= \frac{v_3 v_1}{x_{13}} \sin(\theta_3 - \theta_1) + \frac{v_3 v_2}{x_{23}} \sin(\theta_3 - \theta_2) - p_3 \\
0 &= \frac{v_3^2}{x_{13}} + \frac{v_3^2}{x_{23}} - \frac{v_3 v_1}{x_{13}} \cos(\theta_3 - \theta_1) - \frac{v_3 v_2}{x_{23}} \cos(\theta_3 - \theta_2) - q_3
\end{aligned} \tag{4.5}$$

where the unknowns are v_3, θ_3 and θ_2. It is important to note that the resulting set of equations is intrinsically nonlinear since the load at bus 3 is specified as a constant power consumption.

Equations (4.5) originate from the models of generators and loads. These models are defined based on common practice, as follows.

1. At high voltage level, loads represents an equivalent of sub-transmission systems or distribution networks. The equivalent load power consumptions can be generally well approximated as voltage dependent monomials:

$$\begin{aligned}
p_L &= p_{L0} v^{a_p} \\
q_L &= q_{L0} v^{a_q}
\end{aligned} \tag{4.6}$$

   where p_{L0} and q_{L0} are the active and reactive power consumption, respectively, at the nominal voltage. Moreover, the voltage of the equivalent load bus is typically regulated through an under load tap changer (see Chapter 14). Thus, if the voltage is maintained constant, the load can be modelled in steady-state as a constant active and reactive power consumption.

2. Synchronous generator turbine governors and automatic voltage controls are able to regulate the generated active power and the voltage at the machine bus, respectively (see Chapter 16). In steady-state, these controls can be modelled as constant p and v at the generator bus.
3. Less intuitive is the physical meaning of the slack bus, which is actually quite artificial. A slack bus is, in principle, a generator as any PV ones. However, since one active power cannot be assigned, the slack generator can only specify the voltage magnitude. Then, for similarity with all other buses where there are two unknowns, the slack generator bus is also used as reference voltage angle. Further insights and critics on the standard slack generator model are provided in Subsection 4.4.9 and in Section 10.2.2 of Chapter 10.
4. Transmission lines and transformers are generally modelled as lumped π-circuits with constant parameters (see Chapter 11).

These assumptions yield the *classical* power flow model. This model is nowadays so widely accepted that practically all commercial tools implement power flow equations using the assumptions discussed above. Main differences about power flow methods implemented in practice concern how variable limits (e.g., generator reactive power and load voltage limits) are handled. Another important issue is how discrete variables are handled (e.g., tap ratio of under load tap changer transformers). While these topics are detailed in Chapters 10 and 11, this chapter only focuses on power flow solution algorithms. However, it is important to note that the standard power flow model is not established by "law". For example, there is no specific reason for assigning the reference angle at the slack generator, or for not using load and generator voltage dependent models. A very common error is to confuse common (and reasonable) model assumptions with the power flow problem itself. Furthermore, whenever possible, the solution method should not rely on the model. Last but not least, these assumptions are not unchangeable just because they are used in common commercial software packages.

In the remainder of this chapter, the power flow analysis is formulated as a general problem of finding a physical[5] solution of a set of nonlinear equations. Device models are exploited only if the algorithm used for solving the power flow problem relies on them. However, one has to be aware that relying on a specific device model is a drawback of the solution method, not an advantage.

Finally, before going into the mathematical matter, let me question the power flow problem itself. Saint Agustin wrote that the doubt is the first step on the way to the truth. Thus, to question everything, especially very basic concepts, is always a good habit. The power flow analysis is actually an

[5] In this context, physical means *acceptable*. Since power flow equations are not linear, they have, generally, more than one solution. Only the solutions for which system variables are within admissible limits are of interest. Non-physical solutions of the power flow problem are further discussed in Chapter 5.

artificial problem. Determining the bus voltage profile based on bus power injections (positive or negative) is the correct approach only if one knows power injections. Actually, when scheduling the power productions based on forecasted load demand, generator powers and voltage are not known. On the other hand, during normal operations, bus voltages magnitudes as well as power injections can be measured. However, the power flow problem is the starting point for a variety of important further analysis such as, for example, those discussed in the following chapters of this part. For this reason, the solution of power flow problem is a crucial step in power system analysis.

4.2 Taxonomy of Power Flow Problems

The origins of the formulation of the power flow problem and the solution based on the Newton's method date back to the late sixties [310]. Since then, a huge variety of studies have been presented about the solution of the power flow problem, addressing starting initial guess [296], computational efficiency [54, 97, 176, 275, 297, 299, 324], ill-conditioned cases and robustness [27, 29, 94, 149, 150, 270, 304, 315, 317], multiple solutions [135, 221], and unsolvable cases [219, 220].

It is relevant to classify the power flow problems into the following categories:

1. *Well-conditioned case.* The power flow solution exists and is reachable using a flat initial guess (e.g., all load voltage magnitudes equal to 1 and all bus voltage angles equal to 0) and a standard Newton's method. This case is the most common situation.
2. *Ill-conditioned case.* The solution of the power flow problem does exist, but standard solvers fail to get this solution starting from a flat initial guess. This situation is due to the fact that the region of attraction of the power flow solution is narrow or far from the initial guess. In this case, the failure of standard power flow procedure is due to the instability of the numerical method, not of the power flow equations. Robust power flow methods have proved to be efficacious for solving ill-conditioned cases.
3. *Bifurcation point.* The solution of the power flow exists but it is either a saddle-node bifurcation or a limit-induced bifurcation [39].

   a. Saddle-node bifurcations are associated with the maximum loading condition of a system. The solution cannot be obtained using standard or robust power flow methods, since the power flow Jacobian matrix is singular at the solution point.
   b. Limited-induced bifurcations are associated with a physical limit of the system, such as the shortage of generator reactive power. Although limit-induced bifurcation can in some cases lead to the voltage collapse of the system, the solution point is typically a well-conditioned case and does not show convergence issues.

Several continuation techniques [5, 39] and optimal power flow problems [36, 112, 148] have been proposed for determining bifurcation points (for additional references see also Chapters 5). These methods allow defining the distance between the present power flow solution and the bifurcation points and thus are useful for assessing the voltage stability of the system [39]. However, encountering a case study whose solution is exactly a bifurcation point is quite uncommon in practice.

4. *Unsolvable case.* The solution of the power flow problem does not exist. Typically, the issue is that the loading level of the network is too high. As in the case of the bifurcation points, a continuation method or an optimal power flow problem allow defining the maximum loading level that the system can supply. An alternative method to analyze unsolvable cases is given in [219, 220]. As shown in [219], robust power flow methods provide a solution close to the feasibility boundary rather than diverge.

The continuation power flow analysis discussed in Chapter 5 provides a general and powerful approach to both assess bifurcation points and handle unsolvable cases. Thus, this chapter only focuses on solvable and ill-conditioned cases.

4.3 Classical Power Flow Equations

As discussed in the previous section, there is no particular reason (but historical ones) for reducing the power flow model to power flows in transmission lines, constant PQ loads and constant PV or slack generators. Generally speaking, the power flow problem consists in finding the zero of a set of nonlinear equations starting from an adequate initial guess. Thus, the most general form of the power flow equations is a set of DAE in steady-state, as follows:

$$\begin{aligned} \mathbf{0} &= \boldsymbol{f}(\boldsymbol{x}, \boldsymbol{y}) \\ \mathbf{0} &= \boldsymbol{g}(\boldsymbol{x}, \boldsymbol{y}) \end{aligned} \tag{4.7}$$

where differential equations $\boldsymbol{f}$ model dynamic devices such as, for example, under load tap changers,[6] and algebraic equations $\boldsymbol{g}$ define the power balance at network buses.

However, in general, most dynamic devices are initialized after solving the power flow problem (e.g., synchronous machines and regulators).[7] Thus, the most common formulation of the power flow equations is reduced to the algebraic part of (4.7):

$$\mathbf{0} = \boldsymbol{g}(\boldsymbol{y}) \tag{4.8}$$

[6] A discussion about regulating transformers is given in Subsection 11.2.2 of Chapter 11.

[7] Subsection 9.1.1 of Chapter 7 discusses in detail this topic.

Equations (4.8) can be written in several ways. However, the classical formulation of power flow equations, which is intuitively introduced in Section 4.1, is as follows. The vector of currents injected at each node is:

$$\bar{\boldsymbol{i}} = \bar{\boldsymbol{Y}}\bar{\boldsymbol{v}} \tag{4.9}$$

which leads to write (4.8) as the complex power injections at buses:

$$\bar{\boldsymbol{s}} = \bar{\boldsymbol{V}}\bar{\boldsymbol{i}}^* = \bar{\boldsymbol{V}}\bar{\boldsymbol{Y}}^*\bar{\boldsymbol{v}}^* \tag{4.10}$$

where $\bar{\boldsymbol{V}} = \text{diag}(\bar{v}_1, \bar{v}_2, \dots, \bar{v}_{n_b})$ and n_b is the number of network buses. In tensorial form, (4.10) becomes:

$$\bar{s}_h = \bar{v}_h\bar{i}_h^* = \bar{v}_h \sum_{k\in\mathcal{B}} \bar{y}_{hk}^*\bar{v}_k^*, \quad h \in \mathcal{B} \tag{4.11}$$

where $\mathcal{B} = \{1, 2, \dots, n_b\}$ and $\bar{y}_{hk}$ is the element (h, k) of the admittance matrix $\bar{\boldsymbol{Y}}$. From (4.11), one obtains:

$$\bar{s}_h = p_h + jq_h = \bar{v}_h \sum_{k\in\mathcal{B}} (g_{hk} - jb_{hk})\bar{v}_k^*, \quad h \in \mathcal{B} \tag{4.12}$$

where $\bar{y}_{hk} = g_{hk} + jb_{hk}$.

In (4.12), p_h and q_h are neat power injections at the bus h. If at the same bus there is both a PV generator and a PQ load, then the bus is considered a PV whose power injection p_h is defined as the difference of generator and load powers connected to that bus:

$$p_h = p_G - p_L \tag{4.13}$$

The reactive power generated by the PV can be computed only after solving the power flow as

$$q_G = q_h + q_L \tag{4.14}$$

Similar considerations can be done if a PQ load is connected at the same bus as a slack generator.

The product of voltage phasors can be written in polar form as:

$$\bar{v}_h\bar{v}_k^* = v_h e^{j\theta_h} v_k e^{-j\theta_k} = v_h v_k e^{j(\theta_h - \theta_k)} \tag{4.15}$$

Thus, using (4.15), (4.12) becomes:

$$\begin{aligned} p_h &= v_h \sum_{k\in\mathcal{B}} v_k (g_{hk}\cos\theta_{hk} + b_{hk}\sin\theta_{hk}), \quad h \in \mathcal{B} \\ q_h &= v_h \sum_{k\in\mathcal{B}} v_k (g_{hk}\sin\theta_{hk} - b_{hk}\cos\theta_{hk}), \quad h \in \mathcal{B} \end{aligned} \tag{4.16}$$

where $\theta_{hk} = \theta_h - \theta_k$. In the classical power flow formulation, the variables are voltage amplitudes and phases at load buses, reactive powers and voltage phases at generator PV buses and active and reactive power at the slack bus [310]. A synoptic summary of variables and data for each bus type is shown in Table 4.1.

Table 4.1 Variables and parameters for each bus type in the classical power flow problem formulation

Bus type	Variables	Data
Slack generator	p, q	v, θ
PV generator	q, θ	p, v
PQ load	v, θ	p, q

Alternatively, the product of voltage phasors in (4.12) can be written in rectangular form as:

$$\bar{v}_h \bar{v}_k^* = (v_{d,h} + jv_{q,h})(v_{d,k} - jv_{q,k}) \tag{4.17}$$

Thus, using (4.17), (4.12) becomes:

$$\begin{aligned} p_h &= \sum_{k \in \mathcal{B}} [v_{d,h}(g_{hk}v_{d,k} - b_{hk}v_{q,k}) + v_{q,h}(g_{hk}v_{q,k} + b_{hk}v_{d,k})], \quad h \in \mathcal{B} \\ q_h &= \sum_{k \in \mathcal{B}} [v_{q,h}(g_{hk}v_{d,k} - b_{hk}v_{q,k}) - v_{d,h}(g_{hk}v_{q,k} + b_{hk}v_{d,k})], \quad h \in \mathcal{B} \end{aligned} \tag{4.18}$$

In (4.18), bus voltage magnitudes of PV generators do not appears explicitly. Thus, it is necessary to add a set of equations for imposing such voltage magnitudes:

$$0 = v_{d,h}^2 + v_{q,h}^2 - v_{h,\mathrm{PV}}^2, \quad h \in \mathcal{B}_{\mathrm{PV}} \tag{4.19}$$

where $\mathcal{B}_{\mathrm{PV}}$ is the set of PV generator buses and $v_{h,\mathrm{PV}}$ is the desired PV generator voltage. As for the slack generator, (4.19) is not necessary since at the slack generator imposes both the magnitude and the phase angle of the bus voltage. For example, if $\theta_{\mathrm{slack}} = 0$, one has:

$$0 = v_{d,\mathrm{slack}} - v_{\mathrm{slack}}, \qquad 0 = v_{q,\mathrm{slack}} \tag{4.20}$$

From the mathematical viewpoint, equations (4.16) and equations (4.18)-(4.20) are equivalent. From the computational viewpoint, the polar form involves the relatively costly sine and cosine functions while the rectangular form needs the additional equations (4.19). Both formulations show advantages and drawbacks. In some cases, as it is discussed later in the chapter, the solution method drives the choice of the polar or of the rectangular form.

Both (4.16) and (4.18)-(4.20) are nonlinear and have no analytical explicit solution. Not even the simple loss-less system (4.5) can be solved by hand. Actually, the analytical solution of the power flow problem can be found only for a loss-less two-bus system. Thus, one has to use a numerical iterative technique for solving the power flow problem.

4.4 Power Flow Solvers

4.4.1 Jacobi and Gauss-Seidel's Method

The Jacobi's and Gauss-Seidel's methods are iterative techniques for solving a set of linear equations in the form:

$$\boldsymbol{A}\boldsymbol{y} = \boldsymbol{b} \tag{4.21}$$

Especially the Gauss-Seidel's method has been widely used in the last decades for solving the power flow problem since it do not require factorizing the matrix $\boldsymbol{A}$. Nowadays, computational constraints are less binding and other methods are preferred. However, both Jacobi's and Gauss-Seidel's methods still have a didactic value.

Decomposing $\boldsymbol{A}$ in a diagonal component $\boldsymbol{D}$ and a remainder $\boldsymbol{R}$,

$$\boldsymbol{A} = \boldsymbol{D} + \boldsymbol{R} \tag{4.22}$$

where:

$$\boldsymbol{D} = \begin{bmatrix} a_{11} & 0 & \cdots & 0 \\ 0 & a_{22} & \cdots & 0 \\ \vdots & \vdots & \ddots & \vdots \\ 0 & 0 & \cdots & a_{nn} \end{bmatrix}, \quad \boldsymbol{R} = \begin{bmatrix} 0 & a_{12} & \cdots & a_{1n} \\ a_{21} & 0 & \cdots & a_{2n} \\ \vdots & \vdots & \ddots & \vdots \\ a_{n1} & a_{n2} & \cdots & 0 \end{bmatrix} \tag{4.23}$$

Thus, (4.21) can be rewritten as:

$$\boldsymbol{D}\boldsymbol{y} = \boldsymbol{b} - \boldsymbol{R}\boldsymbol{y} \tag{4.24}$$

The Jacobi's method consists in solving iteratively the left-end side of (4.24) using the current values of the elements of the vector $\boldsymbol{y}$ in the right-end side:

$$\boldsymbol{y}^{(i+1)} = \boldsymbol{D}^{-1}(\boldsymbol{b} - \boldsymbol{R}\boldsymbol{y}^{(i)}) \tag{4.25}$$

or, in tensorial form:

$$y_h^{(i+1)} = y_h^{(i)} + \frac{1}{a_{hh}}(b_h - \sum_{k=1}^{n} a_{hk} y_k^{(i)}), \quad h = 1, 2, \ldots, n \tag{4.26}$$

The Gauss-Seidel's method is very similar to the Jacobi's one. In this case, $\boldsymbol{A}$ is decomposed in a lower triangular component $\boldsymbol{L}$ and a strictly upper triangular one $\boldsymbol{U}$,

$$\boldsymbol{A} = \boldsymbol{L} + \boldsymbol{U} \tag{4.27}$$

where:

$$\boldsymbol{L} = \begin{bmatrix} a_{11} & 0 & \cdots & 0 \\ a_{21} & a_{22} & \cdots & 0 \\ \vdots & \vdots & \ddots & \vdots \\ a_{n1} & a_{n2} & \cdots & a_{nn} \end{bmatrix}, \quad \boldsymbol{U} = \begin{bmatrix} 0 & a_{12} & \cdots & a_{1n} \\ 0 & 0 & \cdots & a_{2n} \\ \vdots & \vdots & \ddots & \vdots \\ 0 & 0 & \cdots & 0 \end{bmatrix} \tag{4.28}$$

Thus, (4.21) can be rewritten as:

$$\boldsymbol{L}\boldsymbol{y} = \boldsymbol{b} - \boldsymbol{U}\boldsymbol{y} \tag{4.29}$$

The Gauss-Seidel's method consists in solving iteratively the left-end side of (4.29) using the current values of the elements of the vector $\boldsymbol{y}$ in the right-end side:

$$\boldsymbol{y}^{(i+1)} = \boldsymbol{L}^{-1}(\boldsymbol{b} - \boldsymbol{U}\boldsymbol{y}^{(i)}) \tag{4.30}$$

or, in tensorial form:

$$y_h^{(i+1)} = \frac{1}{a_{hh}}(b_h - \sum_{k=h+1}^{n} a_{hk} y_k^{(i)} - \sum_{k=1}^{h-1} a_{hk} y_k^{(i+1)}), \quad h = 1, 2, \ldots, n \tag{4.31}$$

Both methods stop if the maximum equation mismatch:

$$\max\{|\boldsymbol{A}\boldsymbol{y}^{(i+1)} - \boldsymbol{b}|\} < \epsilon \tag{4.32}$$

or the maximum variable variation:

$$\max\{|\boldsymbol{y}^{(i+1)} - \boldsymbol{y}^{(i)}|\} < \epsilon \tag{4.33}$$

where ϵ is a given tolerance, or the number of iterations is greater than a given limit $i^{\max}$. The main differences between the Jacobi's method and the Gauss-Seidel's one are twofold:

1. In the Gauss-Seidel's method, each variable $y_h^{(i+1)}$ is updated using the previously updated variables $y_k^{(i+1)}$, for all $k < h$. This allows accelerating the convergence of the algorithm. This is why the Gauss-Seidel's method is more used than the Jacobi's one for solving the power flow problem.
2. The previous point yields another relevant difference. Since the Gauss-Seidel's method uses updated variables $y_k^{(i+1)}$ with $k < h$ for computing $y_h^{(i+1)}$, (4.30) cannot be computed using vectorial operations, but has to be necessarily implemented element by element in a for-loop. On the other

hand, (4.25) can be implemented using vectors. This difference is clearly not relevant if using programming languages such as C or FORTRAN. However, in case of using scripting languages, such as Matlab and Python, for-loops are always less efficient than vector algebra. The rationale for this difference in performance is the following. In scripting languages, for-loops implies working on arrays element by element, which is generally not very efficient. On the other hand, vector algebra internally calls efficient C-based routines (e.g., BLAS library), which generally leads to reduce the computing time.

Thus, the choice of the method depends in some measure on the programming language. For system languages, the Gauss-Seidel's method is the best option. For scripting languages, The Jacobi's method can lead to save time in case the time spent in solving extra iterations (with respect to the Gauss-Seidel's method) is compensated by the time saved using vectorial code.

Equation (4.21) is linear, thus is not directly applicable to power flow equations. However, from (4.11), one can write:

$$\sum_{k \in \mathcal{B}} \bar{y}_{hk} \bar{v}_k = \bar{s}_h^* / \bar{v}_h^*, \quad h \in \mathcal{B} \tag{4.34}$$

or, in vectorial form:

$$\bar{\boldsymbol{Y}} \bar{\boldsymbol{v}} = [\bar{\boldsymbol{V}}^*]^{-1} \bar{\boldsymbol{s}}^* \tag{4.35}$$

where $\bar{\boldsymbol{V}} = \mathrm{diag}(\bar{v}_1, \bar{v}_2, \dots, \bar{v}_{n_b})$. Imposing $\boldsymbol{A} = \bar{\boldsymbol{Y}}$, $\boldsymbol{y} = \bar{\boldsymbol{v}}$ and $\boldsymbol{b} = [\bar{\boldsymbol{V}}^*]^{-1} \bar{\boldsymbol{s}}^*$, the Jacobi's and Gauss-Seidel's methods can be straightforwardly implemented.

The convergence criterion generally used for the power flow problem is:

$$\max\{|\bar{\boldsymbol{s}}^{(i+1)} - \bar{\boldsymbol{v}}^{(i+1)} \bar{\boldsymbol{Y}}^* \bar{\boldsymbol{v}}^{*,(i+1)}|\} < \epsilon \tag{4.36}$$

which is generally referred to as *power mismatch*. In common practice, equation (4.36) is split into its real and imaginary components. This convergence criterion is generally preferred to (4.32) due to its direct physical meaning. For example, $\epsilon = 0.001$ means a 0.1% power mismatch that, on a 100 MVA base, corresponds to a 0.1 MVA.

According to the classical power flow model, there are only two kinds of devices that update variables $\bar{\boldsymbol{v}}$, namely PQ loads and PV generators, as follows.

PQ loads: The ith iteration of the Jacobi's method is:

$$\bar{v}_h^{(i+1)} = \bar{v}_h^{(i)} + \frac{1}{\bar{y}_{hh}} \left(\frac{\bar{s}_h^*}{\bar{v}_h^{*,(i)}} - \sum_{k \in \mathcal{B}} \bar{y}_{hk} \bar{v}_k^{(i)} \right), \quad h \in \mathcal{B} \tag{4.37}$$

The ith iteration of the Gauss-Seidel's method is:

$$\bar{v}_h^{(i+1)} = \frac{1}{\bar{y}_{hh}} \left(\frac{\bar{s}_h^*}{\bar{v}_h^{*,(i)}} - \sum_{k>h} \bar{y}_{hk} \bar{v}_k^{(i)} - \sum_{k<h} \bar{y}_{hk} \bar{v}_k^{(i+1)} \right), \quad h \in \mathcal{B}, \ k \in \mathcal{B} \tag{4.38}$$

PV generator: Equations (4.37) and (4.38) hold also for PV generators for the Jacobi's and Gauss-Seidel's method, respectively. However, since reactive power injections are not known at PV generator buses, $\bar{s}_h$ is estimated as follows:

$$\bar{s}_h^{(i)} = p_h + j\Im\{\bar{v}_h^{(i)} \sum_{k \in \mathcal{B}} \bar{y}_{hk}^* \bar{v}_k^{*,(i)}\}, \quad h \in \mathcal{B} \tag{4.39}$$

Equations (4.37) and (4.38) provide both the new magnitude and the new phase of voltage $\bar{v}_h^{(i+1)}$. Since the PV generator imposes the voltage magnitude, only the voltage angle has to be updated.

Script 4.1 Jacobi's and Gauss-Seidel's Methods

The following code fragments implement (4.37) and (4.38) in Python language. It is assumed that these methods are part of a class that implements the PQ load.

```
from cvxopt.base import matrix, spdiag, mul, div, log, exp
from cvxopt.blas import dotu # scalar product
import cmath

def gauss(self, dae, line, pinj, qinj):

    for item in range(self.n):
        a = self.a[item]
        v = self.v[item]
        vy = mul(dae.y[line.v], exp(dae.y[line.a]*1j))
        vl = vy[a]
        k1 = dotu(matrix(line.Y[a, :]), vy)
        k3 = (pinj[a] + qinj[a]*1j)/vl
        vl += (k3.conjugate() - k1)/line.Y[a, a]
        dae.y[v] = abs(vl)
        dae.y[a] = cmath.log(vl/abs(vl)).imag

def jacobi(self, dae, line, pinj, qinj):

    vl = mul(dae.y[self.v] + 0j, exp(dae.y[self.a]*1j))
    vy = mul(dae.y[line.v] + 0j, exp(dae.y[line.a]*1j))
    k1 = div(pinj[self.a] + qinj[self.a]*1j, vl)
    u = matrix(1, (self.n, 1), 'z')
    y = mul(spdiag(u), line.Y[self.a, self.a])*u
    vl += div(k1.H.T - line.Y[self.a, :]*vy, y)
    mod = abs(vl)
    ang = log(div(vl, mod))
```

```
dae.y[self.v] = mod
dae.y[self.a] = ang.imag()
```

In the code above, `self.a` and `self.v` represent the PQ bus voltage angle and magnitude indexes, respectively, of the algebraic variable vector `dae.y`. Other variable names are self-explicative. For example, `line.Y` is the admittance matrix $\bar{\boldsymbol{Y}}$ defined as a sparse matrix `cvxopt.base.spmatrix`. The statement `k1.H.T` means the transpose of the Hermitian (e.g., conjugate transpose) of `k1` and thus yields the transpose of `k1` (see also Appendix A). As discussed above, the Jacobi's iteration does not need a for-loop, while the Gauss-Seidel's iteration does. Thus, it has to be expected that the Jacobi's iteration is executed faster than the Gauss-Seidel's one, especially for large networks with several PQ loads (see Example 4.3).

4.4.2 Newton's Method

The Newton's method (also known as Newton-Raphson's or Newton-Fourier's method) for solving the power flow problem is described in many books and papers (e.g., [310]). It is nowadays the most commonly used algorithm for solving the power flow problem (along with the fast decoupled power flow). However, in the seventies it was considered very computationally expensive due to the need of computing and factorizing the Jacobian matrix at each iteration.

The i-th iteration of the classical Newton's method for (4.8) is as follows:

$$\begin{aligned} \Delta \boldsymbol{y}^{(i)} &= -[\boldsymbol{g}_{\boldsymbol{y}}^{(i)}]^{-1} \boldsymbol{g}^{(i)} \\ \boldsymbol{y}^{(i+1)} &= \boldsymbol{y}^{(i)} + \Delta \boldsymbol{y}^{(i)} \end{aligned} \tag{4.40}$$

where $\boldsymbol{g}^{(i)} = \boldsymbol{g}(\boldsymbol{y}^{(i)})$, $\boldsymbol{g}_{\boldsymbol{y}}^{(i)} = \boldsymbol{g}_{\boldsymbol{y}}(\boldsymbol{y}^{(i)})$, and $\boldsymbol{g}_{\boldsymbol{y}} = \nabla_{\boldsymbol{y}}^T \boldsymbol{g}$ is the Jacobian matrix of the power flow equations. The geometrical interpretation of the Newton's method is well-known. For the actual value $\boldsymbol{y}^{(i)}$, one computes the tangent of $\boldsymbol{g}^{(i)}$ as:

$$\boldsymbol{\tau}(\boldsymbol{y}) = \boldsymbol{g}^{(i)} + \boldsymbol{g}_{\boldsymbol{y}}^{(i)}(\boldsymbol{y} - \boldsymbol{y}^{(i)}) \tag{4.41}$$

Then imposing $\boldsymbol{\tau}(\boldsymbol{y}) = \boldsymbol{0}$ yields the value $\boldsymbol{y}^{(i+1)}$ defined in (4.40). Figure 4.3.a illustrates the Newton's method for a scalar function $g(y)$.

The algorithm ends if the maximum equation mismatch or variable increment satisfies:

$$\max\{|\boldsymbol{g}|\} < \epsilon, \quad \text{or} \quad \max\{|\Delta \boldsymbol{y}|\} < \epsilon \tag{4.42}$$

or if the number of iterations is greater than a given limit $i^{\max}$. In the latter case, the algorithm has likely failed to converge.

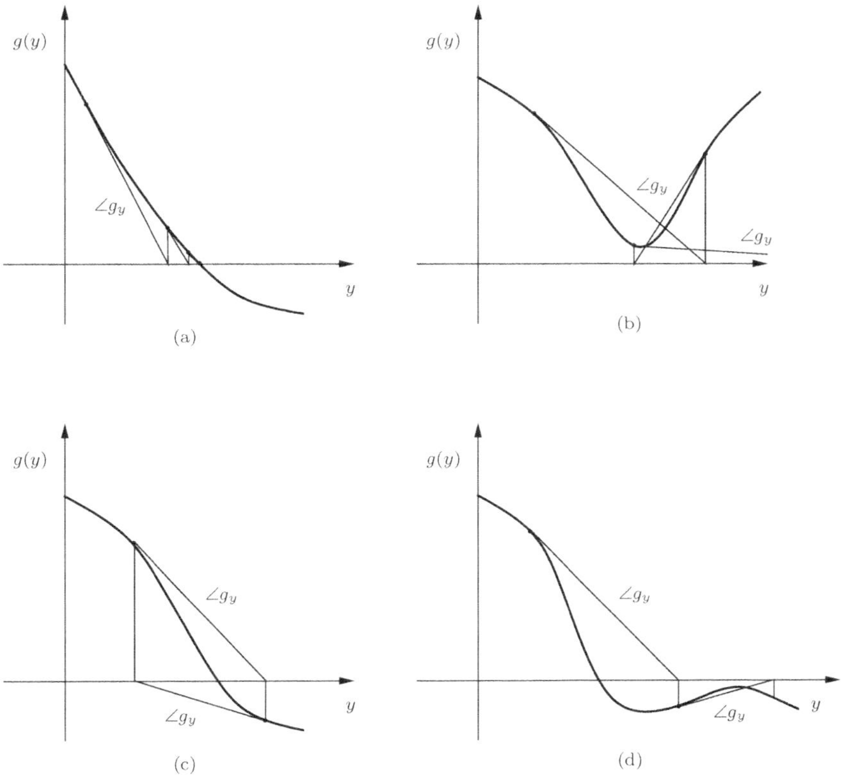

Fig. 4.3 Geometrical interpretation of the Newton's method for a scalar function $g(y)$. (a) well-conditioned case, (b) unsolvable case, (c) and (d) ill-conditioned cases

Script 4.2 Newton's method

The following Python code implements a basic Newton's routine.

```
import system
from cvxopt.base import matrix, spmatrix, sparse, div
from cvxopt.umfpack import symbolic, numeric, solve
from cvxopt.blas import dotu

F = None

def calcInc():

    global F

    exec system.Device.call_pf
    A = sparse(system.DAE.Gy)
    inc = matrix(system.DAE.g)
```

```
    if system.DAE.factorize:
        F = symbolic(A)
        system.DAE.factorize = False

    try:
        N = numeric(A, F)
        solve(A, N, inc)
    except:
        print 'Unexpected symbolic factorization'
        F = symbolic(A)
        solve(A, numeric(A, F), inc)

    return -inc

def powerflow():

    """main power flow routine"""

    # general settings
    iteration = 1
    iter_max = system.Settings.pf_max_iter
    convergence = True
    tol = system.Settings.tol
    system.Settings.error = tol + 1
    err_vec = []

    # main loop
    while system.Settings.error > tol and iteration <= iter_max:

        inc = calcInc()
        system.DAE.y += inc

        system.Settings.error = max(abs(inc))
        err_vec.append(system.Settings.error)

        msg = 'Iteration = %3d   Max. Convergence Error = %8.7f' \
              % (iteration, system.Settings.error)
        print msg

        iteration += 1

        # stop if the error increases too much
        if iteration > 4 and err_vec[-1] > 1000*err_vec[0]:
            print 'The error is increasing too much'
            print 'Convergence is likely not reachable'
            convergence = False
            break

    if iteration > iter_max:
        print 'Reached maximum number of iterations'
        convergence = False
```

The proposed code is fully independent from the system model. As discussed in Script 3.2 of Chapter 3, $\boldsymbol{g}^{(i)}$ and $\boldsymbol{g}_{\boldsymbol{y}}^{(i)}$ are calculated by means of the statement `exec system.Device.call_pf`, which updates the vector `system.DAE.g` the matrix and `system.DAE.Gy` using the current value `system.DAE.y`. The class `system.Device` works as an interface between the power flow routine and the classes that define system devices. For further details on the class `system.Device` see Script 3.2 of Chapter 3.

The function `symbolic` symbolically factorizes the power flow Jacobian matrix. If the number and the position of non-zero elements of the Jacobian matrix do not change, the symbolic factorization is needed only once, which allows saving CPU time. The symbolic re-factorization is needed, for example, each time a PV generator reaches a reactive power production limit. The code also handles an exception in case the Jacobian matrix has changed unexpectedly. This exception generally occurs if the Jacobian matrix is badly conditioned and it is thus a symptom of convergence issues. Finally, the list `err_vec` allows stopping the algorithm in case the variable increment anomalously increases.

4.4.3 Power Flow Jacobian Matrix

For the classical power flow problem (4.16) written in polar form, it is common practice to order variables $\boldsymbol{y}$ so that bus voltage angles are grouped together and come before bus voltage magnitudes:

$$\boldsymbol{y} = [\boldsymbol{\theta}^T, \boldsymbol{v}^T]^T \tag{4.43}$$

In the same vein, equations $\boldsymbol{g}$ are organized grouping together active power mismatches first and then reactive power ones:

$$\boldsymbol{g} = [\boldsymbol{g}_p^T, \boldsymbol{g}_q^T]^T \tag{4.44}$$

With these assumptions, the Jacobian matrix $\boldsymbol{g}_{\boldsymbol{y}}$ can be written as:

$$\boldsymbol{g}_{\boldsymbol{y}} = \begin{bmatrix} \boldsymbol{g}_{p,\boldsymbol{\theta}} & \boldsymbol{g}_{p,\boldsymbol{v}} \\ \boldsymbol{g}_{q,\boldsymbol{\theta}} & \boldsymbol{g}_{q,\boldsymbol{v}} \end{bmatrix} \tag{4.45}$$

where $\boldsymbol{g}_{p,\boldsymbol{\theta}} = \nabla_{\boldsymbol{\theta}}^T \boldsymbol{g}_p$, $\boldsymbol{g}_{p,\boldsymbol{v}} = \nabla_{\boldsymbol{v}}^T \boldsymbol{g}_p$, $\boldsymbol{g}_{q,\boldsymbol{\theta}} = \nabla_{\boldsymbol{\theta}}^T \boldsymbol{g}_q$, and $\boldsymbol{g}_{q,\boldsymbol{v}} = \nabla_{\boldsymbol{v}}^T \boldsymbol{g}_q$.

The expressions for the elements of the Jacobian matrix $\partial \boldsymbol{g}_p / \partial \boldsymbol{y}$ of (4.16) are:

$$\begin{aligned}
\frac{\partial g_{p,h}}{\partial \theta_h} &= -v_h v_k \sum_{k \neq h}^{n_b} (g_{hk} \sin\theta_{hk} - b_{hk}\cos\theta_{hk}) \\
\frac{\partial g_{p,h}}{\partial \theta_k} &= v_h v_k (g_{hk} \sin\theta_{hk} - b_{hk}\cos\theta_{hk}) \\
\frac{\partial g_{p,h}}{\partial v_h} &= 2 v_h g_{hh} + v_k \sum_{k \neq h}^{n_b} (g_{hk} \cos\theta_{hk} + b_{hk}\sin\theta_{hk}) \\
\frac{\partial g_{p,h}}{\partial v_k} &= v_h (g_{hk} \cos\theta_{hk} + b_{hk}\sin\theta_{hk})
\end{aligned} \tag{4.46}$$

and the elements of the Jacobian matrix $\partial \boldsymbol{g}_q / \partial \boldsymbol{y}$ of (4.16) are:

$$\begin{aligned}
\frac{\partial g_{q,h}}{\partial \theta_h} &= -v_h v_k \sum_{k \neq h}^{n_b} (g_{hk} \cos\theta_{hk} + b_{hk}\sin\theta_{hk}) \\
\frac{\partial g_{q,h}}{\partial \theta_k} &= v_h v_k (g_{hk} \cos\theta_{hk} + b_{hk}\sin\theta_{hk}) \\
\frac{\partial g_{q,h}}{\partial v_h} &= -2 v_h b_{hh} - v_k \sum_{k \neq h}^{n_b} (g_{hk} \sin\theta_{hk} - b_{hk}\cos\theta_{hk}) \\
\frac{\partial g_{q,h}}{\partial v_k} &= -v_h (g_{hk} \sin\theta_{hk} - b_{hk}\cos\theta_{hk})
\end{aligned} \tag{4.47}$$

An advantage of the vectorial notation is that (4.46) and (4.47) can be written very compactly as exemplified in the following script.

Script 4.3 Power Flow Jacobian Matrix

The following Python code implements (4.46) and (4.47) without using for-loops:

```
def build_gy(self, dae):

    Vn = exp(dae.y[self.a]*1j)
    Vc = mul(dae.y[self.v] + 0j, Vn)
    Ic = self.Y*Vc
    nb = len(self.a)

    diagVn = spmatrix(Vn, self.a, self.a, (nb, nb), 'z')
    diagVc = spmatrix(Vc, self.a, self.a, (nb, nb), 'z')
    diagIc = spmatrix(Ic, self.a, self.a, (nb, nb), 'z')

    dS = self.Y*diagVn
    dS = diagVc*dS.H.T
    dS += diagIc.H.T*diagVn

    dR = diagIc
    dR -= self.Y*diagVc
```

Table 4.2 Base case power flow results for the IEEE 14-bus system

Bus h	v [pu]	θ [rad]	p_G [pu]	q_G [pu]	p_L [pu]	q_L [pu]
1	1.06	0	2.324	−0.1655	0	0
2	1.045	−0.0870	0.4	0.4356	0.217	0.127
3	1.01	−0.2221	0	0.2507	0.942	0.19
4	1.018	−0.18	0	0	0.478	−0.039
5	1.02	−0.1531	0	0	0.076	0.016
6	1.07	−0.2482	0	0.1273	0.112	0.075
7	1.062	−0.2332	0	0	0	0
8	1.09	−0.2332	0	0.1762	0	0
9	1.056	−0.2607	0	0	0.295	−0.0459
10	1.051	−0.2635	0	0	0.09	0.058
11	1.057	−0.2581	0	0	0.035	0.018
12	1.055	−0.2631	0	0	0.061	0.016
13	1.05	−0.2645	0	0	0.135	0.058
14	1.036	−0.2798	0	0	0.149	0.05
Totals			2.7239	0.8244	2.59	0.5232

```
    dR = diagVc.H.T*dR

    return sparse([[dR.imag(), dR.real()], [dS.real(), dS.imag()]])
```

where `self.a` and `self.v` are the indexes of all bus voltage phase angles and magnitudes, respectively, `nb` is the number of network ac buses and `self.Y` contains the admittance matrix $\bar{\boldsymbol{Y}}$. As explained above, the notation `mat.H.T` indicates the transpose of `mat`. The code above is written as a method of the class that describe transmission lines (see Chapter 11).

Example 4.1 Power Flow Analysis of the IEEE 14-Bus system

The results of the power flow analysis can be conveniently presented in tabular form. For example the base case power flow solution of the IEEE 14-bus system are shown in Table 4.2. Once all bus voltages are known, any other variable of the system can be straightforwardly computed. For example, active and reactive flows as well as losses in transmission lines and transformers are typically included in the power flow report, as shown in Table 4.3.

Example 4.2 Region of Attraction of the Power Flow Solution

Whatever method is used for solving the power flow problem, a *good* initial guess $\boldsymbol{y}^{(0)}$ is needed to start the iterative process. Typically a *flat start* is an acceptable initial guess, i.e., load voltage magnitudes are set to 1 pu and all voltage phase angles are set equal to the reference (e.g., 0 rad) [296].

Table 4.3 Base case branch power flows for the IEEE 14-bus system

Branch	From bus h	To bus k	p_{hk} [pu]	p_{kh} [pu]	q_{hk} [pu]	q_{kh} [pu]	p_{loss} [pu]	q_{loss} [pu]
1	1	2	1.569	−1.526	−0.204	0.2768	0.0430	0.07272
2	1	5	0.7551	−0.7275	0.0386	0.02229	0.0276	0.06084
3	2	3	0.7324	−0.7091	0.0356	0.01602	0.02323	0.05162
4	2	4	0.5613	−0.5445	−0.0155	0.0302	0.0168	0.0147
5	2	5	0.4152	−0.4061	0.0117	−0.0210	0.0090	−0.0093
6	3	4	−0.2329	0.2366	0.0447	−0.0484	0.0037	−0.00363
7	4	5	−0.6116	0.6167	0.1582	−0.142	0.0051	0.0162
8	4	7	0.2807	−0.2807	−0.0968	0.1138	0	0.0170
9	4	9	0.1608	−0.1608	−0.0043	0.0173	0	0.0131
10	5	6	0.4409	−0.4409	0.1247	−0.0805	0	0.0442
11	6	11	0.0735	−0.0730	0.0356	−0.0345	0.0006	0.0012
12	6	12	0.0779	−0.0771	0.0250	−0.0235	0.0007	0.0015
13	6	13	0.1775	−0.1754	0.0722	−0.0680	0.0021	0.0042
14	7	8	0	−0	−0.1716	0.1762	0	0.0046
15	7	9	0.2807	−0.2807	0.0578	−0.0498	0	0.0080
16	9	10	0.0523	−0.0522	0.0422	−0.0419	0.00013	0.0003
17	9	14	0.0943	−0.0931	0.0361	−0.0336	0.00116	0.0025
18	10	11	−0.03785	0.0380	−0.0162	0.0165	0.00013	0.0003
19	12	13	0.0161	−0.0161	0.0075	−0.0075	0	0
20	13	14	0.0564	−0.0559	0.0175	−0.0164	0.0005	0.0011
Totals							0.1339	0.3012

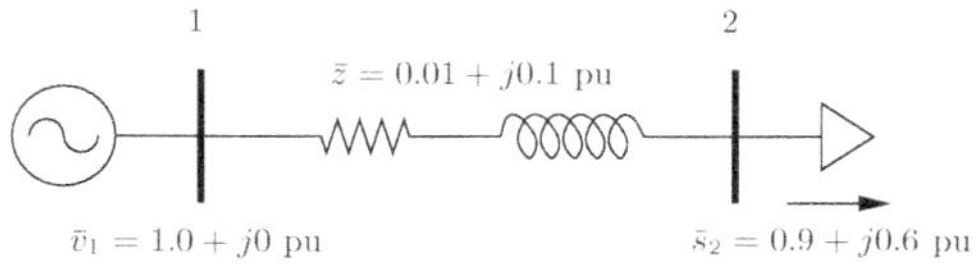

Fig. 4.4 2-bus system

Although the flat start works in the majority of the cases, convergence is never guaranteed.

In theory, the only way to know if a given initial guess is adequate for obtaining a solution $\boldsymbol{y}_0$ of (4.8) is to determine the region of attraction of $\boldsymbol{y}_0$. At this regard, the initial guess can be of three types:

1. The initial guess is inside the region of attraction of the solution $\boldsymbol{y}_0$ and the numerical method converges.
2. The initial guess is outside the region of attraction of the solution $\boldsymbol{y}_0$. Numerical methods typically diverge if one starts with such initial guess.
3. Although the initial guess is within the region of attraction, the numerical method diverges.

This example focuses on initial guesses of the first and second type, whereas the latter case implies some interesting mathematical issues that are discussed in Section 4.5.

Unfortunately, an analytical definition of the region of attraction is not possible. Thus, only numerical methods can be used. A simple way to determine the region of attraction of a given solution $\boldsymbol{y}_0$ is to generate a huge number of initial guesses and solve (4.8) for each initial guess. As example, consider the 2-bus system depicted in Figure 4.4. All power flow data are indicated in this figure. To define the region of attraction of the solution $\bar{v}_{2,0} = 0.9209 - j0.0913$ pu, one can create a grid of initial guesses and solve for each pair $(\theta_2^{(0)}, v_2^{(0)})$ the power flow problem using the Newton's method.

The resulting map is depicted in Figure 4.5. Contour labels indicate the number of iterations needed to get the solution while black regions means that the method does not converge. The region of attraction is quite wide and the standard flat start $\bar{v}_2^{(0)} = 1.0 + j0$ pu falls within the region that requires less iteration to converge. It is important to note that the region of attraction depends on the solution method. Different algorithms are characterized by different region of attractions.

The method discussed above is extremely costly[8] and cannot be used for real size systems. For example, the map of Figure 4.5 was obtained using a grid of 1000×1000 initial guesses, e.g., solving the power flow problem one

[8] The computational burden is similar to that of computing Julia's or Mandelbrot's sets.

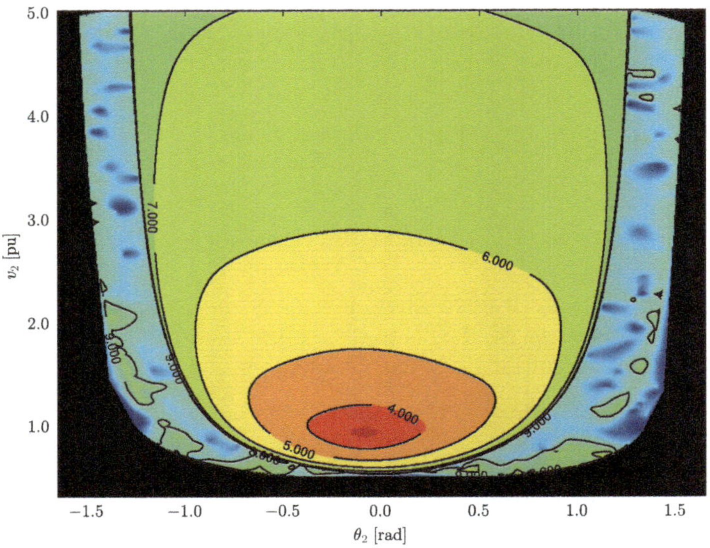

Fig. 4.5 Region of attraction of the Newton's method for a 2-bus system. Contour labels indicate the number of iteration needed to obtain the solution within a tolerance of 10^{-5}

million times. The interested reader can find further discussion on the region of attraction of power flow solutions in [219].

4.4.4 Robust Newton's Method

For well-conditioned cases, the standard Newton's method generally converges in 4-5 iterations. Most books on numerical techniques warn about the possibility that the Newton's technique can cycle around the solution without actually never getting to the solution. Figures 4.3.c and 4.3.d illustrate two ill-conditioned cases by means of a scalar function $g(y)$.

A power flow example that shows a behavior similar to ones depicted in Figures 4.3.c and 4.3.d is quite rare indeed. However, there are idiosyncratic cases for which the Newton's technique fails to converge. A variety of *robust* variations of the basic Newton's method have been proposed in the literature for solving ill-conditioned cases [27, 29, 149, 150, 270, 304, 317]. The majority of these techniques mainly consist in modifying the first equation of (4.40) as follows:

$$\Delta \boldsymbol{y}^{(i)} = -\alpha[\boldsymbol{g}_{\boldsymbol{y}}^{(i)}]^{-1}\boldsymbol{g}^{(i)} \tag{4.48}$$

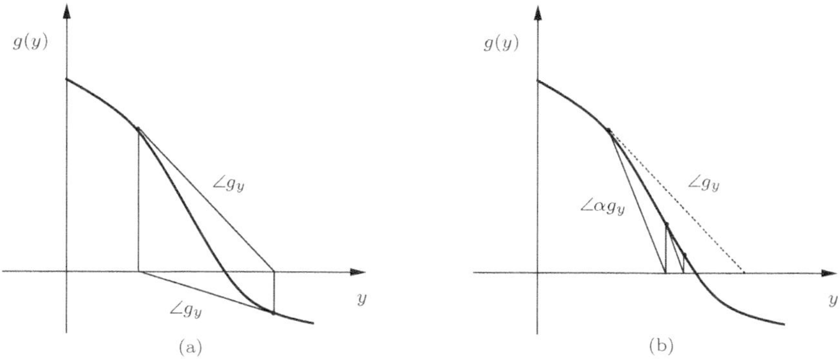

Fig. 4.6 Geometrical interpretation of the robust Newton's method for a scalar function $g(y)$. (a) standard method and (b) robust method

where α is a factor that improves the convergence properties of the iterative process. If α is the result of an optimization process, α is called *optimal multiplier*. The geometrical interpretation of robust Newton's methods is shown in Figure 4.6.

It is important not to confuse ill-conditioned cases with those that are unsolvable since the solution does not exist (see Figure 4.3.b). Robust solvers are useful in case of ill-conditioned systems but do not generally work well for unsolvable cases. Unsolvable cases are better tackled using the continuation power flow technique described in Chapter 5.

At a given iteration i, the optimal value of α is the one that minimizes the maximum power mismatch $\max\{|\boldsymbol{g}(\boldsymbol{y}^{(i+1)})|\}$. Since the maximum power mismatch at the iteration $i+1$ is not known *a priori*, one has to iterate over α. However, it is not necessary to find the optimum. A simple, yet quite robust method is the bisection method, as follows.

1. Set $\alpha \Leftarrow 1$.
2. Compute $\max\{|\boldsymbol{g}(\boldsymbol{y}^{(i+1)})|\}$.
3. If $\max\{|\boldsymbol{g}(\boldsymbol{y}^{(i+1)})|\} \leq \max\{|\boldsymbol{g}(\boldsymbol{y}^{(i)})|\}$, continue with the next iteration, otherwise set $\alpha \Leftarrow 0.5 \cdot \alpha$ and go back to step 2.

For unsolvable cases, $\alpha \to 0$, thus one has to fix a minimum value for α.

Script 4.4 Robust Newton's method

The following Python code illustrates this simple robust method:

```
inc = alpha*calcInc()
error = max(abs(inc))

if error > error_old:
    alpha *= 0.5
```

```
    if alpha < tol:
        fm_disp('The optimal multiplier is too small.')
        iteration = iter_max + 1
        break
else:
    system.DAE.y += inc
    error_old = error
    alpha = 1.0
```

In the previous code, it is assumed that `calcInc()` returns $\Delta \boldsymbol{y}^{(i)}$. Clearly, the previous code has to be inserted within the main power flow loop.

4.4.5 Iwamoto's Method

More sophisticate methods attempt to estimate $\max\{|\boldsymbol{g}(\boldsymbol{y}^{(i+1)})|\}$. For the sake of example, this subsection describes the Iwamoto's method, that is one of the firstly proposed robust power flow methods [149]. With this aim, consider the Taylor's expansion of (4.8) at the ith iteration:

$$\boldsymbol{g}(\boldsymbol{y}) = \boldsymbol{g}^{(i)} + \boldsymbol{g}_{\boldsymbol{y}}^{(i)} \Delta \boldsymbol{y}^{(i)} + \boldsymbol{g}(\Delta \boldsymbol{y}^{(i)}) \tag{4.49}$$

In (4.49) the correction vector $\Delta \boldsymbol{y}^{(i)}$ is not known. In order to optimize the length of $\Delta \boldsymbol{y}^{(i)}$, a factor α is included in (4.49), as follows:

$$\boldsymbol{g}(\boldsymbol{y}) = \boldsymbol{g}^{(i)} + \boldsymbol{g}_{\boldsymbol{y}}^{(i)} \alpha \Delta \boldsymbol{y}^{(i)} + \boldsymbol{g}(\alpha \Delta \boldsymbol{y}^{(i)}) \tag{4.50}$$

Assuming the rectangular form (4.18) of power flow equations:

$$\boldsymbol{g}(\alpha \Delta \boldsymbol{y}^{(i)}) = \alpha^2 \boldsymbol{g}(\Delta \boldsymbol{y}^{(i)}) \tag{4.51}$$

Thus (4.49) is a quadratic equation with respect to α:

$$\boldsymbol{d}(\alpha) = \boldsymbol{c}_0 + \boldsymbol{c}_1 \alpha + \boldsymbol{c}_2 \alpha^2 \tag{4.52}$$

where:

$$\boldsymbol{c}_0 = \boldsymbol{g}^{(i)}, \quad \boldsymbol{c}_1 = \boldsymbol{g}_{\boldsymbol{y}}^{(i)} \Delta \boldsymbol{y}^{(i)}, \quad \boldsymbol{c}_2 = \boldsymbol{g}(\Delta \boldsymbol{y}^{(i)}) \tag{4.53}$$

It is relevant to note that, from (4.40), one has $\boldsymbol{c}_1 = -\boldsymbol{c}_0$. The optimal value of α is determined minimizing the following cost function:

$$\kappa(\alpha) = \frac{1}{2} \boldsymbol{d}(\alpha)^T \boldsymbol{d}(\alpha) \tag{4.54}$$

In this case, the KKT conditions simply give:

$$\frac{\partial \kappa}{\partial \alpha} = 0 \quad \Rightarrow \quad g_0 + g_1 \alpha + g_2 \alpha^2 + g_3 \alpha^3 = 0 \tag{4.55}$$

where:

$$g_0 = \boldsymbol{c}_0^T \boldsymbol{c}_1, \quad g_1 = \boldsymbol{c}_1^T \boldsymbol{c}_1 + 2\boldsymbol{c}_0^T \boldsymbol{c}_2, \quad g_2 = 3\boldsymbol{c}_1^T \boldsymbol{c}_2, \quad g_3 = 2\boldsymbol{c}_2^T \boldsymbol{c}_2 \tag{4.56}$$

Since (4.55) is a cubic scalar polynomial, the Cardan's formula provides the analytical solution:

$$\alpha = a_1 + \sqrt[3]{a_2 + a_3} + \sqrt[3]{a_2 - a_3} \tag{4.57}$$

where:

$$a_1 = -\frac{g_2}{3g_3} \qquad a_2 = a_1^3 + \frac{g_2 g_1 - 3g_3 g_0}{6g_3^2} \tag{4.58}$$
$$a_3 = \sqrt{a_2^2 + (a_4 - a_1^2)^3} \qquad a_4 = \frac{g_1}{3g_3}$$

An issue of the Iwamoto's method is that the optimal multiplier α decreases as $\boldsymbol{y}$ converges to the solution, thus the Iwamoto's method typically converges slowly.

4.4.6 Inexact and Dishonest Newton's Methods

One of the most relevant drawbacks of the Newton's method is the need of factorizing the full Jacobian matrix at each iteration. From the computational point of view, the factorization of a matrix is an order N^3 operation, i.e., the computational weight increases with the cube of the size N of the matrix. The computational effort can be reduced to $N^{1.5}$ if using sparse matrices techniques, which allows saving a considerable time for large systems (e.g., thousands of buses). However, the Jacobian matrix factorization remains the most critical issue of the Newton's method (about 85% of the total CPU time for networks with thousands of buses). Thus, in the literature, there are a variety of proposals for reducing as much as possible the computational effort of the Jacobian matrix factorization.

A family of methods based on the Generalized Minimal Residual (GMRES) method are [54, 97, 275]. The GMRES is a particular case of Krylov's subspace approaches and attempts to minimize (4.21) by minimizing the residual:

$$\boldsymbol{r}(\boldsymbol{y}) = \boldsymbol{b} - \boldsymbol{A}\boldsymbol{y} \tag{4.59}$$

For the power flow problem, $\boldsymbol{b} = \boldsymbol{g}^{(i)}$ and $\boldsymbol{A} = \boldsymbol{g}_{\boldsymbol{y}}^{(i)}$. Since $\boldsymbol{A}$ and $\boldsymbol{b}$ are not constant, the residual $\boldsymbol{r}$ in (4.59) has to be minimized at each iteration. GMRES-based methods differ from the Gauss-Seidel's method in that the latter does not compute the Jacobian matrix. Without entering into mathematical details, GMRES-based methods can be used for setting up the so-called inexact Newton's methods. The term *inexact* comes from the fact that the power flow Jacobian matrix factorization is not computed exactly

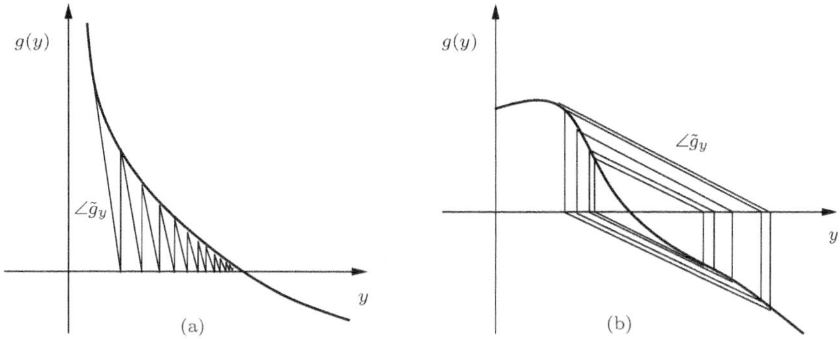

Fig. 4.7 Geometrical interpretation of the dishonest Newton's method for a scalar function $g(y)$. (a) well-conditioned case and (b) infinite-cycle case

(for example using the LU factorization) but approximated with the GMRES using an acceptable accuracy. The higher the accuracy, the smaller the number of iterations but also the higher the CPU time for approximating the Jacobian matrix factorization. Thus, GMRES-inexact Newton's methods require a fine tuning that is generally obtained by preconditioning the matrix $\boldsymbol{A}$ in (4.59).

Pushing inexact methods to the limit, one can approximate the Jacobian matrix factorization using always the same factorization for each iteration of the Newton's method. In other words, one can rewrite (4.40) as follows:

$$\begin{aligned} \Delta \boldsymbol{y}^{(i)} &= -[\tilde{\boldsymbol{g}}_{\boldsymbol{y}}]^{-1} \boldsymbol{g}^{(i)} \\ \boldsymbol{y}^{(i+1)} &= \boldsymbol{y}^{(i)} + \Delta \boldsymbol{y}^{(i)} \end{aligned} \tag{4.60}$$

where $\tilde{\boldsymbol{g}}_{\boldsymbol{y}}^{-1}$ is the inverse of $\boldsymbol{g}_{\boldsymbol{y}}$ computed for a given vector $\tilde{\boldsymbol{y}}$ and maintained constant during all iterations. In this way, the number of iterations is generally quite high (tens versus 4-5 of the standard Newton's method). However, since $\tilde{\boldsymbol{g}}_{\boldsymbol{y}}$ is factorized only once, CPU times generally decreases for large systems. The method (4.60) is called *dishonest* Newton's method. The standard Newton's method can be used in conjunction with the dishonest one. For example $\boldsymbol{g}_{\boldsymbol{y}}$ can be factorized at the first two or three iterations and then left constant for the remainders. It is also important to note that the dishonest method does not guarantee convergence (see Figure 4.7).

4.4.7 Fast Decoupled Power Flow

A widely used variant of dishonest methods is the so-called Fast Decoupled Power Flow (FDPF). This technique was originally proposed in [299] and has been further developed and generalized in several variations. Most common variants are the XB and BX methods presented in [324].

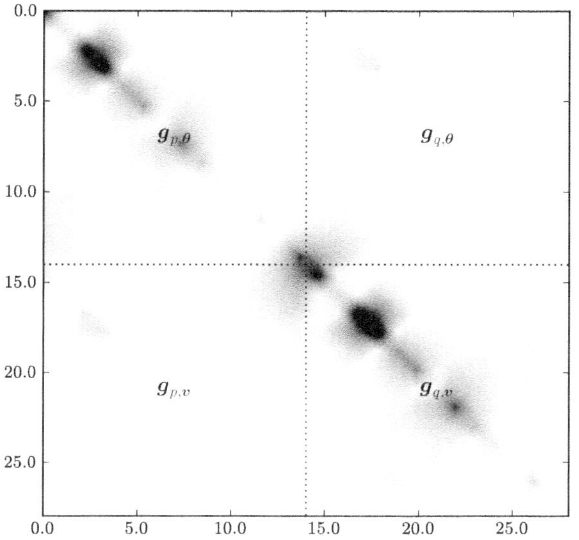

Fig. 4.8 Pictorial representation of the power flow Jacobian matrix for the IEEE 14-bus system. The darker the area the higher the absolute value of the elements of the Jacobian matrix

The FDPF method can be used only if algebraic variables are voltage magnitudes and phases, i.e., the FDPF can be used only with the classical power flow model in polar form (4.16). Recalling (4.45), the power flow Jacobian matrix $\boldsymbol{g_y}$ can be decomposed in four sub-matrices, namely $\boldsymbol{g}_{p,\boldsymbol{\theta}}$, $\boldsymbol{g}_{p,\boldsymbol{v}}$, $\boldsymbol{g}_{q,\boldsymbol{\theta}}$, and $\boldsymbol{g}_{q,\boldsymbol{v}}$. The FDPF approach consists in approximating these sub-matrices as constant through some *smart* assumptions. Thus the main difference with the dishonest method lies in the fact that $\boldsymbol{g_y}$ is not approximated using a given value $\boldsymbol{y}$, but computed *a priori*.

The main observation that has led to the FDPF method is that there is a clear decoupling between the (p, θ) and (q, v) variables. A direct consequence is that the sensitivity of active power mismatches with respect to voltage magnitudes (i.e., $\boldsymbol{g}_{p,\boldsymbol{v}}$) and of reactive power mismatches with respect to voltage phases are relatively small (i.e., $\boldsymbol{g}_{q,\boldsymbol{\theta}}$). This is pictorially illustrated in Figure 4.8. In this figure, the darker the region the higher the absolute value of matrix elements. Since the darker region is diagonal, one can expect that $\boldsymbol{\theta}$ and $\boldsymbol{v}$ are quite decoupled.

The basic assumptions of the FDPF method are:

$$\begin{aligned} \boldsymbol{g}_{p,\boldsymbol{\theta}} &\approx \boldsymbol{B}' & \qquad \boldsymbol{g}_{p,\boldsymbol{v}} &= \boldsymbol{0} \\ \boldsymbol{g}_{q,\boldsymbol{\theta}} &= \boldsymbol{0} & \qquad \boldsymbol{g}_{q,\boldsymbol{v}} &\approx \boldsymbol{B}'' \end{aligned} \tag{4.61}$$

where $\boldsymbol{B}'$ and $\boldsymbol{B}''$ are simplified admittance matrices, as follows:

1. Line charging, shunts and transformer tap ratios are neglected when computing $\boldsymbol{B}'$;
2. Phase shifters are neglected and line charging and shunts are doubled when computing $\boldsymbol{B}''$.

The XB and BX variants differ only in further simplifications of the $\boldsymbol{B}'$ and $\boldsymbol{B}''$ matrices respectively, as follows:

XB: line resistances and shunt conductances are neglected when computing $\boldsymbol{B}'$;

BX: line resistances and shunt conductances are neglected when computing $\boldsymbol{B}''$.

For loss-less systems the XB and BX variants coincide. All assumptions are valid for HV transmission systems, where line and transformer resistances are generally one order of magnitude smaller than reactances. For distribution systems (e.g., medium or low voltage systems), where series resistances are similar or higher than reactances, the FDPF does not generally work well.[9]

The FDPF algorithm consists in solving two linear systems at each iteration, as follows:

$$\begin{aligned} \boldsymbol{\theta}^{(i+1)} &= \boldsymbol{\theta}^{(i)} - [\boldsymbol{B}']^{-1}[\boldsymbol{V}^{(i)}]^{-1}\boldsymbol{g}_p(\boldsymbol{v}^{(i)}, \boldsymbol{\theta}^{(i)}) \\ \boldsymbol{v}^{(i+1)} &= \boldsymbol{v}^{(i)} - [\boldsymbol{B}'']^{-1}[\boldsymbol{V}^{(i)}]^{-1}\boldsymbol{g}_q(\boldsymbol{v}^{(i)}, \boldsymbol{\theta}^{(i+1)}) \end{aligned} \tag{4.62}$$

where $\boldsymbol{V}^{(i)} = \mathrm{diag}(v_1^{(i)}, v_2^{(i)}, \ldots, v_{n_b}^{(i)})$. To reduce the iteration number, the angles $\boldsymbol{\theta}^{(i+1)}$ are used for computing the reactive power mismatches $\boldsymbol{g}_q$. Since $\boldsymbol{B}'$ and $\boldsymbol{B}''$ are constant, these matrices can be factorized only once. However, $\boldsymbol{B}''$ does not change only in case reactive power limits of PV generators are not binding. Otherwise, $\boldsymbol{B}''$ has to be re-built and re-factorized each time a PV generator switches to a constant PQ load.

As in case of dishonest Newton's methods, the FDPF does not ensure to get a solution, but the FDPF method generally shows better convergence properties than the dishonest one. It has also been observed that the FDPF is able to converge for some ill-conditioned case for which the standard Newton method fails [299].

[9] This is one of the reasons why some commercial tools distinguish between the power flow analysis for transmission and distribution systems. However, network parameters do not make any difference if one uses a general Newton method. Rather, the main relevant difference between transmission and distribution systems is the network topology. Transmission system are generally meshed, while distribution systems show a radial configuration. For the interested reader a selection of methods that takes advantage of the radial topology has been developed for distribution systems can be found in [71, 121, 191, 279, 357].

Script 4.5 Fast-Decoupled Power Flow

The Python implementation of a the FDPF algorithm is as follows:

```
from cvxopt.base import matrix, div
from cvxopt.umfpack import symbolic, numeric, solve

def fdpf():

    # general settings
    iteration = 1
    iter_max = system.Settings.pf_max_iter
    convergence = True
    tol = system.Settings.tol
    system.Settings.error = tol + 1
    err_vec = []

    # main loop
    while system.Settings.error > tol and iteration <= iter_max:

        if iteration == 1: # initialize variables
            sw = system.SW._geta()
            sw.sort(reverse = True)
            no_sw = system.Bus.a[:]
            no_swv = system.Bus.v[:]
            for item in sw:
                no_sw.pop(item)
                no_swv.pop(item)
            pv = system.SW._geta(True) + system.PV._geta(True)
            ngen = sum(system.SW.n) + sum(system.PV.n)
            pv.sort(reverse = True)
            no_g = system.Bus.a[:]
            no_gv = system.Bus.v[:]
            for item in pv:
                no_g.pop(item)
                no_gv.pop(item)
            Bp = system.Line.Bp[no_sw, no_sw]
            Bpp = system.Line.Bpp[no_g, no_g]
            Fp = numeric(Bp, symbolic(Bp))
            Fpp = numeric(Bpp, symbolic(Bpp))
            exec system.Device.call_fdpf

        # P-theta
        da = -matrix(div(system.DAE.g[no_sw], system.DAE.y[no_swv]))
        solve(Bp, Fp, da)
        system.DAE.y[no_sw] += da
        exec system.Device.call_fdpf
        normP = max(abs(system.DAE.g[no_sw]))

        # Q-V
        dV = -matrix(div(system.DAE.g[no_gv], system.DAE.y[no_gv]))
        solve(Bpp, Fpp, dV)
        system.DAE.y[no_gv] += dV
        exec system.Device.call_fdpf
```

```
        normQ = max(abs(system.DAE.g[no_gv]))

        inc = matrix([normP, normQ])
        system.DAE.y += inc

        system.Settings.error = max(abs(inc))
        err_vec.append(system.Settings.error)

        msg = 'Iteration = %3d   Max. Convergence Error = %8.7f' \
              % (iteration, system.Settings.error)
        print msg

        iteration += 1

        # stop if the error increases too much
        if iteration > 4 and err_vec[-1] > 1000*err_vec[0]:
            print 'The error is increasing too much'
            print 'Convergence is likely not reachable'
            convergence = False
            break

    if iteration > iter_max:
        print 'Reached maximum number of iterations'
        convergence = False
```

where the variables `SW` and `PV` indicate the slack and PV generator classes, respectively, the method `_geta()` returns the indexes of bus voltage phase angles, and the expression `exec system.Device.call_fdpf` implements equations $\boldsymbol{g}$. Function `symbolic` and `numeric` are executed only once at the first iteration for factorizing $\boldsymbol{B}'$ and $\boldsymbol{B}''$.

Example 4.3 Comparison of Methods for Power Flow Analysis

In this example, several methods for power flow analysis are compared, namely the standard Newton's method, the Jacobi and the Gauss-Seidel's methods, a simple GMRES-based approach, the dishonest Newton's method, the XB and the BX version of the FDPF. The test systems are the IEEE 14-bus system, the IEEE 118-bus system, a real-world 1228-bus system and the 11856-bus system which is available at [359].[10] These systems are chosen so that the bus number spans four orders of magnitude and there is one order of magnitude between one system and the following one.

[10] The 11856-bus system is formed putting together the IEEE 57-bus system 208 times. The resulting test system is thus composed of 208 independent networks and the complete power flow Jacobian matrix is formed by 208 diagonal blocks. However, if one does not take advantage of the structure of the Jacobian matrix, the factorization time depends only on the sparsity of the complete matrix, which is similar to the one of a fully interconnected network.

In order to use similar conditions for all methods, a flat start is used and generator reactive power limits are not enforced. Results are shown in Table 4.4. Relevant remarks are as follows:

1. Methods that involve the computation of the Jacobian matrix, no matter if approximated or kept constant, are generally faster than the Gauss-Seidel's and Jacobi's methods. Thus, the concern that is reported in some relatively old papers about the amount of memory used by Jacobian matrix is not actually an issue, at least for modern personal computers.
2. Overall, the Newton's method is the best method among the ones considered in this example. The rationale behind this fact is twofold:

   a. Sparse matrix factorization algorithms are very efficient. Furthermore, the symbolic factorization provided by the CVXOPT module allows saving additional CPU time.
   b. One of the most critical bottlenecks of scripting languages is matrix memory allocation, slicing and manipulation in general. This is because any operation done on floating-point multi-dimensional arrays are not done by the scripting language itself but sent to some C-library through a hidden-layer interface (*bridge*). This bridge is where the interpreter wastes time.

   In conclusion, in scripting languages the less the array manipulations the better. For these reasons, FDPF approaches are not faster than the Newton's method.
3. As previously discussed, the Jacobi's method is faster than the Gauss-Seidel's one even if it requires more iterations. This happens because, in scripting languages, array operations are faster than scalar operations within a for-loop. However, the Jacobi's method cannot be competitive with the Newton's method for the reasons discussed in the previous point.
4. The GMRES method used in the simulations below is the version implemented in the module `scipy.sparse.linalg` [334]. No preconditioning is used. This is the slowest method among those considered in this example. Moreover, the simulation hangs up for the two biggest systems, probably due to array interface issues in the `scipy` module. In any case, also the Matlab implementation of the GMRES algorithm is generally slower than the LU factorization of sparse matrices.
5. With regard to the dishonest method, the Jacobian matrix is kept constant starting from the second iteration. As expected the number of iterations is higher than for the Newton's method and also of the FDPF. However, CPU time is generally competitive.
6. The two versions of the FDPF show similar results for the considered systems. The BX version is slightly better (one iteration less for the 118 and 11856-bus systems) than the XB one, thus confirming the experimental results discussed in [324].

Table 4.4 Comparison of a variety of methods for power flow analysis

	Newton		Jacobi		Gauss-Seidel	
Bus #	Iter. #	time [s]	Iter. #	time [s]	Iter. #	time [s]
14	4	0.0050	76	0.0217	56	0.0288
118	5	0.0287	580	0.505	388	2.738
1228	5	0.210	454	5.120	224	112.4
11856	4	3.15	340	399.0	173	9112

	GMRES		Dishonest		FDPF-BX	
Bus #	Iter. #	time [s]	Iter. #	time [s]	Iter. #	time [s]
14	4	0.4339	7	0.0040	6	0.0053
118	7	53.53	15	0.0183	6	0.0117
1228	n.a.	n.a.	26	0.207	12	0.160
11856	n.a.	n.a.	10	3.820	5	5.174

4.4.8 DC Power Flow

All methods discussed so far provide, with more or less efficiency, the exact solution of the problem (4.8). Besides the computational burden, the nonlinearity of (4.8) implies three relevant drawbacks:

1. Possible difficulties in finding the solution (e.g., ill-conditioned cases).
2. In case of no convergence, impossibility to determine if the solution does not exist or cannot be reached.
3. Existence of multiple solutions.

These issues, while quite stimulating for researchers, seriously concern practitioners. A common solution consists in simplifying $\boldsymbol{g}(\boldsymbol{y})$ in order to obtain a set of linear equations. The result is the so-called *dc power flow*.

The basic assumption of the dc power flow model is to focus only on the (p, θ) world and completely neglect the (q, v) one. The hypotheses are:

1. All voltage magnitudes are assumed constant and equal to 1.0 pu and reactive powers are neglected.
2. Line resistance and charging are neglected when computing the simplified admittance matrix $\boldsymbol{B}$. The matrix $\boldsymbol{B}$ coincides with the susceptance matrix $\boldsymbol{B}'$ of the XB variant of the FDPF if all transformer tap ratios and phase shifter angles are 1 pu and 0 rad, respectively.
3. Bus voltage phases are considered *small*, so that $\sin\theta_{hk} \approx \theta_{hk}$ and $\cos\theta_{hk} \approx 1.0$.

The resulting system equations are:

$$\boldsymbol{p} = \boldsymbol{B}\boldsymbol{\theta} + \boldsymbol{p}^{\phi} \tag{4.63}$$

where $\boldsymbol{p}$ are the power injections at buses and $\boldsymbol{\theta}$ are unknowns and $\boldsymbol{p}^{\phi}$ are the power shifts introduced by phase shifting transformers. Equation (4.63) accounts for each bus of the network but the reference one. In other words, if the network contains n_b buses, then $\boldsymbol{B}$ is a square matrix of order $n_b - 1$. Introducing the incidence matrix $\boldsymbol{C}$, (4.63) can be rewritten as:

$$\boldsymbol{p} = \boldsymbol{X}^{-1}\boldsymbol{C}\boldsymbol{\theta} + \boldsymbol{C}^T\boldsymbol{X}^{-1}\boldsymbol{\phi} \tag{4.64}$$

where $\boldsymbol{X} = \mathrm{diag}(m_1 x_1, m_2 x_2, \ldots, m_{n_c} x_{n_c})$ are the series reactances multiplied by the off-nominal tap ratio of each network branch,[11] n_c is the number of branches that compose the transmission network, $\boldsymbol{\phi}$ is the $n_c \times 1$ vector n_c of phase shifts of each network branch. The incidence matrix $\boldsymbol{C}$ is a $n_c \times (n_b - 1)$ matrix that accounts for network topology. Further insights of the incidence matrix $\boldsymbol{C}$ are given in Subsection 11.3.1 of Chapter 11.

Equation (4.63) has a formal similarity with a purely resistive linear dc system:

$$\boldsymbol{i}_{\mathrm{dc}} = \boldsymbol{G}_{\mathrm{dc}}\boldsymbol{v}_{\mathrm{dc}} \tag{4.65}$$

where $\boldsymbol{i}_{\mathrm{dc}}$ are currents injected at nodes, $\boldsymbol{v}_{\mathrm{dc}}$ are the node voltages and $\boldsymbol{G}_{\mathrm{dc}}$ is the conductance nodal matrix. Hence the name "dc power flow". However, this notation is misleading, since it implies that a dc system is necessarily linear. More precise definitions are *linear power flow* or *approximated active power flow*.

The dc power flow is an easy way out to avoid the issues inherent the nonlinear nature of power flow equations. The voltage phase angle error introduced by the dc power flow can vary between 1% and 20%. Errors typically increase as the loading level increases (see Figure 4.9 in Example 4.4). Furthermore, neglecting voltage magnitude variations and reactive power flows is an arguable assumption since voltage magnitudes and reactive powers play an important role in the behavior of ac power systems (e.g., voltage stability issues [39]). Chapters 5 and 6 show that it is possible to tackle and often to successfully solve nonlinearity issues at the price of involving adequate mathematical tools.

Example 4.4 Accuracy of the DC Power Flow

The most important drawback of the dc power flow approach is the inaccuracy of the results. The error on voltage magnitudes is generally within 10% and the error of bus voltage phase angles can be up to 20%. This may lead to a wrong estimation of power flows in transmission lines and transformers. Figure 4.9 shows the mean and maximum angle errors introduced by the dc power flow with respect to the exact solution provided by the Newton's method for the IEEE 14-bus system. Angle errors are plotted versus the

[11] Strictly speaking, the correct i^{th} element of matrix $\boldsymbol{X}$ is $m_i \frac{r_i^2 + x_i^2}{x_i^2}$, where r_i is the series resistance of the ith branch element. However, for $r_i < x_i/3$ the error introduced by approximating the correct value with $m_i x_i$ is lower than 1%.

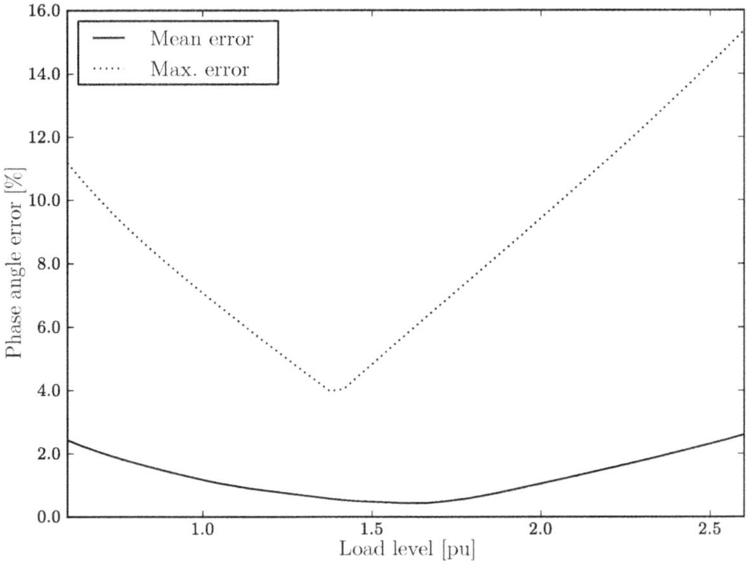

Fig. 4.9 Mean voltage phase angle error of the dc power flow as a function of the loading level for the IEEE 14-bus system

loading level (1 means base case). While the mean error is relatively low, the maximum error vary considerably depending on the loading level.

In order to estimate the error introduced in transmission line active power flows, consider the following simple calculation. The active power flowing in a loss-less transmission line connecting buses h and k through a series reactance x_{hk} is:

$$p_{hk} = \frac{v_h v_k}{x_{hk}} \sin \theta_{hk}$$

Assuming that $v_h = v_k = 1.0$ pu, $\theta_{hk} = 0.1$ rad and $x_{hk} = 0.1$ pu, one has:

$$p_{hk} \approx 1.0 \text{ pu}$$

If each voltage is affected by a 5% error and the angle difference by a 10% error, and assuming the worst-case scenario in which all errors have the same sign:

$$\tilde{p}_{hk} \approx \frac{(v_h + \Delta v_h)(v_k + \Delta v_k)}{x_{hk}} (\theta_{hk} + \Delta\theta_{hk}) = 1.21 \text{ pu}$$

that leads to an error of more than 20%. In most engineering applications, an error of 20% is a sufficient argument for dropping a method, especially taking into account that electrical measures of active power flows can be easily obtained with an error of at most 5%. However, surprisingly enough, the dc power flow is widely used in power system planning and operation and in economic dispatch.

4.4.9 Single and Distributed Slack Bus Models

The concept of single slack bus used in the classical power flow formulation is arguable for the following rationale. In the single slack bus model, an unique generator has to take care of system losses or, at least, of the entire power loss increment in case a contingency occurs (e.g., a transmission line outage). This is not the real behavior of the system, since turbine governors of synchronous machines vary the power production of *all* generators in case of power unbalance (see also Chapter 16). Furthermore, no physical generator has an unlimited capacity or can absorb a negative active power, while this is possible for slack generators. Thus, unless the slack bus is an equivalent model of a big network,[12] the single slack bus model is not realistic. The distributed slack bus model allows avoiding this inconsistency.

The distributed slack bus model is based on a generalized power center concept and consists in distributing losses among all generators [23]. In mathematical terms, this is obtained by substituting the slack generator active power injection p_h in (4.12) by a scalar variable k_G and by rewriting all p_h at generator buses (slack bus included) as follows:

$$p_h = (1 + k_G \gamma_h) p_{G0,h} - p_{Lh}, \quad h \in \mathcal{B}_G \tag{4.66}$$

where $p_{G0,h}$ is the loss net generator power, $\mathcal{B}_G$ is the set of static generator buses, and γ_h is a factor that allows weighting the participation of each generator to network losses.

The loss participation factors γ_h can be fixed to 1, i.e., each generator contributes proportionally to its generated power, or computed based on the droop of the turbine governors as discussed Chapter 16. Other methods for defining γ_h values are based on synchronous machine inertiæ, frequency control participation factors and also economic dispatch [116, 118, 364]. The single slack bus formulation is a particular case of the distributed one, being $\gamma_h = 0$ for all generators except for the slack bus that has $\gamma_{\text{slack}} = 1$ (see also Subsections 10.2.2 and 10.2.1).

Equations (4.40) are still valid if using the distributed slack bus model. The same cannot be said for the Gauss-Seidel, FDPF and dc power flow approaches. The only change in (4.40) is the form of $\boldsymbol{g_y}$. In particular, the following partial derivative has to be added to (4.46):

$$\frac{\partial g_{ph}}{\partial k_G} = -\gamma_h p_{G0,h}, \quad h = \mathcal{B}_G \tag{4.67}$$

[12] *Big* means with a power capacity much greater than those of physical generators modelled in the network.

Example 4.5 Distributed Slack Bus Power Flow for the IEEE 14-Bus System

Table 4.5 shows the results of the power flow with distributed slack bus model for the IEEE 14-bus test system. The loss participation factors of generators 1 and 2 are $\gamma_1 = \gamma_2 = 1$. Furthermore, $p_{G0,1} = 2.3$ pu for the slack bus. In this case, the final generated powers are very similar to the single slack bus model. This result is a consequence of the fact that $p_{G0,1} \gg p_{G0,2}$.

Table 4.5 Power flow results for the IEEE 14-bus system with distributed slack bus model

Bus h	v [pu]	θ [rad]	p_G [pu]	q_G [pu]	p_L [pu]	q_L [pu]
1	1.06	0	2.32	−0.1647	0	0
2	1.045	−0.0868	0.4035	0.4342	0.217	0.127
3	1.01	−0.2219	0	0.2507	0.942	0.19
4	1.018	−0.1799	0	0	0.478	−0.039
5	1.02	−0.153	0	0	0.076	0.016
6	1.07	−0.2481	0	0.1273	0.112	0.075
7	1.062	−0.233	0	0	0	0
8	1.09	−0.233	0	0.1762	0	0
9	1.056	−0.2606	0	0	0.295	−0.0459
10	1.051	−0.2634	0	0	0.09	0.058
11	1.057	−0.258	0	0	0.035	0.018
12	1.055	−0.263	0	0	0.061	0.016
13	1.05	−0.2644	0	0	0.135	0.058
14	1.036	−0.2797	0	0	0.149	0.05
Totals			2.724	0.8237	2.59	0.5232

4.5 A General Framework for Power Flow Solvers

Let us consider a set of autonomous ODE:

$$\dot{\boldsymbol{y}} = \tilde{\boldsymbol{f}}(\boldsymbol{y}) \tag{4.68}$$

The simplest method of numerical integrating (4.68) is the explicit Euler's method, as follows:

$$\begin{aligned} \Delta \boldsymbol{y}^{(i)} &= \Delta t \tilde{\boldsymbol{f}}(\boldsymbol{y}^{(i)}) \\ \boldsymbol{y}^{(i+1)} &= \boldsymbol{y}^{(i)} + \Delta \boldsymbol{y}^{(i)} \end{aligned} \tag{4.69}$$

where Δt is a given step length.

The analogy between (4.40) and (4.69) is straightforward if one defines:

$$\tilde{\boldsymbol{f}}(\boldsymbol{y}) = -[\boldsymbol{g_y}]^{-1}\boldsymbol{g}(\boldsymbol{y}) \tag{4.70}$$

Equations (4.40) can thus be viewed as the ith step of the Euler's forward method where $\Delta t = 1$ [127]. Furthermore, robust Newton's methods (4.48) are nothing but the ith step of the Euler's integration method where $\Delta t = \alpha$. In other words, the computation of the optimal multiplier α corresponds to a variable step forward Euler's method.

Equations (4.68) and (4.70) leads to:

$$\dot{\boldsymbol{y}} = -[\boldsymbol{g_y}]^{-1}\boldsymbol{g}(\boldsymbol{y}) \tag{4.71}$$

which is known as *continuous Newton's method.*

The equilibrium point $\boldsymbol{y}_0$ of (4.71) is

$$\boldsymbol{0} = \tilde{\boldsymbol{f}}(\boldsymbol{y}_0) = -[\boldsymbol{g_y}|_0]^{-1}\boldsymbol{g}(\boldsymbol{y}_0) \tag{4.72}$$

Thus, assuming that $\boldsymbol{g_y}$ is not singular, $\boldsymbol{y}_0$ is also the solution of (4.8). Assuming a non-singular Jacobian matrix for the power flow equations is an implicit hypothesis of any power flow analysis based on the factorization of $\boldsymbol{g_y}$ (see also the discussion in Section 5.4 of the Chapter 5).

4.5.1 Stability of the Continuous Newton's Method

Differentiating (4.70) with respect to $\boldsymbol{y}$ leads to:

$$\begin{aligned}\tilde{\boldsymbol{f}}_{\boldsymbol{y}} &= \nabla_{\boldsymbol{y}}^T \tilde{\boldsymbol{f}}(\boldsymbol{y}) \\ &= -[\boldsymbol{g_y}]^{-1}\boldsymbol{g_y} - (\nabla_{\boldsymbol{y}}^T([\boldsymbol{g_y}]^{-1}))\boldsymbol{g}(\boldsymbol{y}) \\ &= -\boldsymbol{I}_{n_y} - (\nabla_{\boldsymbol{y}}^T([\boldsymbol{g_y}]^{-1}))\boldsymbol{g}(\boldsymbol{y})\end{aligned} \tag{4.73}$$

where $\boldsymbol{I}_{n_y}$ is the identity matrix of order n_y. Since the equilibrium point $\boldsymbol{y}_0$ is a solution for $\boldsymbol{g}(\boldsymbol{y}_0) = 0$, one has:

$$\tilde{\boldsymbol{f}}_{\boldsymbol{y}}|_0 = -\boldsymbol{I}_{n_y} \tag{4.74}$$

To prove equation (4.74) let us define the following quantities, based on tensor notation:

$\tilde{f}_i$ element i of the vector function $\tilde{\boldsymbol{f}}(\boldsymbol{y})$.
g_k element k of the vector function $\boldsymbol{g}(\boldsymbol{y})$.
a_{ik} element (i, k) of the matrix $[\boldsymbol{g_y}]^{-1}$.
$\tilde{f}_{i,j}$ partial derivative of $\tilde{f}_i$ with respect to the variable y_j.
$g_{k,j}$ partial derivative of g_k with respect to the variable y_j.
$a_{ik,j}$ partial derivative of a_{ik} with respect to the variable y_j.

For simplicity, the dependence of $\tilde{f}_{i,j}$, $g_{k,j}$ and $a_{ik,j}$ on $\boldsymbol{y}$ is omitted. Equations (4.70) and (4.73) can be rewritten as follows:

$$\tilde{f}_i = -\sum_{k=1}^{n_y} a_{ik} \cdot g_k \tag{4.75}$$

$$\tilde{f}_{i,j} = -\sum_{k=1}^{n_y} a_{ik} \cdot g_{k,j} - \sum_{k=1}^{n_y} a_{ik,j} \cdot g_k \tag{4.76}$$

Since the matrix $[a_{ik}]$ is the inverse of $\boldsymbol{g_y}$, then

$$\sum_{k=1}^{n_y} a_{ik} \cdot g_{k,j} = \begin{cases} 1 & \text{if} \quad i = j \\ 0 & \text{if} \quad i \neq j \end{cases} \tag{4.77}$$

Thus, (4.76) can be written in the compact form:

$$\tilde{f}_{i,j} = -\delta_{ij} - \sum_{k=1}^{n} a_{ik,j} \cdot g_k \tag{4.78}$$

where δ_{ij} is the well-known Kronecker's operator. Furthermore, if $g_k = 0 \; \forall k = 1, \ldots, n_y$ (which is verified at the solution $\boldsymbol{y}_0$), then one obtains the final expression:

$$\tilde{f}_{i,j} = -\delta_{ij} \tag{4.79}$$

that is the tensor version of (4.74).

This result is straightforward for a scalar $g(y)$, i.e. for $y \in \mathbb{R}$ and $g \in \mathbb{R}$, as follows:

$$\dot{y} = \tilde{f}(y) = -\frac{g(y)}{g_y(y)} \tag{4.80}$$

$$\begin{aligned} \Rightarrow \quad \tilde{f}_y(y) &= -\frac{g_y(y)}{g_y(y)} + \frac{g_{yy}(y)}{g_y^2(y)} g(y) \\ &= -1 + \frac{g_{yy}(y)}{g_y^2(y)} g(y) \end{aligned} \tag{4.81}$$

thus $\tilde{f}_y(y_0) = -1$ if $g(y_0) = 0$ and $g_y(y_0) \neq 0$.

Equation (4.74) implies that all eigenvalues of $\tilde{\boldsymbol{f}}_{\boldsymbol{y}}$ at the solution point are equal to -1. Thus, (4.74) means that the solution of (4.8), if exists, is asymptotically stable. The reachability of this solution depends on the starting point $\boldsymbol{y}(t_0) = \boldsymbol{y}^{(0)}$, which has to be within the *region of attraction* or *stability region* of the equilibrium point $\boldsymbol{y}_0$.

The continuous Newton's method is expected to show better ability to converge than other methods previously discussed if the initial guess is within the region of attraction. A case study that confirms this hypothesis is given in [196].

Example 4.6 Runge-Kutta's Formula for Solving the Power Flow Problem

It is well-known that the forward Euler's method, even with variable time step, can be numerically unstable. Reference [127] suggests that, given the analogy between the power flow equations (4.8) and the ODE (4.68), any well-assessed numerical method can be used for integrating (4.68). It is thus intriguing to use some efficient integration method for solving (4.8). Since the computation of $\tilde{\boldsymbol{f}} = -[\boldsymbol{g_y}]^{-1}\boldsymbol{g}$ implies the inversion of the power flow Jacobian matrix, only explicit integration methods are suitable and computationally efficient, since one does not need to compute the Jacobian matrix of $\tilde{\boldsymbol{f}}$. Explicit numerical integration methods, including a variety of Runge-Kutta's formulæ are described in Subsection 8.3.1 of Chapter 8.

It is worth noting that any order and any version of explicit integration method can be used, and any of these methods is expected to be numerically more stable than the Euler's forward method. For example, Figure 4.10 compares the convergence behavior of continuous Newton's method solved using the classical 4th order Runge-Kutta's formula (8.35) (see Example 8.2

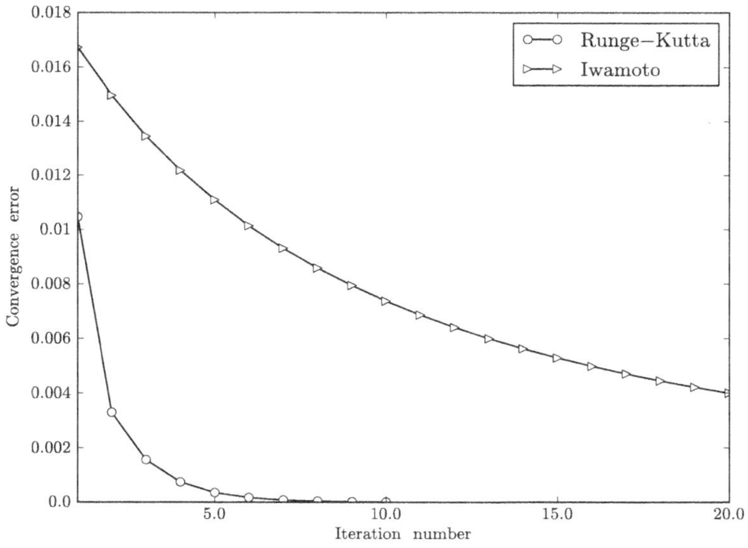

Fig. 4.10 Convergence behavior of Runge-Kutta's 4th order formula and the Iwamoto's method for the IEEE 14-bus system

of Chapter 8) and the Iwamoto's method for the IEEE 14-bus system. The use of a robust power flow solver algorithm is not required in this case since this is a well-conditioned case. However, the results confirm that the Runge-Kutta's formula converges faster than a variable step Euler's method (e.g., Iwamoto's method).

Script 4.6 Runge-Kutta's Formula for Solving the Power Flow Problem

The Python implementation of the generic i^{th} iteration of the Runge-Kutta's 4th order formula (8.35) provided in Chapter 8 looks as follows:

```
yold = system.DAE.y

k1 = alfa*calcInc()
system.DAE.y = yold + 0.5*k1

k2 = alfa*calcInc()
system.DAE.y = yold + 0.5*k2

k3 = alfa*calcInc()
system.DAE.y = yold + k3

k4 = alfa*calcInc()

# compute RK4 increment of variables
inc = (k1 + 2*(k2+k3) + k4)/6

# to estimate RK error, use the RK2:Dy and RK4:Dy.
csi = max(abs(abs(k2) - abs(inc)))
if csi > 0.01:
    deltat = max([0.985*deltat, 0.75])
else:
    deltat = min([1.015*deltat, 0.75])

system.DAE.y = yold + inc
```

where it is assumed that `calcInc()` returns $\tilde{\boldsymbol{f}}(\boldsymbol{y}^{(i)})$. The code above has to be inserted in a while-loop similar to that discussed in Script 4.2.

4.6 Summary

This section summarizes most relevant concepts related to power flow analysis.

Power flow problem vs. classical circuit analysis: The power flow analysis differs from classical circuit analysis in what concerns input data. Typical circuit input data are generator voltage phasors and load impedances. Thus, circuit analysis problem formulation is based on current injections that result in a linear problem if all parameters are constant. On the other hand, typical power flow data are load power consumptions and generator active power supplies. Thus, the power flow problem formulation is based on power injections that result in a nonlinear set of algebraic equations. The intrinsic nonlinearity of the power flow problem makes its solution a challenge.

System model: The classical power flow problem formulation considers only constant PV generators and constant PQ loads. Furthermore, one generator, the slack, is defined as slack and is used as phase angle reference. These models are deduced based on practical assumptions and operation of transmission systems. However, the classical power flow problem formulation can be improved in a variety of ways. For example, the distributed slack model is a better description of the behavior of the system than the standard single slack model. Other improvements of the classical models concerns the inclusion of detailed models of regulating transformers and dynamic devices. The entire Part III is devoted to provide further insights on detailed power flow models.

Power flow solvers: The classical formulation of the power flow problem can be solved only numerically. With this aim, several methods are available, including but not limited to: Gauss-Seidel's, Jacobi's, Newton's, FDPF, and dc power flow methods. The Newton's method is available in several versions: inexact (e.g., GMRES), dishonest and robust.

Dc power flow: The dc power flow formulation avoids the issues originated by the nonlinearity of the classical power flow problem. The dc formulation allows simplifying the equations so that only active power mismatches and bus voltage phase angles are considered. The resulting problem is linear but completely neglects reactive power flows and bus voltage magnitude deviations. For this reasons, the error introduced by the dc power flow model are not negligible.

General framework for power flow analysis solvers: Power flow algorithms based on the Newton's method and its variants can be viewed as particular implementations of a general class of solvers, namely the continuous Newton's method. This approach shows that the map of an iterative method is formally equivalent to find the equilibrium point of a set of differential equations. The equilibrium point of the equivalent ODE system is the solution of the power flow problem.

Chapter 5
Continuation Power Flow Analysis

This chapter describes a variety of techniques used for determining the point of collapse of power flow equations with particular emphasis on continuation power flow analysis. Section 5.1 introduces the maximum loading condition problem using a didactic 2-bus system. Section 5.2 defines the general model for the maximum loading condition. Section 5.3 describes direct methods for computing saddle-node bifurcation and limit-induced bifurcation points while Section 5.4 describes the continuation power flow technique. Section 5.4 also explains the conceptual differences, advantages and drawbacks of continuation (or homotopy) methods with respect to direct ones. Subsections 5.4.4 and 5.4.5 discuss the analogy between the continuous Newton's method and the homotopy approach and the $N-1$ contingency analysis, respectively. Finally, Section 5.5 summarizes the main concepts of the continuation power flow analysis.

5.1 Background

A relevant problem of power system analysis is the determination of the maximum load power consumption that a network can supply. With this aim, consider the 2-bus test system depicted in Figure 5.1. A similar example considering voltage dependent load characteristics can be found in Example 14.1 of Chapter 14.

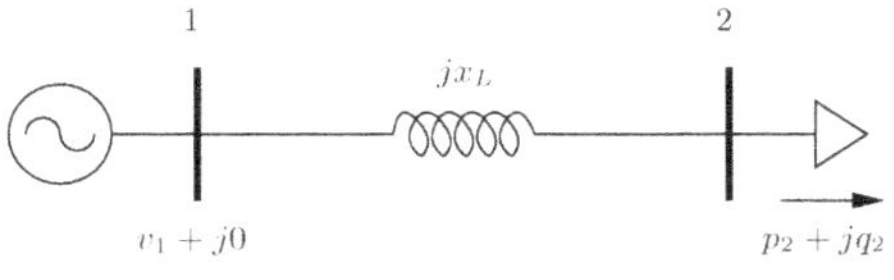

Fig. 5.1 2-bus system

F. Milano: Power System Modelling and Scripting, Power Systems, pp. 103–130.
springerlink.com

The power flow equations that describe this system are:

$$\begin{aligned} -p_2 &= \frac{v_2 v_1^{\text{ref}}}{x_L} \sin \theta_2 \\ -q_2 &= \frac{v_2^2}{x_L} - \frac{v_2 v_1^{\text{ref}}}{x_L} \cos \theta_2 \end{aligned} \tag{5.1}$$

where we assume that the slack generator voltage at bus 1 is the phase reference. After some simple manipulations, one obtains:

$$\begin{aligned} p_2^2 &= \frac{v_2^2 (v_1^{\text{ref}})^2}{x_L^2} \sin^2 \theta_2 \\ q_2^2 + \frac{v_2^4}{x_L^2} + 2q_2 \frac{v_2^2}{x_L} &= \frac{v_2^2 (v_1^{\text{ref}})^2}{x_L^2} \cos^2 \theta_2 \\ \Rightarrow \quad 0 &= p_2^2 + q_2^2 + \frac{v_2^4}{x_L^2} + 2q_2 \frac{v_2^2}{x_L} - \frac{v_2^2 (v_1^{\text{ref}})^2}{x_L^2} \end{aligned} \tag{5.2}$$

The load bus voltage magnitude v_2 can be written as a function of the load power p_2:

$$v_2 = \sqrt{-(q_2 x_L - (v_1^{\text{ref}})^2/2) \pm \sqrt{(q_2 x_L - (v_1^{\text{ref}})^2/2)^2 - x_L^2 (p_2^2 + q_2^2)}} \tag{5.3}$$

Assuming, without loss of generality, that the load has a constant power factor, i.e., $q_2 = p_2 \tan \phi_2$, (5.3) becomes:

$$v_2 = \sqrt{-a \pm \sqrt{a^2 - x_L^2 p_2^2 (1 + \tan^2 \phi_2)}} \tag{5.4}$$

where:

$$a = p_2 \tan \phi_2 x_L - \frac{(v_1^{\text{ref}})^2}{2} \tag{5.5}$$

Equation (5.4) is depicted in Figure 5.2 and is known as *PV curve* or *nose curve*, due to its characteristic shape. It may be of interest to note that, in order to plot the PV curve of Figure 5.2, using (5.4) is not the best choice because the function $v_2(p_2)$ is not defined on all $\mathbb{R}$ and is not biunivocal. To use the other way round, i.e., the function $p_2(v_2)$ results easier:

$$p_2 = \frac{v_2^2}{x_L} \left(\frac{-\tan \phi_2 + \sqrt{\tan^2 \phi_2 - \left(1 - \frac{(v_1^{\text{ref}})^2}{v_2^2}\right)}}{1 + \tan^2 \phi_2} \right) \tag{5.6}$$

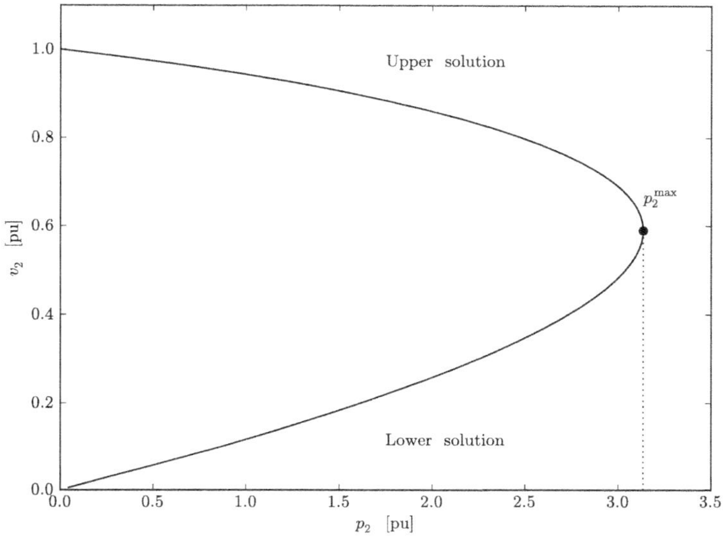

Fig. 5.2 PV curve for the 2-bus system

Some relevant remarks on (5.4) are:

1. The system is characterized by a maximum value of the load power, say $p_2^{\max}$, which is known as *maximum loading condition*.
2. For $p_2 > p_2^{\max}$ the power flow equations (5.1) have no solution. For this reason, the power flow solution for $p_2 = p_2^{\max}$ is known as *point of collapse*. In physical terms, this means that the system cannot supply a load whose power is $p_2 > p_2^{\max}$. Thus, $p_2^{\max}$ is the maximum power that can be transmitted by the network and it can be considered a limit of the transmission system.
3. For $p_2 < p_2^{\max}$, there are two values of v_2 that solve (5.1). However, only the solution with the higher v_2 value (*upper solution*) is physically acceptable. The other value (*lower solution*) has only a mathematical interest.
4. The shape of the PV curve is independent of the load power factor, as well as of system parameters. In other words, any network of any size and complexity shows a similar relationship between bus voltage magnitudes and load powers. PV curves are inherent the structure of classical power flow equations. As a matter of fact, as shown in (4.16) or (4.18), these have a quadratic dependence on bus voltages.

The PV curve depicted in Figure 5.2 was obtained assuming that the slack generator at bus 1 can supply any amount of active and reactive power. If the hypothesis about the reactive power is removed, there can be a value of

p_2 for which the reactive power generated by the slack bus is the maximum one, say $q_1^{\max}$. Since the system has no other reactive power source, the load power cannot be increased any further. If the generator reactive power limit is reached before the transmission system limit, the former yields the point of collapse or, which is the same, the $p_2^{\max}$ value.

In order to determine the collapse point due to reactive power limit for the 2-bus system, assume that the slack is modelled as a constant QV generator, where $q_1 = q_1^{\max}$. The resulting power flow equations are:

$$\begin{aligned} -p_2 &= \frac{v_2 v_1}{x_L} \sin\theta_2 \\ -q_2 &= \frac{v_2^2}{x_L} - \frac{v_2 v_1}{x_L}\cos\theta_2 \\ q_1^{\max} &= \frac{v_1^2}{x_L} - \frac{v_2 v_1}{x_L}\cos(-\theta_2) \end{aligned} \tag{5.7}$$

In this case, the voltage magnitude v_1 is a variable. Using the second and the third equations of (5.7), one has:

$$v_1 = \sqrt{x_L q_1^{\max} + v_2^2 + x_L p_2 \tan\phi_2} \tag{5.8}$$

Then, substituting (5.8) in (5.2), the expression of $p_2(v_2)$ becomes:

$$p_2 = \frac{\frac{v_2^2}{x_L}\tan\phi_2 + \sqrt{(\frac{v_2^2}{x_L}\tan\phi_2)^2 + 4v_2^2 q_1^{\max}(1+\tan^2\phi_2)/x_L}}{2(1+\tan^2\phi_2)} \tag{5.9}$$

If (5.6) and (5.9) intersect in the upper part of (5.6), then the generator reactive power limit occurs before the transmission system limit. This situation is illustrated in Figure 5.3. The interpretation of Figure 5.3 is as follows: as far as $q_1 < q_1^{\max}$, the system is described by (5.6); then, at $q_1 = q_1^{\max}$, the slack bus model changes from constant $v\theta$ to constant qv and the system behavior is described by the equation (5.9).

The determination of PV curves and, in particular, of the maximum loading condition, has great relevance in security analysis. In fact, the knowledge of the maximum loading condition allows defining the distance of the current working condition to the collapse. If this distance is too small, the system operator has to take corrective actions to provide a minimum security margin, i.e. a minimum distance of the current operating point to the collapse.

Unfortunately, analytical formulæ such as (5.6) or (5.9) cannot be found for a generic system. Even for the 2-bus system considered so far, including a resistance in the transmission line prevents from obtaining an explicit expression for $p_2(v_2)$. Thus, a general numerical method for determining the maximum loading condition is desirable. The following sections describe systematic approaches to tackle this problem.

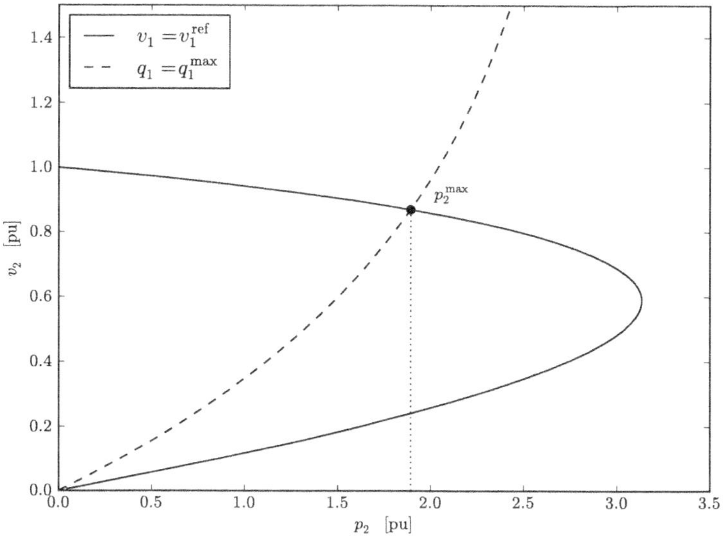

Fig. 5.3 PV curve for the 2-bus system considering generator reactive power limits

5.2 System Model

In order to generalize the concepts presented in the previous section, it is necessary to define a power system model suitable for encountering the maximum loading condition. With this aim, the power flow model is modified as follows:

$$\mathbf{0} = \boldsymbol{g}(\boldsymbol{y}, \mu) \tag{5.10}$$

where $\boldsymbol{y}$ are, as usual, the algebraic variables and $\mu \in \mathbb{R}$ is a *loading parameter*, i.e., a scalar independent parameter that multiplies all generator and load powers, as follows:

$$\begin{aligned} \boldsymbol{p}_G &= (\mu \boldsymbol{I}_{n_G} + k_G \boldsymbol{\Gamma}) \boldsymbol{p}_{G0} \\ \boldsymbol{p}_L &= \mu \boldsymbol{p}_{L0} \\ \boldsymbol{q}_L &= \mu \boldsymbol{q}_{L0} \end{aligned} \tag{5.11}$$

where $\boldsymbol{I}_{n_G}$ is the identity matrix of order n_G, $\boldsymbol{\Gamma} = \mathrm{diag}(\gamma_1, \gamma_2, \dots, \gamma_{n_G})$ are generator loss participation factors, k_G is a scalar variable used for accomplishing the distributed slack bus model as discussed in Subsection 4.4.9 of Chapter 4, and $\boldsymbol{p}_{G0}$, $\boldsymbol{p}_{L0}$ and $\boldsymbol{q}_{L0}$ are the "base case" or initial generator and load powers, respectively.

An alternative loading model that can be found in the literature is:

$$\begin{aligned} \boldsymbol{p}_G &= (\boldsymbol{I}_{n_G} + \tilde{\mu}\boldsymbol{I}_{n_G} + k_G\boldsymbol{\Gamma})\boldsymbol{p}_{G0} \\ \boldsymbol{p}_L &= (1 + \tilde{\mu})\boldsymbol{p}_{L0} \\ \boldsymbol{q}_L &= (1 + \tilde{\mu})\boldsymbol{q}_{L0} \end{aligned} \tag{5.12}$$

where $\tilde{\mu}$ expresses the distance of the base case to the stressed operating condition and $\tilde{\mu}^{\max}$ is thus the distance to the point of collapse.

Another loading model is as follows:

$$\begin{aligned} \boldsymbol{p}_G &= \boldsymbol{p}_{G0} + (\breve{\mu}\boldsymbol{I}_{n_G} + k_G\boldsymbol{\Gamma})\boldsymbol{p}_{S0} \\ \boldsymbol{p}_L &= \boldsymbol{p}_{L0} + \breve{\mu}\boldsymbol{p}_{D0} \\ \boldsymbol{p}_L &= \boldsymbol{p}_{L0} + \breve{\mu}\boldsymbol{p}_{D0} \end{aligned} \tag{5.13}$$

where $\boldsymbol{p}_{S0}$, $\boldsymbol{p}_{D0}$ and $\boldsymbol{q}_{D0}$ are called generator (or supply) and load (or demand) power *directions* since they define a custom path along which the system is loaded through the parameter $\breve{\mu}$. The three models (5.11), (5.12) and (5.13) are equivalent and do not introduce conceptual differences. Thus, in the following, unless explicitly stated, model (5.11) is used.

5.3 Direct Methods

Probably, the most intuitive approach for solving the maximum loading condition problem is to use a *direct method*. The idea is to encounter a set of equations that defines the point of collapse of the power system modelled as (5.10). As observed in Section 5.1, system equations change depending on the type of maximum loading condition: transmission system or reactive power limit. Thus one has to formulate a different set of equations for each type of maximum loading condition.

In the literature, this task has been assessed with the help of bifurcation theory which allows defining a taxonomy of the points of collapse, which are, in turns, *bifurcation points*. A bifurcation point is a solution of (5.10) that satisfies certain mathematical conditions. In particular, classical power flow equations can show only two kinds of bifurcation points, namely, the saddle-node bifurcation and limit-induced bifurcation.

It is important to note that bifurcation theory applies to dynamical systems (i.e., ordinary differential equations), while the power flow equations (5.10) are algebraic. Thus, strictly speaking, stability concepts and bifurcation theory cannot be applied to (5.10). However, it is common practice to associate the maximum loading condition with bifurcation points and, hence, to infer the *static voltage stability region* of the power system through the analysis of algebraic power flow equations.

Using stability concepts proper of dynamical systems for a set of algebraic equations can be justified in mathematical terms if one considers that (5.10) can be rewritten as:

$$\boldsymbol{T}_{\epsilon}\dot{\boldsymbol{y}} = \boldsymbol{g}(\boldsymbol{y}, \mu(t)) \tag{5.14}$$

where $\boldsymbol{T}_{\epsilon} = \text{diag}(T_{\epsilon 1}, T_{\epsilon 2}, \dots, T_{\epsilon n_y})$ is a diagonal matrix composed of "small" time constants. Furthermore, the time dependent variation of $\mu(t)$ is assumed to be "slow", so that system dynamics are considered steady-state and systems devices are approximated by means of the models used in the classical power flow analysis. These assumptions agree with the definition of algebraic variables given in Section 1.4 of Chapter 1. Further insights on (5.14) are given in Section 8.6 of Chapter 8.

5.3.1 Saddle-Node Bifurcation

The Saddle-Node Bifurcation (SNB) is the formal mathematical notation of the transmission system limit that is described in Section 5.1. Figure 5.2 shows that the tangent to the curve $v_2(p_2)$ at the maximum loading point is vertical, i.e., $dv_2/dp_2 \to \infty$ at the point of collapse.

The conditions for a SNB point are as follows:

$$\begin{aligned} \boldsymbol{g}(\boldsymbol{y}, \mu) &= \boldsymbol{0} \\ \boldsymbol{g}_{\boldsymbol{y}}\boldsymbol{\nu} &= \boldsymbol{0} \\ \|\boldsymbol{\nu}\|_2 &= 1 \end{aligned} \tag{5.15}$$

or

$$\begin{aligned} \boldsymbol{g}(\boldsymbol{y}, \mu) &= \boldsymbol{0} \\ \boldsymbol{g}_{\boldsymbol{y}}^T\boldsymbol{w} &= \boldsymbol{0} \\ \|\boldsymbol{w}\|_2 &= 1 \end{aligned} \tag{5.16}$$

where $\boldsymbol{v}$ and $\boldsymbol{w}$ are the right and the left eigenvectors, respectively, and the operator $\|\cdot\|_2$ indicates the Euclidean norm. The solution of (5.15) and (5.16) can be obtained by means of a Newton's method, which requires to compute and factorize the Jacobian matrix. For example, the complete Jacobian matrix of (5.15) is:

$$\begin{bmatrix} \boldsymbol{g}_{\boldsymbol{y}} & 0 & \boldsymbol{g}_{\mu} \\ \boldsymbol{g}_{\boldsymbol{yy}}\boldsymbol{\nu} & \boldsymbol{g}_{\boldsymbol{y}} & 0 \\ 0 & \partial\|\boldsymbol{\nu}\|_2/\partial\boldsymbol{\nu} & 0 \end{bmatrix} \tag{5.17}$$

The Euclidean norm $|\boldsymbol{\nu}|$ reduces the sparsity of the Jacobian matrix computed at each iteration of the Newton's method, but allows avoiding refactorizations (as happens in the case of ∞-norm) and is numerically more stable than the 1-norm. Since (5.17) requires the calculations of the Hessian matrix $\boldsymbol{g_{yy}}$, direct methods are computationally expensive. Furthermore, if the number of buses is n_b, the variables in (5.15) or (5.16) are $2n_b + 1$.

Unfortunately, even assuming that modern computers and adequate functions for handling sparse matrices are able to reduce the computational burden, problems (5.15) and (5.16) are hard to solve. The main issue is to find a *good* initial guess for the eigenvectors $\boldsymbol{\nu}$ or $\boldsymbol{w}$. This is not an easy task since the eigenvector elements vary drastically close to the bifurcation point. Thus, in order to find a good initial guess, one has to start relatively close to the bifurcation point.

Example 5.1 Saddle-Node Bifurcation of the 2-Bus System

This example provides the analytical expression of the SNB point of the 2-bus system of Figure 5.1. System data are: $v_1^{\rm ref} = 1$ pu, $x_L = 0.1$ pu, and $\cos\phi_2 = 0.9$ lag. Due to the simplicity of the problem, the solution can be find analytically. The value of $p_2^{\max}$ is:

$$p_2^{\max} = \frac{(v_1^{\rm ref})^2}{2x_L}(-\tan\phi_2 + \sqrt{1+\tan^2\phi_2}) = 3.1339 \text{ pu} \tag{5.18}$$

Equation (5.18) can be obtained by imposing that the argument of the inner square root of (5.4) is zero:

$$(p_2^{\max}\tan\phi_2 x_L - (v_1^{\rm ref})^2/2)^2 - x_L^2(p_2^{\max})^2(1+\tan^2\phi_2) = 0 \tag{5.19}$$

And, substituting (5.18) in (5.4), one obtains:

$$v_2^M = \sqrt{-a} = \sqrt{\frac{(v_1^{\rm ref})^2}{2} - p_2^{\max}\tan\phi_2 x_L} = 0.5901 \text{ pu} \tag{5.20}$$

The same solution can be found observing that the second equation of (5.16) implies that the Jacobian matrix $\boldsymbol{g_y}$ has to be singular at the SNB point.[1] This is another way of imposing the the tangent of $v_2(p_2)$ is vertical at the SNB. Since in this simple case the power flow Jacobian matrix size is 2×2, one can compute analytically the determinant. The Jacobian matrix of (5.1) is:

$$\boldsymbol{g_y} = \frac{1}{x_L}\left[\begin{array}{c|c} v_1^{\rm ref} v_2\cos\theta_2 & v_1^{\rm ref}\sin\theta_2 \\ \hline v_1^{\rm ref} v_2\sin\theta_2 & 2v_2 - v_1^{\rm ref}\cos\theta_2 \end{array}\right] \tag{5.21}$$

Hence, imposing that the determinant is zero leads to:

$$2\frac{v_1^{\rm ref} v_2}{x_L}\cos\theta_2 - \frac{v_1^{\rm ref} v_2}{x_L} = 0 \tag{5.22}$$

[1] In fact $\boldsymbol{\nu}$ cannot be zero because of the constraint $\|\boldsymbol{\nu}\|_2 = 1$.

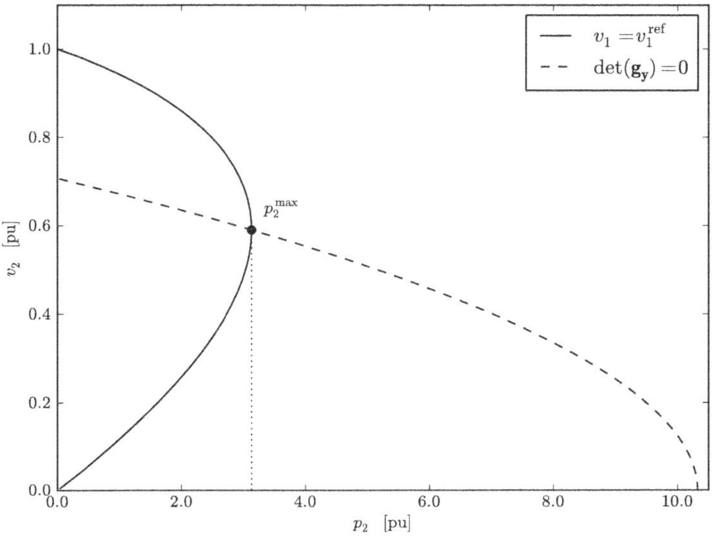

Fig. 5.4 Graphical illustration of the direct method for computing the SNB of the 2-bus system

The trigonometric function $\cos\theta_2$ can be substituted for the expression deduced by the second of (5.1):

$$\cos\theta_2 = \frac{1}{v_1^{\rm ref} v_2}\left(v_2^2 + x_L p_2 \tan\phi_2\right) \tag{5.23}$$

Thus (5.22) and (5.23) lead to:

$$p_2 = \frac{(v_1^{\rm ref})^2 - 2v_2^2}{2x_L \tan\phi_2} \tag{5.24}$$

The intersection of (5.6) and (5.24) is the SNB point (see Figure 5.4).

5.3.2 Limit-Induced Bifurcation

Limit-Induced Bifurcations (LIBs) allow formally defining maximum loading conditions due to generator reactive power limits. LIB points are the solution of two sets of equations:

$$\begin{aligned} \mathbf{0} &= \boldsymbol{g}_1(\boldsymbol{y}, \mu) \\ \mathbf{0} &= \boldsymbol{g}_2(\boldsymbol{y}, \mu) \end{aligned} \tag{5.25}$$

where $\boldsymbol{g}_1$ and $\boldsymbol{g}_2$ differ in the model of some devices. For example, let us assume that in $\boldsymbol{g}_1$ a certain generator is modelled as a constant pv and that in $\boldsymbol{g}_2$ the same generator is modelled as a constant pq, with $q = q^{\max}$. Then, the LIB point is the intersection of the two sets of power flow equations $\boldsymbol{g}_1$ and $\boldsymbol{g}_2$.

In practice, it is not necessary to duplicate all equations, since at the LIB point $(\boldsymbol{y}, \mu)$ is the same for the two systems. It suffices to solve:

$$\begin{aligned} \mathbf{0} &= \boldsymbol{g}(\boldsymbol{y}, \mu) \\ 0 &= \breve{g}(\boldsymbol{y}, \mu) \end{aligned} \tag{5.26}$$

where $\boldsymbol{g} \equiv \boldsymbol{g}_1$ and $\breve{g} \equiv \boldsymbol{g}_2$ is a scalar equation that imposes the desired additional constraint. For example, assuming that the limit is a generator reactive power, $\boldsymbol{g}$ can model the generator at bus h as a constant pv and $\breve{g}$ is:

$$\breve{g} = q_{Gh} - q_{Gh}^{\max} \tag{5.27}$$

In principle, one can impose more than one scalar constraint in (5.26). As a matter of fact, not all LIBs lead to the maximum loading condition (see Example 5.4). Moreover, constraints can be of any kind, e.g., voltage limits, transmission line thermal limits, etc. However, in the literature, LIBs generally concern only generator reactive power limits, since only these limits can lead to a point of collapse.

In mathematical terms, a LIB is not characterized by a Jacobian matrix singularity. The Jacobian matrix is only discontinuous at the bifurcation point. This property is used in continuation power flow analysis for distinguishing between SNB and LIB points.

Example 5.2 Limit-Induced Bifurcation of the 2-Bus System

For the 2-bus example of Figure 5.1, the LIB point can be determined analytically. In particular, the intersection of (5.6) and (5.9) can be determined solving the following system:

$$-p_2 = \frac{v_2 v_1^{\text{ref}}}{x_L} \sin\theta_2 \tag{5.28}$$

$$-q_2 = \frac{v_2^2}{x_L} - \frac{v_2 v_1^{\text{ref}}}{x_L} \cos\theta_2 \tag{5.29}$$

$$q_1^{\max} = \frac{(v_1^{\text{ref}})^2}{x_L} - \frac{v_2 v_1^{\text{ref}}}{x_L} \cos(-\theta_2) \tag{5.30}$$

From (5.30) and (5.29):

$$q_2 = \frac{(v_1^{\text{ref}})^2 - v_2^2}{x_L} - q_1^{\max} \tag{5.31}$$

Then, substituting (5.31) in (5.2), one obtains:

$$(b-1)v_2^4 + dv_2^2 + dc^2 = 0 \tag{5.32}$$

where:

$$\begin{aligned} b &= \frac{1+\tan^2\phi_2}{\tan^2\phi_2} \\ c &= x_L q_1^{\max} - (v_1^{\mathrm{ref}})^2 \\ d &= (1-2b)(v_1^{\mathrm{ref}})^2 + 2x_L(b-1)q_1^{\max} \end{aligned} \tag{5.33}$$

The only physical solution of the bi-quadratic equation (5.32) is:

$$v_2^M = \sqrt{\frac{-d-\sqrt{d^2-4b(b-1)c^2}}{2(b-1)}} = 0.8708 \text{ pu} \tag{5.34}$$

Finally, $p_2^{\max} = p_2(v_2^M)$ can be determined indifferently from (5.6) or (5.9), that yield $p_2^{\max} = 1.8928$ pu. These results are obtained assuming the following system data: $q_1^{\max} = 1.5$ pu, $x_L = 0.1$ pu, and $\cos\phi_2 = 0.9$ lag.

5.3.3 Nonlinear Programming

Another class of direct approaches consists in formulating the maximum loading condition as a nonlinear programming problem. This approach has proved to be robust [35, 36, 37, 259]. Furthermore, the formulation as an optimization problem allows including a variety of constraints and setting up complex models [198, 260]. Due to the relevance of nonlinear programming in power system analysis, the entire Chapter 6 is dedicated to this topic. This section only provides a simple yet relevant example.

Example 5.3 Optimization Problem Equivalent to the Saddle-Node Direct Method

This example proves that (5.16) is formally equivalent to an optimization problem. This proof firstly appeared in [36]. Consider the following problem:

$$\begin{aligned} &\underset{\boldsymbol{y},\,\mu}{\text{Minimize}} \quad -\mu \\ &\text{subject to} \quad \boldsymbol{g}(\boldsymbol{y},\mu) = \boldsymbol{0} \end{aligned} \tag{5.35}$$

The Lagrangian function associated with problem (5.35) is:

$$\mathscr{L}(\boldsymbol{y},\mu,\boldsymbol{\rho}) = -\mu + \boldsymbol{g}^T(\boldsymbol{y},\mu)\boldsymbol{\rho} \tag{5.36}$$

whose the Karush-Kuhn-Tucker's optimality conditions are:

$$\begin{aligned} \boldsymbol{g}(\boldsymbol{y}, \mu) &= \mathbf{0} \\ \boldsymbol{g}_{\boldsymbol{y}}^T \boldsymbol{\rho} &= \mathbf{0} \\ \boldsymbol{g}_{\mu}^T \boldsymbol{\rho} &= 1 \end{aligned} \tag{5.37}$$

Thus, the global minimizer of (5.35) is a solution of (5.16). In fact, the left eigenvector is equal to the vector of dual variables ($\boldsymbol{w} = \boldsymbol{\rho}$) and the condition $\boldsymbol{g}_{\mu}^T \boldsymbol{\rho} = 1$ is equivalent to $\|\boldsymbol{w}\|_2 = 1$ since its effect is to impose that the vector of dual variables is not zero ($\boldsymbol{\rho} \neq \mathbf{0}$).

The analogy between direct methods and nonlinear programming problems can be extended to LIBs given that proper inequalities and complementarity constraints are included in (5.35). The interested reader can find a rigorous proof in [21].

5.4 Homotopy Methods

The common feature of all direct methods is to find a single solution, i.e., the maximum loading condition. Since the objective is to find the maximum loading condition, one may argue that direct methods provide just the required information. Actually, this is not completely true. The nose curves shown in Figures 5.2 and 5.3 provide much more information than the mere bifurcation point ($v_2^M, p_2^{\max}$). In fact, nose curves provide a *continuous set* of solutions, one of which is the maximum loading condition. This continuous set allows understanding the behavior of the system likely better than the single bifurcation point. Moreover, important drawbacks of the direct methods discussed above except for the nonlinear programming approach are:

1. The solution can be difficult to obtain as a good initial guess is required. This is especially true for determining SNB points.
2. Each bifurcation point is treated separately. Since one does not know *a priori* whether a LIB or a SNB yields the maximum loading condition, taking into account all possible bifurcations (a real-world system can show tens of LIBs before getting to the maximum loading condition) can be a lengthy process.

Finally, with regard to the nonlinear programming approach, it has to be noted that the computational burden of an optimal power flow problem of a real-size system is not negligible.

To overcome the drawbacks of direct methods, consider the following strategy. Starting from an initial solution (e.g., base case) of (5.10), the loading parameter μ is increased by a small amount, say $\Delta\mu$, and then (5.10) is solved again for the new value $\mu + \Delta\mu$. The operation is repeated as long as the solution of (5.10) cannot be found. This procedure is known as *embedding algorithm* [147]. In case (5.10) represent power flow equations, this procedure yields a series of power flow solutions, of which the last one is supposedly the maximum loading condition.

The critical point of the embedding algorithm is the word *supposedly*. In fact, there is no guarantee that the procedure ends at the maximum loading condition. Each solution is obtained by means of an iterative algorithm (e.g., Newton's method) and thus failing to obtain the convergence can be due either to numerical issues or to the fact that the solution actually does not exists. This issue is unsolvable, even using some robust technique as those described in Chapter 4. Other critical issues of the embedding algorithm are:

1. At SNB points, the Jacobian matrix is singular. Thus the Newton's method surely diverges for cases close enough to SNB points.
2. The number of steps and the magnitude of $\Delta\mu$ are not known *a priori*. In any case, $\Delta\mu$ has to be sufficiently small to ensure that the Newton's method applied to the next step converges. This fact prevents from using parallel computing as the initial guess of each step is the solution of the previous step. Thus, the computational burden of embedding algorithms is generally high.

A solution to all issues above is provided by *homotopy methods* [218]. The homotopy approach consists in defining a *homotopy map* $\boldsymbol{\psi}$ based on the original system equations $\boldsymbol{g}(\boldsymbol{y})$, as follows:

$$\boldsymbol{\psi}(\boldsymbol{y}, \mu) = \boldsymbol{g}(\boldsymbol{y}, \mu) \tag{5.38}$$

where the main difference between $\boldsymbol{g}$ and $\boldsymbol{\psi}$ is that in the latter μ is a variable of the system, thus $\boldsymbol{g} : \mathbb{R}^{n_y} \mapsto \mathbb{R}^{n_y}$ and $\boldsymbol{\psi} : \mathbb{R}^{n_y+1} \mapsto \mathbb{R}^{n_y}$. According to the definitions given in Section 1.4 of Chapter 1, μ ($\mu \in \mathbb{R}$) is an independent variable (i.e., $\boldsymbol{\mu} = [\mu]$) and $\boldsymbol{\psi} : \mathbb{R}^{n_y} \times \mathbb{R}^{n_\mu} \mapsto \mathbb{R}^{n_y}$ where $n_\mu = 1$. In homotopy methods, μ is called *continuation parameter*. However, in the following, μ is called *loading level*, thus referring to its physical meaning rather to the mathematical one.

Equation (5.38) may seem trivial since $\boldsymbol{\psi}$ and $\boldsymbol{g}$ coincide. This is because (5.38) is a *forced* or *natural parameter* homotopy, i.e., μ is a parameter of $\boldsymbol{g}$. It is important to note that, in general, homotopy maps can be defined also using an external parameter that has no physical meaning and no relation with $\boldsymbol{g}(\boldsymbol{y})$. In this case, the homotopy is called *free-running* or *artificial parameter* homotopy and has the general form:

$$\boldsymbol{\psi}(\boldsymbol{y}, \mu) = (1 - \mu)\boldsymbol{r}(\boldsymbol{y}) + \mu\boldsymbol{g}(\boldsymbol{y}) \tag{5.39}$$

where $\mu \in [0, 1]$ and $\boldsymbol{r}(\boldsymbol{y})$ is an arbitrary smooth function. However, in the following, only forced homotopy methods are considered.

Parametrizing $\boldsymbol{y}$ and μ by means of an arc length s, one has:

$$\boldsymbol{\psi}(\boldsymbol{y}(s), \mu(s)) = \boldsymbol{0} \tag{5.40}$$

and differentiating (5.40) with respect to s:

$$\frac{d\boldsymbol{\psi}}{ds} = \begin{bmatrix} \boldsymbol{\psi}_{\boldsymbol{y}} \dfrac{d\boldsymbol{y}}{ds} \\ \boldsymbol{\psi}_{\mu} \dfrac{d\mu}{ds} \end{bmatrix} \tag{5.41}$$

with:

$$\left\| \left(\frac{d\boldsymbol{y}}{ds}, \frac{d\mu}{ds} \right) \right\|_2 = 1 \tag{5.42}$$

and initial conditions:

$$\boldsymbol{y}(0) = \boldsymbol{y}^{(0)}, \qquad \mu(0) = \mu^{(0)} \tag{5.43}$$

Equation (5.41) is used for mapping $\boldsymbol{\psi}$ along a path parametrized by s. For a current point $(\boldsymbol{y}^{(i)}, \mu^{(i)})$, the next point $(\boldsymbol{y}^{(i+1)}, \mu^{(i+1)})$ can be found using a predictor-corrector method. In particular, the predictor step can be as follows:

$$\begin{aligned} \tilde{\boldsymbol{y}}^{(i+1)} &= \boldsymbol{y}^{(i)} + \Delta \boldsymbol{y}^{(i)} = \boldsymbol{y}^{(i)} + k \frac{d\boldsymbol{y}}{ds} \\ \tilde{\mu}^{(i+1)} &= \mu^{(i)} + \Delta \mu^{(i)} = \mu^{(i)} + k \frac{d\mu}{ds} \end{aligned} \tag{5.44}$$

where k is an adequate step size. The corrector step consists in ensuring that the point $(\boldsymbol{y}^{(i+1)}, \mu^{(i+1)})$ satisfies the condition $\boldsymbol{\psi}(\boldsymbol{y}^{(i+1)}, \mu^{(i+1)}) = 0$. A possible choice of the corrector step is to solve an optimization problem, as follows:

$$\begin{aligned} \underset{\boldsymbol{y},\ \mu}{\text{Minimize}} \quad & \|(\boldsymbol{y}^{(i+1)} - \tilde{\boldsymbol{y}}^{(i+1)}, \mu^{(i+1)} - \tilde{\mu}^{(i+1)})\|_2 \\ \text{subject to} \quad & \boldsymbol{\psi}(\boldsymbol{y}^{(i+1)}, \mu^{(i+1)}) = \boldsymbol{0} \end{aligned} \tag{5.45}$$

Although elegant, the solution of (5.45) for each step of the path can be cumbersome. In practical applications, it is thus preferred to substitute the optimization problem (5.45) for:

$$\begin{aligned} \boldsymbol{g}(\boldsymbol{y}, \mu) &= \boldsymbol{0} \\ \varrho(\boldsymbol{y}, \mu, \tilde{\boldsymbol{y}}^{(i+1)}, \tilde{\mu}^{(i+1)}) &= \boldsymbol{0} \end{aligned} \tag{5.46}$$

where $\varrho : \mathbb{R}^{n_y+1} \mapsto \mathbb{R}$ is the *continuation function*. How to choose ϱ depends on the application. Subsection 5.4.3 discusses typical expressions of ϱ used in the continuation power flow analysis.

The mathematical construction above may appear artificial. However, the advantage of homotopy methods is the robustness. The convergence is guaranteed by a series of theorems that the interested reader can find in [218]. Moreover, the homotopy approach has been successfully applied to bifurcation analysis [276]. If the function ϱ is properly set up, the Jacobian matrix of (5.46), i.e.,

$$\begin{bmatrix} \boldsymbol{g}_{\boldsymbol{y}} & \boldsymbol{g}_{\mu} \\ \varrho_{\boldsymbol{y}} & \varrho_{\mu} \end{bmatrix} \tag{5.47}$$

is not singular at the SNB point. This property is important for ensuring the convergence of the Newton's method close to the SNB point.

5.4.1 Continuation Power Flow

Although the very first appearance of homotopy methods in a power system conference dates back to the seventies [307], the proposal of the technique known as *Continuation Power Flow* (CPF) was conceived in the nineties [5]. Currently, the most authoritative reference about continuation power flow analysis and its usage for voltage stability assessment is [39].

For the interested reader, a system programming-based implementation of general homotopy methods can be found in [344]. Simulation results of this software package for power system problems, as well as an in-depth discussion of homotopy methods can be found in [147]. Finally, an implementation of homotopy methods for power system analysis is presented in [56], whereas C- and Matlab-based implementations of the CPF analysis can be found in [42] and [194], respectively.

The CPF method described in following subsections consists in a predictor step realized by the computation of the tangent vector and a corrector step that can be obtained either by means of a local parametrization or a perpendicular intersection.

5.4.2 Predictor Step

In the CPF analysis, the arc length that parametrizes (5.40) is defined as $s \equiv \mu$. Thus, for a generic step i and for the solution $(\boldsymbol{y}^{(i)}, \mu^{(i)})$, the following relation for the homotopy map $\boldsymbol{\psi}$ applies:

$$\boldsymbol{\psi}(\boldsymbol{y}^{(i)}(\mu^{(i)}), \mu^{(i)}) = \boldsymbol{0} \quad \Rightarrow \quad \left.\frac{d\boldsymbol{\psi}}{d\mu}\right|_i = \boldsymbol{0} = \boldsymbol{\psi}_{\boldsymbol{y}}|_i \left.\frac{d\boldsymbol{y}}{d\mu}\right|_i + \boldsymbol{\psi}_{\mu}|_i \tag{5.48}$$

Hence, the tangent vector for the solution $(\boldsymbol{y}^{(i)}, \mu^{(i)})$ can be approximated by:

$$\boldsymbol{\tau}^{(i)} = \left.\frac{d\boldsymbol{y}}{d\mu}\right|_i \approx \frac{\Delta \boldsymbol{y}^{(i)}}{\Delta \mu^{(i)}} \tag{5.49}$$

From (5.48) and (5.49), one has:

$$\begin{aligned} \boldsymbol{\tau}^{(i)} &= -\boldsymbol{\psi}_{\boldsymbol{y}}^{-1}|_i \boldsymbol{\psi}_{\mu}|_i \\ \Delta \boldsymbol{y}^{(i)} &= \boldsymbol{\tau}^{(i)} \Delta \mu^{(i)} \end{aligned} \tag{5.50}$$

A step size control k has to be chosen for determining the increment $\Delta \boldsymbol{y}^{(i)}$ and $\Delta\mu^{(i)}$ that appear in (5.44), along with the normalization (5.42) that avoids large steps when $\|\boldsymbol{\tau}^{(i)}\|_2$ increases:

$$\Delta\mu^{(i)} \triangleq \frac{k}{\|\boldsymbol{\tau}^{(i)}\|_2} \qquad \Delta \boldsymbol{y}^{(i)} \triangleq \frac{k\boldsymbol{\tau}^{(i)}}{\|\boldsymbol{\tau}^{(i)}\|_2} \tag{5.51}$$

Figure 5.5 provides a pictorial representation of the predictor step.

The sign of the step size k determines whether to increase or decrease μ. In order to obtain a complete nose curve, $k > 0$ in the upper part and $k < 0$ in the lower one. It is easy to know if the current point is in the upper or in the lower part of the nose curve, since the determinant of the Jacobian matrix sign before and after SNB or critical LIB points changes.

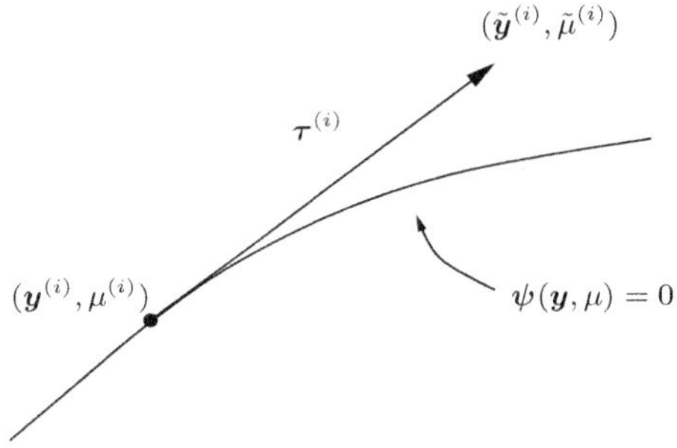

Fig. 5.5 Tangent predictor

For the sake of completeness, it is worth observing that predictor steps other than the tangent vector can be implemented. For example, an alternative predictor step is based on the secant [56]. The secant allows approximating the tangent $\boldsymbol{\tau}^{(i)}$ once two solutions $(\boldsymbol{y}^{(i-1)}, \mu^{(i-1)})$ and $(\boldsymbol{y}^{(i)}, \mu^{(i)})$ are known:

$$\boldsymbol{\tau}^{(i)} = \left.\frac{d\boldsymbol{y}}{d\mu}\right|_i \approx \boldsymbol{y}^{(i)} - \boldsymbol{y}^{(i-1)} \tag{5.52}$$

Predictions based on the secant have been observed to reach the maximum loading condition faster than those based on the tangent vector [39]. *Faster* means that the secant predictor requires less steps to reach the maximum loading point. However, the secant method may provide an inadequate prediction in case of sharp corners or if the solutions $i-1$ and i are too far apart (see Figure 5.6). An efficient method to accelerate the convergence of the CPF analysis to the point of collapse is the Tangent Vector Index (TVI) proposed in [289].

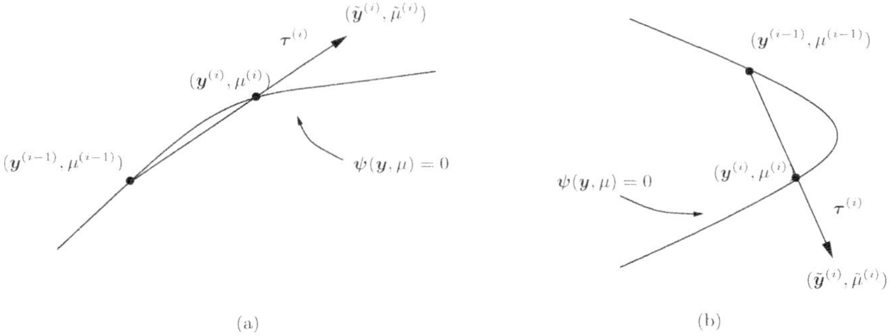

Fig. 5.6 Secant predictor: (a) smooth function, and (b) sharp corner

Script 5.1 CPF Predictor Step

The following Python script implements a predictor step based on the tangent vector.

```
import system
from cvxopt.base import matrix, spmatrix, sparse, log
from cvxopt.umfpack import symbolic, numeric, solve
from cvxopt.lapack import gees

def predictor(k, Jsign):

    exec system.Device.call_pf
    exec system.Device.call_gmu

    B = sparse(system.DAE.Gy)
    dy_dmu = -matrix(system.DAE.Gmu)

    if system.DAE.factorize:
        G = symbolic(B)
    try:
        solve(B, numeric(B, G), dy_dmu)
    except:
        G = symbolic(B)
        solve(B, numeric(B, G), dy_dmu)

    Jsign_old = Jsign
    Jsign = gees(matrix(B, tc = 'z'), select = negative)

    if Jsign != Jsign_old:
        ksign = -1
        Sflag = False

    norm = sum(dz_dmu**2)**0.5
    if not norm: norm = 1.0
    d_mu = ksign*k/norm
    d_y = d_mu*dy_dmu
    inc = matrix([d_y, d_mu])
```

```
def negative(x):

    return (x.real < 0.0)
```

The interface `system.Device` computes $\boldsymbol{g}$ and $\boldsymbol{g_y}$ and $\boldsymbol{g_\mu}$ of all devices that compose the system. The functioning of the class `system.Device` is described in Script 3.2 of Chapter 3. Furthermore, observe the use of the symbolic factorization of the power flow Jacobian matrix. The symbolic re-factorization is needed whenever a LIB occurs.

An important point of the predictor step is how to decide if the sign of the step k has to be changed. As previously discussed, the sign of the determinant of the Jacobian matrix $\boldsymbol{g_y}$ changes if the path has reached the lower part of the nose curve. Unfortunately computing the determinant is an expensive operation and cannot be used for real-size systems.

A simple method to compute the determinant of a matrix is using the LU factorization. In fact, one has:

$$\begin{aligned} & \boldsymbol{A} = \boldsymbol{L}\boldsymbol{U} \\ \Rightarrow \quad & \det(\boldsymbol{A}) = \det(\boldsymbol{L})\det(\boldsymbol{U}) = 1 \cdot \det(\boldsymbol{U}) = \prod_{h=1}^{n} u_{hh} \end{aligned} \tag{5.53}$$

where u_{hh} are the diagonal elements of the matrix $\boldsymbol{U}$. However, efficient sparse matrix algorithms generally provides a variant of the LU factorization, namely the LU factorization with partial pivoting, or LUP factorization. In this case:

$$\boldsymbol{A} = \boldsymbol{L}\boldsymbol{U}\boldsymbol{P} \tag{5.54}$$

where $\boldsymbol{P}$ is a permutation matrix. Thus, the determinant of the matrix $\boldsymbol{A}$ depends also on $\boldsymbol{P}$ whose number of permutations has to be taken into account to define the sign of the determinant of $\boldsymbol{A}$.

The proposed code uses an ordered Schur's factorization through the function `gees` provided by the module `cvxopt`. The function `gees` returns the number of eigenvalues selected by a user-defined "ordering routine". In this case, the ordering routine is the function `negative`. The rationale of this procedure is as follows. The determinant of a matrix can be also computed as the product of all eigenvalues λ_i of that matrix:

$$\det(\boldsymbol{A}) = \prod_{h=1}^{n} \lambda_h \tag{5.55}$$

Thus, if the number of negative eigenvalues increases or decreases by one, the sign of the determinant changes. The proposed approach allows also determining if the number of eigenvalues that change sign is more than one (e.g., in case a Hopf bifurcation occurs). In particular, it would work also if this number is even, while the method based on the LU factorization would fail in this case. However, the Schur's factorization is computationally expensive for large matrices and the LU factorization-based calculation of the determinant sign has to be preferred if Hopf bifurcations are not of interest.[2]

5.4.3 Corrector Step

As discussed above, the most commonly used corrector step is (5.46), which is a set of $n_y + 1$ equations in $n_y + 1$ variables. The solution of (5.46) is the new point $(\boldsymbol{y}^{(i+1)}, \mu^{(i+1)})$. This solution must be in the bifurcation manifold (i.e., the nose curve) of $\boldsymbol{\psi}$ and ϱ is an additional equation to guarantee a non-singular set at the bifurcation point. As for the choice of ϱ, there are several options. Common continuation equations are the *perpendicular intersection* and the *local parametrization*.

In case of perpendicular intersection, whose pictorial representation is reported in Fig. 5.7, the expression of ϱ becomes [136]:

$$\varrho(\boldsymbol{y}, \mu) = \begin{bmatrix} \Delta \boldsymbol{y}^{(i)} \\ \Delta \mu^{(i)} \end{bmatrix}^T \begin{bmatrix} \boldsymbol{y} - (\boldsymbol{y}^{(i)} + \Delta \boldsymbol{y}^{(i)}) \\ \mu - (\mu^{(i)} + \Delta \mu^) \end{bmatrix} = \begin{bmatrix} \Delta \boldsymbol{y}^{(i)} \\ \Delta \mu^{(i)} \end{bmatrix}^T \begin{bmatrix} \boldsymbol{y} - \tilde{\boldsymbol{y}}^{(i)} \\ \mu - \tilde{\mu^{(i)}} \end{bmatrix} \tag{5.56}$$

whereas for the local parametrization, either the parameter μ or a variable y_k is forced to be an assigned value [5, 56]:

$$\varrho(\boldsymbol{y}, \mu) = \mu - \mu^{(i)} - \Delta \mu^{(i)} = \mu - \tilde{\mu}^{(i)} \tag{5.57}$$

or

$$\varrho(\boldsymbol{y}, \mu) = y_k - y_k^{(i)} - \Delta y_k^{(i)} = y_k - \tilde{y}_k^{(i)} \tag{5.58}$$

The choice of the variable to be fixed depends on the bifurcation manifold of $\boldsymbol{\psi}$, as depicted in Fig. 5.8.

Script 5.2 CPF Corrector Step

The following Python script implements a corrector step with both perpendicular intersection and local parametrization corrector steps. The system (5.46) is solved using a Newton's method.

---

[2] The CVXOPT module does not currently provide a method for computing the determinant and does not return the LU factorization as a transparent Python object. Hence, a solution is to use the `det` function of the `numpy.linalg` module, as follows:

```
Jsign = numpy.linalg.det(matrix(B))
```

Since the function `det` internally uses the LAPACK package and the LU factorization, the computation of the determinant is relatively efficient. However, this requires factorizing the matrix `B` twice per each predictor step. Thus, the best solution is to modify the method `solve cvxopt.umfpack` module so that it returns the value (or the sign) of the determinant of the argument matrix to be factorized. This is possible only modifying the source C code of the CVXOPT package.

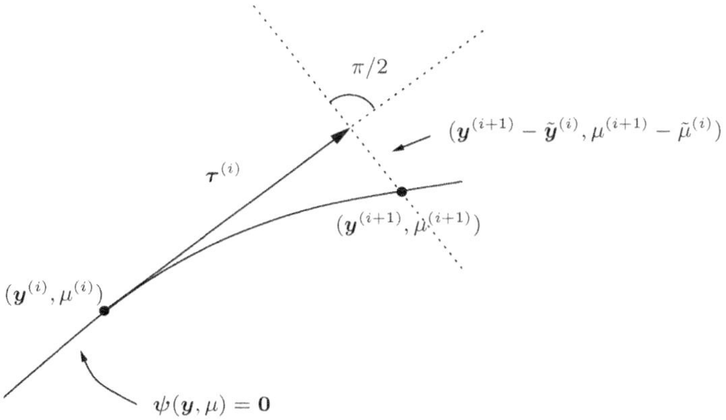

Fig. 5.7 Perpendicular intersection corrector

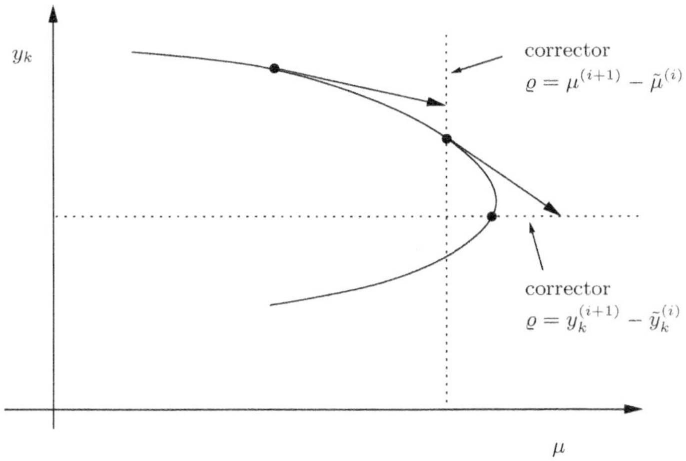

Fig. 5.8 Local parametrization corrector

```
import system
from cvxopt.base import matrix, spmatrix, sparse, log
from cvxopt.umfpack import symbolic, numeric, solve

def corrector(inc, y_old, v_bus):

    iter_corr = 0
    system.Settings.error = system.Settings.tol + 1

    while system.Settings.error > system.Settings.tol:

        if iter_corr > system.Settings.maxiter: break

        # call component functions
        exec system.Device.call_pf
```

```
exec system.Device.call_gmu

Mmu[0, 0] = 0.0
if system.CPF.method == 'perpendicular intersection':
     My = inc[n:n+m].T
     rho = My*(system.DAE.y - y_old - My.T)
else: # local parametrization
     My[0, v_bus] = 1.0
     rho = system.DAE.y[v_bus] - inc[v_bus] - y_old[v_bus]

A = sparse([[system.DAE.Gy,  My ],
            [system.DAE.Gmu, Mmu]])
inc_corr = matrix([system.DAE.g, rho])

if system.DAE.factorize:
    F = symbolic(A)
    if iteration > 1: system.DAE.factorize = False
try:
    solve(A, numeric(A, F), inc_corr)
except:
    F = symbolic(A)
    solve(A, numeric(A, F), inc_corr)

system.DAE.y  -= inc_corr[:m]
system.DAE.mu -= inc_corr[-1]
iter_corr += 1

system.Settings.error = max(abs(inc_corr))
```

The following observations are relevant: (i) the interface `Device` is used for computing $\boldsymbol{g}$ and $\boldsymbol{g}_{\boldsymbol{y}}$ and $\boldsymbol{g}_{\mu}$ of all devices that compose the system; (ii) the function `symbolic` is used for symbolically factorizing the power flow Jacobian matrix (the symbolic re-factorization is needed whenever a LIB occurs); and (iii) in case of local parametrization, the variable y_k used in the continuation function ϱ is the bus voltage magnitude of a PQ load with index `v_bus`.

Example 5.4 Continuation Power Flow Analysis for the IEEE 14-Bus System

Figure 5.9 shows some results of the CPF analysis for the IEEE 14-bus system. The nose curves are obtained using a distributed slack bus model and neglecting reactive power limits of PV generators. In this case, the maximum loading condition is due to a SNB.

Figure 5.10 shows the results of the continuation power flow analysis enforcing PV generator reactive power limits. The figure shows both bus voltage magnitude at load buses, and the voltage magnitude and reactive power at bus 6. The voltage at bus 6 is constant until the PV generator reactive power is lower than its limits (i.e., 0.24 pu). Then, the reactive power is maintained equal to 0.24 pu and the bus voltage is let free to vary. The LIB caused by the reactive power saturation of the PV generator at bus 6 is not critical

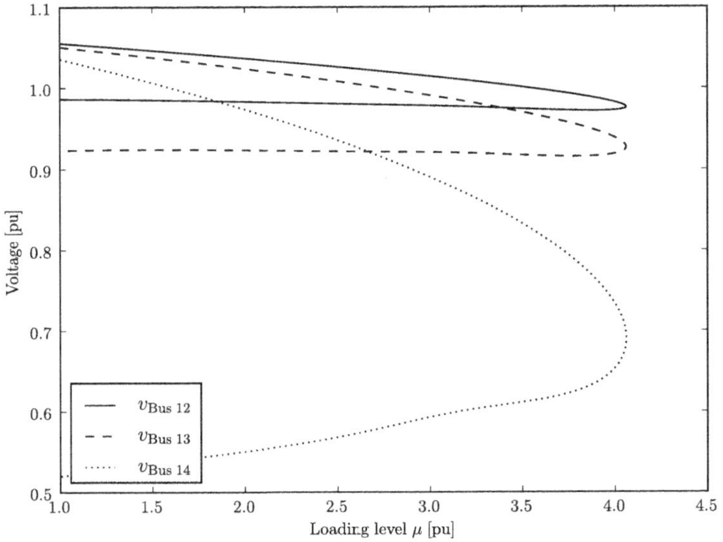

Fig. 5.9 Nose curve for the IEEE 14-bus system without PV reactive power limits

since other generators have still an available reserve of reactive power. As a matter of fact, the maximum loading condition is a SNB because the slack generator is assumed to have an unlimited reactive power reserve.

Figure 5.11 shows the results of the continuation power flow analysis enforcing both PV and slack generator reactive power limits (it is assumed $q_{G1}^{\max} = 1.0$ pu). In this case, the maximum loading condition is a LIB due to the slack generator maximum reactive power. In [332], a LIB is classified as *static* if it leads to the maximum loading condition and *dynamic* otherwise. This taxonomy can be misleading, since no dynamic device is actually considered. A simpler notation is *critical* for static LIBs and *non-critical* for dynamic ones.[3]

[3] In [325], only critical LIBs are called LIB, while non-critical LIBs are called *breaking points*. This definition bases its rationale on the fact that stability is not endangered for non-critical LIBs. In fact, a common definition of *local bifurcation* states that a bifurcation occurs when a parameter change causes the stability of an equilibrium (or fixed point) to change. In continuous systems, this corresponds to the real part of an eigenvalue of an equilibrium passing through zero. According to this definition, breaking points are not bifurcations. However, according to a more qualitative (or philosophical) definition, a bifurcation is a point of branching or forking where a non-linear system qualitatively shows new types of behavior. In this respect, while a change of stability of a non-linear system always implies the occurrence of a bifurcation, the occurrence of a bifurcation does not necessarily imply a change of stability.

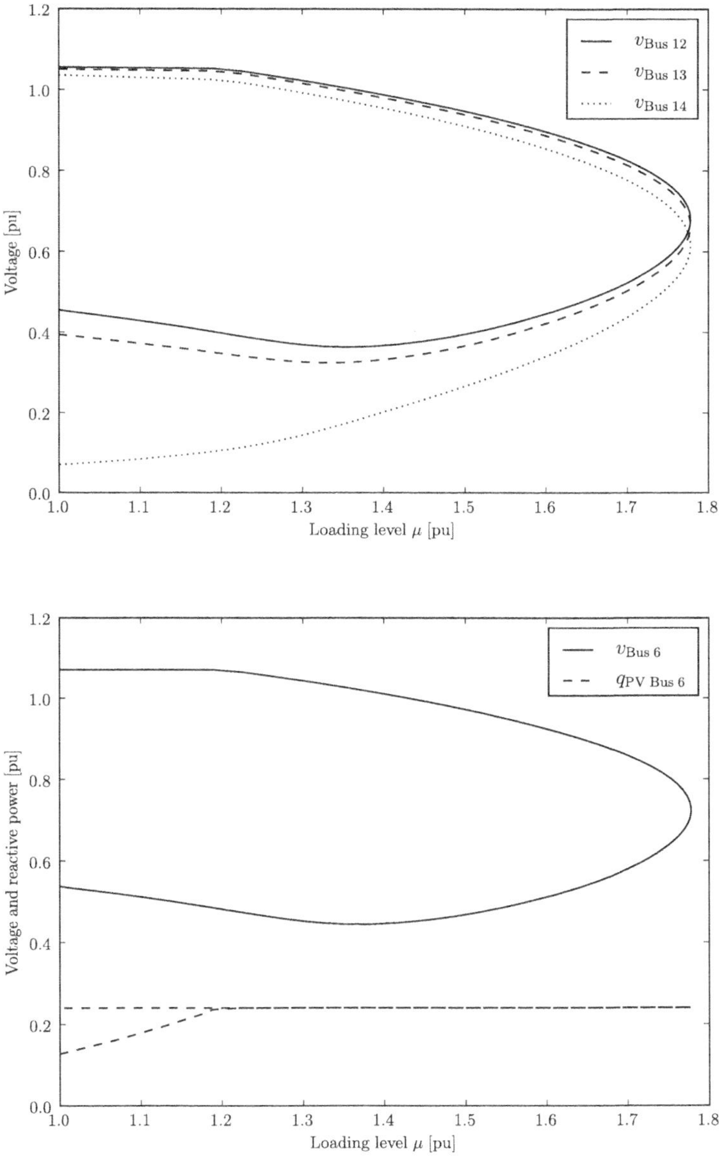

Fig. 5.10 Nose curve for the IEEE 14-bus system enforcing PV generator reactive power limits

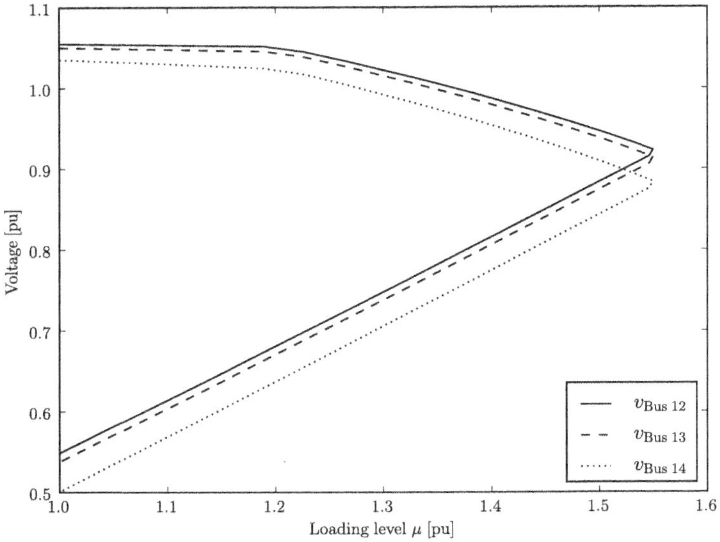

Fig. 5.11 Nose curve for the IEEE 14-bus system enforcing PV and slack generator reactive power limits

Any other limit can be checked during the CPF analysis. For example, bus voltage magnitude limits and transmission line and transformer thermal limits. However, these limits are conceptually different than generator reactive power limits because the former do not directly lead to a point of collapse. However, voltage and thermal limit violations have to be avoided to prevent cascade line tripping phenomena.

5.4.4 Continuous Newton's Method and Homotopy

This subsection discusses an analogy between the homotopy method proposed by Davidenko in [72] and the continuous Newton's method presented in Chapter 4. Differentiating the power flow equations (5.10) at an equilibrium point, one has:

$$\mathbf{0} = \boldsymbol{g}_{\boldsymbol{y}} d\boldsymbol{y} + \boldsymbol{g}_{\mu} d\mu \tag{5.59}$$

Equation (5.59) leads to the homotopy method proposed by Davidenko for computing the variation of $\boldsymbol{y}$ as a function of the parameter μ [72]:

$$\frac{d\boldsymbol{y}}{d\mu} = -[\boldsymbol{g}_{\boldsymbol{y}}]^{-1} \boldsymbol{g}_{\mu} \tag{5.60}$$

where $\boldsymbol{g}_\mu = \nabla^T_\mu \boldsymbol{g}$.[4] The Davidenko's method fails at turning points (e.g. saddle-node bifurcations) because of the singularity of $\boldsymbol{g}_{\boldsymbol{y}}$. More details on homotopy techniques can be found in [276].

Equations (5.60) is equivalent to a set of ODE, where the integration variable is μ. It is relevant to note the similarity of the continuous Newton's method (4.70) and (5.60). In fact, let us define the function $\boldsymbol{h}(\boldsymbol{y}, t)$ as follows:

$$\mathbf{0} = \boldsymbol{h}(\boldsymbol{y}, t) = e^t \boldsymbol{g}(\boldsymbol{y}) \tag{5.61}$$

where e represents the natural exponential. Then, differentiating $\boldsymbol{h}$, one has:

$$\begin{aligned} \mathbf{0} &= \boldsymbol{h}_{\boldsymbol{y}} d\boldsymbol{y} + \boldsymbol{h}_t dt \\ &= e^t \boldsymbol{g}_{\boldsymbol{x}} d\boldsymbol{y} + e^t \boldsymbol{g} dt \end{aligned} \tag{5.62}$$

Thus, (4.70) can be rewritten as:

$$\frac{d\boldsymbol{y}}{dt} = -[\boldsymbol{h}_{\boldsymbol{y}}]^{-1} \boldsymbol{h}_t \tag{5.63}$$

Equation (5.63) shows that t can be viewed as the continuation parameter for the function $\boldsymbol{h}(\boldsymbol{y}, t)$. As for the Davidenko's method, the continuous Newton's method fails at turning points where the power flow Jacobian matrix is singular.

The main difference between (5.60) and (5.63) is that μ is an internal parameter (i.e., the homotopy is forced) while t is an external one (i.e., the homotopy is free-running). Thus, only the final equilibrium point of (5.63) is physically relevant, while the values of $\boldsymbol{y}$ in intermediate iterations lack of physical meaning.

5.4.5 N-1 Contingency Analysis

On the use of the CPF analysis there is sometime some confusion. The most frequent critic that is moved to the CPF model is that the path used for increasing load powers cannot be known *a priori*. Another critic is about the use of a scalar parameter μ that increases *all* load and generator powers.

Actually, these critics are not justified since the basic assumption of continuation power flow analysis is that the load increase is *virtual*. In other words, the information provided by the CPF is how distant from the point of collapse is the current loading condition. Thus, the loading level μ is not a

[4] Equation (5.60) can be also obtained from (4.70). As a matter of fact, differentiating (4.68) and imposing $d\dot{\boldsymbol{y}} = \mathbf{0}$ leads to:

$$\mathbf{0} = d\tilde{\boldsymbol{f}} = -\boldsymbol{I}_n d\boldsymbol{y} - [\boldsymbol{g}_{\boldsymbol{y}}]^{-1} \boldsymbol{g}_\mu d\mu$$

where it is implicitly assumed that $\boldsymbol{g}_{\boldsymbol{y}}$ does not depend on μ, as discussed in Section 5.2.

real load increase, but rather a measure of the security margin of the current loading condition.

The usefulness of the CPF analysis is clear if applied to contingency analysis. System operators have to ensure that for each credible contingency (e.g., line or generator outage) the system is able to supply the current load. This is also referred as N-1 contingency criterion. The CPF provides a tool to evaluate the impact of each contingency in terms of the maximum loading level $\mu^{\max}$. If a contingency leads to a $\mu^{\max} < 1$, then the contingency is not feasible.

System operators generally fix a minimum loading margin to ensure that the current loading condition has a minimum distance to the point of collapse. Thus, for each contingency, $\mu^{\max} \geq \mu^{\rm sm}$ must hold, where $\mu^{\rm sm}$ is the required security margin. For example, if $\mu^{\rm sm} = 1.1$, the system has to ensure a 10% loading margin. If a certain contingency is characterized by $\mu^{\max} < \mu^{\rm sm}$, then that contingency is classified as *critical* and corrective actions are recommended.

Example 5.5 N-1 Contingency Analysis for the IEEE 14-Bus System

Figure 5.12 and Table 5.1 illustrate the N-1 contingency analysis for the IEEE 14-bus test system. Results are obtained considering only generator reactive power limits. The table shows the maximum loading level $\mu^{\max}$ correspondent

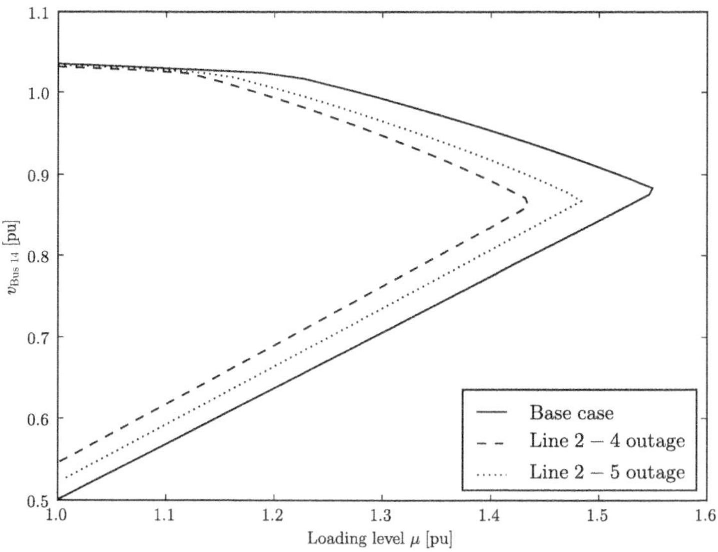

Fig. 5.12 Nose curves for the IEEE 14-bus system considering a variety of line outages

Table 5.1 N-1 contingency analysis for the IEEE 14-bus system

Branch #	From bus h	To bus k	Outage type	$\mu^{\max}$ [pu]
1	1	2	*Unfeasible*	**0.9930**
2	1	5	Feasible	1.3223
3	2	3	*Critical*	**1.2622**
4	2	4	Feasible	1.4428
5	2	5	Feasible	1.4876
6	3	4	Feasible	1.5243
7	4	5	Feasible	1.4578
8	4	7	Feasible	1.3310
9	4	9	Feasible	1.5289
10	5	6	Feasible	1.3081
11	6	11	Feasible	1.5489
12	6	12	Feasible	1.5499
13	6	13	Feasible	1.5239
14	7	8	Feasible	1.4623
15	7	9	Feasible	1.4476
16	9	10	Feasible	1.5499
17	9	14	Feasible	1.5318
18	10	11	Feasible	1.5554
19	12	13	Feasible	1.5572
20	13	14	Feasible	1.5501

to the outage of each line of the network. The outage is considered feasible if the maximum loading condition associated to line outage is $\mu^{\max} > 1$ ($\mu = 1$ is the base case loading condition). According to this condition, only line 1-2 outage is unfeasible. Moreover, assuming for the sake of example that the required security margin is $\mu^{\rm sm} = 1.3$ (i.e., 30% loading margin), the line 2-3 outage is critical.

5.5 Summary

This section summarizes most relevant concepts related to continuation power flow analysis.

Maximum loading condition: The calculation of the maximum loading condition of a power system can be formulated as the problem of computing the bifurcation points of classical power flow equations. Although bifurcation theory applies to dynamical systems, the use of static power flow equations is justified by assuming *slow* load variations and *fast* dynamics of power flow variables.

Point of collapse and bifurcation points: The only bifurcations of interest in power flow analysis are those that lead to a point of collapse, i.e., the maximum loading level beyond which power flow equations have no solution. These are the saddle-node and the limit-induced bifurcations. The saddle-node bifurcation is associated with the transmission system capacity and is intrinsic of the quadratic form of power flow equations. The limit-induced bifurcation refers to the reactive power reserve of synchronous generators. While the saddle-node bifurcation is always a point of collapse, limit-induced bifurcations can be critical or non-critical. Only critical limit-induced bifurcations are points of collapse.

Direct methods: Direct methods allows calculating *directly* the maximum loading condition of the system. The simplest direct methods are formulated as a set of nonlinear equations that include power flow equations and other constraints aimed to impose the conditions of either the saddle-node or the limited-induced bifurcation. Such direct methods are difficult to solve and have no practical interest. Another class of direct methods is based on nonlinear programming techniques. These lead to the formulation of the maximum loading condition as an optimal power flow problem. Due to its several implications that go far beyond the determination of the maximum loading condition, the entire Chapter 6 is dedicated the discussion of the optimal power flow analysis.

Homotopy methods: Homotopy methods are a class of algorithms that allow efficiently and reliably solving the problem of calculating the maximum loading condition of power flow equations. Homotopy methods consist in defining a homotopy map and a continuation equation whose Jacobian matrix is not singular at bifurcation points, hence the numerical robustness. In particular, the continuation power flow analysis is a forced homotopy method and consists in a predictor and in a corrector step. The predictor can be obtained through a tangent vector or the secant, whereas the corrector can be obtained through a perpendicular intersection or a local parametrization.

N-1 contingency analysis: The maximum loading condition can be considered as a measure of the distance to the collapse of the current operating point and not the other way round, i.e., the amount of load that the system can feed before collapsing. Thus, the most relevant application of the CPF technique is the N-1 contingency analysis. Given a set of contingencies (e.g., line outages), the CPF analysis provides the maximum loading level corresponding to each contingency. If a contingency is characterized by a maximum lading level lower than the current operating condition, that contingency is unfeasible. If a contingency is characterized by a maximum lading level lower than a given margin (typically fixed by the system operator), that contingency is critical. In both cases, the system operator has to take corrective actions to improve system security.

Chapter 6
Optimal Power Flow Analysis

This chapter describes the optimal power flow problem. Section 6.1 provides the background of the OPF problem and justifies the need for numerical methods. Section 6.2 provides a general nonlinear programming model for the OPF problem. A variety of examples are also provided in this section. Section 6.3 describes two solver methods for tackling the OPF problem, namely the generalized reduced gradient and the primal-dual interior-point methods. For the latter method, the Python implementation and numerical results are also provided. Finally, Section 6.4 summarizes common parameters of the interior point method.

6.1 Background

Several issues in power system analysis can be formulated as an optimization problem. For example, in centralized power systems typical objectives are minimizing generation cost and/or transmission system losses. In recent years, most national power grids have been restructured so that the centralized management has been substituted by a decentralized electricity market. In a competitive environment, the objective function is typically the maximization of the social benefit. Other common objectives are to minimize emissions, to maximize system security, etc.

The common constraint of most power system optimization problems are the power flow equations that are discussed in Chapter 4. As a simple example, consider again the didactic 3-bus example system discussed in Section 4.1. For clarity, this system is depicted again in Figure 6.1.

F. Milano: Power System Modelling and Scripting, Power Systems, pp. 131–153.
springerlink.com

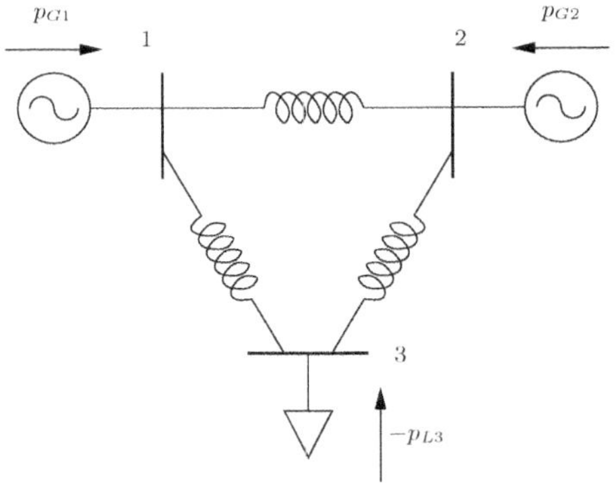

Fig. 6.1 3-bus system

As a simple example of optimization problem, consider the classical problem of generator cost minimization. Since this 3-bus system is loss-less, one has:

$$\begin{array}{ll} \underset{p_{G1},\ p_{G2}}{\text{Minimize}} & c_{G1}(p_{G1}) + c_{G2}(p_{G2}) \qquad (6.1) \\ \text{subject to} & p_{G1} + p_{G2} - p_{L3} = 0 \qquad (6.2) \end{array}$$

where c_{G1} and c_{G2} are generator cost functions and can be assumed quadratic. For example, $c_{G1}(p_{G1})$ is:

$$c_{G1}(p_{G1}) = a_1 + b_1 p_{G1} + \frac{c_1}{2} p_{G1}^2 \tag{6.3}$$

A similar expression holds for $c_{G2}(p_{G2})$.

The solution of (6.1)-(6.2) is straightforward if one defines the Lagrangian function and imposes the Karush-Kuhn-Tacker's (KKT) optimality conditions. The Lagrangian function is:

$$\mathscr{L}(p_{G1}, p_{G2}, \rho) = c_{G1}(p_{G1}) + c_{G2}(p_{G2}) - \rho(p_{G1} + p_{G2} - p_{L3}) \tag{6.4}$$

Hence the KKT condition are:

$$\begin{aligned} 0 &= \frac{\partial \mathscr{L}}{\partial p_{G1}} = c_1 p_{G1} - \rho \\ 0 &= \frac{\partial \mathscr{L}}{\partial p_{G2}} = c_2 p_{G2} - \rho \\ 0 &= \frac{\partial \mathscr{L}}{\partial \rho} = p_{G1} + p_{G2} - p_{L3} \end{aligned} \tag{6.5}$$

where ρ is the dual variable or Lagrangian multiplier associated with the power balance equation (6.2). Thus, the solution is:

$$p_{G1} = \frac{c_2}{c_1 + c_2} p_{L3} \tag{6.6}$$
$$p_{G2} = \frac{c_1}{c_1 + c_2} p_{L3}$$
$$\rho = \frac{c_1 c_2}{c_1 + c_2} p_{L3}$$

The dimension of the dual variable ρ is a cost per power unit and per hour. For this reason, ρ is also called marginal cost of the system. In a restructured power system, generator costs have become "secret" and only offers are known. However, as it often happens, this is only a new name for the same mathematical quantity. If one assumes that functions $c_{G1}(p_{G1})$ and $c_{G2}(p_{G2})$ are offers, the dual variable ρ takes the meaning of *market clearing price*, i.e., the amount that each generator of the system has to be paid for each unit of produced power. In any case, the solution procedure of (6.1)-(6.2) does not change.

Problem (6.1)-(6.2) is linear and does not contain inequalities. There are virtually infinite ways of complicating this problem. Actually, optimization problems are a kind of Swiss-army knife in power system analysis and can be used for tackling a huge variety of power system issues. The scope of this chapter is not to enumerate all possible formulations but rather to provide the tools for solving an as general as possible class of optimization problems. With this aim, following sections focus only nonlinear programming (NLP) techniques and, among all possible NLP methods, only the reduced gradient method and the interior point methods are discussed.

6.2 Optimal Power Flow Model

The very first formulations of the optimal power flow problem taking into account both power flow equations and economic dispatch were presented at the beginning of the sixties [48, 290] and further important developments were given in the following decade. At that time, there was no distinction between the solution method and the problem formulation. On the contrary, solution methods were suited to a specific problem formulation. For example, reduced gradient method in [80], Powell's and Fletcher-Powell's methods [265], Hessian method in [266], sequential programming [255], differential injection method in [49], and linear programming [301]. By the middle of the seventies, the optimal power flow problem was a mature tool for assessing both power system security and economic dispatch [50, 120, 267, 309]. Since then, a huge variety of methods/models have been proposed so that survey papers have been necessary (e.g., [134, 300]).

In the last decade, power system restructuring has led to a renewed interest in mathematical programming, since it provides the adequate tools for tackling electricity markets [277]. From the programming point of view, in the last decade, the trend has been to take implementation details and problem formulation separated. This separation has been made possible by the availability of mature software packages that provide a general purpose meta-language for mathematical programming (e.g., GAMS [31] and AMPL [99]). In this chapter, modelling and implementation are also discussed separately.

The system model that is used throughout this chapter is a constrained nonlinear programming problem in the following general form:

$$\begin{aligned} \underset{\boldsymbol{z}}{\text{Minimize}} \quad & \varphi(\boldsymbol{z}) \\ \text{subject to} \quad & \boldsymbol{g}(\boldsymbol{z}) = \boldsymbol{0} \\ & \boldsymbol{h}(\boldsymbol{z}) \leq \boldsymbol{0} \end{aligned} \tag{6.7}$$

where $\boldsymbol{z} \in \mathbb{R}^{n_z}$, $\varphi(\boldsymbol{z})$ is the objective function ($\varphi(\boldsymbol{z}) : \mathbb{R}^{n_z} \mapsto \mathbb{R}$), $\boldsymbol{g}(\boldsymbol{z})$ are the equality constraints ($\boldsymbol{g}(\boldsymbol{z}) : \mathbb{R}^{n_z} \mapsto \mathbb{R}^{n_g}$), and $\boldsymbol{h}(\boldsymbol{z})$ are the inequality constraints ($\boldsymbol{h}(\boldsymbol{z}) : \mathbb{R}^{n_z} \mapsto \mathbb{R}^{n_h}$), and $n_g < n_z$. The latter condition allows (6.7) having $(n_z - n_g)$ *degrees of freedom*.[1] The functions $\varphi(\boldsymbol{z})$, $\boldsymbol{g}(\boldsymbol{z})$ and $\boldsymbol{h}(\boldsymbol{z})$ are assumed to be smooth, i.e., continuous and differentiable at least two times for $\boldsymbol{z} \in \mathbb{R}^{n_z}$.

According to the definitions given in Section 1.4 of Chapter 1, the vector $\boldsymbol{z}$ is formed by algebraic variables $\boldsymbol{y}$ and controllable parameters $\boldsymbol{\eta}$. Thus, $\boldsymbol{z} = [\boldsymbol{y}^T, \boldsymbol{\eta}^T]^T$ and $n_z = n_y + n_\eta$.

In order to properly describe the solution methods for the problem (6.7), some definitions are required.

1. A point $\boldsymbol{z}^* \in \mathbb{R}^{n_z}$ is a *local minimizer* if $\boldsymbol{g}(\boldsymbol{z}^*) = \boldsymbol{0}$ and $\boldsymbol{h}(\boldsymbol{z}^*) \leq \boldsymbol{0}$ and there exists an $\epsilon > 0$ such that $\varphi(\boldsymbol{z}^*) \leq \varphi(\boldsymbol{z})$ $\forall \boldsymbol{x} \in \mathbb{R}^{n_z}$, with $\boldsymbol{g}(\boldsymbol{z}) = \boldsymbol{0}$ and $\boldsymbol{h}(\boldsymbol{z}) \leq \boldsymbol{0}$ and $|\boldsymbol{z} - \boldsymbol{z}^*| < \epsilon$.
2. A point $\boldsymbol{z}^r \in \mathbb{R}^{n_z}$ is said to be a *regular point* of the constraints $\boldsymbol{g}(\boldsymbol{z})$ and $\boldsymbol{h}(\boldsymbol{z})$ if satisfies the conditions $\boldsymbol{g}(\boldsymbol{z}) = \boldsymbol{0}$ and $\boldsymbol{h}(\boldsymbol{z}) \leq \boldsymbol{0}$ and if the gradients $\boldsymbol{g}_{\boldsymbol{z}^r}$ and $\boldsymbol{h}_{\boldsymbol{z}^r}$ are linearly independent.

The former definitions allows formulating the first-order KKT optimality conditions, as follows. If $\boldsymbol{z}^*$ is both a local minimizer of (6.7) and a regular point of the constraints $\boldsymbol{g}(\boldsymbol{z})$ and $\boldsymbol{h}(\boldsymbol{z})$, then there exist vectors $\boldsymbol{\rho} \in \mathbb{R}^{n_g}$ and $\boldsymbol{\pi} \in \mathbb{R}^{n_h}$, with $\boldsymbol{\pi} \geq \boldsymbol{0}$, such that:

$$\begin{aligned} \varphi_{\boldsymbol{z}}(\boldsymbol{z}^*) + \boldsymbol{\rho}^T \boldsymbol{g}_{\boldsymbol{z}}(\boldsymbol{z}^*) + \boldsymbol{\pi}^T \boldsymbol{h}_{\boldsymbol{z}}(\boldsymbol{z}^*) &= \boldsymbol{0} \\ \boldsymbol{\pi}^T \boldsymbol{h}(\boldsymbol{z}^*) &= \boldsymbol{0} \end{aligned} \tag{6.8}$$

The vectors $\boldsymbol{\rho}$ and $\boldsymbol{\pi}$ are called *dual variables* or *multipliers*. The first-order optimality conditions (6.8) can be conveniently expressed in terms of the *Lagrangian* function. The Lagrangian of the constrained problem (6.7) is:

$$\mathscr{L}(\boldsymbol{z}, \boldsymbol{\rho}, \boldsymbol{\pi}) = \varphi(\boldsymbol{z}) + \boldsymbol{\rho}^T \boldsymbol{g}(\boldsymbol{z}) + \boldsymbol{\pi}^T \boldsymbol{h}(\boldsymbol{z}) \tag{6.9}$$

[1] This definition is actually borrowed from mechanical engineering and is not used in mathematical programming.

Thus, the first-order optimality conditions (6.8) become:

$$\mathscr{L}_{\boldsymbol{z}}(\boldsymbol{z}, \boldsymbol{\rho}, \boldsymbol{\pi}) = \mathbf{0} \tag{6.10}$$

$$\mathscr{L}_{\boldsymbol{\rho}}(\boldsymbol{z}, \boldsymbol{\rho}, \boldsymbol{\pi}) = \mathbf{0} \tag{6.11}$$

$$\boldsymbol{h}(\boldsymbol{z}) \leq \mathbf{0} \tag{6.12}$$

$$\boldsymbol{\Pi} \boldsymbol{h}(\boldsymbol{z}) = \mathbf{0} \tag{6.13}$$

$$\boldsymbol{\pi} \geq \mathbf{0} \tag{6.14}$$

where $\Pi = \text{diag}(\pi_1, \pi_2, \ldots, \pi_{n_h})$, the conditions (6.11) and (6.12) ensure the *primal feasibility*; the conditions (6.10) and (6.14) ensure the *dual feasibility*; and (6.13) are the *complementarity conditions*. It is important to note that the first-order optimality conditions allow characterizing only regular points. Non-regular points require specific conditions and are not considered in what follows since non-regular points are quite uncommon in power system problems.

The system (6.10)-(6.14) is a set of nonlinear equations that include both equalities and inequalities. A constraint is said to be *binding* (or *active*) if it is equal to zero. By definition, equalities $\boldsymbol{g}$ are always binding, while an inequality is binding only if $h_k = 0$. It is worth observing that if a constraint is not binding for a given local minimizer, that constraint can be removed from (6.7), thus reducing the problem to:

$$\begin{aligned} \underset{\boldsymbol{z}}{\text{Minimize}} \quad & \varphi(\boldsymbol{z}) \\ \text{subject to} \quad & \boldsymbol{g}(\boldsymbol{z}) = \mathbf{0} \\ & \tilde{\boldsymbol{h}}(\boldsymbol{z}) = \mathbf{0} \end{aligned} \tag{6.15}$$

where $\tilde{\boldsymbol{h}}$ ($\tilde{\boldsymbol{h}} \subset \boldsymbol{h}$) is the set of binding constraints. Unfortunately, one knows which inequalities are binding only after founding a local minimizer $\boldsymbol{z}^*$ of the original problem (6.7). However (6.15) can be useful for extracting some properties of the solution $\boldsymbol{z}^*$.

If $\boldsymbol{h}(\boldsymbol{z})$ is a null vector, then the solution of the optimization problem (6.7) reduces to the solution of a set of nonlinear equalities. However, in general, $\boldsymbol{h}(\boldsymbol{z})$ is not null, which considerably complicates the solution of (6.10)-(6.14). With this aim, it may be useful to transform the initial optimization problem (6.7) introducing a vector of non-negative slack variables $\boldsymbol{s} \in \mathbb{R}^{n_h}$, as follows:[2]

$$\begin{aligned} \underset{\boldsymbol{z}}{\text{Minimize}} \quad & \varphi(\boldsymbol{z}) \\ \text{subject to} \quad & \boldsymbol{g}(\boldsymbol{z}) = \mathbf{0} \\ & \boldsymbol{s} + \boldsymbol{h}(\boldsymbol{z}) = \mathbf{0}, \quad \boldsymbol{s} \geq \mathbf{0} \end{aligned} \tag{6.16}$$

[2] The advantages of this transformation are clarified in Subsection 6.3.2.

The first-order optimality conditions of (6.16) are:

$$\mathscr{L}_{\boldsymbol{z}}(\boldsymbol{z}, \boldsymbol{\rho}, \boldsymbol{\pi}, \boldsymbol{s}) = \mathbf{0} \tag{6.17}$$
$$\mathscr{L}_{\boldsymbol{\rho}}(\boldsymbol{z}, \boldsymbol{\rho}, \boldsymbol{\pi}, \boldsymbol{s}) = \mathbf{0} \tag{6.18}$$
$$\mathscr{L}_{\boldsymbol{\pi}}(\boldsymbol{z}, \boldsymbol{\rho}, \boldsymbol{\pi}, \boldsymbol{s}) = \mathbf{0} \tag{6.19}$$
$$\boldsymbol{\Pi} \boldsymbol{s} = \mathbf{0} \tag{6.20}$$
$$\boldsymbol{s} \geq \mathbf{0} \tag{6.21}$$
$$\boldsymbol{\pi} \geq \mathbf{0} \tag{6.22}$$

The latter problem is somewhat easier to solve than (6.10)-(6.14). Analogously to the non-transformed problem, the conditions (6.18), (6.19) and (6.21) ensure the primal feasibility; the conditions (6.17) and ((6.22)) ensure the dual feasibility; and (6.20) are the complementarity conditions.

The concept of *degenerated constraints* completes this brief list of definitions. A constraint is degenerate if it is binding and its associated multiplier is null. For example, an inequality constraint is degenerate if $h_k = 0$ and $\pi_k = 0$, and is non-degenerate otherwise. In physical optimization problems and in particular in optimal power flow problems, binding constraints are generally non-degenerate. This observation eases the solution of the first-order optimality conditions. Assuming non-degenerate binding constraints (6.10)-(6.14) become:

$$\begin{aligned}
&\mathscr{L}_{\boldsymbol{z}}(\boldsymbol{z}, \boldsymbol{\rho}, \boldsymbol{\pi}) = \mathbf{0} \\
&\mathscr{L}_{\boldsymbol{\rho}}(\boldsymbol{z}, \boldsymbol{\rho}, \boldsymbol{\pi}) = \mathbf{0} \\
&\text{if } h_k < 0 \quad \Rightarrow \quad \pi_k = 0 \\
&\text{if } h_k = 0 \quad \Rightarrow \quad \pi_k > 0
\end{aligned} \tag{6.23}$$

and the first-order optimality conditions (6.17)-(6.22) of the transformed problem (6.16) become:

$$\begin{aligned}
&\mathscr{L}_{\boldsymbol{z}}(\boldsymbol{z}, \boldsymbol{\rho}, \boldsymbol{\pi}, \boldsymbol{s}) = \mathbf{0} \\
&\mathscr{L}_{\boldsymbol{\rho}}(\boldsymbol{z}, \boldsymbol{\rho}, \boldsymbol{\pi}, \boldsymbol{s}) = \mathbf{0} \\
&\mathscr{L}_{\boldsymbol{\pi}}(\boldsymbol{z}, \boldsymbol{\rho}, \boldsymbol{\pi}, \boldsymbol{s}) = \mathbf{0} \\
&\text{if } s_k > 0 \quad \Rightarrow \quad \pi_k = 0 \\
&\text{if } s_k = 0 \quad \Rightarrow \quad \pi_k > 0
\end{aligned} \tag{6.24}$$

Mathematical programming is a broad branch of mathematics and this section is not intended to provide a comprehensive treatise. The interested reader can find useful the following references [24, 51, 95, 96, 106, 175].

Example 6.1 Standard Optimal Power Flow Problem

A typical, relatively general OPF-based problem can be represented using the following nonlinear constrained optimization problem:

$$
\begin{array}{lll}
\underset{\boldsymbol{z}}{\text{Minimize}} & \varphi = -(\sum_{h \in \mathcal{D}} c_L(\boldsymbol{p}_L) - \sum_{h \in \mathcal{S}} c_G(\boldsymbol{p}_G)) & \\
\text{subject to} & \boldsymbol{g}(\boldsymbol{\theta}, \boldsymbol{v}, \boldsymbol{q}_G, \boldsymbol{p}_G, \boldsymbol{p}_L) = \boldsymbol{0} & \rightarrow \text{Power flow equations} \\
 & \boldsymbol{p}_G^{\min} \leq \boldsymbol{p}_G \leq \boldsymbol{p}_G^{\max} & \rightarrow \text{Generator } p \text{ limits} \\
 & \boldsymbol{q}_G^{\min} \leq \boldsymbol{q}_G \leq \boldsymbol{q}_G^{\max} & \rightarrow \text{Generator } q \text{ limits} \\
 & \boldsymbol{p}_L^{\max} \leq \boldsymbol{p}_L \leq \boldsymbol{p}_L^{\max} & \rightarrow \text{Load } p \text{ limits} \\
 & \mid \boldsymbol{\phi}_{ij}(\boldsymbol{\theta}, \boldsymbol{v}) \mid \leq \boldsymbol{\phi}_{ij}^{\max} & \rightarrow \text{Flow limits} \\
 & \mid \boldsymbol{\phi}_{ji}(\boldsymbol{\theta}, \boldsymbol{v}) \mid \leq \boldsymbol{\phi}_{ji}^{\max} & \\
 & \boldsymbol{v}^{\min} \leq \boldsymbol{v} \leq \boldsymbol{v}^{\max} & \rightarrow \text{Voltage limits}
\end{array}
\tag{6.25}
$$

where $\boldsymbol{z} = (\boldsymbol{\theta}, \boldsymbol{v}, \boldsymbol{q}_G, \boldsymbol{p}_G, \boldsymbol{p}_L)$, $\boldsymbol{c}_G$ and $\boldsymbol{c}_L$ are vectors of functions of the generator and load powers, respectively; $\boldsymbol{q}_G$ stand for the generator reactive powers; $\boldsymbol{v}$ and $\boldsymbol{\theta}$ represent the bus phasor voltages; and $\boldsymbol{p}_G$ and $\boldsymbol{p}_L$ represent bounded generator and load limits; and $\boldsymbol{\phi}_{ij}$ and $\boldsymbol{\phi}_{ji}$ represent the active powers (or apparent powers or currents) flowing through the lines in both directions. In the security context, power transfer limits are usually determined based only on power flow based voltage stability studies [107] and can be determined using the $N-1$ contingency analysis that is described in Subsection 5.4.5 of Chapter 5.

In spite of its simplicity, problem (6.25) can tackle a variety of important problems.

1. If $\boldsymbol{c}_G$ are generation cost functions, $\boldsymbol{c}_L = \boldsymbol{0}$ and $\boldsymbol{p}_L^{\min} = \boldsymbol{p}_L^{\max}$, (6.25) allows solving the classical economic dispatch ensuring security limits such as voltage limits and transmission line thermal limits.
2. If $\boldsymbol{c}_G(\boldsymbol{p}_G) = \boldsymbol{p}_G$ and $\boldsymbol{c}_G(\boldsymbol{p}_L) = \boldsymbol{p}_L$, (6.25) allows minimizing power system losses.
3. If $\boldsymbol{c}_G(\boldsymbol{p}_G) = -\boldsymbol{p}_G$ and $\boldsymbol{c}_L = \boldsymbol{0}$, (6.25) allows maximizing the power production. This problem is useful to evaluate the allowable penetration of energy resources in the power system (e.g., renewable and distributed generation).
4. If $\boldsymbol{c}_G(\boldsymbol{p}_G)$ and $\boldsymbol{c}_L(\boldsymbol{p}_L)$ have the meaning of *offers* and *bids*, respectively, rather then costs, then the objective function becomes the *social benefit* and (6.25) allows solving the security constrained market dispatch [353]. The demand is said to be *inelastic* if $\boldsymbol{p}_L^{\min} = \boldsymbol{p}_L^{\max}$ (which is the common case), *elastic* if $\boldsymbol{p}_L^{\min} < \boldsymbol{p}_L^{\max}$.

Example 6.2 Maximization of the Distance to Voltage Collapse

The following optimization problem is implemented to represent system security through the use of voltage stability conditions, based on what was proposed in [43, 46, 47]:

$$
\begin{aligned}
\underset{\boldsymbol{z}}{\text{Minimize}} \quad & \varphi = -\mu & (6.26)\\
\text{subject to} \quad & \boldsymbol{g}(\boldsymbol{\theta}, \boldsymbol{v}, \boldsymbol{q}_G, \boldsymbol{p}_G, \boldsymbol{p}_L) = \boldsymbol{0} & \rightarrow \text{PF equations}\\
& \boldsymbol{g}^c(\boldsymbol{\theta}^c, \boldsymbol{v}^c, \boldsymbol{q}_G^c, k_G^c, \mu, \boldsymbol{p}_G, \boldsymbol{p}_L) = \boldsymbol{0} & \rightarrow \text{Max load PF equations}\\
& \mu^{\min} \leq \mu & \rightarrow \text{Loading level}\\
& \boldsymbol{p}_G^{\min} \leq \boldsymbol{p}_G \leq \boldsymbol{p}_G^{\max} & \rightarrow \text{Generator } p \text{ limits}\\
& \boldsymbol{p}_L^{\min} \leq \boldsymbol{p}_L \leq \boldsymbol{p}_L^{\max} & \rightarrow \text{Load } p \text{ limits}\\
& \boldsymbol{\phi}_{ij}(\boldsymbol{\theta}, \boldsymbol{v}) \leq \boldsymbol{\phi}_{ij}^{\max} & \rightarrow \text{Flow limits}\\
& \boldsymbol{\phi}_{ji}(\boldsymbol{\theta}, \boldsymbol{v}) \leq \boldsymbol{\phi}_{ji}^{\max} &\\
& \boldsymbol{\phi}_{ij}(\boldsymbol{\theta}^c, \boldsymbol{v}^c) \leq \boldsymbol{\phi}_{ij}^{\max} &\\
& \boldsymbol{\phi}_{ji}(\boldsymbol{\theta}^c, \boldsymbol{v}^c) \leq \boldsymbol{\phi}_{ji}^{\max} &\\
& \boldsymbol{q}_G^{\min} \leq \boldsymbol{q}_G \leq \boldsymbol{q}_G^{\max} & \rightarrow \text{Generator } q \text{ limits}\\
& \boldsymbol{q}_G^{\min} \leq \boldsymbol{q}_G^c \leq \boldsymbol{q}_G^{\max} &\\
& \boldsymbol{v}^{\min} \leq \boldsymbol{v} \leq \boldsymbol{v}^{\max} & \rightarrow \text{Voltage limits}\\
& \boldsymbol{v}^{\min} \leq \boldsymbol{v}^c \leq \boldsymbol{v}^{\max} &
\end{aligned}
$$

where $\boldsymbol{z} = (\mu, \boldsymbol{\theta}, \boldsymbol{v}, \boldsymbol{q}_G, \boldsymbol{\theta}^c, \boldsymbol{v}^c, \boldsymbol{q}_G^c, \boldsymbol{p}_G, \boldsymbol{p}_L)$.

In this case, a second set of power flow equations and constraints with a superscript c is introduced to represent the system at the limit or *critical* conditions associated with the loading margin μ that drives the system to its maximum loading condition. The critical power flow equations $\boldsymbol{g}^c$ can present a line outage. The maximum or critical loading point could be either associated with a thermal or bus voltage limit or a voltage stability limit (collapse point) corresponding to a system singularity (saddle-node bifurcation) or system controller limits like generator reactive power limits (limit induced bifurcation) [39, 259]. Thus, for the current and maximum loading conditions, the generator and load powers are defined as follows:

$$
\begin{aligned}
\boldsymbol{p}_G^c &= (1 + \mu + k_G^c)\boldsymbol{p}_G\\
\boldsymbol{p}_L^c &= (1 + \mu)\boldsymbol{p}_L
\end{aligned}
$$

where k_G^c represents a scalar variable which distributes system losses associated *only* with the solution of the critical power flow equations in proportion to the power injections obtained in the solution process (distributed slack bus model). It is assumed that the losses corresponding to the maximum loading level defined by μ are distributed among all generators.

For the sake of example, consider the Lagrangian function $\mathscr{L}$ associated to problem (6.26) with all inequalities transformed into equalities through the vector of slack variables $\boldsymbol{s}$ as in (6.16).

$$
\begin{aligned}
\mathscr{L} = \varphi &- \boldsymbol{\rho}^T \boldsymbol{g}(\boldsymbol{\theta}, \boldsymbol{v}, \boldsymbol{q}_G, \boldsymbol{p}_G, \boldsymbol{p}_L) \\
&- \boldsymbol{\rho}_c^T \boldsymbol{g}^c(\boldsymbol{\theta}^c, \boldsymbol{v}^c, \boldsymbol{q}_G^c, \mu, \boldsymbol{p}_G, \boldsymbol{p}_L) \\
&- \pi_{\mu^{\min}}(\mu - \mu^{\min} - s_{\mu^{\min}}) \\
&- \boldsymbol{\pi}_{p_G^{\max}}^T(\boldsymbol{p}_G^{\max} - \boldsymbol{p}_G - \boldsymbol{s}_{p_G^{\max}}) \\
&- \boldsymbol{\pi}_{p_G^{\min}}^T(\boldsymbol{p}_G - \boldsymbol{p}_G^{\min} - \boldsymbol{s}_{p_G^{\min}}) \\
&- \boldsymbol{\pi}_{p_L^{\max}}^T(\boldsymbol{p}_L^{\max} - \boldsymbol{p}_L - \boldsymbol{s}_{p_L^{\max}}) \\
&- \boldsymbol{\pi}_{p_L^{\min}}^T(\boldsymbol{p}_L - \boldsymbol{p}_L^{\min} - \boldsymbol{s}_{p_L^{\min}}) \\
&- \boldsymbol{\pi}_{\phi_{ij}^{\max}}^T(\boldsymbol{\phi}_{ij}^{\max} - \boldsymbol{\phi}_{ij} - \boldsymbol{s}_{\phi_{ij}^{\max}}) \\
&- \boldsymbol{\pi}_{\phi_{ji}^{\max}}^T(\boldsymbol{\phi}_{ji}^{\max} - \boldsymbol{\phi}_{ji} - \boldsymbol{s}_{\phi_{ji}^{\max}}) \\
&- \boldsymbol{\pi}_{\phi_{ij}^{c}\max}^T(\boldsymbol{\phi}_{ij}^{\max} - \boldsymbol{\phi}_{ijc} - \boldsymbol{s}_{\phi_{ij}^{c}\max}) \\
&- \boldsymbol{\pi}_{\phi_{ji}^{c}\max}^T(\boldsymbol{\phi}_{ji}^{\max} - \boldsymbol{\phi}_{jic} - \boldsymbol{s}_{\phi_{ji}^{c}\max}) \\
&- \boldsymbol{\pi}_{q_G^{\max}}^T(\boldsymbol{q}_G^{\max} - \boldsymbol{q}_G - \boldsymbol{s}_{q_G^{\max}}) \\
&- \boldsymbol{\pi}_{q_G^{\min}}^T(\boldsymbol{q}_G - \boldsymbol{q}_G^{\min} - \boldsymbol{s}_{q_G^{\min}}) \\
&- \boldsymbol{\pi}_{q_G^{c}\max}^T(\boldsymbol{q}_G^{\max} - \boldsymbol{q}_G^c - \boldsymbol{s}_{q_G^{c}\max}) \\
&- \boldsymbol{\pi}_{q_G^{c}\min}^T(\boldsymbol{q}_G^c - \boldsymbol{q}_G^{\min} - \boldsymbol{s}_{q_G^{c}\min}) \\
&- \boldsymbol{\pi}_{v^{\max}}^T(\boldsymbol{v}^{\max} - \boldsymbol{v} - \boldsymbol{s}_{v^{\max}}) \\
&- \boldsymbol{\pi}_{v^{\min}}^T(\boldsymbol{v} - \boldsymbol{v}^{\min} - \boldsymbol{s}_{v^{\min}}) \\
&- \boldsymbol{\pi}_{v^{c}\max}^T(\boldsymbol{v}^{\max} - \boldsymbol{v}^c - \boldsymbol{s}_{v^{c}\max}) \\
&- \boldsymbol{\pi}_{v^{c}\min}^T(\boldsymbol{v}^c - \boldsymbol{v}^{\min} - \boldsymbol{s}_{v^{c}\min})
\end{aligned} \tag{6.27}
$$

where $\boldsymbol{\rho}$ and $\boldsymbol{\rho}^c \in \mathbb{R}^{n_g}$, and all the other $\boldsymbol{\pi}$ ($\pi_k > 0$, $\forall k$) correspond to the Lagrangian multipliers. The $\boldsymbol{s}$ variables have to satisfy the non-negativity condition $\boldsymbol{s} > \boldsymbol{0}$.

6.3 Nonlinear Programming Solvers

As indicated in the previous section, the problem of finding a local minimizer of (6.7) or (6.16) is equivalent to solve (6.10)-(6.14) or (6.17)-(6.22), respectively. These are sets of nonlinear equalities and inequalities. The main challenges from the solution method viewpoint are twofold:

1. The inequalities constraints (6.12) and (6.14) or (6.21) and (6.22) complicate considerably the solution process.
2. The conditions (6.10) or (6.17) contain the Jacobian matrices $\varphi_{\boldsymbol{z}}$ and $\boldsymbol{g}_{\boldsymbol{z}}$ and $\boldsymbol{h}_{\boldsymbol{z}}$. Thus any solution method that involves the calculation of $\mathscr{L}_{\boldsymbol{zz}}$ (such as the any Newton's method), implies setting up the Hessian matrices $\varphi_{\boldsymbol{zz}}$ and $\boldsymbol{g}_{\boldsymbol{zz}}$ and $\boldsymbol{h}_{\boldsymbol{zz}}$. In case of power flow equations, calculating $\boldsymbol{g}_{\boldsymbol{zz}}$ is not a trivial task.

In the following sections, only two solution methods are described, namely the reduced gradient method and the primal-dual interior point method.

6.3.1 Generalized Reduced Gradient Method

The generalized reduced gradient (GRG) has been one of first methods used in power system analysis [80]. The GRG method is also used in well-assessed solvers such as CONOPT [81]. This method works for constrained nonlinear problems and resembles the solution approach of the simplex method used for linear programming [96]. The main idea of the GRG method is to divide the variables $\boldsymbol{z}$ into two subsets, one of *basic* (or dependent) variables and one of *non-basic* (or independent) variables. In mathematical terms, basic variables are those variables that are unequivocally determined once the vector of non-basic variables is assigned. To define basic and non-basic variables is generally easy in physical problems. As a matter of fact, according to Sections 1.4 and 6.2, $\boldsymbol{z} = [\boldsymbol{y}^T, \boldsymbol{\eta}^T]^T$. Thus $\boldsymbol{y}$ are the basic variables and $\boldsymbol{\eta}$ are the non-basic ones. For example, in the standard optimal power flow problem, $\boldsymbol{y}$ are bus voltages, while $\boldsymbol{\eta}$ are the generator active powers.

In order to describe the reduced gradient method, consider for simplicity the following optimization problem:

$$\begin{array}{ll} \underset{\boldsymbol{y},\boldsymbol{\eta}}{\text{Minimize}} & \varphi(\boldsymbol{y},\boldsymbol{\eta}) \\ \text{subject to} & \boldsymbol{g}(\boldsymbol{y},\boldsymbol{\eta}) = \boldsymbol{0} \end{array} \tag{6.28}$$

How to handle inequalities is explained later on. The reduced gradient $\boldsymbol{r}(\boldsymbol{\eta})$ of (6.7), with $\boldsymbol{r}(\boldsymbol{\eta}) : \mathbb{R}^{n_\eta} \mapsto \mathbb{R}^{n_\eta}$, is defined as:

$$\boldsymbol{r}(\boldsymbol{\eta}) = \frac{d\varphi}{d\boldsymbol{\eta}} = \varphi_{\boldsymbol{y}} + \frac{d\boldsymbol{y}}{d\boldsymbol{\eta}}\varphi_{\boldsymbol{\eta}} \tag{6.29}$$

Differentiating $\boldsymbol{g}(\boldsymbol{y},\boldsymbol{\eta})$, and assuming that the current point $(\boldsymbol{y}^{(i)},\boldsymbol{\eta}^{(i)})$ is feasible and satisfies $\boldsymbol{g}(\boldsymbol{y}^{(i)},\boldsymbol{\eta}^{(i)}) = \boldsymbol{0}$, one has:

$$\boldsymbol{g}_{\boldsymbol{y}}|_i d\boldsymbol{y} + \boldsymbol{g}_{\boldsymbol{\eta}}|_i d\boldsymbol{\eta} = \boldsymbol{0} \tag{6.30}$$

Thus, (6.29) can be rewritten as:

$$\boldsymbol{r}(\boldsymbol{\eta}) = \varphi_{\boldsymbol{y}} - \boldsymbol{g}_{\boldsymbol{y}}^{-1}\boldsymbol{g}_{\boldsymbol{\eta}}\varphi_{\boldsymbol{\eta}} \tag{6.31}$$

The reduced gradient is used a direction along which finding a *small* move from the current value of $\boldsymbol{\eta}^{(i)}$ that is able to decrease the objective function φ. For the current feasible point $(\boldsymbol{y}^{(i)}, \boldsymbol{\eta}^{(i)})$, the step size $\Delta\eta_k^{(i)}$ for $k = 1, \ldots, n_\eta$ is:

$$\Delta\eta_k^{(i)} = \begin{cases} 0 & \text{if} \quad \eta_k^{(i)} = 0 \quad \text{and} \quad r_k(\boldsymbol{\eta}) > 0 \\ -r_k(\boldsymbol{\eta}) & \text{otherwise} \end{cases} \tag{6.32}$$

Then, the basic variable step size $\Delta\boldsymbol{y}^{(i)}$ is computed as:

$$\Delta\boldsymbol{y}^{(i)} = -\boldsymbol{g}_{\boldsymbol{y}}^{-1}\boldsymbol{g}_{\boldsymbol{\eta}}\Delta\boldsymbol{\eta}^{(i)} \tag{6.33}$$

Thus, the *projection move* is:

$$\tilde{\boldsymbol{z}}^{(i+1)} = \boldsymbol{z}^{(i)} + \Delta\boldsymbol{z}^{(i)} \tag{6.34}$$

where $\Delta\boldsymbol{z}^{(i)} = [[\Delta\boldsymbol{y}^{(i)}]^T, [\Delta\boldsymbol{\eta}^{(i)}]^T]^T$.

Due to the nonlinearity of constraints $\boldsymbol{g}$, the projection move provides a new point that does not satisfy $\boldsymbol{g}(\tilde{\boldsymbol{y}}^{(i+1)}, \tilde{\boldsymbol{\eta}}^{(i+1)}) = \boldsymbol{0}$. It is thus necessary to apply a *restoration move* that moves the current point back to the constraint boundary. A possibility is to use a linear approximation of the constraints $\boldsymbol{g}(\boldsymbol{z}^{(i+1)})$:

$$\boldsymbol{g}(\boldsymbol{z}^{(i+1)}) \approx \boldsymbol{g}(\tilde{\boldsymbol{z}}^{(i+1)}) + \boldsymbol{g}_{\boldsymbol{z}}(\boldsymbol{z}^{(i+1)} - \tilde{\boldsymbol{z}}^{(i+1)}) \tag{6.35}$$

Since equations $\boldsymbol{g}(\boldsymbol{z})$ are not linear and the Jacobian matrix $\boldsymbol{g}_{\boldsymbol{z}}$ is not square, the correction $\boldsymbol{z}^{(i+1)}$ can be found using a Newton's method and iterating the following equation until $\max\{\text{abs}(\boldsymbol{g}(\boldsymbol{z}^{(i+1)}))\}$ is sufficiently small:

$$\boldsymbol{z}^{(i+1)} = \tilde{\boldsymbol{z}}^{(i+1)} - \boldsymbol{g}_{\boldsymbol{z}}(\boldsymbol{g}_{\boldsymbol{z}}^T\boldsymbol{g}_{\boldsymbol{z}})^{-1}\boldsymbol{g}(\boldsymbol{z}^{(i+1)}) \tag{6.36}$$

The whole reduced gradient procedure ends if $\max\{\text{abs}(\Delta\boldsymbol{z}^{(i)})\} < \epsilon$ or if the maximum number of moves are completed.

In case the optimization problem contains inequalities, the procedure described above uses the vector of all binding constraints, e.g., one has to substitute $\boldsymbol{g}$ with $\boldsymbol{g}_a = [\boldsymbol{g}^T, \tilde{\boldsymbol{h}}^T]^T$, where $\tilde{\boldsymbol{h}}$ are the binding inequalities at the step i. The main difficulty is to find a projection move that does not violates inactive constraints. With this aim, the projection move (6.34) is modified using the Haug and Arora's procedure [125], which is a combination of the projection and the restoration modes:

$$\boldsymbol{z}^{(i+1)} = \alpha^*\boldsymbol{z}^{(i)} - \boldsymbol{g}_{a,\boldsymbol{z}}(\boldsymbol{g}_{a,\boldsymbol{z}}^T\boldsymbol{g}_{a,\boldsymbol{z}})^{-1}\boldsymbol{g}_a(\boldsymbol{z}^{(i+1)}) \tag{6.37}$$

where α^* is defined by defining a given reduction γ in the objective function:

$$\alpha^* = \frac{\gamma\varphi(\boldsymbol{z}^{(i)})}{(\Delta\boldsymbol{z}^{(i)})^T\varphi_{\boldsymbol{z}}(\boldsymbol{z}^{(i)})} \tag{6.38}$$

After solving the restoration move, new constraints may become binding, and the vector $\boldsymbol{g}_a$ has to be updated before solving the next projection step.

Example 6.3 Continuation Power Flow as Reduced Gradient Method

This example states the formal analogy between the reduced gradient method described above and the homotopy predictor-corrector method described in Section 5.4 of Chapter 5. A similar proof was originally presented in [37] and further formalized in [21].

The analogy can be shown straightforwardly by observing that in the continuation power flow analysis, the non-basic variable is the loading level μ and that the objective function is $\varphi = -\mu$. Since $\varphi_{\boldsymbol{y}} = \mathbf{0}$ and $\varphi_\mu = -1$, the reduced gradient (6.31) becomes:

$$\boldsymbol{r}(\mu) = \boldsymbol{g}_{\boldsymbol{y}}^{-1}\boldsymbol{g}_\mu \tag{6.39}$$

Hence, the reduced gradient $\boldsymbol{r}$ coincides with the tangent vector $\boldsymbol{\tau}$ used for the predictor step (e.g., projection move) discussed in Section 5.4. As for the restoration move discussed above, it is just another version of the correctors steps described the same Section 5.38. Moreover, the PV generator reactive power limits discussed in Example 5.4 are inequality constraints. Whenever a reactive power limit (e.g., $q_{Gh}^{\max}$) is reached, the constraint $q_{Gh} \leq q_{Gh}^{\max}$ becomes binding and the vector of $\boldsymbol{y}$ is updated to include the voltage magnitude v_h at the PV generator bus. Alternatively, if v_h is already defined as a variable, the constraint $v_h = v_{Gh}^{\mathrm{ref}}$ is removed from $\boldsymbol{g}$.

6.3.2 Interior Point Method

Although Interior Points Methods (IPMs) have been formalized in late sixties [95], the nineties are the dawn of most IPM-based applications for power system analysis [8, 111, 148, 209, 311, 352]. IPM-based OPF problems proved to be robust, especially in large networks, as the number of iterations increase slightly with the number of constraints and network size.

In particular, in [251, 311, 312], the authors present a comprehensive investigation of the use of primal-dual IPM for nonlinear problems, and describe the application of Newton's direction and Mehrotra's predictor-corrector to the OPF. The latter allows reducing the number of iterations to obtain the final solution. Both methods are described in this section.

The main idea of the primal-dual IPM discussed in [95] is the introduction of *logarithmic barrier function* that allows incorporating inequality constraints in the objective function. In this way, inequalities are implicitly taken into account. The objective function modified by means of the logarithmic barrier function is:

$$\hat{\varphi}(\boldsymbol{z}, \hat{\mu}) = \varphi(\boldsymbol{z}) - \hat{\mu} \sum_{k=1}^{n_h} \ln(-h_k(\boldsymbol{z})) \tag{6.40}$$

where $\hat{\mu} > 0$ is the *barrier parameter*. The logarithmic function ensures that $\boldsymbol{h}(\boldsymbol{z}) < \boldsymbol{0}$. In order to effectively minimize the objective function, during the iterative process of the IPM, $\hat{\mu}$ is decreased monotonically to zero.

Applying the logarithmic function to the transformed problem (6.16), one obtains:

$$\begin{aligned} \underset{\boldsymbol{z}}{\text{Minimize}} \quad & \varphi(\boldsymbol{z}) - \hat{\mu} \sum_{k=1}^{n_h} \ln(s_k) \\ \text{subject to} \quad & \boldsymbol{g}(\boldsymbol{z}) = \boldsymbol{0} \\ & \boldsymbol{s} + \boldsymbol{h}(\boldsymbol{z}) = \boldsymbol{0}, \quad \boldsymbol{s} > \boldsymbol{0} \end{aligned} \tag{6.41}$$

The logarithmic terms impose strict positivity on the slack variables. First order optimality conditions for (6.41) are:

$$\begin{aligned} & \varphi_{\boldsymbol{z}} + \boldsymbol{\rho}^T \boldsymbol{g}_{\boldsymbol{z}}(\boldsymbol{z}) + \boldsymbol{\pi}^T \boldsymbol{h}_{\boldsymbol{z}}(\boldsymbol{z}) = \boldsymbol{0} \\ & \boldsymbol{g}(\boldsymbol{z}) = \boldsymbol{0} \\ & \boldsymbol{s} + \boldsymbol{h}(\boldsymbol{z}) = \boldsymbol{0} \\ & \boldsymbol{\pi} - \hat{\mu} \boldsymbol{S}^{-1} \boldsymbol{e} = \boldsymbol{0} \end{aligned} \tag{6.42}$$

where $\boldsymbol{S} = \text{diag}(s_1, s_2, \ldots, s_{n_h})$ and $\boldsymbol{e} = [1, 1, \ldots, 1]^T$. The complementarity constraints can be rewritten as:

$$\boldsymbol{S}\boldsymbol{\pi} - \hat{\mu}\boldsymbol{e} = \boldsymbol{0} \tag{6.43}$$

where $\hat{\mu}\boldsymbol{e}$ with $\hat{\mu} > 0$ is a perturbation of the standard complementarity conditions.

The primal-dual IPM consists in the following steps.

1. *Initial guess.* Set a starting point $i \Leftarrow 0$, $\boldsymbol{z}^{(0)}$, $\boldsymbol{s}^{(0)}$, $\boldsymbol{\pi}^{(0)}$ and $\hat{\mu}^{(0)}$. The initial guess must satisfy the strict positivity condition.
2. *Computing variable directions.* Compute the Jacobian matrix of the first-order optimality conditions (6.42) and compute variable directions.
3. *Updating variables.* Update primal and dual variables using a step length on the directions computed in the previous step.
4. *Reducing the barrier parameter.* A new barrier parameter $\hat{\mu}^{(i+1)}$ is updated based on the current slack and dual variables $\boldsymbol{s}^{(i)}$ and $\boldsymbol{\pi}^{(i)}$, respectively.
5. *Convergence test.* Check if the new point is a local minimizer. If yes the algorithm ends, otherwise set $i \Leftarrow i + 1$, update the barrier parameter $\hat{\mu}^{(i)}$ and go back to Step 2.

Each step is briefly described in the following subsections.

Initial Guess

IPM methods do not requires that the initial point $\boldsymbol{z}$ is a feasible point, however the strict positivity conditions $\boldsymbol{s} > \mathbf{0}$ and $\boldsymbol{\pi} > \mathbf{0}$ must be satisfied, otherwise the method does not converge. Some heuristics can also help obtain the convergence. In [311], the following initialization are proposed:

1. Primal variables $\boldsymbol{z}$ can be obtained as the solution of a power flow problem or computing the middle point between the upper and the lower limit for the bounded variables.
2. The slack variables $\boldsymbol{s}$ are initialized to satisfy the strict positivity constraint. Rewriting inequalities as:

$$\boldsymbol{h}^{\min} \leq \hat{\boldsymbol{h}}(\boldsymbol{z}) \leq \boldsymbol{h}^{\max} \tag{6.44}$$

   slack variables associated with lower limits, say $\boldsymbol{s}_{\min}$ are obtained as:

$$\boldsymbol{s}_{\min}^{(0)} = \min\{\max\{\gamma \boldsymbol{h}^{\Delta}, \hat{\boldsymbol{h}}(\boldsymbol{z}^{(0)}) - \boldsymbol{h}^{\min}\}, (1-\gamma)\boldsymbol{h}^{\Delta}\} \tag{6.45}$$

   where $\boldsymbol{h}^{\Delta} = \boldsymbol{h}^{\max} - \boldsymbol{h}^{\min}$ and $\gamma = 0.25$. Then, slack variables associated with upper limits are set as:

$$\boldsymbol{s}_{\max}^{(0)} = \boldsymbol{h}^{\Delta} - \boldsymbol{s}_{\min}^{(0)} \tag{6.46}$$

3. The dual variables $\boldsymbol{\pi}^{(0)}$ are given by:

$$\boldsymbol{\pi}^{(0)} = \hat{\mu}^{(0)}[\boldsymbol{S}^{(0)}]^{-1}\boldsymbol{e} \tag{6.47}$$

4. The dual variables $\rho_k^{(0)}$ are set to 1 if associated with an active power constraint, 0 otherwise.

Computing Variable Directions

At each step i of the IPM method, variable directions are used for following the path of minimizers parametrized by $\hat{\mu}^{(i)}$. The most common method to compute directions is the Newton's method which consists in solving the following linear system obtained from (6.42) and (6.43):

$$\mathscr{L}_{\xi\xi}\begin{bmatrix}\Delta \boldsymbol{z}\\ \Delta \boldsymbol{\rho}\\ \Delta \boldsymbol{\pi}\\ \Delta \boldsymbol{s}\end{bmatrix} = \begin{bmatrix}\mathscr{L}_{\boldsymbol{zz}} & \boldsymbol{g}_{\boldsymbol{z}}^T & \boldsymbol{h}_{\boldsymbol{z}}^T & \mathbf{0}\\ \boldsymbol{g}_{\boldsymbol{z}} & \mathbf{0} & \mathbf{0} & \mathbf{0}\\ \boldsymbol{h}_{\boldsymbol{z}} & \mathbf{0} & \mathbf{0} & \boldsymbol{I}_{n_h}\\ \mathbf{0} & \mathbf{0} & \boldsymbol{S} & \boldsymbol{\Pi}\end{bmatrix}\begin{bmatrix}\Delta \boldsymbol{z}\\ \Delta \boldsymbol{\rho}\\ \Delta \boldsymbol{\pi}\\ \Delta \boldsymbol{s}\end{bmatrix} = \begin{bmatrix}\mathscr{L}_{\boldsymbol{z}}\\ \mathscr{L}_{\boldsymbol{\rho}}\\ \mathscr{L}_{\boldsymbol{\pi}}\\ \mathscr{L}_{\boldsymbol{s}}\end{bmatrix} \tag{6.48}$$

where $\boldsymbol{\xi} = [\boldsymbol{z}^T, \boldsymbol{\rho}^T, \boldsymbol{\pi}^T, \boldsymbol{s}^T]^T$ and the super-script i has been omitted for simplicity. The size of the system (6.48) can be reduced to a $2n_z \times 2n_z$ system by approximating $\mathscr{L}_{\boldsymbol{\pi}} \approx \mathbf{0}$ and $\mathscr{L}_{\boldsymbol{s}} \approx \mathbf{0}$ as follows:

$$\begin{bmatrix} \hat{\mathscr{L}}_{zz} & \boldsymbol{g}_z^T \\ \boldsymbol{g}_z & \boldsymbol{0} \end{bmatrix} \begin{bmatrix} \Delta \boldsymbol{z} \\ \Delta \boldsymbol{\rho} \end{bmatrix} = \begin{bmatrix} \mathscr{L}_z \\ \mathscr{L}_\rho \end{bmatrix} \tag{6.49}$$

that provides $\Delta \boldsymbol{z}$ and $\Delta \boldsymbol{\rho}$, plus the following direct equations:

$$\begin{aligned} \Delta \boldsymbol{s} &= -\boldsymbol{h}_z \Delta \boldsymbol{z} \\ \Delta \boldsymbol{\pi} &= -\boldsymbol{S}^{-1} \boldsymbol{\Pi} \Delta \boldsymbol{s} \end{aligned} \tag{6.50}$$

where:

$$\hat{\mathscr{L}}_{zz} = \mathscr{L}_{zz} + \boldsymbol{h}_z^T \boldsymbol{S}^{-1} \boldsymbol{\Pi} \boldsymbol{h}_z \tag{6.51}$$

Newton directions obtained from (6.48) or (6.49) are generally sufficient to lead to convergence. However, sparse matrix factorization is the most time consuming operation of the whole algorithm. Thus, it is worth looking for methods that allows reducing the iterations and, consequently, the number of factorizations. In this vein, the Mehrotra's predictor-corrector method is a good option. The Mehrotra's method consists in computing variable directions in two steps but only needs one factorization of the matrices in (6.48) or (6.49), thus leading to a computational burden similar to the standard Newton's directions. Details on the Mehrotra's method can be found in [190].

Predictor Step: The predictor step is obtained as follows:

$$\mathscr{L}_{\xi\xi} \begin{bmatrix} \Delta \boldsymbol{z}_p \\ \Delta \boldsymbol{\rho}_p \\ \Delta \boldsymbol{\pi}_p \\ \Delta \boldsymbol{s}_p \end{bmatrix} = \begin{bmatrix} \mathscr{L}_z \\ \mathscr{L}_\rho \\ \mathscr{L}_\pi \\ -\boldsymbol{S}\boldsymbol{\pi} \end{bmatrix} \tag{6.52}$$

The prediction provided by (6.52) is also called the *affine-scaling* direction. Using this direction is possible to estimate a new barrier parameter value $\hat{\mu}_p^{(i)}$. How to update the barrier parameter is explained in the following subsection.

Corrector Step: The corrector step is obtained as follows:

$$\mathscr{L}_{\xi\xi} \begin{bmatrix} \Delta \boldsymbol{z} \\ \Delta \boldsymbol{\rho} \\ \Delta \boldsymbol{\pi} \\ \Delta \boldsymbol{s} \end{bmatrix} = \begin{bmatrix} \mathscr{L}_z \\ \mathscr{L}_\rho \\ \mathscr{L}_\pi \\ -\boldsymbol{S}\boldsymbol{\pi} + \hat{\mu}_p^{(i)} \boldsymbol{e} - \Delta \boldsymbol{S}_p \Delta \boldsymbol{\pi}_p \end{bmatrix} \tag{6.53}$$

where the term $\hat{\mu}_p^{(i)} \boldsymbol{e}$ is called *centering direction* and helps the current point keep away from the boundary of the feasible region, and $\Delta \boldsymbol{S}_p \Delta \boldsymbol{\pi}_p$ is called *corrector direction* and in some measure compensates nonlinearity not taken into account in the affine-scaling direction.

Since the matrix $\mathscr{L}_{\xi\xi}$ in (6.52) and (6.53) is the same, only one factorization is needed and thus the corrector step does not suppose a relevant extra computing time. Nevertheless, the variable directions obtained using the Mehrotra's method allow reducing the number of iterations with respect to Newton's directions.

Updating Variables

The new primal and dual variables are computed based on the previously computed directions:

$$
\begin{aligned}
\boldsymbol{z}^{(i+1)} &= \boldsymbol{z}^{(i)} + \alpha_P^{(i)} \Delta \boldsymbol{z} \\
\boldsymbol{s}^{(i+1)} &= \boldsymbol{s}^{(i)} + \alpha_P^{(i)} \Delta \boldsymbol{s} \\
\boldsymbol{\rho}^{(i+1)} &= \boldsymbol{\rho}^{(i)} + \alpha_D^{(i)} \Delta \boldsymbol{\rho} \\
\boldsymbol{\pi}^{(i+1)} &= \boldsymbol{\pi}^{(i)} + \alpha_D^{(i)} \Delta \boldsymbol{\pi}
\end{aligned}
\tag{6.54}
$$

where $\alpha_P^{(i)} \in (0, 1]$ and $\alpha_D^{(i)} \in (0, 1]$ are the *step length* parameters for the primal and dual variables, respectively. The maximum values of the step lengths can be estimated using the following heuristic rules:

$$
\begin{aligned}
\alpha_P^{(i)} &= \min\{1, \gamma \min_k \{-s_k^{(i)}/\Delta s_k \text{ if } \Delta s_k < 0\}\} \\
\alpha_Q^{(i)} &= \min\{1, \gamma \min_k \{-\pi_k^{(i)}/\Delta \pi_k \text{ if } \Delta \pi_k < 0\}\}
\end{aligned}
\tag{6.55}
$$

where $\gamma \in (0, 1)$ is a *safety factor* that ensure that the next point satisfies the strict positivity condition. A typical value for the safety factor is $\gamma = 0.99995$. In NLP problems, such as the optimal power flow, primal and dual variables are interdependent due to the dual feasibility conditions. In this case, to use the same step length for both primal and dual variables can help obtain convergence:

$$
\alpha_P^{(i)} = \alpha_Q^{(i)} \leftarrow \min\{\alpha_P^{(i)}, \alpha_Q^{(i)}\}
\tag{6.56}
$$

However, separate step lengths have proved to work well in [111].

Reducing the Barrier Parameter

The barrier parameter $\hat{\mu}^{(i)}$ has to be updated (and hopefully reduced) at each iteration. The new value of the barrier parameter is computed based on the *complementarity gap* $\hat{\varrho}^{(i)}$ that is the residual of the complementarity conditions:

$$
\hat{\varrho}^{(i)} = [\boldsymbol{s}^{(i)}]^T \boldsymbol{\pi}^{(i)}
\tag{6.57}
$$

The complementarity gap $\hat{\varrho}^{(i)} \to 0$ as the primal variables approach a local minimizer $\boldsymbol{z} \to \boldsymbol{z}^*$. Then, the new barrier parameter is computed as:

$$
\hat{\mu}^{(i+1)} = \sigma^{(i+1)} \frac{\hat{\varrho}^{(i)}}{n_h}
\tag{6.58}
$$

where $\sigma^{(i+1)} \in (0,1)$ is the *centering parameter* that can be evaluated as $\sigma^{(i+1)} = \min\{0.99\sigma^{(i)}\}$, with $\sigma^{(0)} = 0.2$.[3] For the Mehrotra's predictor step, the barrier parameter $\hat{\mu}_p^{(i)}$ is computed in a similar way as for the Newton's direction-based method. Firstly, the predictor complementarity gap is computed as:

$$\hat{\varrho}_p^{(i)} = [\boldsymbol{s} + \alpha_P^p \Delta \boldsymbol{s}_p]^T (\boldsymbol{\pi} + \alpha_Q^p \Delta \boldsymbol{\pi}_p) \tag{6.59}$$

where α_P^p and α_Q^p are the primal and dual predictor step lengths, respectively. Then the predictor barrier parameter is given by:

$$\hat{\mu}_p^{(i)} = \min\left\{ \left(\frac{\hat{\varrho}_p^{(i)}}{\hat{\varrho}^{(i)}} \right)^2, 0.2 \right\} \frac{\hat{\varrho}_p^{(i)}}{n_h} \tag{6.60}$$

Finally, the barrier parameter following the Mehrotra's corrector step is computed using (6.57) and (6.58).

Convergence Test

The convergence test has to satisfy primal feasibility, dual feasibility and complementarity conditions:

$$\begin{aligned} \max\{\max\{\boldsymbol{h}(\boldsymbol{z})\}, \|\boldsymbol{g}(\boldsymbol{z})\|_\infty\} &\le \epsilon_1 \\ \frac{\|\varphi_{\boldsymbol{z}} + \boldsymbol{\rho}^T \boldsymbol{g}_{\boldsymbol{z}} + \boldsymbol{\pi}^T \boldsymbol{h}_{\boldsymbol{z}}\|_\infty}{1 + \|\boldsymbol{z}\|_2 + \|\boldsymbol{\rho}\|_2 + \|\boldsymbol{\pi}\|_2} &\le \epsilon_1 \\ \frac{\hat{\varrho}}{1 + \|\boldsymbol{z}\|_2} &\le \epsilon_2 \end{aligned} \tag{6.61}$$

where the index i is omitted for simplicity. Alternative convergence conditions are:

$$\begin{aligned} \hat{\mu}^{(i)} &\le \epsilon_\mu \\ \|\Delta \boldsymbol{z}^{(i)}\|_\infty &\le \epsilon_2 \\ \|\boldsymbol{g}(\boldsymbol{z}^{(i)})\|_\infty &\le \epsilon_1 \end{aligned} \tag{6.62}$$

Finally, a convergence test on the objective function is as follows:

$$\Delta\varphi(\boldsymbol{z}^{(i)}) = \frac{|\varphi(\boldsymbol{z}^{(i)}) - \varphi(\boldsymbol{z}^{(i-1)})|}{1 + |\varphi(\boldsymbol{z}^{(i)})|} \le \epsilon_2 \tag{6.63}$$

If all convergence tests are satisfied, the current point $\boldsymbol{z}^{(i)}$ is a local minimizer of (6.41). Typical tolerances are $\epsilon_1 = 10^{-4}$, $\epsilon_2 = 10^{-2}\epsilon_1$ and $\epsilon_\mu = 10^{-12}$.

---

[3] If $\sigma^{(i)} = 1$, the first-order optimality conditions define a centering direction, i.e., a step towards a point at the barrier trajectory. If $\sigma^{(i)} = 0$, the direction is a pure Newton's one.

Script 6.1 Interior Point Method

The following Python code implement the IPM described above including both Newton's direction method and the Mehrotra's predictor-corrector step. As usual, the `system.Device` class provides the interface between the IPM algorithm and system devices as discussed in Script 3.2 of Chapter 3. The matrices `system.DAE.Oz` and `system.DAE.Hes` are $\varphi_{\mathbf{z}}$ and $\mathscr{L}_{\mathbf{zz}}$, respectively.

```
import system
from cvxopt.base import matrix, spmatrix, sparse, spdiag, div, mul, log
from cvxopt.umfpack import solve, symbolic, numeric
from cvxopt.blas import dotu

def opf():

    # setup dimensions, variables, vectors and matrices
    exec system.Device.setup_opf
    Zz = spmatrix([], [], [], (system.DAE.ng, system.DAE.ng), 'd')

    # parameters
    iteration = 1
    mui = system.OPF.sigma/system.DAE.nh
    msg = 'Iter. = %3d,  mui = %s, |dz| = %s, |g(z)| = %s, |dOF| = %s'

    # initial guess of primal, dual and slack variables
    exec system.Device.call_opfguess
    exec system.Device.call_obj
    system.DAE.s += 1e-6  # avoid zero slack variables
    system.DAE.pi = div(mui, system.DAE.s)
    system.DAE.rho[:system.Bus.n] = matrix(1, (system.Bus.n, 1), 'd')

    # primal dual interior-point method
    while 1:

        # compute g(z), h(z, s), and Jacobian and Hessian matrices
        exec system.Device.call_opf
        system.DAE.h += system.DAE.s

        s = mul(system.DAE.pi, system.DAE.s) - mui
        Lz = system.DAE.Oz + (system.DAE.Gz.T)*system.DAE.rho + \
             (system.DAE.Hz.T)*system.DAE.pi
        Hp = spdiag(div(system.DAE.pi, system.DAE.s))

        # reduced system
        Lr = sparse([[system.DAE.Hes + system.DAE.Hz.T*(Hp*system.DAE.Hz), \
                      system.DAE.Gz], [system.DAE.Gz.T, Zz]])
        if iteration <= 2: F = symbolic(Lr)
        try:
            C = numeric(Lr, F)
        except ValueError:
            # unexpected refactorization of Langragian Jacobian matrix
            F = symbolic(Lr)
            C = numeric(Lr, F)
```

```
if system.OPF.method == 'Newton':

    Hs = spdiag(div(1.0, system.DAE.s))
    Dz = -matrix([Lz + system.DAE.Hz.T*(Hp*system.DAE.h - Hs*s), \
                  system.DAE.g])
    solve(Lr, C, Dz)
    Ds = -system.DAE.h - system.DAE.Hz*Dz[:system.DAE.nz]
    Dp = -Hs*s - Hp*Ds

elif system.OPF.method == 'Mehrotra':

    # predictor step
    Dz = -matrix([Lz + system.DAE.Hz.T*(Hp*system.DAE.h - \
                  system.DAE.pi), system.DAE.g])
    solve(Lr, C, Dz)
    Ds = -system.DAE.h - system.DAE.Hz*Dz[:system.DAE.nz]
    Dp = -system.DAE.pi - Hp*Ds

    # centering correction
    alpha_P, alpha_D = step_length(Ds, Dp)
    cgap_p = dotu(system.DAE.s + alpha_P*Ds, \
                   system.DAE.pi + alpha_D*Dp)
    cgap = dotu(system.DAE.s, system.DAE.pi)
    mui = min((cgap_p/cgap)**2, 0.2)*cgap_p/system.DAE.nh
    s = system.DAE.pi + div(mul(Ds, Dp) - mui, system.DAE.s)

    # corrector step
    Dz = -matrix([Lz + system.DAE.Hz.T*(Hp*system.DAE.h - s), \
                  system.DAE.g])
    solve(Lr, C, Dz)
    Ds = -system.DAE.h - system.DAE.Hz*Dz[:system.DAE.nz]
    Dp = -s - Hp*Ds

# update primal and dual variables
alpha_P, alpha_D = step_length(Ds, Dp)
system.DAE.z   += alpha_P*Dz[:system.DAE.nz]
system.DAE.s   += alpha_P*Ds
system.DAE.rho += alpha_D*Dz[system.DAE.nz:]
system.DAE.pi  += alpha_D*Dp

# objective function
obj_old = system.DAE.obj
exec system.Device.call_obj
system.DAE.obj -= mui*sum(log(system.DAE.s + system.OPF.eps_mu))

# centering parameter, complementarity gap and barrier parameter
system.OPF.sigma = max(0.99*system.OPF.sigma, 0.1)
cgap = dotu(system.DAE.s, system.DAE.pi)
mui = min(abs(system.OPF.sigma*cgap/system.DAE.nh), 1.0)

# convergence tests
test1  = mui <= system.OPF.eps_mu
norma2 = max(abs(Dz))
```

```
        test2  = norma2 <= system.OPF.eps2
        norma3 = system.Settings.error = max(abs(system.DAE.g))
        test3  = norma3 <= system.OPF.eps1
        norma4 = abs(system.DAE.obj - obj_old)/(1 + abs(system.DAE.obj))
        test4  = norma4 <= system.OPF.eps2
        print msg % (iteration, mui, norma2, norma3, norma4)
        if test1 and test2 and test3 and test4: break

        iteration += 1
        if iteration > system.OPF.max_iter: break

def step_length(Ds, Dp):

    ratio1 = [1.0/system.OPF.gamma] + \
             [-s/d for s, d in zip(system.DAE.s, Ds) if d < 0]
    ratio2 = [1.0/system.OPF.gamma] + \
             [-s/d for s, d in zip(system.DAE.pi, Dp) if d < 0]

    alpha_P = system.OPF.gamma*min(ratio1)
    alpha_D = system.OPF.gamma*min(ratio2)

    return alpha_P, alpha_D
```

Example 6.4 Optimal Power Flow Analysis for the IEEE 14-Bus System

This example shows the results of the primal-dual nonlinear IPM applied to the IEEE 14-bus system. The demand is considered inelastic and equal to the base case while generator costs are quadratic and, in order to provide more "freedom degrees", also synchronous compensators are assumed to be standard generators with an active power output. The cost function data and variable limits are taken from [363] and are given in Appendix D. At the local minimizer, the objective function is 8081.55 €/h. Tables 6.1, 6.2, 6.3 and 6.4 show generator active and reactive powers, bus voltages and power flow results, respectively. As expected, dual variables are positive only if a limit is binding. Transmission line limits are not imposed.

Table 6.1 Optimal power flow results for the IEEE 14-bus system: power supplies

Bus #	$\pi_{p_G^{\min}}$	$p_G^{\min}$ [MW]	p_G [MW]	$p_G^{\max}$ [MW]	$\pi_{p_G^{\min}}$
1	0.0000	0.0000	194.3276	200.0000	0.0000
2	0.0000	0.0000	36.7192	140.0000	0.0000
3	0.0000	0.0000	28.7453	100.0000	0.0000
6	0.2663	0.0000	0.0000	100.0000	0.0000
8	0.0000	0.0000	8.4948	100.0000	0.0000

Table 6.2 Optimal power flow results for the IEEE 14-bus system: generator reactive powers

Bus #	$\pi_{q_G^{\min}}$	$q_G^{\min}$ [MVAr]	q_G [MVAr]	$q_G^{\max}$ [MVAr]	$\pi_{q_G^{\min}}$
1	0.0938	0.0000	0.0000	10.0000	0.0000
2	0.0000	−40.0000	23.6860	50.0000	0.0000
3	0.0000	0.0000	24.1265	40.0000	0.0000
6	0.0000	−6.0000	11.5448	24.0000	0.0000
8	0.0000	−6.0000	8.2719	24.0000	0.0000

Table 6.3 Optimal power flow results for the IEEE 14-bus system: bus voltages

Bus #	$\pi_{v^{\min}}$	$v^{\min}$ [pu]	v [pu]	$v^{\max}$ [pu]	$\pi_{v^{\min}}$	θ [rad]
1	0.0000	0.9400	1.0600	1.0600	5.8375	0.0000
2	0.0000	0.9400	1.0408	1.0600	0.0000	−0.0702
3	0.0000	0.9400	1.0156	1.0600	0.0000	−0.1732
4	0.0000	0.9400	1.0145	1.0600	0.0000	−0.1512
5	0.0000	0.9400	1.0164	1.0600	0.0000	−0.1296
6	0.0000	0.9400	1.0600	1.0600	0.5522	−0.2215
7	0.0000	0.9400	1.0464	1.0600	0.0000	−0.1953
8	0.0000	0.9400	1.0600	1.0600	0.7107	−0.1818
9	0.0000	0.9400	1.0437	1.0600	0.0000	−0.2268
10	0.0000	0.9400	1.0391	1.0600	0.0000	−0.2309
11	0.0000	0.9400	1.0460	1.0600	0.0000	−0.2285
12	0.0000	0.9400	1.0448	1.0600	0.0000	−0.2362
13	0.0000	0.9400	1.03995	1.0600	0.0000	−0.2371
14	0.0000	0.9400	1.0239	1.0600	0.0000	−0.2491

In the OPF report, the LMP (Locational Marginal Prices) column are dual variables $\boldsymbol{\rho}_p$, while the NCP column indicates Nodal Congestion Prices that are computed as follows. Using the decomposition formula for LMPs proposed in [353], one has:

$$\mathrm{NCP} = \boldsymbol{g}_{\boldsymbol{y}}^{-1} \hat{\boldsymbol{h}}_{\boldsymbol{y}}^{T} (\boldsymbol{\pi}_{\boldsymbol{y}^{\max}} - \boldsymbol{\pi}_{\boldsymbol{y}^{\min}}) \tag{6.64}$$

where $\hat{\boldsymbol{h}}$ represents the inequality constraints as defined in (6.44), and $\boldsymbol{\pi}_{\boldsymbol{y}^{\max}}$ and $\boldsymbol{\pi}_{\boldsymbol{y}^{\min}}$ are the dual variables associated with the lower and upper limits of such inequality constraints.

Figure 6.2 shows the behavior of the convergence tests of the IPM method during the 8 iterations required to find the local minimizer with the given tolerance. The quantities used for the convergence tests are $\hat{\mu}^{(i)}$, $\|\Delta \boldsymbol{z}^{(i)}\|_{\infty}$, $\|\boldsymbol{g}(\boldsymbol{z}^{(i)})\|_{\infty}$, and $\Delta\varphi(\boldsymbol{z}^{(i)})$. Each quantity is normalized with respect to the

Table 6.4 Optimal power flow results for the IEEE 14-bus system: bus power injections

Bus #	p [MW]	q [MVAr]	ρ_p [$/MWh]	ρ_q [$/MVArh]	NCP [$/MWh]
1	194.3276	0.0000	36.7238	−0.0938	0.0000
2	15.0192	10.9860	38.3596	0.00005	0.6220
3	−65.4547	5.1265	40.5749	0.00005	1.5524
4	−47.8000	3.9000	40.1902	0.1199	1.3823
5	−7.6000	−1.6000	39.6608	0.2076	1.1809
6	−11.2000	4.0448	39.7337	0.00002	1.9533
7	0.0000	−0.0000	40.1715	0.1197	1.7654
8	8.4948	8.2719	40.1699	0.00002	1.6526
9	−29.5000	4.0968	40.1662	0.1961	2.0468
10	−9.0000	−5.8000	40.3178	0.3089	2.1032
11	−3.5000	−1.8000	40.1554	0.2282	2.0627
12	−6.1000	−1.6000	40.3791	0.2124	2.1386
13	−13.5000	−5.8000	40.5755	0.3535	2.1729
14	−14.9000	−5.0000	41.1974	0.5710	2.3410

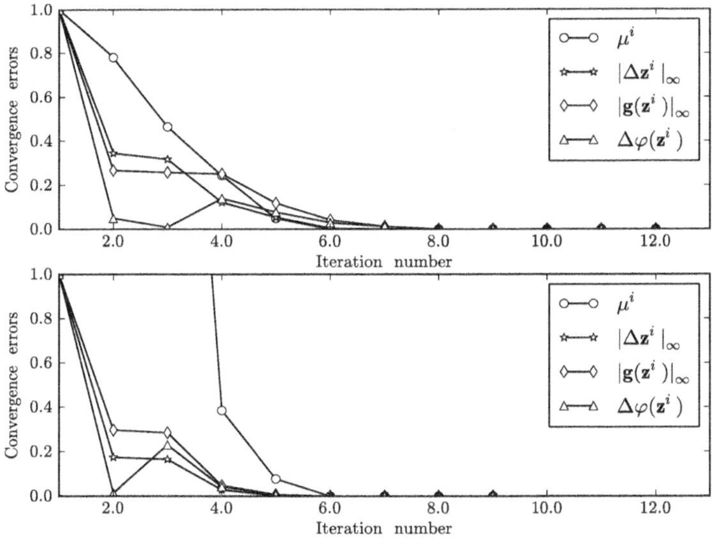

Fig. 6.2 Convergence behavior of IPM for the IEEE 14-bus system. The upper plot is obtained using the Newton's direction method, the lower plot using the Mehrotra's predictor-corrector method

value of the first iteration. As expected, the Mehrotra's predictor-corrector method converges in less iterations than the Newton's direction method.

6.4 Summary of IPM Parameters

This section summarizes most relevant parameters related to the interior point method.

Solver method: The methods described in this chapter are the GRG and the IPM. Of course, several other methods for nonlinear programming are available [24]. It is important to note that, since the OPF problem is non-convex,[4] in general only local minima can be found. Furthermore, due to idiosyncrasies of nonlinearity, it is difficult to say *a priori* if a solver method performs well for a given NLP problem.

Tolerance ϵ_1: Tolerance used for checking the convergence of equalities $\boldsymbol{g}(\boldsymbol{z}^{(i)})$. Typical value is $\epsilon_1 = 10^{-4}$.

Tolerance ϵ_2: Tolerance used for checking the convergence of variable increments $\Delta \boldsymbol{z}^{(i)}$. Typical value is $\epsilon_2 = 10^{-2}\epsilon_1$.

Tolerance ϵ_μ: Threshold of the barrier parameter $\hat{\mu}^{(i)}$ used for defining the convergence of the IPM. Typical value is $\epsilon_\mu = 10^{-12}$.

Initial guess type: The initial guess can be obtained as a solution of a preliminary power flow analysis or as a flat start (e.g., middle value between the upper and lower limits).

Safety factor γ: The safety factor $\gamma \in (0, 1)$ is used for estimating a *good* step length and ensuring the strict positivity condition. A typical value for the safety factor is 0.99995.

Maximum number of iterations: The maximum number of iterations before stopping the solver method if convergence is not reached. For IPM, a reasonable iteration limit is 50.

Method used for computing variable directions: Various methods are available for computing variable directions for the IPM. The most common is the Newton's direction. The Mehrotra's predictor-corrector method provides more precise directions with a negligible additional computational effort. Higher order methods (such as composite Newton's method) can be also used [312].

Centering parameter $\sigma^{(i)}$: The centering factor $\sigma^{(i)} \in (0, 1)$. The initial value of the centering factor can be fixed to $\sigma^{(0)} = 0.2$.

---

[4] In fact, if $\hat{\boldsymbol{h}}(\boldsymbol{z}) \leq \boldsymbol{h}^{\max}$ is convex, then $-\hat{\boldsymbol{h}}(\boldsymbol{z}) \leq -\boldsymbol{h}^{\min}$ is concave, and *vice versa*.

Chapter 7
Eigenvalue Analysis

This chapter describes various aspects of eigenvalues analysis. Section 7.1 provides the background of modal analysis using a simple one-machine infinite-bus example. Section 7.2 describes the small signal stability analysis for dynamical power systems and outlines the properties of equilibrium points, including bifurcation points commonly shown by power systems, participation factors and the Z-domain transformation. Section 7.3 describes a variety of methods for computing a reduced set of eigenvalues and introduces the singular value decomposition. Section 7.4 describes the modal analysis applied to the power flow Jacobian matrix. Finally, Section 7.5 summarizes most relevant concepts related to eigenvalue analysis.

7.1 Background

Consider the one-machine infinite-bus system of Figure 7.1. Assuming that the machine dynamics are described by the classical model, system equations are as follows:

$$\dot{\delta} = \Omega_b(\omega - \omega_s) \tag{7.1}$$

$$\dot{\omega} = \frac{1}{2H}(p_m - p_e(\delta)) \tag{7.2}$$

where Ω_b is the synchronous frequency in rad/s, ω_s is the synchronous speed ($\omega_s = 1$ pu), H is the rotor inertia constant, p_m is the mechanical power, $p_e(\delta)$ is the electrical power defined as:

$$p_e = \frac{ev}{x_{\rm eq}} \sin\delta \tag{7.3}$$

and $x_{\rm eq}$ is the equivalent series reactance composed of the internal machine reactance, of the transmission line reactance and of the Thevenin reactance of the infinite bus, i.e., $x_{\rm eq} = x'_d + x_L + x_{\rm Th}$.

F. Milano: Power System Modelling and Scripting, Power Systems, pp. 155–178.
springerlink.com

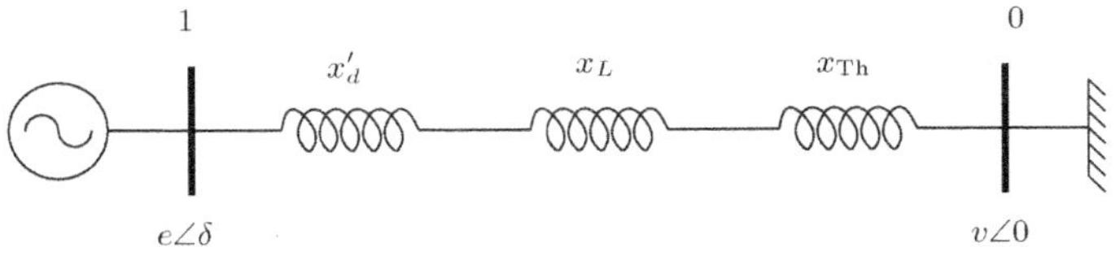

Fig. 7.1 OMIB system

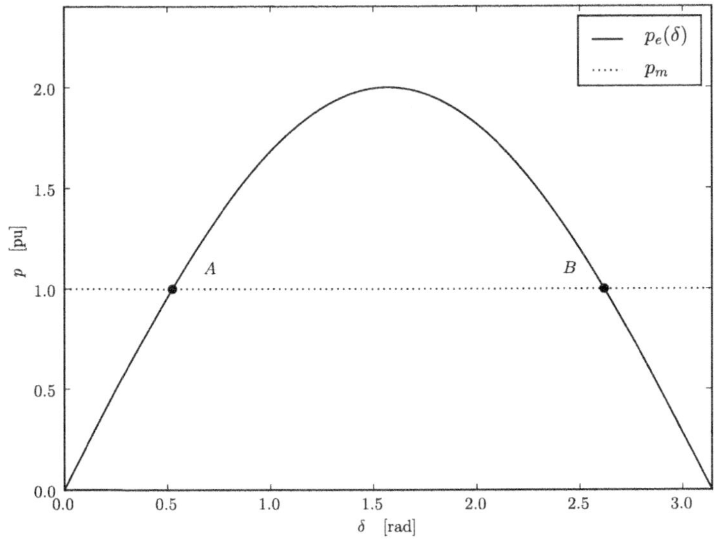

Fig. 7.2 Equilibrium points of the OMIB system

As it is well known, for $\delta \in [0, \pi]$, the dynamical system (7.1)-(7.2) has two equilibrium points. Figure 7.2 shows these equilibrium points assuming $e = v = p_m = 1.0$ pu and $x_{\mathrm{eq}} = 0.5$ pu.

Given the vector of state variables $\boldsymbol{x} = (\delta, \omega)$, the equilibrium points are:

$$\boldsymbol{x}_A = (0.5236, 1)$$
$$\boldsymbol{x}_B = (2.6180, 1)$$

The most important property of an equilibrium point is whether it is stable or unstable. For the simple OMIB system, the stability of the points $\boldsymbol{x}_A$ and $\boldsymbol{x}_B$ can be easily determined using *virtual variations*, as follows.

Point $\boldsymbol{x}_A$: Assume that the system is steady-state at the equilibrium point $\boldsymbol{x}_A$ and that the angle δ is perturbed by a small positive quantity, say

$\partial\delta > 0$. Then, $p_e(\delta_A + \partial\delta) > p_m$ and, from (7.1), $\dot{\omega} < 0$, which leads to decrease ω. If ω decreases, $\omega < \omega_A = 1$ and, from (7.2), $\dot{\delta} < 0$ that leads to decrease δ. In conclusion, the system responds to an increase of δ by decreasing it. Similarly, the system responds to a decrease of δ by increasing it. Hence, the point $\boldsymbol{x}_A$ is a *sink* as it attracts the trajectories of the system that are sufficiently close to it. Such point is also commonly called a *stable* equilibrium point.

Point $\boldsymbol{x}_B$: A specular reasoning can be done for the equilibrium point $\boldsymbol{x}_B$. Assume that the system is steady-state at the equilibrium point $\boldsymbol{x}_B$ and that the angle δ is perturbed by a small positive quantity, say $\partial\delta > 0$. Then, $p_e(\delta_B + \partial\delta) < p_m$ and, from (7.1), $\dot{\omega} > 0$, which leads to increase ω. If ω increases, $\omega > \omega_B = 1$ and, from (7.2), $\dot{\delta} > 0$, which leads to increase δ. In conclusion, the system responds to an increase of δ by further increasing it. Similarly, the system responds to a decrease of δ by further decreasing it. Hence, the point $\boldsymbol{x}_B$ is a *source* as it repulses the trajectories of the system that are sufficiently close to it. Such point is also commonly called an *unstable* equilibrium point.

Unfortunately, the procedure above cannot be applied to a system with hundreds of state variables. There is thus the need of a systematic procedure that allows defining the stability of an equilibrium point. This systematic approach is the eigenvalue analysis.

Assume a continuous, nonlinear and smooth ODE system:

$$\dot{\boldsymbol{x}} = \boldsymbol{f}(\boldsymbol{x}) \tag{7.4}$$

where $\boldsymbol{x} \in \mathbb{R}^{n_x}$ and $\boldsymbol{f} : \mathbb{R}^{n_x} \mapsto \mathbb{R}^{n_x}$. An equilibrium point of (7.4) is a point $\boldsymbol{x}_0$ that satisfies $\boldsymbol{f}(\boldsymbol{x}_0) = \boldsymbol{0}$. Then, the solution $\boldsymbol{\lambda}$ of the following system:

$$\det(\boldsymbol{f}_{\boldsymbol{x}}|_0 - \boldsymbol{\lambda}\boldsymbol{I}_{n_x}) = 0 \tag{7.5}$$

are the eigenvalues (or *characteristic roots* [132] or *latent roots* [185]) of the state matrix $\boldsymbol{f}_{\boldsymbol{x}}|_0$.

An equilibrium point (or stationary solution) $\boldsymbol{x}_0$ is called *hyperbolic* or *non-degenerate* when the state matrix $\boldsymbol{f}_{\boldsymbol{x}}|_0$ has no eigenvalue with zero real part. Otherwise the equilibrium is called *non-hyperbolic* or *degenerate*. The well-known Lyapunov's first stability method states that the properties of the eigenvalues $\boldsymbol{\lambda}$ can describe the behavior of the nonlinear system (7.4) around a hyperbolic equilibrium point $\boldsymbol{x}_0$ [178]. In particular one has:

1. If $\Re\{\lambda_h\} < 0, \forall h = 1, 2, \ldots, n_x$, then the equilibrium point $\boldsymbol{x}_0$ is stable.
2. If $\exists\ \Re\{\lambda_h\} > 0$ for some $h = 1, 2, \ldots, n_x$, then the equilibrium point $\boldsymbol{x}_0$ is unstable.

If the equilibrium is non-hyperbolic, i.e., $\exists\ \Re\{\lambda_h\} = 0$ for some $h = 1, 2, \ldots, n$, the eigenvalues are not adequate to define the properties of the equilibrium point $\boldsymbol{x}_0$ and further analysis is needed. The latter statement can be explained with a simple example. Let us consider the following scalar ODE system:

$$\dot{x} = x^2 \tag{7.6}$$

that has the equilibrium $x_0 = 0$. Linearizing (7.6) at x_0, one has:

$$\Delta\dot{x} = 2x_0 \cdot \Delta x = 0 \cdot \Delta x \tag{7.7}$$

where $2x_0$ is both the state matrix and the unique eigenvalue of the system. As it can be noted, (7.7) provides no relevant information about the stability of the equilibrium point since $\Delta\dot{x} = 0$ notwithstanding any variation of Δx.[1]

Equilibrium points for which at least one eigenvalue has zero real part are called *bifurcation points*. The only bifurcation points that are of practical interest are *generic* ones, i.e., bifurcations that are likely to appear in a physical system. The concept of genericity also implies *robustness*, i.e., the bifurcation persists for any parameter perturbation. For example, the bifurcation point $x_0 = 0$ of (7.6) is generic.[2]

For the OMIB system, the eigenvalue analysis reduces to find the roots of the following equation:

$$\det(\boldsymbol{A}_S - \boldsymbol{\lambda}\boldsymbol{I}_2) = 0 \tag{7.8}$$

where $\boldsymbol{A}_S = \boldsymbol{f_x}$ is the state matrix of (7.1)-(7.2) computed at an equilibrium point.

Point $\boldsymbol{x}_A$: The state matrix at the equilibrium point $\boldsymbol{x}_A$ is:

$$\boldsymbol{f_x}|_A = \left[\begin{array}{c|c} 0 & \Omega_b \\ \hline -\dfrac{1}{2H}\dfrac{ev}{x_{\mathrm{eq}}}\cos\delta_A & 0 \end{array}\right] \tag{7.9}$$

Thus, (7.8) yields:

$$\lambda_{1,2} = \sqrt{-\Omega_b \frac{1}{2H}\frac{ev}{x_{\mathrm{eq}}}\cos\delta_A} = \pm j5.8317 \tag{7.10}$$

where $\Omega_b = 314.16$ rad/s (i.e., the system rated frequency is 50 Hz) and the mechanical starting time is $H = 8$ MWs/MVA. Since the pair of complex eigenvalues is purely imaginary, the system trajectories around the equilibrium point are undamped and $\boldsymbol{x}_A$ is a bifurcation point. However,

[1] Observe that the approach based on virtual variation used for studying (7.1) works fine if one varies x_0 by δx in (7.6).

[2] In fact, the differential equation:

$$\dot{x} = x^2 + \mu$$

is the *germ* or the *normal form* of the saddle-node bifurcation.

this is an ideal condition. By including a damping (i.e., a viscous friction) coefficient D in (7.2), e.g.:

$$\dot{\omega} = \frac{1}{2H}(p_m - p_e(\delta) - D(\omega - \omega_s)) \tag{7.11}$$

the pair of complex eigenvalues shows a negative real part, hence the bifurcation is not robust. The equilibrium point $\boldsymbol{x}_A$ is stable (or *weakly* stable as reported in some books), which confirms the qualitative analysis above based on virtual perturbations.

Point $\boldsymbol{x}_B$: Drawing the eigenvalue analysis for the point $\boldsymbol{x}_B$, one has:

$$\lambda_{1,2} = \sqrt{-\Omega_b \frac{1}{2H}\frac{ev}{x_{\rm eq}}\cos\delta_B} = \pm 5.8317 \tag{7.12}$$

One of the characteristics roots is positive, hence the system is unstable. In this case, the instability cannot be eliminated by simply including a damping coefficient in (7.2).

The solution of (7.5) is a problem conceptually different from the ones discussed so far, e.g., finding a solution (only *one*) of a set of nonlinear functions. The solution of (7.5) consists in finding *all* the roots (e.g., zeros) of a polynomial function. Since Abel proved in 1826 that there is no explicit solution for generic polynomials of order greater than 4, (7.5) has to be necessarily solved numerically. Some matrix-based methods are discussed in the following sections.

7.2 Small Signal Stability Analysis

The system used for the small signal stability analysis is a set of differential algebraic equations, in the form:

$$\begin{aligned} \dot{\boldsymbol{x}} &= \boldsymbol{f}(\boldsymbol{x}, \boldsymbol{y}) \\ \boldsymbol{0} &= \boldsymbol{g}(\boldsymbol{x}, \boldsymbol{y}) \end{aligned} \tag{7.13}$$

where $\boldsymbol{x}$ is the vector of the state variables and $\boldsymbol{y}$ the vector of the algebraic variables, $\boldsymbol{f}$ is the vector of differential equations, and $\boldsymbol{g}$ is the vector of algebraic equations. Small signal stability analysis studies the properties of *equilibria* or *stationary points* $(\boldsymbol{x}_0, \boldsymbol{y}_0)$ that satisfies:

$$\begin{aligned} \boldsymbol{0} &= \boldsymbol{f}(\boldsymbol{x}_0, \boldsymbol{y}_0) \\ \boldsymbol{0} &= \boldsymbol{g}(\boldsymbol{x}_0, \boldsymbol{y}_0) \end{aligned} \tag{7.14}$$

through an eigenvalue analysis of the state matrix $\boldsymbol{A}_S$ of the system.

The state matrix $\boldsymbol{A}_S$ is obtained by manipulating the complete Jacobian matrix $\boldsymbol{A}_C$, that is defined by the linearization of the DAE system equations (7.13) at the equilibrium point:

$$\begin{bmatrix} \Delta\dot{\boldsymbol{x}} \\ \boldsymbol{0} \end{bmatrix} = \begin{bmatrix} \boldsymbol{f_x} & \boldsymbol{f_y} \\ \boldsymbol{g_x} & \boldsymbol{g_y} \end{bmatrix} \begin{bmatrix} \Delta\boldsymbol{x} \\ \Delta\boldsymbol{y} \end{bmatrix} = \boldsymbol{A}_C \begin{bmatrix} \Delta\boldsymbol{x} \\ \Delta\boldsymbol{y} \end{bmatrix} \tag{7.15}$$

The state matrix $\boldsymbol{A}_S$ is obtained by eliminating the algebraic variables and, thus, it is implicitly assumed that $\boldsymbol{g_y}$ is not singular (i.e., absence of singularity-induced bifurcations):

$$\boldsymbol{A}_S = \boldsymbol{f_x} - \boldsymbol{f_y}\boldsymbol{g_y}^{-1}\boldsymbol{g_x} \tag{7.16}$$

There have been an extensive research on singularity-induced bifurcations [22, 113, 256]. It is now recognized that the singularity of $\boldsymbol{g_y}$ at an equilibrium point is a *folding* of the manifold of algebraic variables. This is not actually a stability issue, but rather a modelling one. If $\boldsymbol{g_y}$ is singular at a given equilibrium, then the dynamic of some of the algebraic variables y_h cannot be considered infinitely fast (e.g., its time constant cannot be considered zero). In other words, recalling the discussion given in Section 1.4 of Chapter 1, the sets of variables $\boldsymbol{\xi}_i$ and $\boldsymbol{\xi}_f$ have to be revised and some of the $\boldsymbol{\xi}_f$ have to be passed to the set of state variables $\boldsymbol{\xi}_i$. The best candidates to be switched to state variables are the subset of $\boldsymbol{\xi}_f$ that most participate to the zero eigenvalue of $\boldsymbol{g_y}$.

The matrix $\boldsymbol{D} = \boldsymbol{f_y}\boldsymbol{g_y}^{-1}\boldsymbol{g_x}$ can be considered a *degradation matrix* since it degrades the stability of the matrix $\boldsymbol{f_x}$. In fact, the eigenvalues of $\boldsymbol{f_x}$ are typically all negative, as in $\boldsymbol{f_x}$, each device and associated controls are basically decoupled from the others. The matrix $\boldsymbol{D}$ couples each device through the network and, hence, through variables $\boldsymbol{y}$, which can be considered *aggregation variables* (see also Section 8.2 of Chapter 8). Since $\boldsymbol{A}_C$ may show poorly damped or even positive real part eigenvalues, an effective technique for improving or obtaining stability is to reduce, by means of adequate controllers, the effect of $\boldsymbol{D}$ and, in turn, to decouple dynamic devices.

Script 7.1 Small-Signal Stability Analysis

The Python implementation of the standard small-signal stability analysis is rather simple. Provided that the Jacobian matrices of the DAE system are available, a possible implementation is as follows.

```
import system
from numpy.linalg import eigvals
from cvxopt.umfpack import linsolve

def state_matrix():

    Gyx = matrix(system.DAE.Gx)
    linsolve(system.DAE.Gy, Gyx)
    return system.DAE.Fx - system.DAE.Fy*Gyx
```

```
def eigs():

    As = state_matrix()
    return eigvals(As)
```

Clearly, the efficiency of the implementation strongly depends on the algorithm used for computing all eigenvalues.

Example 7.1 Eigenvalues of the IEEE 14-Bus System in the *S*-Domain

Figure 7.3 shows the eigenvalue analysis for the IEEE 14-bus system whose dynamic data are given in Appendix D. For the sake of illustration, only the eigenvalues with $\Re\{\lambda_h\} > -10$ are shown in Figure 7.3. The figure also shows two dotted lines indicating the locus with 5% damping. Given a complex eigenvalue $\alpha \pm j\beta$, the damping ζ is defined as:

$$\zeta = -\frac{\alpha}{\sqrt{\alpha^2 + \beta^2}} \tag{7.17}$$

with $\zeta \in [-1, 1]$. The slope s of the line with a given damping ζ is given by:

$$s = \pm\sqrt{\frac{1 - \zeta^2}{\zeta^2}} \tag{7.18}$$

The damping is an important measure of the quality of the transient response of the system. A poorly damped system oscillates for a relatively long time after a contingency, which has to be avoided. Figure 7.3 shows that a pair of complex eigenvalues is poorly damped, thus some corrective actions are advisable for improving the system response. Further discussion on oscillation damping is given in Section 16.3 of Chapter 16.

7.2.1 Bifurcation Points

In most stability applications, it is of interest determining whether the system equilibrium is a bifurcation point or not, i.e., whether some eigenvalues of $\boldsymbol{A}_S$ have zero real part.

At practical effects, there are only two cases that have to be taken into account:[3]

1. One eigenvalue is $\lambda_k = 0$. This condition generally implies the occurrence of a *saddle-node bifurcation* [6, 35, 37, 76].

[3] Eigenvalue conditions for saddle-node and Hopf bifurcations are necessary but not sufficient. Proper transversality conditions that impose the dependence of the critical eigenvalues on system parameters complete the definitions of these bifurcation points. Transversality conditions are omitted in the chapter for the sake of simplicity. The interested reader can find rigorous definitions in [276].

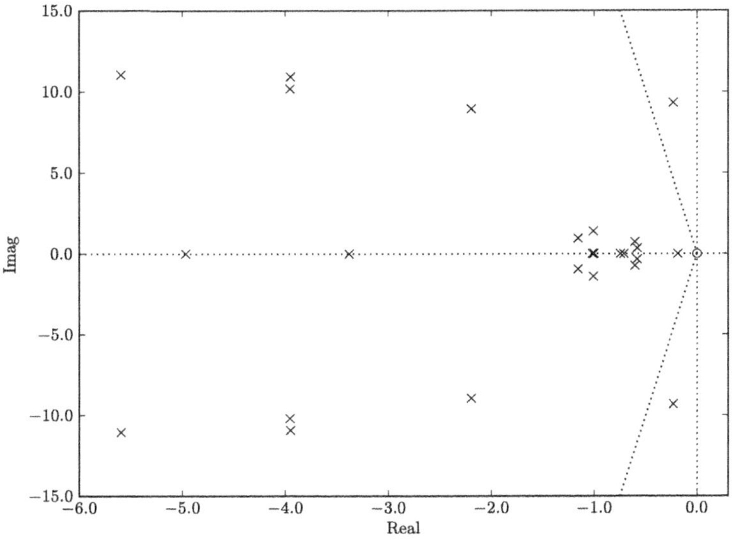

Fig. 7.3 Relevant eigenvalues in the S-domain for the IEEE 14-bus system

2. A pair of complex eigenvalues as zero real part, i.e., $\lambda_{h,k} = \pm j\beta$. This condition generally implies the occurrence of a *Hopf bifurcation* [44, 45, 77, 207, 361].

For saddle-node bifurcations, the determinant $\det(\boldsymbol{A}_S) = 0$, which directly derives from the matrix property:

$$\det(\boldsymbol{A}_S) = \prod_{h=1}^{n_x} \lambda_h \tag{7.19}$$

For Hopf bifurcations, the determinant of $\boldsymbol{A}_S$ is not informative. However, if $\alpha \pm j\beta$ is a pair of complex eigenvalues of $\boldsymbol{A}_S$ with associated eigenvectors $\boldsymbol{\nu}_r \pm j\boldsymbol{\nu}_i$, then the following expression holds [45]:

$$\begin{aligned} (\boldsymbol{A}_S - \alpha \boldsymbol{I}_n)\boldsymbol{\nu}_r + \beta \boldsymbol{\nu}_i &= \boldsymbol{0} \\ (\boldsymbol{A}_S - \alpha \boldsymbol{I}_n)\boldsymbol{\nu}_i - \beta \boldsymbol{\nu}_r &= \boldsymbol{0} \end{aligned} \tag{7.20}$$

Then the modified state matrix:

$$\boldsymbol{A}_{MS} = \begin{bmatrix} \boldsymbol{A}_S & +\beta \boldsymbol{I}_{n_x} \\ -\beta \boldsymbol{I}_{n_x} & \boldsymbol{A}_S \end{bmatrix} \tag{7.21}$$

is singular at the bifurcation point. It is interesting to note that $\boldsymbol{A}_{MS}$ is singular for both saddle-node ($\beta = 0$) and Hopf bifurcations ($\beta \neq 0$).

Since the computation of the determinant is computationally inefficient, various alternative methods can be used, for example (i) computing the eigenvalue with smallest magnitude (see also Subsection 7.3), (ii) computing the minimum singular value [39], and (iii) computing the LU factorization of $\boldsymbol{A}_S$ or $\boldsymbol{A}_{MS}$ (see also the discussion given in Section 5.4 of Chapter 5).

It has also been observed that using the complete Jacobian matrix $\boldsymbol{A}_C$ can be more efficient than $\boldsymbol{A}_S$ [45]. In fact, $\boldsymbol{A}_C$ does not requires the computation of the inverse of $\boldsymbol{g}_y$ and is generally sparser than $\boldsymbol{A}_S$. Thus sparse matrix algorithms work generally faster for $\boldsymbol{A}_C$ than for $\boldsymbol{A}_S$ even though the size of $\boldsymbol{A}_C$ is greater than that of $\boldsymbol{A}_S$. At this regard, the following remarks are relevant:

1. For saddle-node bifurcations, from the Leibniz's formula, one has:

$$\begin{aligned} \det(\boldsymbol{A}_C) &= \det\left(\begin{bmatrix} \boldsymbol{f}_x & \boldsymbol{f}_y \\ \boldsymbol{g}_x & \boldsymbol{g}_y \end{bmatrix}\right) \qquad (7.22) \\ &= \det\left(\begin{bmatrix} \boldsymbol{f}_x - \boldsymbol{f}_y\boldsymbol{g}_y^{-1}\boldsymbol{g}_x & \boldsymbol{0} \\ \boldsymbol{g}_y^{-1}\boldsymbol{g}_x & \boldsymbol{I}_{n_y} \end{bmatrix}\right)\det\left(\begin{bmatrix} \boldsymbol{I}_{n_x} & \boldsymbol{f}_y \\ \boldsymbol{0} & \boldsymbol{g}_y \end{bmatrix}\right) \\ &= \det(\boldsymbol{g}_y)\det(\boldsymbol{f}_x - \boldsymbol{f}_y\boldsymbol{g}_y^{-1}\boldsymbol{g}_x) \\ &= \det(\boldsymbol{g}_y)\det(\boldsymbol{A}_S) \end{aligned}$$

   where $\det(\boldsymbol{g}_y) \neq 0$, since by hypothesis $\boldsymbol{g}_y$ is invertible. Hence $\det(\boldsymbol{A}_C) = 0$ iff $\det(\boldsymbol{A}_S) = 0$.
2. Generalizing equation (7.20) for the complete matrix $\boldsymbol{A}_C$ leads to:

$$\begin{bmatrix} \boldsymbol{f}_x & \boldsymbol{f}_y \\ \boldsymbol{g}_x & \boldsymbol{g}_y \end{bmatrix} \begin{bmatrix} \boldsymbol{\nu}_{x,r} \pm j\boldsymbol{\nu}_{x,i} \\ \boldsymbol{\nu}_{y,r} \pm j\boldsymbol{\nu}_{y,i} \end{bmatrix} = (\alpha \pm j\beta) \begin{bmatrix} \boldsymbol{\nu}_{x,r} \pm j\boldsymbol{\nu}_{x,i} \\ \boldsymbol{0} \end{bmatrix} \qquad (7.23)$$

   Thus, the modified complete matrix $\boldsymbol{A}_{MC}$ becomes:

$$\boldsymbol{A}_{MC} = \begin{bmatrix} \boldsymbol{f}_x & \boldsymbol{f}_y & +\beta\boldsymbol{I}_{n_x} & \boldsymbol{0} \\ \boldsymbol{g}_x & \boldsymbol{g}_y & \boldsymbol{0} & \boldsymbol{0} \\ -\beta\boldsymbol{I}_{n_x} & \boldsymbol{0} & \boldsymbol{f}_x & \boldsymbol{f}_y \\ \boldsymbol{0} & \boldsymbol{0} & \boldsymbol{g}_x & \boldsymbol{g}_y \end{bmatrix} \qquad (7.24)$$

Example 7.2 Synchronous Reference Zero Eigenvalue for the IEEE 14-Bus System

In Example 7.1, Figure 7.3 shows one zero eigenvalue indicated as a circle at the intersection of the real and imaginary axes. According to the discussion above, a zero eigenvalue always requires special care.

In most situations, a zero eigenvalue implies a saddle-node bifurcation. However, in this case the bifurcation is due to the model of synchronous generators. Although the details of synchronous machine models are discussed in Chapter 15, the explanation of this issue can be explained using the simple classical model used in Section 7.1. In (7.1), the reference angle $\theta_0 = 0$ is that of the infinite bus. The machine rotor angle δ is implicitly a relative angle, i.e., $\delta - \theta_0$. Thus, as discussed in Section 7.1, the state matrix of the system does not show zero eigenvalues. Consider now the same two bus system of Figure 7.1, but assuming that the machines connected at both buses are modelled using a classical generator model. Thus, one has:

$$\begin{aligned} \dot{\delta}_1 &= \Omega_b(\omega_1 - \omega_s) \\ \dot{\omega}_1 &= \frac{1}{2H_1}(p_m - p_e(\delta_1, \delta_2)) \\ \dot{\delta}_2 &= \Omega_b(\omega_2 - \omega_s) \\ \dot{\omega}_2 &= \frac{1}{2H_2}(-p_m + p_e(\delta_1, \delta_2)) \end{aligned} \tag{7.25}$$

where

$$p_e = \frac{e_1 e_2}{x_{\text{eq}}} \sin(\delta_1 - \delta_2) = k \sin(\delta_1 - \delta_2) \tag{7.26}$$

The state matrix of the system is:

$$\boldsymbol{f_x} = \left[\begin{array}{c|c|c|c} 0 & \Omega_b & 0 & 0 \\ \hline -k\cos(\delta_1 - \delta_2) & 0 & k\cos(\delta_1 - \delta_2) & 0 \\ \hline 0 & 0 & 0 & \Omega_b \\ \hline k\cos(\delta_1 - \delta_2) & 0 & -k\cos(\delta_1 - \delta_2) & 0 \end{array}\right] \tag{7.27}$$

where the first and third columns are linearly dependent and, thus, $\det(\boldsymbol{f_x}) = 0$ for any equilibrium point.[4] Actually, the issue is not the stability of the system, but the model of the synchronous machines that refer to a fictitious ideal synchronous reference. In other words, the reference angle of the machines is not a physical angle of the network. Hence, one of the angles of the machines is redundant, as in the ODE system (7.25) rotor angles only appear as $\delta_1 - \delta_2$. By defining the angle and speed differences $\delta_{12} = \delta_1 - \delta_2$ and $\omega_{12} = \omega_1 - \omega_2$, (7.25) becomes:

$$\begin{aligned} \dot{\delta}_{12} &= \Omega_b(\omega_1 - \omega_2) \\ \dot{\omega}_{12} &= \frac{1}{2H_{\text{eq}}}(p_m - p_e(\delta_{12})) \end{aligned} \tag{7.28}$$

where $H_{\text{eq}} = H_1 H_2/(H_1 + H_2)$. The transformed ODE system (7.28) does not show the zero eigenvalues of the original system (7.25). The transformation

[4] In particular, since damping is not considered, the state matrix shows two zero eigenvalues. One can easily check that, by adding a damping coefficient, the state matrix shows an unique zero eigenvalue.

used in (7.28) cannot be promptly generalized for a multi-machine system, mainly for two reasons:

1. One of the synchronous machines has to be chosen as reference machine. This implies that the model of a machine changes depending if it chosen as reference or not.
2. In order to maintain a consistent model, the reference machine has to be connected through ac branches to all machines of the network. This cannot be ensured in case of network topological changes, (e.g., line outages) or if the system is composed of regions connected only through HVDC systems.

In conclusion, in multi-machine systems, each generator rotor angle and speed is referred to a fictitious ideal synchronous machine which is not physically connected to the system. As a consequence, the state matrix shows a zero eigenvalue that is not a symptom of instability but rather an intrinsic characteristic of the machine model.

7.2.2 Participation Factors

Along with eigenvalues $\boldsymbol{\lambda}$, it is relevant to compute the participation factors, that are evaluated in the following way [269]. Let $\boldsymbol{N}$ and $\boldsymbol{W}$ be the right and the left eigenvector matrices respectively, such that $\boldsymbol{\Lambda} = \boldsymbol{W}\boldsymbol{A}_S\boldsymbol{N}$ and $\boldsymbol{W} = \boldsymbol{N}^{-1}$, then the participation factor p_{ij} of the ith state variable to the jth eigenvalue can be defined as:

$$p_{ij} = \frac{w_{ij}\nu_{ji}}{\boldsymbol{w}_j^T\boldsymbol{\nu}_j} \tag{7.29}$$

In case of complex eigenvalues, the amplitude of each element of the eigenvectors is used:

$$p_{ij} = \frac{|w_{ij}||\nu_{ji}|}{\sum_{k=1}^{n_x}|w_{jk}||\nu_{kj}|} \tag{7.30}$$

Script 7.2 Participation Factors

The following Python code implements the function for computing participation factors of a given matrix `As`. The function `gesv` accomplishes the operation $\boldsymbol{N}^{-1}\boldsymbol{W}$ and works for both real and complex matrices. The participation factor matrix `pf` is normalized and small eigenvalues are rounded to zero. The latter operation allows quickly detecting bifurcation points.[5] In the code the eigenvalues are called `mu` because, for an idiosyncrasy of Python, `lambda` is a language reserved word.

[5] In fact, it takes a little while to recognize that $-1.209803e^{-15}$ is practically zero, especially if this number appears in a long eigenvalue list.

```
import system
from numpy.linalg import eig
from cvxopt.lapack import gesv
from cvxopt.base import matrix, spmatrix

def compute_eigs(As):

    mu, N = eig(matrix(As))
    N = matrix(N)
    n = len(mu)
    idx = range(n)
    W = matrix(spmatrix(1.0, idx, idx, (n, n), v.typecode))
    gesv(N, W)
    pf = mul(abs(W.T), abs(V))
    b = matrix(1.0, (1, n))
    WN = b * pf
    pf = pf.T

    for item in idx:
        mur = mu[item].real
        mui = mu[item].imag
        mu[item] = complex(round(mur, 5), round(mui, 5))
        pf[item, :] /= WN[item]
```

Example 7.3 Eigenvalue Participation Factors for the IEEE 14-Bus System

Table 7.1 shows the full eigenvalue list for the IEEE 14-bus system, as well as the most associated state variables to each eigenvalue, the undamped natural angular frequency ω_0, the damping ratio ζ and the damped natural angular frequency ω_d.

The damped frequency has the following meaning. Let $\alpha \pm j\beta$ be a pair of complex conjugate eigenvalues. The frequency ω_0 is defined as:

$$\omega_0 = \sqrt{\alpha^2 + \beta^2} \tag{7.31}$$

while the damping is:

$$\zeta = -\frac{\alpha}{\omega_0} \tag{7.32}$$

where $\zeta \in [-1, 1]$. The damping is positive if the mode is stable (i.e. $\alpha < 0$). Thus, one has:

$$\begin{aligned} \alpha &= -\zeta\omega_0 \\ \beta &= \sqrt{1-\zeta^2}\,\omega_0 \end{aligned} \tag{7.33}$$

The frequency ω_0 is also called the frequency of resonance, or undamped frequency. However, the frequency that can be observed during the transient, namely the damped frequency, depends on the damping ζ, as follows:

$$\omega_d = \sqrt{1-\zeta^2}\,\omega_0 = \beta \tag{7.34}$$

The frequency of resonance coincides with the damped frequency only if $\alpha = 0$.

Table 7.1 Eigenvalues and most associated state variables for the IEEE 14-bus system

Eigen. λ_h	Mostly associated state variable	$\Re\{\lambda_h\}$	$\Im\{\lambda_h\}$	Damped Freq. ω_d	Freq. ω_0	Damping ζ [%]
1	v_m AVR 1	−1000	0	0	0	100
2	v_m AVR 2	−1000	0	0	0	100
3	v_m AVR 3	−1000	0	0	0	100
4	v_m AVR 5	−1000	0	0	0	100
5	v_m AVR 4	−1000	0	0	0	100
6	v_{r1} AVR 1, ψ''_d Syn 1	−45.22	9.375	1.492	7.351	97.92
7	v_{r1} AVR 1, ψ''_d Syn 1	−45.22	−9.375	1.492	7.351	97.92
8	v_{r1} AVR 2	−50	0	0	0	100
9	v_{r1} AVR 3	−49.99	0	0	0	100
10	v_{r1} AVR 5	−49.9	0	0	0	100
11	v_{r1} AVR 4	−49.93	0	0	0	100
12	ψ''_q Syn 1	−37.37	0	0	0	100
13	ψ''_d Syn 2	−33.8	0	0	0	100
14	ψ''_d Syn 3	−31.23	0	0	0	100
15	ψ''_d Syn 4	−29.59	0	0	0	100
16	ψ''_d Syn 5	−25.79	0	0	0	100
17	ψ''_q Syn 2	−20.93	0	0	0	100
18	ψ''_q Syn 3	−18.16	0	0	0	100
19	ω Syn 4, δ Syn 4	−5.592	11.05	1.758	1.97	45.17
20	ω Syn 4, δ Syn 4	−5.592	−11.05	1.758	1.97	45.17
21	e'_q Syn 1, v_f AVR 1	−0.2317	9.32	1.483	1.484	2.485
22	e'_q Syn 1, v_f AVR 1	−0.2317	−9.32	1.483	1.484	2.485
23	ω Syn 2, δ Syn 2	−3.949	10.93	1.74	1.85	33.97
24	ω Syn 2, δ Syn 2	−3.949	−10.93	1.74	1.85	33.97
25	ω Syn 3, δ Syn 3	−3.954	10.19	1.622	1.74	36.17
26	ω Syn 3, δ Syn 3	−3.954	−10.19	1.622	1.74	36.17
27	ω Syn 3, δ Syn 3	−2.194	8.95	1.424	1.467	23.81
28	ω Syn 3, δ Syn 3	−2.194	−8.95	1.424	1.467	23.81
29	ψ''_q Syn 4	−11.44	0	0	0	100
30	ψ''_q Syn 5	−9.864	0	0	0	100
31	e'_d Syn 2	−4.968	0	0	0	100
32	e'_d Syn 3	−3.381	0	0	0	100
33	v_f AVR 5, e'_q Syn 5	−1.008	1.398	0.2225	0.2743	58.5
34	v_f AVR 5, e'_q Syn 5	−1.008	−1.398	0.2225	0.2743	58.5
35	v_f AVR 4, e'_q Syn 4	−1.156	0.9523	0.1516	0.2384	77.19
36	v_f AVR 4, e'_q Syn 4	−1.156	−0.9523	0.1516	0.2384	77.19
37	δ Syn 1	0	0	0	0	100
38	e'_q Syn 3, v_f AVR 3	−0.6064	0.7361	0.1172	0.1518	63.58
39	e'_q Syn 3, v_f AVR 3	−0.6064	−0.7361	0.1172	0.1518	63.58
40	ω Syn 1	−0.1862	0	0	0	100
41	e'_q Syn 2, v_f AVR 2	−0.584	0.3424	0.05449	0.1077	86.27
42	e'_q Syn 2, v_f AVR 2	−0.584	−0.3424	0.05449	0.1077	86.27
43	e'_d Syn 4	−0.7128	0	0	0	100
44	e'_d Syn 5	−0.7463	0	0	0	100
45	v_{r2} AVR 1	−1.017	0	0	0	100
46	v_{r2} AVR 2	−1.009	0	0	0	100
47	v_{r2} AVR 3	−1.004	0	0	0	100
48	v_{r2} AVR 4	−1.003	0	0	0	100
49	v_{r2} AVR 5	−1.004	0	0	0	100

While the meaning of each state variable is clarified in Part III, in this example, it is relevant to note just the way the most associated state variables are determined. Since the full participation factor matrix would occupy too much space, Table 7.2 shows a selection of eigenvalues, say eigenvalues 19, 20 and 23-28. Generally, only few state variables participate with a $p_{ij} > 0.1$ to a given eigenvalue, which eases the choice of control actions.

Table 7.2 Selection of eigenvalue participation factors for the IEEE 14-bus system

State variable	$p_{k,19}$	$p_{k,20}$	$p_{k,23}$	$p_{k,24}$	$p_{k,25}$	$p_{k,26}$	$p_{k,27}$	$p_{k,28}$
δ Syn 1	0.0002	0.0002	0.0024	0.0024	0.0043	0.0043	0.0827	0.0827
ω Syn 1	0.0002	0.0002	0.0024	0.0024	0.0043	0.0043	0.0831	0.0831
e'_q Syn 1	0.0003	0.0003	0.0022	0.0022	0.0021	0.0021	0.0480	0.0480
ψ''_d Syn 1	0.0001	0.0001	0.0008	0.0008	0.0007	0.0007	0.0123	0.0123
ψ''_q Syn 1	0.0001	0.0001	0.0008	0.0008	0.0011	0.0011	0.0130	0.0130
δ Syn 2	0.0040	0.0040	0.2158	0.2158	0.0550	0.0550	0.0516	0.0516
ω Syn 2	0.0041	0.0041	0.2168	0.2168	0.0553	0.0553	0.0518	0.0518
e'_q Syn 2	0.0001	0.0001	0.0023	0.0023	0.0006	0.0006	0.0008	0.0008
ψ''_d Syn 2	0.0002	0.0002	0.0052	0.0052	0.0012	0.0012	0.0012	0.0012
e'_d Syn 2	0.0010	0.0010	0.0470	0.0470	0.0117	0.0117	0.0066	0.0066
ψ''_q Syn 2	0.0019	0.0019	0.0848	0.0848	0.0193	0.0193	0.0098	0.0098
δ Syn 3	0.0025	0.0025	0.0045	0.0045	0.1480	0.1480	0.1654	0.1654
ω Syn 3	0.0025	0.0025	0.0045	0.0045	0.1487	0.1487	0.1660	0.1660
e'_q Syn 3	0	0	0	0	0	0	0.0005	0.0005
ψ''_d Syn 3	0	0	0.0001	0.0001	0	0	0.0007	0.0007
e'_d Syn 3	0.0009	0.0009	0.0014	0.0014	0.0454	0.0454	0.0400	0.0400
ψ''_q Syn 3	0.0016	0.0016	0.0025	0.0025	0.0747	0.0747	0.0591	0.0591
δ Syn 4	0.2618	0.2618	0.0207	0.0207	0.0443	0.0443	0.0241	0.0241
ω Syn 4	0.2637	0.2637	0.0208	0.0208	0.0446	0.0446	0.0242	0.0242
e'_q Syn 4	0	0	0	0	0	0	0.0001	0.0001
ψ''_d Syn 4	0.0001	0.0001	0.0001	0.0001	0.0001	0.0001	0.0003	0.0003
e'_d Syn 4	0.0047	0.0047	0.0003	0.0003	0.0008	0.0008	0.0004	0.0004
ψ''_q Syn 4	0.2334	0.2334	0.0151	0.0151	0.0305	0.0305	0.0113	0.0113
δ Syn 5	0.0759	0.0759	0.1289	0.1289	0.1148	0.1148	0.0373	0.0373
ω Syn 5	0.0764	0.0764	0.1296	0.1296	0.1156	0.1156	0.0375	0.0375
e'_q Syn 5	0	0	0	0	0	0	0.0001	0.0001
ψ''_d Syn 5	0	0	0.0001	0.0001	0	0	0.0002	0.0002
e'_d Syn 5	0.0013	0.0013	0.0019	0.0019	0.0018	0.0018	0.0005	0.0005
ψ''_q Syn 5	0.0627	0.0627	0.0860	0.0860	0.0724	0.0724	0.0158	0.0158
v_m AVR 1	0	0	0	0	0	0	0.0004	0.0004
v_f AVR 1	0.0003	0.0003	0.0024	0.0024	0.0022	0.0022	0.0452	0.0452
v_{r1} AVR 1	0.0001	0.0001	0.0006	0.0006	0.0005	0.0005	0.0088	0.0088
v_{r2} AVR 1	0	0	0	0	0	0	0.0007	0.0007
v_m AVR 2	0	0	0	0	0	0	0	0
v_f AVR 2	0	0	0	0	0	0	0.0001	0.0001
v_{r1} AVR 2	0	0	0	0	0	0	0	0
v_{r2} AVR 2	0	0	0	0	0	0	0	0
v_m AVR 3	0	0	0	0	0	0	0	0
v_f AVR 3	0	0	0	0	0	0	0.0001	0.0001
v_{r1} AVR 3	0	0	0	0	0	0	0	0
v_{r2} AVR 3	0	0	0	0	0	0	0	0
v_m AVR 4	0	0	0	0	0	0	0	0
v_f AVR 4	0	0	0	0	0	0	0.0001	0.0001
v_{r1} AVR 4	0	0	0	0	0	0	0	0
v_{r2} AVR 4	0	0	0	0	0	0	0	0
v_m AVR 5	0	0	0	0	0	0	0	0
v_f AVR 5	0	0	0	0	0	0	0.0001	0.0001
v_{r1} AVR 5	0	0	0	0	0	0	0	0
v_{r2} AVR 5	0	0	0	0	0	0	0	0

7.2.3 Analysis in the Z-Domain

The state matrix in (7.16) leads to the computation of the eigenvalues in the S-domain, i.e., the system is stable if $\Re\{\lambda_h\} < 0, \forall h = 1, 2, \ldots, n_x$. To compute the eigenvalues in the Z-domain can lead to some numeric advantages as discussed in the following subsection. The Z-domain can also ease the visualization of stiff systems since, in the Z-domain, if the system is stable, all the eigenvalues are inside the unit circle [147]. For obtaining the Z-domain eigenvalues, a bi-linear transformation is performed:

$$\boldsymbol{A}_Z = (\boldsymbol{A}_S + \chi \boldsymbol{I}_{n_x})(\boldsymbol{A}_S - \chi \boldsymbol{I}_{n_x})^{-1} \tag{7.35}$$

where χ is a weighting factor that, based on heuristic considerations, can be set $\chi = 8$. Computing $\boldsymbol{A}_Z$ is more expensive than $\boldsymbol{A}_S$ but using $\boldsymbol{A}_Z$ can be useful for fastening the determination of the maximum amplitude eigenvalue (e.g., by means of a power method), especially in case of unstable equilibrium points with only one eigenvalue outside the unit circle.

Example 7.4 Eigenvalues of the IEEE 14-Bus System in the *Z*-Domain

Figure 7.4 shows the eigenvalue analysis for the IEEE 14-bus tests system in the Z-domain. The figure also shows two dotted curves indicating the

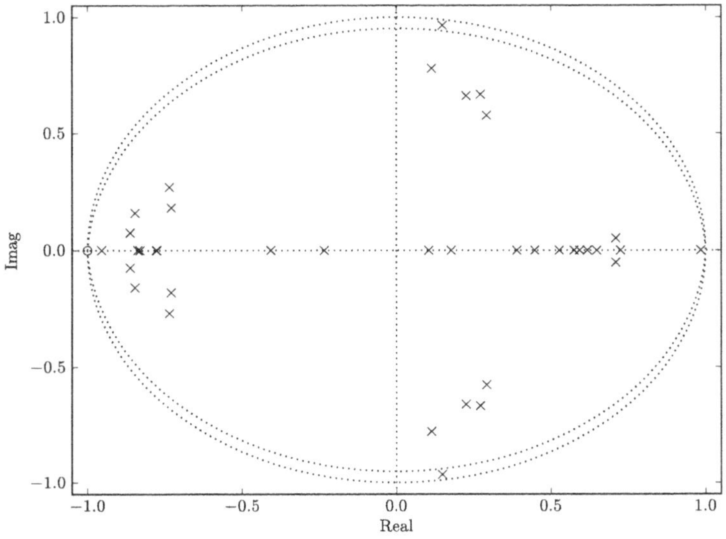

Fig. 7.4 Eigenvalues of the IEEE 14-bus system in the Z-domain

circumference with unitary radius (stability limit in the Z-domain) and the locus of eigenvalues with a damping $\zeta = 5\%$. The latter curve can be computed as follows.

From (7.33) and (7.35), it can be deduced that the transformed value $\alpha_Z \pm j\beta_Z$ in the Z-domain of a given pair of complex eigenvalues $\alpha \pm j\beta$ is:

$$\alpha_Z \pm j\beta_Z = \frac{-\zeta\omega_0 \pm \sqrt{1-\zeta^2}\omega_0 + \chi}{-\zeta\omega_0 \pm \sqrt{1-\zeta^2}\omega_0 - \chi} \tag{7.36}$$

Thus, imposing a fixed damping ζ_c and parametrizing (7.36) with ω_0, each value of ω_0 yields a point in the Z-domain pertaining to the locus with $\zeta = \zeta_c$.

As expected from the analysis presented in Example 7.1, Figure 7.4 shows a pair of complex eigenvalues poorly damped. Figure 7.4 also shows an eigenvalue equal to -1, which is the equivalent in the Z-domain of a zero eigenvalue in the S-domain.

7.3 Computing the Eigenvalues

Most common methods for computing all eigenvalues of a matrix are the QR algorithm, the Arnoldi's iteration, and, if the matrix is Hermitian, the Lanczos' method [74, 313]. However, computing of all eigenvalues generally requires the use of the Gram-Schmidth's orthonormalization method and, thus, can be a lengthy process if the dynamic order of the system is high.

To reduce the computational effort, it is possible to compute only a few eigenvalues with a particular property, i.e., largest or smallest magnitude, largest or smaller real or imaginary part. All or some of these options may be already available in some scientific-oriented scripting language such as Matlab, but have currently to be implemented in general-purpose scripting languages such as Python.

7.3.1 *Power Method*

A very simple method that allows determining the eigenvalue with greatest absolute value, although may show a slow convergence rate, is the *power method*, which works as follows.

Given a matrix $\boldsymbol{A}$ and an initial vector $\boldsymbol{\nu}^{(0)}$, a generic iteration of the power method is given by:

$$\boldsymbol{\nu}^{(i+1)} = \frac{\boldsymbol{A}\boldsymbol{\nu}^{(i)}}{\|\boldsymbol{A}\boldsymbol{\nu}^{(i)}\|_2} \tag{7.37}$$

If $\boldsymbol{A}$ has an eigenvalue λ_k that is strictly greater than all other eigenvalues of $\boldsymbol{A}$, i.e., $\lambda_k > \lambda_h, \forall h \neq k$ (in this case λ_k is called the *dominant* eigenvalue), and $\boldsymbol{\nu}^{(0)}$ has a non-zero component in the direction of the eigenvector $\boldsymbol{\nu}_k$

associated with λ_k, then $\boldsymbol{\nu}^{(i)} \rightarrow \boldsymbol{\nu}_k$ for $i \rightarrow \infty$. Finally, the eigenvalue λ_k is computed as:

$$\lambda_k = \frac{\boldsymbol{\nu}_k^T \boldsymbol{A} \boldsymbol{\nu}_k}{\nu_k^T \nu_k} \tag{7.38}$$

The rationale of the power method is relatively simple and is worth being briefly outlined. Assume that the initial vector $\boldsymbol{\nu}^{(0)}$ can be written as the linear combination of all eigenvectors $\boldsymbol{\nu}_h$ of the matrix $\boldsymbol{A}$:

$$\boldsymbol{\nu}^{(0)} = c_1\boldsymbol{\nu}_1 + c_2\boldsymbol{\nu}_2 + \cdots + c_k\boldsymbol{\nu}_k + \cdots + c_n\boldsymbol{\nu}_n \tag{7.39}$$

where, by hypothesis, $c_k \neq 0$. Assume also that $\boldsymbol{A}$ is diagonalizable and can be written as $\boldsymbol{N\Lambda N}^{-1}$.[6] Then, at the ith iteration:

$$\begin{aligned}
\boldsymbol{\nu}^{(i)} &= \frac{\boldsymbol{A}^{(i)}\boldsymbol{\nu}^{(0)}}{\|\boldsymbol{A}^{(i)}\boldsymbol{\nu}^{(0)}\|_2} \qquad (7.40)\\
&= \frac{(\boldsymbol{N\Lambda N}^{-1})^{(i)}\boldsymbol{\nu}^{(0)}}{\|(\boldsymbol{N\Lambda N}^{-1})^{(i)}\boldsymbol{\nu}^{(0)}\|_2}\\
&= \frac{\boldsymbol{N\Lambda}^{(i)}\boldsymbol{N}^{-1}\boldsymbol{\nu}^{(0)}}{\|\boldsymbol{N\Lambda}^{(i)}\boldsymbol{N}^{-1}\boldsymbol{\nu}^{(0)}\|_2}\\
&= \left(\frac{\lambda_k}{|\lambda_k|}\right)^{(i)} \frac{c_k}{|c_k|} \frac{\boldsymbol{\nu}_k + \frac{1}{c_k}\boldsymbol{N}\left(\frac{1}{\lambda_k}\boldsymbol{\Lambda}\right)^{(i)}\boldsymbol{b}}{\|\boldsymbol{\nu}_k + \frac{1}{c_k}\boldsymbol{N}\left(\frac{1}{\lambda_k}\boldsymbol{\Lambda}\right)^{(i)}\boldsymbol{b}\|_2}
\end{aligned}$$

where

$$\boldsymbol{b} = c_1\boldsymbol{e}_1 + \cdots + c_{k-1}\boldsymbol{e}_{k-1} + c_{k+1}\boldsymbol{e}_{k+1} + \cdots + c_n\boldsymbol{e}_n \tag{7.41}$$

From observing that:

$$\lim_{i\to\infty}\left(\frac{1}{\lambda_k}\boldsymbol{\Lambda}\right)^{(i)} = \begin{bmatrix} \frac{\lambda_1^{(i)}}{\lambda_k^{(i)}} & & & & \\ & \ddots & & & \\ & & \frac{\lambda_k^{(i)}}{\lambda_k^{(i)}} & & \\ & & & \ddots & \\ & & & & \frac{\lambda_n^{(i)}}{\lambda_k^{(i)}} \end{bmatrix} \rightarrow \begin{bmatrix} 0 & & & & \\ & \ddots & & & \\ & & 1 & & \\ & & & \ddots & \\ & & & & 0 \end{bmatrix} \tag{7.42}$$

Then, it follows that:

$$\lim_{i\to\infty}\frac{1}{c_k}\boldsymbol{N}\left(\frac{1}{\lambda_k}\boldsymbol{\Lambda}\right)^{(i)}\boldsymbol{b} \rightarrow 0 \tag{7.43}$$

[6] A similar proof holds if $\boldsymbol{A}$ is decomposed into its Jordan's canonical form.

7.3.2 Inverse Iteration

The inverse method is a variant of the power method described above and allows finding the eigenvalue λ_k and the associated eigenvector $\boldsymbol{\nu}_k$ if a *good* estimation $\tilde{\lambda}_k$ of the eigenvalue is known.

By way of introduction, observe that to find the minimum eigenvalue of a given matrix $\boldsymbol{A}$, it suffices to apply the power method iteration (7.37) to $\boldsymbol{A}^{-1}$. In fact, the eigenvalues of $\boldsymbol{A}^{-1}$ are the inverse of the eigenvalues of $\boldsymbol{A}$. Clearly, this algorithm works only if the matrix $\boldsymbol{A}$ is not singular.

The inverse iteration consists in applying the power method iteration (7.37) to the matrix $(\boldsymbol{A} - \tilde{\lambda}_k \boldsymbol{I}_n)^{-1}$. In fact, the eigenvalues of $(\boldsymbol{A} - \tilde{\lambda}_k \boldsymbol{I}_n)^{-1}$ are $(\lambda_1 - \tilde{\lambda}_k)^{-1}, \ldots, (\lambda_k - \tilde{\lambda}_k)^{-1}, \ldots, (\lambda_k - \tilde{\lambda}_n)^{-1}$. Thus, if $\tilde{\lambda}_k$ is sufficiently close to λ_h, then $(\lambda_k - \tilde{\lambda}_k)^{-1}$ is the biggest eigenvalue of $(\boldsymbol{A} - \tilde{\lambda}_k \boldsymbol{I}_n)^{-1}$, which is the necessary condition for the power method to converge.

7.3.3 Rayleigh's Iteration

The Rayleigh's iteration is an improvement of the inverse iteration. One chooses an initial value $\tilde{\lambda}_k^{(0)}$, then at each iteration both the eigenvector $\boldsymbol{\nu}^{(i+1)}$ and the eigenvalue estimation $\tilde{\lambda}_k^{(i+1)}$ are computed, as follows:

$$\boldsymbol{\nu}^{(i+1)} = \frac{(\boldsymbol{A} - \lambda^{(i)} \boldsymbol{I}_n)^{-1} \boldsymbol{\nu}^{(i)}}{\|(\boldsymbol{A} - \lambda^{(i)} \boldsymbol{I}_n)^{-1} \boldsymbol{\nu}^{(i)}\|_2} \tag{7.44}$$

$$\lambda^{(i+1)} = \frac{(\boldsymbol{\nu}^{(i+1)})^T \boldsymbol{A} \boldsymbol{\nu}^{(i+1)}}{\nu^{(i+1)})^T \boldsymbol{\nu}^{(i+1)}} \tag{7.45}$$

where (7.45) is called the *Rayleigh's quotient*. The convergence characteristics of this method are generally better (i.e., cubically) than the inverse iteration.

Example 7.5 Inverse and Rayleigh's Iterations for the IEEE 14-Bus System

For the sake of example, consider the matrix $\boldsymbol{A}_Z$ defined in (7.35), $\boldsymbol{\nu}^{(0)} = [1, 1, \ldots, 1]$ and $\tilde{\lambda}_k^{(0)} = -0.9$ and a tolerance $\epsilon = 10^{-5}$. The inverse iteration converges to the eigenvalue -0.9545 in 13 iterations, while the Rayleigh's iteration in 5.

From the scripting viewpoint, both inverse and Rayleigh's iterations require practically the same code. For example, the inverse iteration is as follows:

```
from cvxopt.umfpack import linsolve
from cvxopt.base import matrix
from cvxopt.blas import dotu

b = matrix(1, (system.DAE.nx, 1), 'd')
```

```
mold = 99999
m0 = -0.9
iteration = 0
while 1:
    linsolve(As - m0*In, b)
    nor = (dotu(b, b))**0.5
    b = b/nor
    m = dotu(b, As*b)/dotu(b, b)
    if abs(m - mold) < 1e-5:
        break
    mold = m
    iteration += 1
```

To obtain the Rayleigh's iteration, it suffices to substitute the 12th line of the code above with:

```
m0 = m = dotu(b, As*b)/dotu(b, b)
```

7.4 Power Flow Modal Analysis

Beside small-signal stability analysis, eigenvalues and eigenvectors can also be used for assessing sensitivities. In particular, an interesting approach is evaluating the modal analysis for the power flow Jacobian matrix [104, 211, 354].

Let us consider the classical power flow model defined in Section 4.3 (e.g., constant PQ loads and constant PV generators). After solving the power flow analysis, the Jacobian matrix can be easily computed. If the Newton's method is used, the Jacobian matrix is available as byproduct of the solution algorithm. The eigenvalue analysis is performed on a reduced matrix, as follows. Recalling (4.45), the Jacobian matrix can be divided into four sub-matrices:

$$\boldsymbol{g_y} = \begin{bmatrix} \boldsymbol{g}_{p,\boldsymbol{\theta}} & \boldsymbol{g}_{p,\boldsymbol{v}} \\ \boldsymbol{g}_{q,\boldsymbol{\theta}} & \boldsymbol{g}_{q,\boldsymbol{v}} \end{bmatrix} \tag{7.46}$$

In case of the classical power flow model, one can associate a physical meaning to each sub-matrix, since load and generator powers are constant. In fact, consider the linearization of the power flow equations with constant power injections:

$$\begin{bmatrix} \Delta\boldsymbol{p} \\ \Delta\boldsymbol{q} \end{bmatrix} = \begin{bmatrix} \boldsymbol{g}_{p,\boldsymbol{\theta}} & \boldsymbol{g}_{p,\boldsymbol{v}} \\ \boldsymbol{g}_{q,\boldsymbol{\theta}} & \boldsymbol{g}_{q,\boldsymbol{v}} \end{bmatrix} \begin{bmatrix} \Delta\boldsymbol{\theta} \\ \Delta\boldsymbol{v} \end{bmatrix} \tag{7.47}$$

The basic assumption of [104] is to consider that $\Delta\boldsymbol{p} \approx \boldsymbol{0}$. This is reasonable if one is interested only in the relationship between reactive powers and bus voltage magnitudes. Furthermore, recalling the assumptions of the fast-decoupled power flow (see Section 4.4.7 of Chapter 4), the $p\theta$ world is quite decoupled from the qv one. Thus, one can define a reduced power flow Jacobian matrix as follows:

$$\boldsymbol{J}_{\text{LF}} = \boldsymbol{g}_{q,\boldsymbol{v}} - \boldsymbol{g}_{q,\boldsymbol{\theta}}\boldsymbol{g}_{p,\boldsymbol{\theta}}^{-1}\boldsymbol{g}_{p,\boldsymbol{v}} \tag{7.48}$$

where it is assumed that $\boldsymbol{g}_{p,\boldsymbol{\theta}}$ is non-singular. The sensitivity analysis follows from the observation that:

$$\Delta \boldsymbol{q} = \boldsymbol{J}_{\mathrm{LF}} \Delta \boldsymbol{v} \tag{7.49}$$

hence:

$$\Delta \boldsymbol{v} = \boldsymbol{J}_{\mathrm{LF}}^{-1} \Delta \boldsymbol{q} = \boldsymbol{N} \boldsymbol{\Lambda}^{-1} \boldsymbol{N}^{-1} \Delta \boldsymbol{q} \tag{7.50}$$

and defining the modal reactive power and voltage variations as:

$$\begin{aligned} \Delta \boldsymbol{q}_m &= \boldsymbol{N}^{-1} \Delta \boldsymbol{q} \\ \Delta \boldsymbol{v}_m &= \boldsymbol{N}^{-1} \Delta \boldsymbol{v} \end{aligned} \tag{7.51}$$

it follows that the modal sensitivity for each eigenvalue λ_k is:

$$\frac{dv_{m,k}}{dq_{m,k}} = \frac{1}{\lambda_k}, \quad k \in \mathcal{B}_{PQ} \tag{7.52}$$

where $\mathcal{B}_{PQ}$ is the set of PQ load buses. If $\lambda_h > 0$, then an increase in the injected reactive power leads to a bus voltage increase, which is the normal situation for systems with inductive branches and loads. If $\lambda_k < 0$, the voltage decreases if the reactive power injected at the bus increases. This is considered an unstable behavior (at least for inductive systems), and actually it is typical of power flow solutions of the lower part of the nose curve (see Section 5.1 of Chapter 5). A special case is $\lambda_k = 0$ that can be viewed as an infinite sensitivity of the voltage with respect to the reactive power. Actually, this case corresponds to a saddle-node bifurcation.

Example 7.6 Power Flow Modal Analysis for the IEEE 14-Bus System

Figure 7.5 and Table 7.3 show the results of the modal analysis of the power flow Jacobian matrix as well as the participation factors for the IEEE 14-bus test system. In this case study, only static power flow data are considered, i.e., loads are modelled as constant PQ and generators as constant PV or slack.

Figure 7.5 and Table 7.3 only show 9 eigenvalues. In fact, the 14-bus system system has 5 generators that maintain constant the voltage at the bus where they are connected. As expected, all eigenvalues are positive, thus indicating that $\partial v / \partial q > 0$ at all load buses.

7.4.1 Singular Value Decomposition

The Singular Value Decomposition (SVD) is an important kind of matrix factorization and has several applications, for example, for computing the

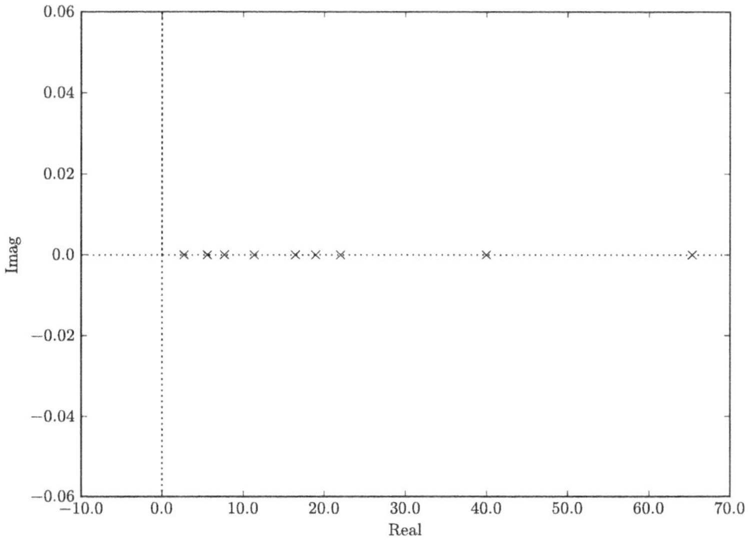

Fig. 7.5 Eigenvalues of the power flow Jacobian matrix for the IEEE 14-bus system

pseudo-inverse, for matrix approximation, and for determining the rank of a matrix. The SVD consists in a factorization of a matrix $\boldsymbol{A}$ in the form:

$$\boldsymbol{A} = \boldsymbol{U}\boldsymbol{\Sigma}\boldsymbol{V}^H \tag{7.53}$$

where $\boldsymbol{U}$ is an unitary (but not diagonal) matrix, $\boldsymbol{\Sigma}$ is a diagonal matrix whose diagonal elements are the *singular values* (non-negative real numbers) and $\boldsymbol{V}^H$ is the Hermitian matrix (conjugate transpose) of $\boldsymbol{V}$, which is also a unitary matrix.

The singular values can be interpreted as "gain controls" that multiply the input signals filtered by the orthonormal $\boldsymbol{V}$ and that pass these signals to the orthonormal $\boldsymbol{U}$ that generates the output signals.

Although the main applications of the SVD can be found in signal processing and statistics, the feature that is relevant in the context of small signal stability is that the computation of the SVD or better of the minimum singular value of a matrix is much more efficient than the corresponding eigenvalue computation [313]. Moreover, the following relevant property of the determinant:

$$\begin{aligned} \det(\boldsymbol{A}) = \det(\boldsymbol{N}\boldsymbol{\Lambda}\boldsymbol{N}^{-1}) &= \det(\boldsymbol{U}\boldsymbol{\Sigma}\boldsymbol{V}^H) \\ \Rightarrow \quad \det(\boldsymbol{A}) = \det(\boldsymbol{\Lambda}) &= \det(\boldsymbol{\Sigma}) \end{aligned} \tag{7.54}$$

has an interesting application in case the $\det(\boldsymbol{A}) = 0$. In fact, if $\boldsymbol{A}$ is singular, there exists a zero singular value. Since all singular values are non-negative,

Table 7.3 Power flow modal analysis and participation factors for the IEEE 14-bus system

Eigenvalue	$p_{h,4}$	$p_{h,5}$	$p_{h,7}$	$p_{h,9}$	$p_{h,10}$
65.3424	0.5416	0.4517	0.0066	0.0001	0
39.9528	0	0.0006	0.1531	0.6147	0.2153
21.9828	0.0756	0.1517	0.4942	0.0030	0.2216
18.9217	0.0005	0.0007	0.0002	0.0002	0.0046
16.4317	0.2835	0.3223	0.0202	0.0476	0.1614
2.7060	0.0082	0.0040	0.0699	0.1999	0.2394
5.5693	0.0024	0.0013	0.0166	0.0314	0.1157
7.6621	0	0	0.0001	0.0001	0.0379
11.3351	0.0881	0.0677	0.2392	0.1030	0.0041

Eigenvalue	$p_{h,11}$	$p_{h,12}$	$p_{h,13}$	$p_{h,14}$
65.3424	0	0	0	0
39.9528	0.0076	0	0.0001	0.0085
21.9828	0.0534	0	0.0001	0.0003
18.9217	0.0021	0.1781	0.7652	0.0485
16.4317	0.1530	0.0024	0.0057	0.0040
2.7060	0.1108	0.0190	0.0324	0.3164
5.5693	0.1281	0.3392	0.1636	0.2017
7.6621	0.1168	0.4512	0.0306	0.3634
11.3351	0.4282	0.0101	0.0023	0.0573

$\sigma_k = 0$ is the minimum singular value of $\boldsymbol{A}$. On the other hand, if one looks for the minimum singular value and $\min\{\sigma_k\} > 0$, then $\boldsymbol{A}$ is certainly non-singular. Thus, if one is interested only in knowing if $\boldsymbol{A}$ is singular or not, computing the minimum singular value is a numerically efficient option. This property has been used in voltage stability studies for determining the distance to saddle-node bifurcations [39, 45].

Example 7.7 Minimum Singular Value Index for the IEEE 14-Bus System

Example 5.4 in Chapter 5 shows the continuation power flow analysis for the IEEE 14-bus system using a distributed slack bus model and neglecting reactive power limits of PV generators. In this case, the maximum loading condition is due to a SNB. As discussed above, at the SNB point, the minimum singular value of the power flow Jacobian matrix is zero. Thus, the minimum singular value of the Jacobian matrix can be used as an index for evaluating the proximity to the point of collapse.

Figure 7.6 shows the behavior of the minimum singular value of the power flow Jacobian matrix. As expected, as the loading level μ increases, the minimum singular value decreases. It is numerically quite difficult to find exactly

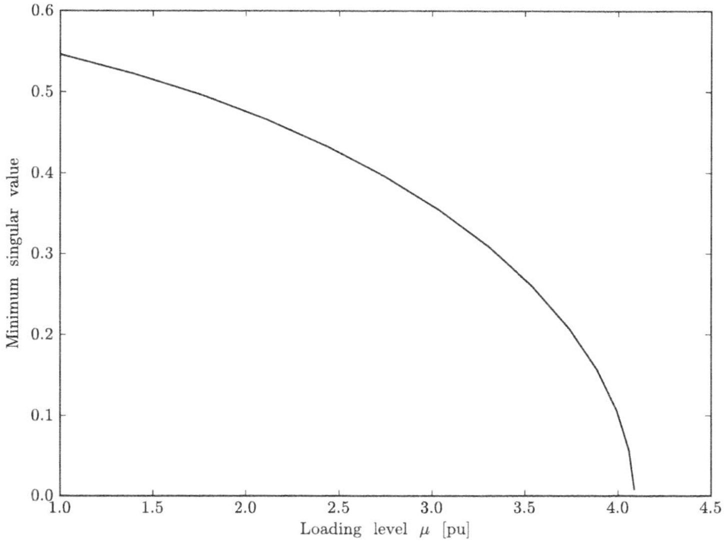

Fig. 7.6 Minimum singular value of the power flow Jacobian matrix computed during the CPF analysis for the IEEE 14-bus system

the saddle-node bifurcation point, hence, the curve shown in Figure 7.6 only gets very close to zero.

7.5 Summary

This section summarizes most relevant concepts related to small-signal stability analysis.

Solver method: There are several methods for computing the eigenvalues of a matrix. Methods that compute all eigenvalues are based on the QR algorithm or on some of its variants, such as the Arnoldi's iteration. In this chapter, three simple methods for computing a reduced number of eigenvalues are described, namely the power method , the inverse iteration and the Rayleigh's iteration. The latter is quite efficient if a good estimation of the eigenvalues of interest is known. In practical applications, it is not necessary to compute all eigenvalues since only the smallest ones are of interest. Furthermore, if one is only interested in knowing if a matrix has a zero eigenvalue or not, computing the minimum singular value problem is generally more efficient than finding the minimum eigenvalue.

Matrix type: Typically, the matrix used for studying the small-signal stability analysis is the state matrix. For ODE systems, the Jacobian $\boldsymbol{f_x}$ coincides with the state matrix. However, for DAE systems, computing the state matrix requires factorizing the algebraic Jacobian matrix $\boldsymbol{g_y}$. To avoid this step, it is also possible to study the complete Jacobian matrix of the DAE system, which has the advantage of being sparser that the state matrix. Finally, the power flow Jacobian matrix is useful for assessing the sensitivities between reactive powers and bus voltages.

Domain type: The typical eigenvalue analysis uses the matrix as is. This leads to the eigenvalues in the S-domain. Using the Z-domain transformation leads to a change of coordinates so that all eigenvalues with negative real part fall inside a unitary circle, while positive real part eigenvalues fall outside the unitary circle. The Z-domain transformation requires the factorization of the state matrix but can be efficient if used in conjunction with a method that computes only a reduced set of eigenvalues since the module of unstable eigenvalues can be easily estimated.

Participation factors: Participation factors are computed using right and left eigenvector matrices and allow defining the participation of each system state variable to each system mode. This information is relevant for synthesizing control systems and for defining sensitivities.

Chapter 8
Time Domain Analysis

This chapter describes numerical integration methods for transient stability analysis. Section 8.1 provides a qualitative justification of the need for numerical integration and describes intrinsic limitations of Lyapunov's direct method. Section 8.2 describes two common models for time domain analysis, namely the current-injection and the power-injection models. Section 8.3 outlines a variety of explicit and implicit numerical methods, paying particular attention to the accuracy and the stability of these methods. Section 8.4 provides a complete numerical integration routine and discusses related issues such as the choice of the step length, disturbances and stop criteria, including the SIME method. Sections 8.5 and 8.6 briefly describes numerical methods for electro-magnetic as well as long-term transients, respectively. Finally Section 8.7 summarizes most relevant concepts given in this Chapter.

8.1 Background

A basic tool of stability analysis is the integration of the initial-value problem

$$\dot{\boldsymbol{x}} = \boldsymbol{f}(\boldsymbol{x}, t), \qquad \boldsymbol{x}(t_0) = \boldsymbol{x}_0 \tag{8.1}$$

where $\boldsymbol{f}(\boldsymbol{x}, t)$ is a set of ordinary differential equations (ODE). The solution of (8.1) is a trajectory $\varphi(\boldsymbol{x}_0, t)$ or, simply, $\boldsymbol{x}(t)$.

As discussed in Chapter 1, the most convenient power system model for transient stability analysis is a set of differential algebraic equations (DAE). Thus, the initial value problem (8.1) becomes:

$$\begin{aligned} \dot{\boldsymbol{x}} &= \boldsymbol{f}(\boldsymbol{x}, \boldsymbol{y}, t), \qquad & \boldsymbol{x}(t_0) &= \boldsymbol{x}_0 \\ \boldsymbol{0} &= \boldsymbol{g}(\boldsymbol{x}, \boldsymbol{y}, t), \qquad & \boldsymbol{y}(t_0) &= \boldsymbol{y}_0 \end{aligned} \tag{8.2}$$

In theory, (8.2) can be transformed in (8.1) if algebraic variables $\boldsymbol{y}$ are explicited using equations $\boldsymbol{g}$.

F. Milano: Power System Modelling and Scripting, Power Systems, pp. 179–218.
springerlink.com

$$\boldsymbol{y} = \tilde{\boldsymbol{g}}(\boldsymbol{x}, t) \tag{8.3}$$

Using (8.3), the DAE system (8.2) can be rewritten as:

$$\dot{\boldsymbol{x}} = \boldsymbol{f}(\boldsymbol{x}, \tilde{\boldsymbol{g}}(\boldsymbol{x}, t), t), \qquad \boldsymbol{x}(t_0) = \boldsymbol{x}_0 \tag{8.4}$$

that is formally the same problem as (8.1).

In the discussion above there are two relevant issues:

1. The initial value problems (8.1) or (8.4) do not generally have an explicit solution due to the nonlinearity of $\boldsymbol{f}$. In other words, to find an analytical expression for the trajectory $\varphi(\boldsymbol{x}, t)$ is generally impossible.
2. The implicit function theorem guarantees the existence of $\tilde{\boldsymbol{g}}$ if $\boldsymbol{g}_{\boldsymbol{y}}$ is not singular. However, to find an analytical expression for (8.3) is generally not possible due to the nonlinearity of $\boldsymbol{g}$.

To overcome these issues, the only solution is to use a numerical integration method. The numerical solution approximates $\boldsymbol{x}(t)$ using a series of discrete values t_i of the independent variable t:

$$\boldsymbol{x}(t_0),\ \boldsymbol{x}(t_1),\ \boldsymbol{x}(t_2),\ \dots \tag{8.5}$$

or, with a compact notation:

$$\boldsymbol{x}_0,\ \boldsymbol{x}_1,\ \boldsymbol{x}_2,\ \dots \tag{8.6}$$

There is a huge variety of numerical integration methods. The most intuitive one is the *explicit forward Euler's method*. At a generic time t_{i+1}, the scheme of the explicit Euler's method is:

$$\boldsymbol{x}_{i+1} = \boldsymbol{x}_i + \Delta t \boldsymbol{f}(\boldsymbol{x}_i, t) \tag{8.7}$$

where Δt is the step length that can be fixed or varied from step to step. Although it worked well for computing planet orbits, the explicit Euler's method can be improved in terms of both accuracy and numerical stability. A small selection of methods that yield an improvement over the basic Euler's scheme is presented in the following Section 8.3.

Assuming that a suitable numerical integration method is used, the importance of time domain simulations for power system analysis is twofold:

1. Assessing the electro-magnetic behavior of power system devices. Relevant studies are transients following line switching operations, symmetrical and asymmetrical faults or imbalanced conditions and power electronics converters [345]. This approach is called *electro-magnetic transient analysis*.
2. Assessing the electro-mechanical response of power system networks following a large disturbance such as line outages or short circuits. This approach is called *transient stability analysis*.

Since the time scales of electro-magnetic and electro-mechanical transients differ at least one order of magnitude (see Figure 1.6 of Chapter 1), the two studies have followed separate development and are generally tackled by different specific software tools. Only in recent years it has been recognized the need for interfacing Electro-Magnetic Transient (EMT) programs and Transient Stability (TS) ones [151]. Although this book focuses on transient stability analysis, HVDC systems and the increasing penetration of renewable energy sources and energy storage systems, most of which in dc (e.g., photovoltaic and fuel cells and battery energy storage systems) lead to the need of interfacing several small dc systems with a large ac network. Thus, Section 8.5 briefly discusses the EMT approach.

The main goal of transient stability analysis is to determine the effects of large disturbances on the dynamic response of a given power system. This problem is conceptually different from the small-signal stability analysis. In fact, the system before and after the disturbance generally shows a stable equilibrium point. In other words, pre- and post-disturbance equilibria exist and all the eigenvalues of the state matrix have negative real part. The issue is that the trajectory of the system following the disturbance is unstable and never reaches the stable post-disturbance equilibrium point.

Thus, the object of transient stability analysis is to determine whether the system trajectory is stable or not. A variety of mathematical and engineering books have been written on this fascinating topic. In this context, it is relevant to summarize the most important solutions that have been proposed.

1. The first solution is to solve the numerical integration and observe the response of the power system. If the time domain simulation diverges, the system is unstable, otherwise it is stable. This approach has the advantage of being "exact". The accuracy only depends on the numerical integration method and on the system model. On the other hand, the numerical integration is computationally demanding, especially taking into account that one has to solve a simulation for each contingency. Although modern computers can solve the numerical integration for real size power systems very quickly, the computational burden of a full contingency analysis is still high.
2. The second solution is based on the *Lyapunov's direct method*. This method attempts to infer the stability of the ODE system (8.1) by building a function $\vartheta(\boldsymbol{x}, \boldsymbol{x}_0) : \mathbb{R}^{n_x} \mapsto \mathbb{R}$ called *Lyapunov's function* able to "measure" the stability of the system. The total energy of the system is a good Lyapunov function and thus the Lyapunov's function is called *Transient Energy Function* (TEF) in most publications on transient stability. The main advantage of the Lyapunov's direct method is that the large disturbance stability of a multi-variable system is reduced to the study of a scalar function. Thus no numerical integration is needed. The main drawbacks are:

a. There is no general systematic method to define the Lyapunov's function $\vartheta(\boldsymbol{x}, \boldsymbol{x}_0)$. As a matter of fact, the TEF can be easily computed only for comparatively simple systems.
b. The stability region of the TEF has to be computed to provide the stability measure. This can be a difficult task for large systems.
c. For systems with losses (e.g., for the totality of real systems) the stability test provided by the TEF is only sufficient, not necessary.
d. An hypothesis of the Lyapunov's direct method is that the structure of the system must not change after a given initial instant t_i. This can be a limiting hypothesis in case one wants to study the effect of corrective actions (e.g., fast valving of synchronous machine turbines).
e. The multi-swing instability phenomenon cannot be taken into account. The multi-swing instability consists in a system that loses synchronism after the first oscillation following a large disturbance. The origin of this phenomenon is intrinsic of the nonlinear system and it has been conjectured to be caused by an unbounded chaotic motion [173].

It is worth observing that all the drawbacks of the TEF are intrinsic of the Lyapunov's direct method and thus can be hardly solved, unless one invents some new theory able to overcome mathematical issues. On the other hand, the main drawback of the numerical time integration is only to be time consuming. Thanks to the enhancement of micro-processors, the computational burden is more and more an irrelevant constraint. For this reason, in this chapter, only numerical integration methods are described. However, the Lyapunov's direct method is so intriguing that it is difficult to resist to the temptation of trying to solve mathematical issues. As a matter of fact, for over three decades there have been attempts to provide suitable procedures for building the Lyapunov's function [20, 57, 98, 228, 229, 257, 328].

For the sake of example, consider the OMIB system described in Section 7.1 of Chapter 7. The OMIB system can be modelled as a two-order ODE:

$$\dot{\delta} = \Omega_b(\omega - \omega_s) \tag{8.8}$$

$$\dot{\omega} = \frac{1}{2H}\left(p_m - \frac{ev}{x_{\rm eq}}\sin\delta\right) \tag{8.9}$$

where all parameters are defined in Section 7.1. The TEF for this system is the sum of the kinetic and potential energy, $\mathscr{E}_K$ and $\mathscr{E}_P$, respectively, of the synchronous machine, as follows:

$$\begin{aligned}\vartheta(\boldsymbol{x}, \boldsymbol{x}_0) &= \vartheta((\delta, \omega), (\delta_0, \omega_0)) \\ &= \mathscr{E}_K + \mathscr{E}_P \\ &= H\omega^2 - \frac{ev}{x_{\rm eq}}(\cos\delta - \cos\delta_0) - p_m(\delta - \delta_0)\end{aligned} \tag{8.10}$$

where $(\delta_0, \omega_0) = (0.5236, 1)$ is the stable equilibrium point discussed in Section 7.1. A three-phase fault occurs at $t_0 = 0.5$ s at one end of the two parallel transmission lines that connect the synchronous machine to the infinite bus. The fault is cleared at $t_0 + t_c$ by disconnecting the faulted transmission line (see Figure 8.1). The parameters of the system are $e = v = p_m = 1$ pu, $H = 8$ MWs/MVA, $\Omega_b = 314.16$ rad/s. The pre-fault equivalent series reactance between buses 1 and 0 is $x_{\mathrm{eq}} = 0.5$ pu, while the post-fault reactance is $x_{\mathrm{eq}} = 0.6$ pu.

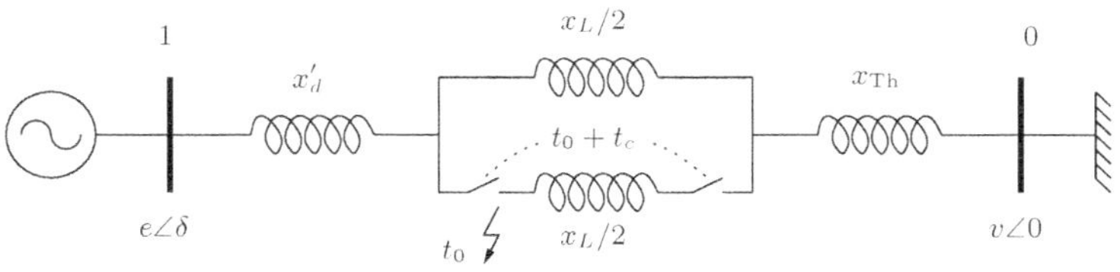

Fig. 8.1 OMIB system with three-phase fault and line outage

A typical didactic problem is to determine the *critical clearing time* t_c, i.e., the maximum time t_c that allows maintaining the synchronism of the synchronous machine. Through numerical integration, the solution can be found using a trial-and-error strategy, as follows.

1. Choose an initial guess for $t_c^{(0)}$.
2. Run the numerical integration.
3. Evaluate $t_c^{(i+1)}$. If the system is stable, t_c is increased, otherwise t_c is decreased. A bisection method can be just fine for choosing the next value of t_c.
4. If $|t_c^{(i+1)} - t_c^{(i)}| < \epsilon$, stop. Otherwise, go back to Step 2.

Figure 8.2 shows some iterations of this procedure. The critical clearing time falls between 0.255 and 0.26 s. Since the time domain simulation provides the state variable trajectory, the value $\delta(t_c)$ of the rotor angle that corresponds to the clearing time t_c is a byproduct of the solution, as shown in Table 8.1.

Table 8.1 Clearing times and angles for the OMIB system

Clearing time t_c [s]	Rotor angle $\delta_c = \delta(t_c)$ [rad]
0.210	0.9619
0.255	1.1686
0.260	1.1939

A similar solution can be obtained through the Lyapunov's direct method. The general procedure of this method is to find the stability boundary of the

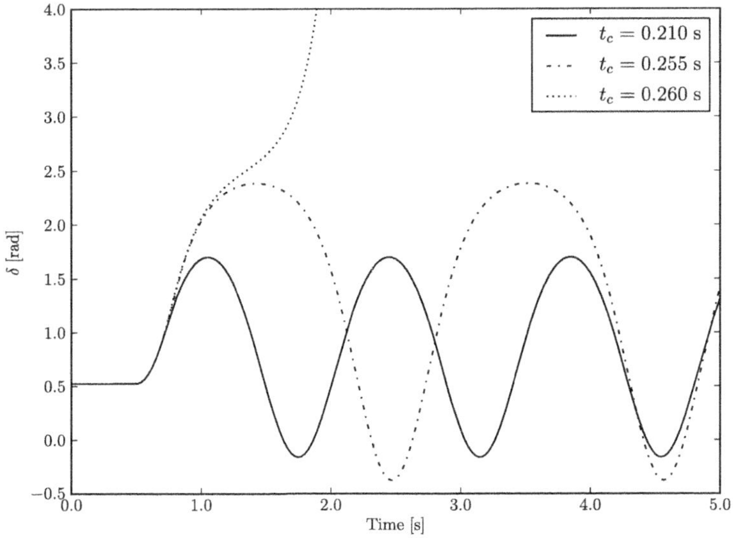

Fig. 8.2 Time domain analysis for the OMIB system

stable equilibrium point. In our case, it is relevant to define the stability region of the post-fault condition. Figure 8.3 shows the potential energy $\mathscr{E}_P(\delta)$ of the post-fault OMIB system (e.g., for $x_{\rm eq} = 0.6$ pu). The minimum of $\mathscr{E}_P(\delta)$ corresponds to the stable equilibrium point $(\tilde{\delta}_0, 1)$, where $\tilde{\delta}_0 = \mathrm{asin}(0.6)$ rad. The stability region is bounded by the maxima of $\mathscr{E}_P(\delta)$, that occur for the unstable equilibrium points $(-\pi - \delta_m, 1)$ and $(\delta_m, 1)$, where $\delta_m = \pi - \tilde{\delta}_0$. Since $\mathscr{E}_{P1} = \mathscr{E}_P(\delta_m) < \mathscr{E}_P(-\pi - \delta_m) = \mathscr{E}_{P2}$, the binding limit is $\mathscr{E}_{P1}$. For loss-less systems, the Lyapunov's direct method states that:

1. If the total system energy at $t_0 + t_c$ is $\mathscr{E} \leq \mathscr{E}_{P1}$, then the system is stable.
2. If the total system energy at $t_0 + t_c$ is greater than $\mathscr{E} > \mathscr{E}_{P1}$, then the system is unstable.

For loss-less systems, the Lyapunov's direct method is both necessary and sufficient.

A simple yet powerful way to apply the Lyapunov's direct method is the equal area criterion (EAC) that is shown in Figure 8.4. The areas in the plane (δ, p) are energies. In particular, the *accelerating area* $A_a = p_m(\delta_c - \delta_0)$ is the energy increase during the fault. In fact, during the fault the electrical power p_e generated by the synchronous machine is zero (the voltage at the short-circuit point is zero). The *decelerating area* $A_d = \int_{\delta_c}^{\delta_m} (p_m - \tilde{p}_e^{\max} \sin\delta) d\delta$ for the post-fault system provides the potential energy that is available for

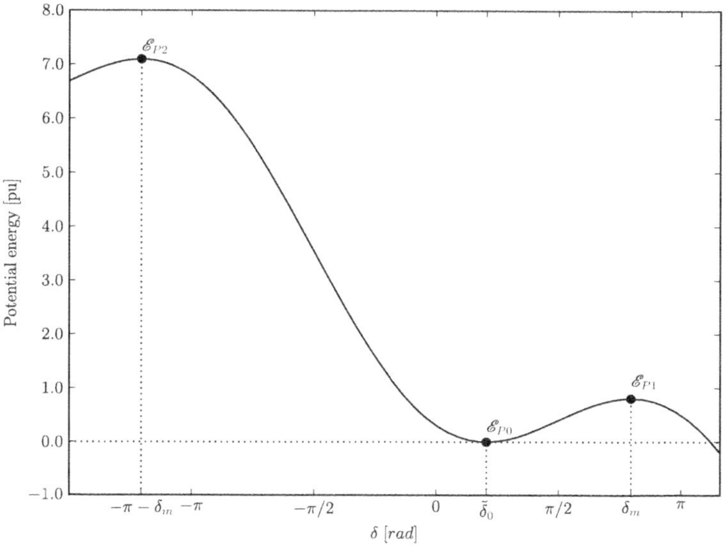

Fig. 8.3 Post-fault potential energy of the OMIB system

compensating the accelerating area A_a. The critical angle δ_c satisfies the condition $A_a = A_d$. In particular one has:

$$\begin{aligned} & A_a = A_d \\ \Rightarrow\ & A_a + A_c = A_d + A_c \\ \Rightarrow\ & p_m(\delta_c - \delta_0) = \int_{\delta_c}^{\delta_m} \tilde{p}_e(\delta)d\delta \\ \Rightarrow\ & \delta_c = 1.1759 \text{ rad} \end{aligned} \tag{8.11}$$

where $\delta_0 = 0.5235$ rad, $\delta_m = 2.4981$ rad and $\tilde{p}_e^{\max} = 1.6667$ pu. The value of the critical angle δ_c confirms the results of the numerical integration.

The following final remarks are relevant.

1. If a damping is included in the machine equations, the Lyapunov's direct method is conclusive only if the energy at $t_0 + t_c$ satisfies $\mathscr{E} \leq \mathscr{E}_{P1}$. If $\mathscr{E} > \mathscr{E}_{P1}$, the Lyapunov direct method is inconclusive or, in mathematical terms, only provides a sufficient stability condition (see Figure 8.5).
2. The critical clearing time is big with respect to typical protection intervention times. Modern protections can detect and open a transmission line in about 4 four cycles (80 ms at 50 Hz) which is well lower then typical values of critical clearing times (about 200 ms). Thus, only in case primary protections fail to clear the fault, there can be a real risk of transient instability. As a consequence, transient instability has become a relatively rare event in the last decades.

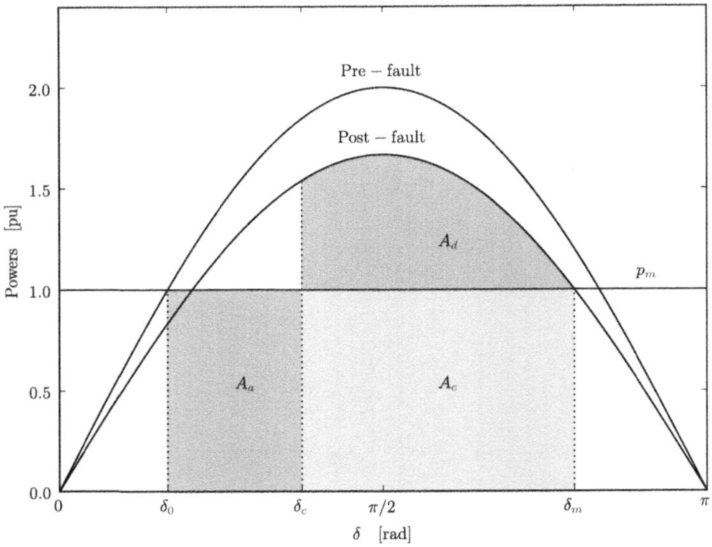

Fig. 8.4 Equal area criterion for the OMIB system

For the reasons above, the interest in TEF has somewhat decreased in recent years. Unfortunately other instability phenomena, included but not limited to voltage and frequency instability, are still a major system operator concern. Numerical integration can tackle any kind of instability phenomena, not just the loss of synchronism of synchronous machines. Thus, numerical integration is and will likely always be the workhorse of any power system stability analysis. The challenge is how to improve efficiency and/or accuracy of numerical methods.

8.2 Power System Model

As discussed in Chapter 1, the power system model used for time domain analysis is a set of nonlinear differential algebraic equations (DAE) with discrete variables:

$$\begin{aligned} \dot{\boldsymbol{x}} &= \boldsymbol{f}(\boldsymbol{x}, \boldsymbol{y}, \boldsymbol{\eta}, \boldsymbol{u}, t) \\ \boldsymbol{0} &= \boldsymbol{g}(\boldsymbol{x}, \boldsymbol{y}, \boldsymbol{\eta}, \boldsymbol{u}, t) \end{aligned} \tag{8.12}$$

If discrete variables $\boldsymbol{u}$ are substituted for if-then rules, (8.12) becomes an *hybrid dynamical system*, e.g., a collection of continuous DAE, one per each

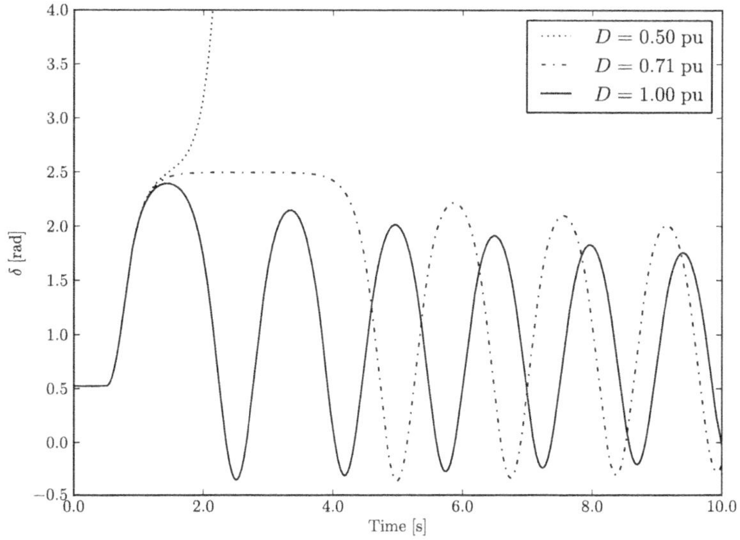

Fig. 8.5 Time domain analysis for the OMIB system with damping. The clearing time is $t_c = 0.26$ s for all simulations

discrete variable change [131].[1] Thus, one can assume that the problem to be solved for each set of discrete variables is a problem similar to (8.2).

In (8.12), differential equations $\boldsymbol{f}$ depends on machine, regulator and load models. On the other hand, the form of algebraic equations in (8.12) can be twofold, namely (i) current-injection model and (ii) power-injection model. These two models are described in the following subsections.

8.2.1 Current-Injection Model

The classical and most common model of algebraic equations for transient stability analysis is the current-injection one [298]. According to this model, the algebraic variables are bus voltage phasors $\bar{\boldsymbol{v}}$ and the algebraic equations express the current injections at network buses:

$$\begin{aligned} \dot{\boldsymbol{x}} &= \boldsymbol{f}(\boldsymbol{x}, \bar{\boldsymbol{v}}) \\ \mathbf{0} &= \bar{\boldsymbol{i}}(\boldsymbol{x}, \bar{\boldsymbol{v}}) - \bar{\boldsymbol{Y}}(\boldsymbol{x})\bar{\boldsymbol{v}} \end{aligned} \tag{8.13}$$

[1] The notation "hybrid system" has not to be confused with *hybrid transient simulator* that is used in the literature on time domain integration of power systems for indicating tools that integrate together EMT and TS analyses [151].

Under certain hypothesis, it is possible that the admittance matrix $\bar{\boldsymbol{Y}}$ does not depend on state variables. For example by modelling dynamic series devices (e.g., regulating transformers or FACTS devices) as current injections at the sending and receiving buses, respectively. In this case the only links among dynamic devices are the algebraic variables $\bar{\boldsymbol{v}}$ through the admittance matrix $\bar{\boldsymbol{Y}}$. For this reason, $\bar{\boldsymbol{v}}$ are also sometimes called *aggregation variables*.

Constant admittance loads allows simplifying (8.13) since, for a constant admittance, the correspondent element of the vector $\bar{\boldsymbol{\imath}}(\boldsymbol{x}, \bar{\boldsymbol{v}})$ is zero.[2] Thus, ordering the vector of bus voltages into a vector of generator bus voltages $\bar{\boldsymbol{v}}_G$ and a vector load bus voltages $\bar{\boldsymbol{v}}_L$, the algebraic equations in (8.13) becomes:

$$
\begin{aligned}
\mathbf{0} &= \bar{\boldsymbol{\imath}}_G(\boldsymbol{x}, \bar{\boldsymbol{v}}) + \bar{\boldsymbol{Y}}_{GG}\bar{\boldsymbol{v}}_G + \bar{\boldsymbol{Y}}_{GL}\bar{\boldsymbol{v}}_L \\
\mathbf{0} &= \bar{\boldsymbol{Y}}_{LG}\bar{\boldsymbol{v}}_G + \bar{\boldsymbol{Y}}_{LL}\bar{\boldsymbol{v}}_L
\end{aligned}
\tag{8.14}
$$

where:

$$
\bar{\boldsymbol{Y}} = \begin{bmatrix} \bar{\boldsymbol{Y}}_{GG} & \bar{\boldsymbol{Y}}_{GL} \\ \bar{\boldsymbol{Y}}_{LG} & \bar{\boldsymbol{Y}}_{LL} \end{bmatrix}
\tag{8.15}
$$

Thus, load bus voltages can be eliminated from (8.14):

$$
\begin{aligned}
\bar{\boldsymbol{v}}_L &= -\bar{\boldsymbol{Y}}_{LL}^{-1}\bar{\boldsymbol{Y}}_{LG}\bar{\boldsymbol{v}}_G \\
\Rightarrow \quad \mathbf{0} &= \bar{\boldsymbol{\imath}}_G(\boldsymbol{x}, \bar{\boldsymbol{v}}) + (\bar{\boldsymbol{Y}}_{GG} - \bar{\boldsymbol{Y}}_{GL}\bar{\boldsymbol{Y}}_{LL}^{-1}\bar{\boldsymbol{Y}}_{LG})\bar{\boldsymbol{v}}_G
\end{aligned}
\tag{8.16}
$$

Finally, by defining a reduced admittance matrix $\bar{\boldsymbol{Y}}_r$ as:

$$
\bar{\boldsymbol{Y}}_r = \bar{\boldsymbol{Y}}_{GG} - \bar{\boldsymbol{Y}}_{GL}\bar{\boldsymbol{Y}}_{LL}^{-1}\bar{\boldsymbol{Y}}_{LG}
\tag{8.17}
$$

the system (8.13) becomes:

$$
\begin{aligned}
\dot{\boldsymbol{x}} &= \boldsymbol{f}(\boldsymbol{x}, \bar{\boldsymbol{v}}_G) \\
\mathbf{0} &= \bar{\boldsymbol{\imath}}_G(\boldsymbol{x}, \bar{\boldsymbol{v}}) + \bar{\boldsymbol{Y}}_r\bar{\boldsymbol{v}}_G
\end{aligned}
\tag{8.18}
$$

The latter model is the most commonly used in transient stability analysis, especially in proprietary software packages. The advantage of this formulation is that the order of algebraic equations is consistently reduced with respect to the full system size since generator buses are much less than load and transit nodes. Furthermore, algebraic equations are linear and most elements of the reduced admittance matrix are constant (although they can vary due to line outages, fault occurrences and load shedding).

The current-injection model (8.18) is a standard *de facto* for transient stability analysis. However, any model, even the most well-accepted one, is

[2] Observe that pure transit nodes are a special case of loads with a zero constant admittance. In the following, pure transit nodes are implicitly considered constant admittance loads.

simply a model, subjected to hypothesis. The most restrictive hypothesis that leads to (8.18) is to assume constant impedance loads. This hypothesis is reasonable only if the time frame is that of transient stability (e.g., few seconds following a short circuit occurrence). In fact, load controls (e.g., tap changer voltage regulation) can be considered frozen for the few seconds following a large disturbance. On the other hand, voltage and frequency stability analyses require detailed models of dynamic loads and their controls [146]. Constant impedance loads are also inadequate for long-term voltage and frequency stability analyses. As a matter of fact, for long-term analysis, tap changer voltage control allows modelling loads as constant power consumptions.

8.2.2 Power-Injection Model

The power injection mode is obtained from (8.13) by multiplying the conjugate of algebraic equations by bus voltages:

$$\begin{aligned} \dot{\boldsymbol{x}} &= \boldsymbol{f}(\boldsymbol{x}, \bar{\boldsymbol{v}}) \\ \boldsymbol{0} &= \bar{\boldsymbol{V}}\bar{\boldsymbol{i}}^*(\boldsymbol{x}, \bar{\boldsymbol{v}}) - \bar{\boldsymbol{V}}\bar{\boldsymbol{Y}}^*(\boldsymbol{x})\bar{\boldsymbol{v}}^* \end{aligned} \tag{8.19}$$

where $\bar{\boldsymbol{V}} = \mathrm{diag}(\bar{v}_1, \bar{v}_2, \ldots, \bar{v}_{n_b})$. The term $\bar{\boldsymbol{V}}\bar{\boldsymbol{Y}}^*(\boldsymbol{x})\bar{\boldsymbol{v}}^*$ are the power flow equations, while $\bar{\boldsymbol{V}}\bar{\boldsymbol{i}}^*(\boldsymbol{x}, \bar{\boldsymbol{v}})$ are the complex powers injected at network buses. Thus, (8.19) can be rewritten as:

$$\begin{aligned} \dot{\boldsymbol{x}} &= \boldsymbol{f}(\boldsymbol{x}, \bar{\boldsymbol{v}}) \\ \boldsymbol{0} &= \bar{\boldsymbol{s}}(\boldsymbol{x}, \bar{\boldsymbol{v}}) - \bar{\boldsymbol{V}}\bar{\boldsymbol{Y}}^*(\boldsymbol{x})\bar{\boldsymbol{v}}^* \end{aligned} \tag{8.20}$$

Equations (8.20) are equivalent to (8.13) but are intrinsically nonlinear and are thus computationally more demanding than (8.13).

Another issue of writing algebraic equations in terms of power injections is numerical. Assume that, as a consequence of a short-circuit, some bus voltage magnitudes become zero. After clearing the fault, voltage magnitudes recover positive values. However, if one uses a Newton's method to solve algebraic equations, one has, at a certain bus h where $v_h = 0$:

$$0 = v_h e^{j\theta_h} \left(\bar{i}_h^*(\boldsymbol{x}, \bar{\boldsymbol{v}}) - \sum_{k=1}^{n} \bar{y}_{hk}^* \bar{v}_k \right) = v_h \kappa(\boldsymbol{z}) \tag{8.21}$$

where $\boldsymbol{z} = [\boldsymbol{x}^T, \boldsymbol{v}^T, \boldsymbol{\theta}^T]^T$. The Newton's equation for (8.21) at a generic iteration i is:

$$v_h^{(i)} \kappa(\boldsymbol{z}^{(i)}) = \kappa^{(i)}(\boldsymbol{z}^{(i)}) \Delta v_h^{(i)} + v_h \left(\sum_{\ell=1}^{n_z} \frac{\partial \kappa^{(i)}}{\partial z_\ell} \Delta z_\ell^{(i)} \right) \tag{8.22}$$

If $v_h^{(i)} = 0$ at a certain iteration i, then (8.22) becomes:

$$0 = \kappa^{(i)}(\boldsymbol{z}^{(i)})\Delta v_h^{(i)} \tag{8.23}$$

which lead to $\Delta v_h^{(i)} = 0$. In other words, if $v_h^{(i)} = 0$ at the iteration i, it will remain zero for all the following iterations. A very small value of $v_h^{(i)}$ would also show a similar numerical issue. Furthermore, $\boldsymbol{v} = \boldsymbol{0}$ is a solution of algebraic equations in (8.20). Thus, one has to carefully avoid that voltages become zero (or very small values) at any iteration, otherwise, the Newton's method is not able to recover voltage magnitudes.

In (8.20), the only algebraic variables are bus voltage magnitudes and phase angles. A more general and flexible model includes additional algebraic variables $\hat{\boldsymbol{y}}$, algebraic equations $\hat{\boldsymbol{g}}$ and controllable parameters $\boldsymbol{\eta}$:

$$\begin{aligned} \dot{\boldsymbol{x}} &= \boldsymbol{f}(\boldsymbol{x}, \hat{\boldsymbol{y}}, \bar{\boldsymbol{v}}, \boldsymbol{\eta}) \\ \boldsymbol{0} &= \hat{\boldsymbol{g}}(\boldsymbol{x}, \hat{\boldsymbol{y}}, \bar{\boldsymbol{v}}, \boldsymbol{\eta}) \\ \boldsymbol{0} &= \bar{\boldsymbol{s}}(\boldsymbol{x}, \hat{\boldsymbol{y}}, \bar{\boldsymbol{v}}, \boldsymbol{\eta}) - \bar{\boldsymbol{V}}\bar{\boldsymbol{Y}}^*(\boldsymbol{x}, \hat{\boldsymbol{y}}, \boldsymbol{\eta})\bar{\boldsymbol{v}}^* \end{aligned} \tag{8.24}$$

The following conclusive remarks are relevant:

1. Both (8.13) and (8.24) are nonlinear. In fact, (8.13) is nonlinear at least in the differential equations of synchronous machines. Thus, if an implicit solution method is used for the numerical integration, both (8.13) and (8.24) requires a Newton's method for solving a set of nonlinear equations at each integration step.
2. Equations (8.24) have the advantage of requiring the same model for power flow, continuation power flow and time domain analyses. In other words, (8.24) allow using an unique structure for all devices (there is no difference between power flow and time domain analysis models of the same device) and writing more compact code (the same algebraic equations are used for both static and dynamic analysis).

Example 8.1 OMIB Differential Algebraic Equations

This example provides the power-injection as well as the current-injection model (8.24) and (8.13), respectively, for the 2-bus system depicted in Figure 8.6. Bus 0 is an infinite bus where the voltage $\bar{v}_0 = v_0\angle\theta_0$ is constant, while the machine is a two-order machine model. Assuming that the voltage at the infinite bus is a parameter, the system power-injection model is:

$$\boldsymbol{f} \Rightarrow \begin{cases} \dot{\delta} = \omega_n(\omega - \omega_s) \\ \dot{\omega} = (p_m - p_e - D(\omega - \omega_s))/2H \end{cases} \tag{8.25}$$

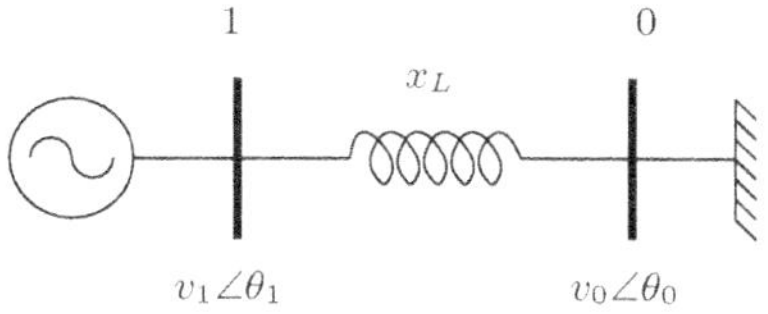

Fig. 8.6 OMIB system

$$\hat{\boldsymbol{g}} \Rightarrow \begin{cases} 0 = (v_q + r_a i_q) i_q + (v_d + r_a i_d) i_d - p_e \\ 0 = v_q + r_a i_q - e'_q + x'_d i_d \\ 0 = v_d + r_a i_d - x'_d i_q \\ 0 = v_1 \sin(\delta - \theta_1) - v_d \\ 0 = v_1 \cos(\delta - \theta_1) - v_q \end{cases}$$

$$\bar{\boldsymbol{s}} \Rightarrow \begin{cases} 0 = v_d i_d + v_q i_q - \frac{v_1 v_0}{x_L} \sin(\theta_1 - \theta_0) \\ 0 = v_q i_d - v_d i_q - \frac{v_1^2}{x_L} + \frac{v_1 v_0}{x_L} \cos(\theta_1 - \theta_0) \end{cases}$$

The current-injection model can be obtained by substituting $\bar{\boldsymbol{s}}$ in (8.25) with the current injection:

$$\bar{\boldsymbol{i}}_G \quad \Rightarrow \quad 0 = (i_d + j i_q) - \frac{1}{j x_L}(\bar{v}_1 - \bar{v}_0) \tag{8.26}$$

In (8.26), the generator current $i_d + j i_q$ can be substituted by explicit functions of $\boldsymbol{x}$ and $\bar{v}_1$. From (8.25), one obtains:

$$\begin{bmatrix} i_d \\ i_q \end{bmatrix} = \frac{1}{{x'_d}^2 + r_a^2} \begin{bmatrix} x'_d & r_a \\ r_a & -x'_d \end{bmatrix} \begin{bmatrix} e'_q - v_q \\ -v_d \end{bmatrix} \tag{8.27}$$

where v_d and v_q are functions of the rotor angle δ and of the bus voltage $\bar{\boldsymbol{v}}_1$. Assuming $r_a \approx 0$, which is a common hypothesis for the classical machine model, (8.27) can be further simplified as:

$$\begin{aligned} i_d &= (e'_q - v_q)/x'_d &= (e'_q - v_1 \cos(\delta - \theta_1))/x'_d \\ i_q &= v_d / x'_d &= v_1 \sin(\delta - \theta_1)/x'_d \end{aligned} \tag{8.28}$$

and the electrical power p_e becomes:

$$p_e = \frac{e'_q v_1}{x'_d} \sin(\delta - \theta_1) \tag{8.29}$$

In summary, the simplified current-injection model is:

$$\boldsymbol{f} \Rightarrow \begin{cases} \dot{\delta} = \omega_n (\omega - \omega_s) \\ \dot{\omega} = (p_m - \frac{e'_q v_1}{x'_d} \sin(\delta - \theta_1) - D(\omega - \omega_s))/2H \end{cases} \tag{8.30}$$

$$\bar{\boldsymbol{i}}_G \Rightarrow \begin{cases} 0 = \frac{1}{x'_d}(e'_q - v_1 \cos(\delta - \theta_1) + j v_1 \sin(\delta - \theta_1)) - \frac{1}{j x_L}(\bar{v}_1 - \bar{v}_0) \end{cases}$$

8.3 Numerical Integration Methods

In order to numerically integrate (8.2), the first issue that has to be solved is how to handle algebraic equations $\boldsymbol{g}$. There are mainly two approaches:

1. *Partitioned-solution approach.* Variables $\boldsymbol{x}$ and $\boldsymbol{y}$ are updated sequentially.
2. *Simultaneous-solution approach.* Variables $\boldsymbol{x}$ and $\boldsymbol{y}$ are solved together in a unique step using a solver such as the Newton's method.

As usual, both approaches have advantages and drawbacks.

In the partitioned approach, since $\boldsymbol{x}$ and $\boldsymbol{y}$ are updated independently, any numerical integration method can be used. However, the sequential approach is typically used combined with *explicit* numerical methods (e.g., Runge-Kutta's formulæ) that do not require computing and factorizing the Jacobian matrix $\boldsymbol{f}_{\boldsymbol{x}}$. On the other hand, the partitioned approach introduces a "delay" between $\boldsymbol{x}$ and $\boldsymbol{y}$. In fact, for a generic step i, while computing $\boldsymbol{x}^{(i+1)}$, algebraic variables are frozen to the old value $\boldsymbol{y}^{(i)}$. Moreover, the state variables $\boldsymbol{x}^{(i+1)}$ are not modified when computing $\boldsymbol{y}^{(i+1)}$. To avoid the delay between $\boldsymbol{x}^{(i)}$ and $\boldsymbol{y}^{(i)}$, one has to iterate over $\boldsymbol{x}^{(i)}$ and $\boldsymbol{y}^{(i)}$ for each time step. This process can lead to numerical instabilities. It has to be noted that solving $\boldsymbol{g} = \boldsymbol{0}$ for updating algebraic variables requires the solution of a nonlinear system, which generally requires computing and factorizing iteratively the Jacobian matrix $\boldsymbol{g}_{\boldsymbol{y}}$. At this aim, to reduce the computational effort, one can use a Newton's dishonest method as described in Subsection 4.4.6 of Chapter 4. In case the Jacobian matrix of algebraic equations is kept constant for multiple integration time steps, the method is called *very dishonest Newton's method* [19].

The simultaneous approach has the advantage that $\boldsymbol{x}^{(i)}$ and $\boldsymbol{y}^{(i)}$ are updated together, thus no delay is introduced. This approach is used in conjunction with *implicit* numerical methods that require, at each time step, the solution of a set of nonlinear equations. This solution is generally obtained through a Newton's method. Thus, the simultaneous approach requires iteratively computing and factorizing an $(n_x + n_y) \times (n_x + n_y)$ Jacobian matrix.

In conclusion, the partitioned approach can be considered faster but less numerically stable than the simultaneous approach. Further details are given in the following subsections.

8.3.1 Explicit Methods

Multi-stage explicit methods can be expressed using a m-stage formula in the form [276]:

$$\begin{aligned} \boldsymbol{x}(t+\Delta t) &= \boldsymbol{x}(t) + \Delta t \sum_{k=0}^{m-1} c_k \boldsymbol{f}^{\{k\}} \\ \hat{\boldsymbol{x}}(t+\Delta t) &= \boldsymbol{x}(t) + \Delta t \sum_{k=0}^{m} \hat{c}_k \boldsymbol{f}^{\{k\}} \end{aligned} \tag{8.31}$$

where t is the current integration time and Δt is the *step length* and:

$$\begin{aligned} \boldsymbol{f}^{\{0\}} &= \boldsymbol{f}(\boldsymbol{x}(t), t_i) \\ \boldsymbol{f}^{\{k\}} &= \boldsymbol{f}(\boldsymbol{x}(t) + \Delta t \sum_{j=0}^{k-1} b_{kj} \boldsymbol{f}^{\{j\}}, t_i + a_k \Delta t) \end{aligned} \tag{8.32}$$

Both $\boldsymbol{x}(t+\Delta t)$ and $\hat{\boldsymbol{x}}(t+\Delta t)$ approximate the exact solution and $\hat{\boldsymbol{x}}(t+\Delta t)$ is an approximation of higher order than $\boldsymbol{x}(t+\Delta t)$. The difference $\boldsymbol{x}(t+\Delta t) - \hat{\boldsymbol{x}}(t+\Delta t)$ allows estimating the error with respect to the exact solution and adjusting the step length Δt.

The m-stage formula (8.31) evaluates the new value of the state variables $\boldsymbol{x}(t+\Delta t)$ using a weighted sum of m values of $\dot{\boldsymbol{x}}$ at suitable points between t and $t+\Delta t$.

A convenient way of visualizing the formulæ (8.31) for a given method is through the Butcher's tableau [34], as follows:

$$\begin{array}{c|ccccc} 0 & & & & & \\ a_1 & b_{10} & & & & \\ a_2 & b_{20} & b_{21} & & & \\ \vdots & \vdots & \vdots & \ddots & & \\ a_m & b_{m0} & b_{m1} & \dots & b_{m,m-1} & \\ \hline & c_0 & c_1 & \dots & c_{m-1} & \\ & \hat{c}_0 & \hat{c}_1 & \dots & \hat{c}_{m-1} & \hat{c}_m \end{array} \tag{8.33}$$

Using (8.31), a huge variety of explicit methods can be defined, including the large family of Runge-Kutta's formulæ.

Multi-step or *predictor-corrector* methods are another class of explicit methods. The general formula of a multi-step method is:

$$\boldsymbol{x}(t+\Delta t) = \boldsymbol{x}(t) + \Delta t \sum_{k=0}^{m} c_k \boldsymbol{f}(\boldsymbol{x}(t-k\Delta t), t-k\Delta t) \tag{8.34}$$

A drawback of multi-step methods is that they are not *self-starting*, since the first m steps has to be known to compute the generic step (8.34) for $t+\Delta t$. The well-known Adams-Bashforth's method, Milne-Simpson's method and Hamming's method belong to the family of multi-step methods [34]. However, these methods have proved to be less accurate and efficient then Runge-Kutta's formulæ, at least for power system applications [163].

Example 8.2 Runge-Kutta's Formulæ

Example 4.6 presented the classical 4th order Runge-Kutta's formula:

$$\begin{aligned} \boldsymbol{f}^{\{0\}} &= \boldsymbol{f}(\boldsymbol{x}(t)) \\ \boldsymbol{f}^{\{1\}} &= \boldsymbol{f}(\boldsymbol{x}(t) + 0.5\Delta t \boldsymbol{f}^{\{0\}}) \\ \boldsymbol{f}^{\{2\}} &= \boldsymbol{f}(\boldsymbol{x}(t) + 0.5\Delta t \boldsymbol{f}^{\{1\}}) \\ \boldsymbol{f}^{\{3\}} &= \boldsymbol{f}(\boldsymbol{x}(t) + \Delta t \boldsymbol{f}^{\{2\}}) \\ \Rightarrow \quad \boldsymbol{x}(t+\Delta t) &= \boldsymbol{x}(t) + \Delta t(\boldsymbol{f}^{\{0\}} + 2\boldsymbol{f}^{\{1\}} + 2\boldsymbol{f}^{\{2\}} + \boldsymbol{f}^{\{3\}})/6 \end{aligned} \tag{8.35}$$

The RK4 is represented by the following Butcher's tableau:

$$\begin{array}{c|cccc} 0 & & & & \\ \frac{1}{2} & \frac{1}{2} & & & \\ \frac{1}{2} & 0 & \frac{1}{2} & & \\ 1 & 0 & 0 & 1 & \\ \hline & \frac{1}{6} & \frac{2}{6} & \frac{2}{6} & \frac{1}{6} \end{array} \tag{8.36}$$

Several more sophisticated schemes have been proposed [117]. For example, the Runge-Kutta-Fehlberg's formula has the following Butcher's tableau:

$$\begin{array}{c|cccccc} 0 & & & & & & \\ \frac{1}{4} & \frac{1}{4} & & & & & \\ \frac{3}{8} & \frac{3}{32} & \frac{9}{32} & & & & \\ \frac{12}{13} & \frac{1932}{2197} & -\frac{7200}{2197} & \frac{7296}{2197} & & & \\ 1 & \frac{439}{216} & -8 & \frac{3680}{513} & -\frac{845}{4104} & & \\ \frac{1}{2} & -\frac{8}{27} & 2 & -\frac{3544}{2565} & \frac{1859}{4104} & -\frac{11}{40} & \\ \hline & \frac{25}{216} & 0 & \frac{1408}{2565} & \frac{2197}{4104} & -\frac{1}{5} & \\ & \frac{16}{135} & 0 & \frac{6656}{12825} & \frac{28561}{56430} & -\frac{9}{50} & \frac{2}{55} \end{array} \tag{8.37}$$

Example 8.3 Modified Euler's Method

The modified Euler's method is the simplest multi-step method and is the only that has been widely used in power system analysis [163, 291]. It is composed of a predictor and a corrector step, as follows:

$$\begin{aligned}\tilde{\boldsymbol{x}}(t+\Delta t) &= \boldsymbol{x}(t) + \Delta t \boldsymbol{f}(\boldsymbol{x}(t), t) \\ \boldsymbol{x}(t+\Delta t) &= \boldsymbol{x}(t) + \frac{1}{2}\Delta t(\boldsymbol{f}(\boldsymbol{x}(t), t) + \boldsymbol{f}(\tilde{\boldsymbol{x}}(t+\Delta t), t))\end{aligned} \tag{8.38}$$

The accuracy of this method can be improved by assigning $\boldsymbol{x}(t) \leftarrow \boldsymbol{x}(t+\Delta t)$ and repeating the two-step formula (8.38).

8.3.2 Implicit Methods

In transient stability analysis, one of the possible issues is that time constant can span various time scales. For example if both transient stability and long term dynamics are considered together, time constants vary between 10^{-2} and 10^3 s (see Figure 1.6 of Chapter 1). Similarly, if one considers both sub-synchronous resonance phenomena and transient stability, the time scale range is between 10^{-4} and 10^1 s. If the time scale range of a ODE problem is "big", the ODE problem is said to be *stiff*.

The behavior of numerical methods on stiff ODE problems can be analyzed by applying these methods to the *test equation*:

$$\dot{x} = kx, \quad k \in \mathbb{C} \tag{8.39}$$

The solution of (8.39) is $x(t) = e^{kt}$ that approaches zero as $t \to \infty$ when $\Re\{k\} < 0$, i.e., the left-half of the complex plane is the stability region of the test equation (8.39). If the numerical method also exhibits this behavior, then the method is said to be *absolute stable* or A-stable according to the Dahlquist's definition [69]. For example, applying the Runge-Kutta's formulæ to the test equation (8.39), one has:

$$x(t+\Delta t) = \chi(k\Delta t)x(t) = \chi^n(k\Delta t)x(t_0) \tag{8.40}$$

where n is the number of steps and $x(t_0)$ is the initial value. The function $\chi(k\Delta t)$ is called *stability function* and must be $|\chi(k\Delta t)| < 1$ to satisfy $x(t) \to 0$ for $n \to \infty$.

An interesting result of Dahlquist's theorems is that an explicit multi-step method cannot be A-stable [69, 165, 316]. Thus, explicit methods are expected to provide poor behavior for stiff ODE problems. On the other hand, implicit methods can be A-stable. For this reason, and because implicit methods allow a simultaneous solution of both state and algebraic variables in DAE problems, implicit methods are of particular relevance for power system analysis.

When using implicit methods, each step of the numerical integration is obtained as the solution of a set of nonlinear equations. Thus, implicit methods are particularly suited for nonlinear DAE systems, since the algebraic equations can be included in the nonlinear system to be solved at each iteration.

For a generic time t, and assumed a step length Δt, one has to solve the following problem [30]:

$$\begin{aligned} \mathbf{0} &= \hat{\boldsymbol{q}}(\boldsymbol{x}(t+\Delta t), \boldsymbol{y}(t+\Delta t), \boldsymbol{f}(t)) \\ \mathbf{0} &= \boldsymbol{g}(\boldsymbol{x}(t+\Delta t), \boldsymbol{y}(t+\Delta t)) \end{aligned} \tag{8.41}$$

where $\boldsymbol{f}$ and $\boldsymbol{g}$ are the differential and algebraic equations and $\hat{\boldsymbol{q}}$ is a function that depends on the implicit numerical method. Equations (8.41) are nonlinear and their solution is obtained by means of a Newton's method, which in turn consists of computing iteratively the increments $\Delta\boldsymbol{x}^{(i)}$ and $\Delta\boldsymbol{y}^{(i)}$ of the state and algebraic variables and updating the actual variables:

$$\begin{aligned} \begin{bmatrix} \Delta\boldsymbol{x}^{(i)} \\ \Delta\boldsymbol{y}^{(i)} \end{bmatrix} &= -[\boldsymbol{A}_c^{(i)}]^{-1} \begin{bmatrix} \hat{\boldsymbol{q}}^{(i)} \\ \boldsymbol{g}^{(i)} \end{bmatrix} \\ \begin{bmatrix} \boldsymbol{x}^{(i+1)}(t+\Delta t) \\ \boldsymbol{y}^{(i+1)}(t+\Delta t) \end{bmatrix} &= \begin{bmatrix} \boldsymbol{x}^{(i)}(t+\Delta t) \\ \boldsymbol{y}^{(i)}(t+\Delta t) \end{bmatrix} + \begin{bmatrix} \Delta\boldsymbol{x}^{(i)} \\ \Delta\boldsymbol{y}^{(i)} \end{bmatrix} \end{aligned} \tag{8.42}$$

where $\boldsymbol{A}_c^{(i)}$ is a matrix depending on the algebraic and state Jacobian matrices of the system. Some examples of implicit method formulæ are given in the following examples. Most techniques described in Chapter 4 can be applied to (8.42). For example, the dishonest or very dishonest Newton's methods can be useful to reduce the number of factorizations of $\boldsymbol{A}_c^{(i)}$ and speed up the simulation.

Example 8.4 Backward Euler's Method

The backward Euler's method is a first order implicit method. It is generally faster but less accurate than the trapezoidal method that is discussed in the next example. At a generic time step $t + \Delta t$ and a generic iteration i, $\boldsymbol{A}_c^{(i)}$ and $\hat{\boldsymbol{q}}^{(i)}$ are as follows:

$$\begin{aligned} \boldsymbol{A}_c^{(i)} &= \begin{bmatrix} \boldsymbol{I}_{n_x} - \Delta t \boldsymbol{f}_{\boldsymbol{x}}^{(i)} & -\Delta t \boldsymbol{f}_{\boldsymbol{y}}^{(i)} \\ \boldsymbol{g}_{\boldsymbol{x}}^{(i)} & \boldsymbol{g}_{\boldsymbol{y}}^{(i)} \end{bmatrix} \\ \hat{\boldsymbol{q}}^{(i)} &= \boldsymbol{x}^{(i)}(t+\Delta t) - \boldsymbol{x}(t) - \Delta t \boldsymbol{f}^{(i)} \end{aligned} \tag{8.43}$$

where $\boldsymbol{I}_{n_x}$ is the identity matrix of the same dimension of the dynamic order of the DAE system and all Jacobian matrices and $\boldsymbol{f}^{(i)}$, are computed at the current point $(\boldsymbol{x}^{(i)}(t+\Delta t), \boldsymbol{y}^{(i)}(t+\Delta t), t+\Delta t)$.

Example 8.5 Trapezoidal Method

The Crank-Nicolson's or trapezoidal method is the workhorse solver for electro-mechanical DAE, and is widely used, in a variety of flavors, in most commercial and non-commercial power system software packages. The implicit version of the trapezoidal method has proved to be very robust and reliable for a variety of stiff ODE and DAE systems. At a generic iteration i, $\boldsymbol{A}_c^{(i)}$ and $\hat{\boldsymbol{q}}^{(i)}$ are as follows:[3]

$$\boldsymbol{A}_c^{(i)} = \begin{bmatrix} \boldsymbol{I}_{n_x} - 0.5\Delta t \boldsymbol{f}_{\boldsymbol{x}}^{(i)} & -0.5\Delta t \boldsymbol{f}_{\boldsymbol{y}}^{(i)} \\ \boldsymbol{g}_{\boldsymbol{x}}^{(i)} & \boldsymbol{g}_{\boldsymbol{y}}^{(i)} \end{bmatrix} \tag{8.44}$$

$$\hat{\boldsymbol{q}}^{(i)} = \boldsymbol{x}^{(i)} - \boldsymbol{x}(t) - 0.5\Delta t(\boldsymbol{f}^{(i)} + \boldsymbol{f}(t))$$

where the notation is the same as in (8.43).

Example 8.6 Rosenbrock's Semi-Implicit Method

Dahlquist's theorems discourage from setting up implicit methods of order greater than 2 [165]. However, an interesting method for improving the accuracy of an implicit method is proposed in [316]. This reference proposes *semi-implicit* methods for solving ODE systems which are A-stable and avoid the need of iterating. However, semi-implicit methods applied to nonlinear DAE systems still require iterating and factorizing the matrix $\boldsymbol{A}_c^{(i)}$ at each iteration.

A well-known semi-implicit method is the one based on Rosenbrock's formulæ. For example, a 4th order Rosenbrock's formula is as follows. At a generic time t and iteration i, the matrix $\boldsymbol{A}_c^{(i)}$ is the same as in (8.44). The variables $\boldsymbol{x}^{(i)}$ and $\boldsymbol{y}^{(i)}$ are determined by means of a 4th order approximation:

$$\begin{aligned}
\hat{\boldsymbol{q}}_1^{(i)} &= [\frac{1}{0.5\Delta t}\boldsymbol{A}_c^{(i)}]^{-1}\boldsymbol{\varphi}(t) \\
\boldsymbol{z}_1^{(i)} &= \boldsymbol{z}(t) + a_{21}\hat{\boldsymbol{q}}_1^{(i)} \\
\hat{\boldsymbol{q}}_2^{(i)} &= [\frac{1}{0.5\Delta t}\boldsymbol{A}_c^{(i)}]^{-1}(\boldsymbol{\varphi}(t + a_{2x}\Delta t) + c_{21}\hat{\boldsymbol{q}}_1^{(i)}/\Delta t) \\
\boldsymbol{z}_2^{(i)} &= \boldsymbol{z}(t) + a_{31}\hat{\boldsymbol{q}}_1^{(i)} + a_{32}\hat{\boldsymbol{q}}_2^{(i)} \\
\hat{\boldsymbol{q}}_3^{(i)} &= [\frac{1}{0.5\Delta t}\boldsymbol{A}_c^{(i)}]^{-1}(\boldsymbol{\varphi}(t + a_{3x}\Delta t) + (c_{31}\boldsymbol{q}_1^{(i)} + c_{32}\hat{\boldsymbol{q}}_2^{(i)})/\Delta t) \\
\hat{\boldsymbol{q}}_4^{(i)} &= [\frac{1}{0.5\Delta t}\boldsymbol{A}_c^{(i)}]^{-1}(\boldsymbol{\varphi}(t + a_{3x}\Delta t) + (c_{41}\hat{\boldsymbol{q}}_1^{(i)} + c_{42}\hat{\boldsymbol{q}}_2^{(i)} + c_{43}\hat{\boldsymbol{q}}_3^{(i)})/\Delta t) \\
\boldsymbol{z}^{(i)} &= \boldsymbol{z}(t) + b_1\hat{\boldsymbol{q}}_1^{(i)} + b_2\boldsymbol{q}_2^{(i)} + b_3\hat{\boldsymbol{q}}_3^{(i)} + b_4\hat{\boldsymbol{q}}_4^{(i)}
\end{aligned} \tag{8.45}$$

---

[3] Butcher's tableaux can be defined also for implicit methods. The only difference with explicit methods is that the coefficient matrix b_{kj} is not necessarily lower triangular. For example, the Butcher's tableau of the trapezoidal method is:

$$\begin{array}{c|cc} 1 & \frac{1}{2} & \frac{1}{2} \\ 0 & 0 & 0 \\ \hline & \frac{1}{2} & \frac{1}{2} \end{array}$$

However, for implicit methods, the determination of each $\boldsymbol{f}^{\{k\}}$ is generally involved and, hence, the Butcher's tableau representation is not practical.

where $\boldsymbol{z} = [\boldsymbol{x}^T, \boldsymbol{y}^T]^T$ and $\boldsymbol{\varphi} = [\boldsymbol{f}^T, \boldsymbol{g}^T]^T$ and the coefficients are:

$$
\begin{array}{llll}
a_{21} = 2 & a_{31} = 48/25 & a_{32} = 6/25 & \\
c_{21} = -8 & c_{31} = 372/25 & c_{32} = 12/5 & \\
c_{41} = -112/125 & c_{42} = -54/125 & c_{43} = -2/5 & \\
b_1 = 19/9 & b_2 = 0.5 & b_3 = 25/108 & b_4 = 125/108 \\
a_{2x} = 1 & a_{3x} = 3/5 & &
\end{array}
$$

For this method, the convergence error is:

$$\epsilon = \max\{\text{abs}(e_1\hat{\boldsymbol{q}}_1 + e_2\hat{\boldsymbol{q}}_2 + e_3\hat{\boldsymbol{q}}_3 + e_4\hat{\boldsymbol{q}}_4)\} \tag{8.46}$$

where:

$$e_1 = 17/54 \qquad e_2 = 7/36 \qquad e_3 = 0 \qquad e_4 = 125/108 \tag{8.47}$$

It is worth noting that (8.45) requires only one factorization of $\boldsymbol{A}_c^{(i)}$. Thus, the Rosenbrock's formula is only slightly more computational demanding than the trapezoidal method.

8.4 Numerical Integration Routine

This section describes a complete routine for numerical integrations. Only the implicit backward Euler's and the trapezoidal methods are considered. However, the routine can be easily modified to take into account any other numerical method.

Figure 8.7 shows the flowchart for a generic implicit numerical method. The integration starts at a given initial point and at a given initial time. Then, for each time step, an inner Newton's method is solved for (8.42). The inner loop stops if the required variable increment is below a tolerance ϵ or if the maximum number of iterations is reached. In the latter case, the step length Δt is decreased. If the inner loop does not converge and the step length Δt is decreased below a minimum threshold, than the simulation stops since a singularity has been likely encountered (this behavior is typical of a collapse point). Otherwise, the simulation goes on the following time step. The external loop stops if $t \geq t_f$, where t_f is the desired final integration time, or if some other stop criterion is verified.

The numerical method is clearly the kernel of the whole routine but other important issues have to be taken into account, namely, (i) the initial condition, (i) how to update the step length Δt, (iii) how to define disturbances, and (iv) the criteria for stopping the simulation. The calculation of initial conditions is described in Subsection 9.1.1 of Chapter 9. Other issues are discussed in the following subsections.

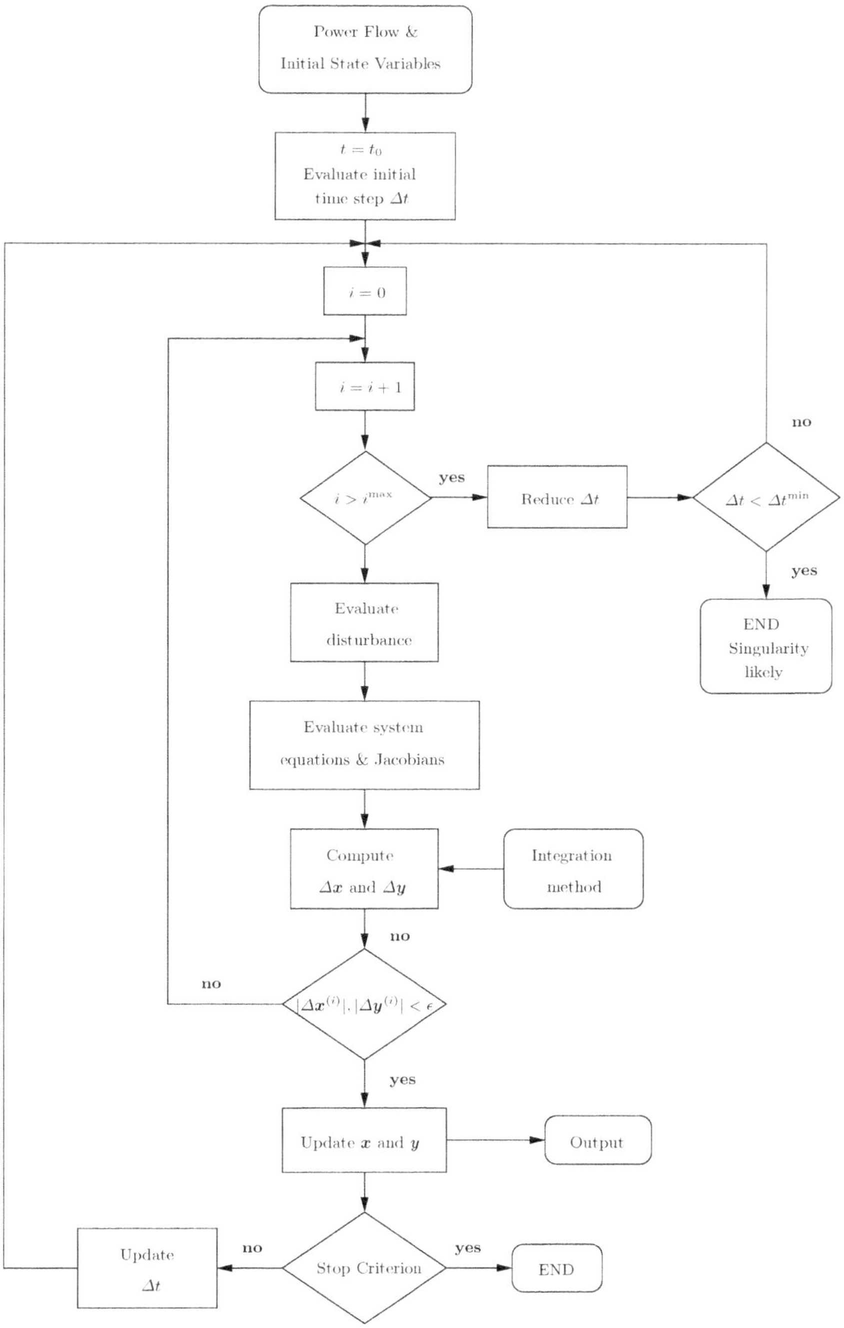

Fig. 8.7 Time domain integration flowchart

8.4.1 Step Length

The step length Δt can be fixed or variable. Increasing the step length is useful for saving time if the numerical integration is going on "smoothly", while decreasing the step length can be useful in the instants after the occurrence of a large disturbance (e.g., short-circuit or line outage).

Multi-stage explicit methods allow estimating the step length. A commonly used formula is:

$$\Delta t \leftarrow \sigma \frac{\epsilon \Delta t}{|\hat{\boldsymbol{x}}(t+\Delta t) - \boldsymbol{x}(t+\Delta t)|} \tag{8.48}$$

where $\sigma \in [0.8, 0.9]$ is a safety factor and ϵ a given tolerance.

A-stable implicit methods are less sensitive to the step length. However, it can be useful to increase Δt if "nothing happens". A heuristic method that has provided good results for the software package PSAT is as follows [195]. The main idea is to use the number of iterations of the inner Newton's method used for solving (8.42) as a "measure" of the difficulty that is facing the numerical method for obtaining the point at $t = \Delta t$. The rules are:

1. If the inner Newton's method has not converged, the step length is defined as:

$$\Delta t \leftarrow 0.5\Delta t \tag{8.49}$$

   Furthermore, if $\Delta t < \Delta t^{\min}$, the routine stops since a singularity has been likely encountered.
2. If the iterations are more than 15, then the following step length is defined as:

$$\Delta t \leftarrow \max\{0.9\Delta t, \Delta t^{\min}\} \tag{8.50}$$

3. If the iterations are less than 10, then the following step length is defined as:

$$\Delta t \leftarrow \min\{1.3\Delta t, \Delta t^{\max}\} \tag{8.51}$$

4. During faults, the time step is defined as:

$$\Delta t \leftarrow \min\{\Delta t, 0.0025 \text{ s}\} \tag{8.52}$$

5. Otherwise the step length is not changed.

The minimum and maximum step lengths $\Delta t^{\min}$ and $\Delta t^{\max}$ are defined at the beginning of the simulation based on the eigenvalues of the system state matrix. In fact, the eigenvalues of the state matrix provide a valuable information on the time constants of the system, as follows:

$$T_k = \frac{1}{|\lambda_k|}, \quad k = 1, 2, \dots, n_x \tag{8.53}$$

where n_x is the dynamic order of the system. The following script further develops this concept.

Script 8.1 Computing the First Time Step

The following script provides some heuristics that are able to estimate the first step length, $\Delta t^{\min}$ and $\Delta t^{\max}$ and, in case of fixed step length, to evaluate if the step length chosen by the user is reasonable.

```
import system
from cvxopt.base import matrix
from cvxopt.umfpack import linsolve

def first_time_step():

    """compute first time step"""

    # estimate the minimum time step
    if not system.DAE.nx:
        freq = 1.0
    elif system.DAE.nx == 1:
        B = matrix(system.DAE.Gx)
        linsolve(system.DAE.Gy, B)
        As = system.DAE.Fx - system.DAE.Fy*B
        freq = abs(As[0,0])
    else:
        freq = 20.0

    if freq > system.Settings.freq:
        freq = float(system.Settings.freq)

    if not freq: freq = 20.0

    # set the minimum time step
    deltaT = abs(system.Settings.tf - system.Settings.t0)
    Tstep = 1/freq
    system.Settings.deltatmax = min(5*Tstep, deltaT/100.0)
    system.Settings.deltat = min(Tstep, deltaT/100.0)
    system.Settings.deltatmin = min(Tstep/64, system.Settings.deltatmax/20)
    if system.Settings.fixt:
        if system.Settings.tstep <= 0:
            print 'Fixed time step is negative or zero'
            print 'Automatic time step has been set'
            system.Settings.fixt = False
        elif system.Settings.tstep < system.Settings.deltatmin:
            print 'Fixed time step is less than estimated minimum time step'
            system.Settings.deltat = system.Settings.tstep
        else:
            system.Settings.deltat = system.Settings.tstep

    return system.Settings.deltat
```

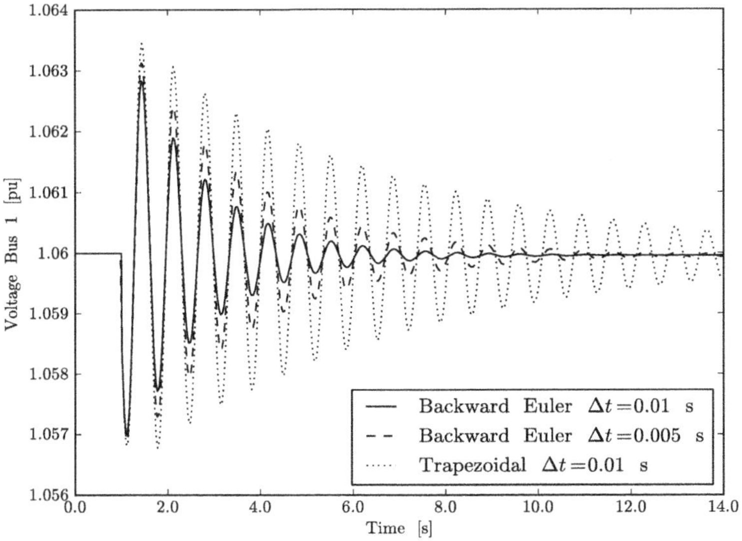

Fig. 8.8 Comparison between the backward Euler's method and the implicit trapezoidal method for the IEEE 14-bus system

Example 8.7 Comparison of Time Domain Integration Methods

Figure 8.8 shows a comparison between the backward Euler's method and the implicit trapezoidal method using a step length $\Delta t = 0.01$ s for the IEEE 14-bus system. The simulation refers to the IEEE 14-bus system and is obtained by applying line 2-4 outage at $t = 1$ s. The backward Euler's method damps oscillations more than the implicit trapezoidal method and, thus, requires a smaller time step to provide precise results.

Figure 8.9 shows a comparison of numerical integration results using the implicit trapezoidal method and different step lengths Δt, namely a fixed step length $\Delta t = 0.01$ s, a fixed step length $\Delta t = 0.10$ s and an adaptive step length as described above. The disturbance is the same as the one used for obtaining the results shown in Figure 8.8. As expected, since the implicit trapezoidal method is A-stable, results are relatively independent from the time step.

8.4.2 Disturbances

In the classical transient stability analysis, whose object is to determine whether synchronous machines go out of step, disturbances of interest are essentially short-circuits and device outages (e.g., lines, generators or loads).

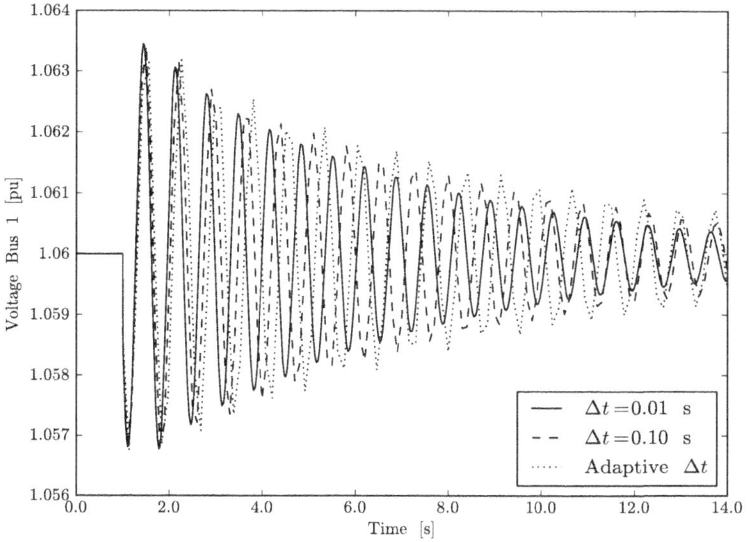

Fig. 8.9 Comparison of numerical integration results using different step lengths for the IEEE 14-bus system

These disturbances are so relevant that have to be handled by means of dedicated classes and functions. These are described in details in Sections 13.1 and 13.2, respectively, of Chapter 13.

Other perturbations, such as load ramps, are more relevant for long-term analysis and, thus, are less commonly included in transient stability packages. A particular case of load ramps is discussed in the following Section 8.6.

Generic disturbances are an issue for most transient stability packages. The main difficulty is that, to be "generic", the time domain integration routine has to be able to support a custom user-defined function. This is quite complicated for software packages built using a system programming language, since the user defined function has to be compiled within the package to work correctly. To overcome this issue, there are mainly three solutions:

1. To allow the user to apply only a predefined set of disturbances. Clearly this is not a solution, but just a way of overcoming the issue. Unfortunately, this solution is more common than not in proprietary software.
2. To define a special syntax for disturbances. The syntax can be in form of data [59], in form of a meta-language [68], or using a system-language function that is compiled at run-time [243].
3. To allow the user to embed any function within the main routine. While relatively complicated for packages based on system programming languages (especially if the source code is not provided), this solution is trivial for

scripting language-based applications. For example, in Python, function names are handled as any other variable type.

As the reader may expect, my favorite solution is the last one. However, this solution has also some drawback. First of all, a user-defined disturbance function must have a fixed syntax to be compatible with the main numerical integration routine. For example, the function header can be of the type:

```
def disturbance(t):
    ...
```

where t is current simulation time. Then, in order to modify the parameters of some device, that device has to be imported and visible in the function scope. The main issue of this approach is that the user has to be familiar with the syntax and the idiosyncrasies of the software package in use. This familiarity is often difficult to obtain. In conclusion, to much flexibility can result in an obstacle rather than a freedom.

8.4.3 Stop Criterion

An important issue to solve when dealing with numerical integration is *when* to stop the simulation. In transient stability, most simulations are performed to define if the system is able to maintain the synchronism after a large disturbance. In practice, a common security analysis consists in defining a set of "credible" contingencies and in running a time domain simulation for each contingency.

A relevant question is: is it possible to anticipate whether the system trajectory is going to be stable or unstable and, thus, to save simulation time by opportunely stopping the numerical integration?

A simple method for determining if the trajectory is going to be unstable is to monitor the rotor angle of synchronous machines. If at a certain time t, the maximum difference between two rotor angles exceeds 2π,[4] then some machine is certainly losing the synchronism and the simulation can be stopped.

However, the previous method does not allow saving time if the simulation is stable. In fact, if the simulation is stable, one has to wait for the final assigned time t_f before stopping the numerical method.

A well-known technique that allows defining the stability or instability of a given trajectory is the SIME method [236], which works as follows. At each step of the numerical integration, the machine rotor angles are sorted and the maximum difference of two consecutive synchronous machine rotor angles is found. Assuming that these angles are δ_i and δ_j, with $\delta_i > \delta_j$, all machines whose rotor angles satisfy $\delta_h \geq \delta_i$ are considered *critical* machines, while all machines whose rotor angles satisfy $\delta_h \leq \delta_j$ are considered *non-critical* machines.

[4] It is assumed that the system data are in pu and thus the phase shift introduced by three-phase transformers do not affect rotor angle values.

Once defined the critical and non-critical machine sets, say $\mathcal{G}_{\rm C}$ and $\mathcal{G}_{\rm NC}$, the equivalent OMIB rotor angle is defined as:

$$\delta^{\rm OMIB} = \frac{1}{H_{\rm C}} \sum_{j=\mathcal{G}_{\rm C}} H_j \delta_j - \frac{1}{H_{\rm NC}} \sum_{j=\mathcal{G}_{\rm NC}} H_j \delta_j \tag{8.54}$$

where the sub-indexes C and NC stand for critical and non-critical, and the equivalent inertia constants are:

$$\begin{aligned} H_{\rm C} &= \sum_{j=\mathcal{G}_{\rm C}} H_j \\ H_{\rm NC} &= \sum_{j=\mathcal{G}_{\rm NC}} H_j \end{aligned} \tag{8.55}$$

Similarly, OMIB electrical and mechanical powers are defined as:

$$\begin{aligned} p_e^{\rm OMIB} &= H^{\rm OMIB} \left[\frac{1}{H_{\rm C}} \sum_{j=\mathcal{G}_{\rm C}} p_{ej} - \frac{1}{H_{\rm NC}} \sum_{j=\mathcal{G}_{\rm NC}} p_{ej} \right] \\ p_m^{\rm OMIB} &= H^{\rm OMIB} \left[\frac{1}{H_{\rm C}} \sum_{j=\mathcal{G}_{\rm C}} p_{mj} - \frac{1}{H_{\rm NC}} \sum_{j=\mathcal{G}_{\rm NC}} p_{mj} \right] \end{aligned} \tag{8.56}$$

where $H^{\rm OMIB} = H_{\rm C} H_{\rm NC}/(H_{\rm C} + H_{\rm NC})$. According to the EAC described in Section 8.1, the equivalent OMIB accelerating power $p_a^{\rm OMIB}$ is:

$$p_a^{\rm OMIB} = p_m^{\rm OMIB} - p_e^{\rm OMIB} \tag{8.57}$$

The following stability conditions hold for the equivalent OMIB:

1. If, at a certain time step t, $p_a^{\rm OMIB} = 0$ and $\dot{p}_a^{\rm OMIB} > 0$ and $\dot{\delta}^{\rm OMIB} > 0$ the system is unstable.[5] In fact, the previous conditions ensure that the system has no further kinetic energy to spend for decelerating the system. Furthermore, $\dot{\delta}^{\rm OMIB} > 0$ implies that the rotor angle is increasing. These are sufficient conditions for defining instability.
2. If, at a certain time step t, $p_a^{\rm OMIB} < 0$ and $\dot{\delta}^{\rm OMIB} \leq 0$ the system is first-swing stable. In fact, in this case, the kinetic energy is enough to stop

[5] The first time derivative of the equivalent OMIB rotor angle $\dot{\delta}^{\rm OMIB}$ can be written as:

$$\begin{aligned} \dot{\delta}^{\rm OMIB} &= \frac{1}{H_{\rm C}} \sum_{j=\mathcal{G}_{\rm C}} H_j \dot{\delta}_j - \frac{1}{H_{\rm NC}} \sum_{j=\mathcal{G}_{\rm NC}} H_j \dot{\delta}_j \\ &= \frac{1}{H_{\rm C}} \sum_{j=\mathcal{G}_{\rm C}} H_j \omega_j - \frac{1}{H_{\rm NC}} \sum_{j=\mathcal{G}_{\rm NC}} H_j \omega_j = \Delta\omega^{\rm OMIB} \end{aligned}$$

Thus, $\dot{\delta}^{\rm OMIB}$ is the speed deviation $\Delta\omega^{\rm OMIB}$ of the equivalent OMIB system.

the critical machine rotor angles and make them to "come back". These stability conditions are only necessary. In fact, the system can later on show multi-swing instability. Only the numerical integration can show if the trajectory is multi-swing stable or not.

3. If $p_a^{\mathrm{OMIB}} > 0$, $\forall t > 0$, then the system is certainly unstable. However, some heuristic is needed to determine when to stop the simulation [236].

The most relevant features of the SIME method are (i) no assumption is made on the original system and (ii) computing the OMIB equivalent is practically inexpensive with respect to the computational burden of the numerical method.

The main assumption of the SIME method is that the two sets of critical and non-critical machines can be considered as an OMIB system. One may argue that there could be a case in which the system separates into three or more groups. Actually, there is no experimental result that shows a system separating in more than two groups since it becomes unstable.[6] Thus, until a case study will prove the contrary, the main assumption of the SIME method can be considered true.

Example 8.8 Application of the SIME Method for the IEEE 14-Bus System

Figure 8.10 shows synchronous machine rotor speeds for the IEEE 14-bus system following a three-phase fault at bus 4. The fault occurs at $t = 1$ s and is cleared at $t = 1.2$ s by line 2-4 outage. The numerical integration provides the first 4 s following the fault clearing. During the fault and in the first instances after the line outage, the system separates into two groups: the first group is composed of machines 1 and 2, and the other one is composed of machines 3, 4 and 5. From observing the plot, it is clear that the system is stable. Nevertheless, the SIME method allows stopping the simulation at about $t = 1.265$ s, i.e., when the stability conditions $p_a^{\mathrm{OMIB}} < 0$ and $\dot{\delta}^{\mathrm{OMIB}} \leq 0$ are satisfied (see Figure 8.11). It is important to note that only the full time domain simulation ensures that the system does not show multi-swing instability.

Script 8.2 Complete Time Domain Integration Algorithm

The following script provides a complete routine for numerical integration implementing the backward Euler's and the trapezoidal implicit methods. The functions `first_time_step()`, `time_step(convergence, iteration, t)` and `anglediff()` allows defining the first step length as well as maximum and minimum step length, updating the step length and computing the rotor angle difference or the stability according to the SIME approach, respectively. The class `system.Varout` is used for storing the simulation results: each instance of the method `store()` append to the output file the current simulation time and the vectors of state and algebraic variables. As

[6] Observe that if the system trajectory is stable, the critical and the non-critical machine groups are arbitrary since there are no critical machines.

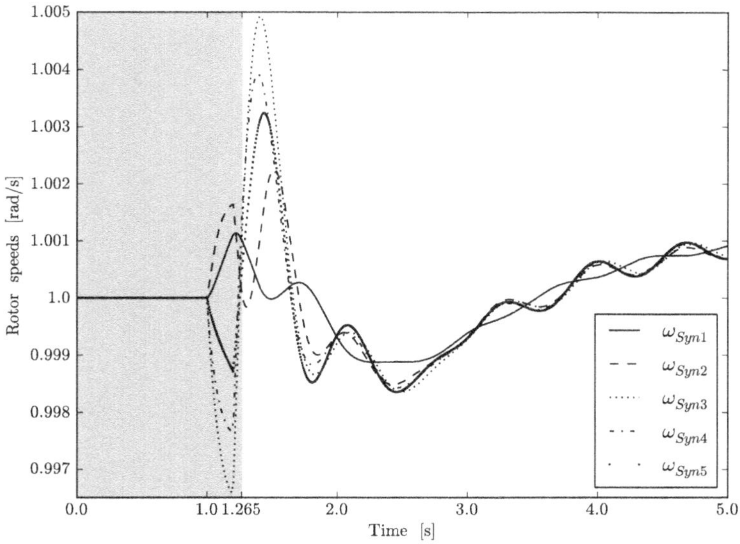

Fig. 8.10 Transient following a three-phase fault at bus 4 for the IEEE 14-bus system. The gray region indicates the part of the simulation that is required by the SIME method to determine that rotor speed trajectories are first-swing stable

usual, the expression `exec system.Device.call_int` is used as interface for computing algebraic and differential equations and Jacobian matrices of all devices defined in the system (see Script 3.2 of Chapter 3).

Two remarks are relevant:

1. State variable hard limits requires a special care. When a state variable hits a hard limit, its value becomes a constant and the correspondent differential equation is "frozen". This has to be taken into account in the matrix $\boldsymbol{A}_c^{(i)}$ and in the vector $\hat{\boldsymbol{q}}^{(i)}$. With this aim, the interface `exec system.Device.call_windup` makes sure that if a state variable, say x_i hits a limit, the ith column and the ith row of $\boldsymbol{A}_c^{(i)}$ as well as the ith element of the vector $\hat{\boldsymbol{q}}^{(i)}$ are set to zero, while the element (i, i) of $\boldsymbol{A}_c^{(i)}$ is set to 1.
2. Special events such as fault occurrences and breaker operations are taken into account so that, regardless the step length Δt, the numerical integration evaluates a point right before and right after that event. In other words, if the event occurs at $t = t_e$, then a point is evaluated at $t_e - \epsilon$ and at $t_e + \epsilon$. This "trick" avoids that a specific event is evaluated with a delay due to the step length.

```
import system
from cvxopt.base import matrix, spmatrix, sparse
from cvxopt.umfpack import linsolve, symbolic, numeric
```

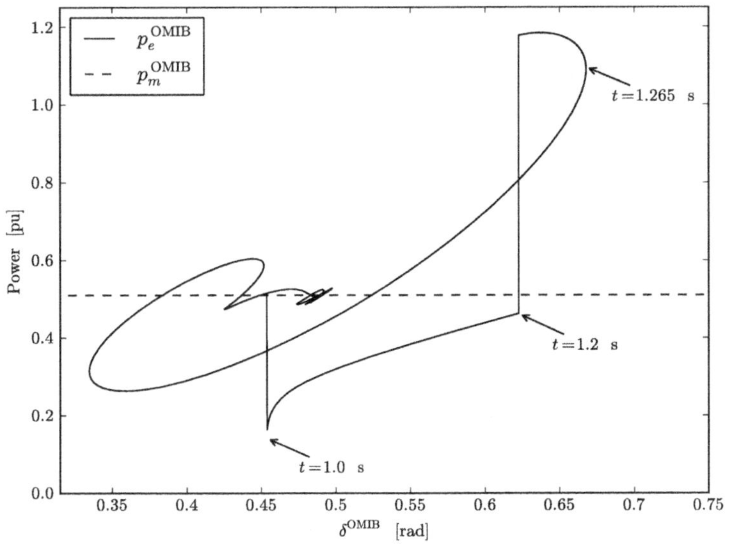

Fig. 8.11 Equivalent OMIB electrical and mechanical powers as a function of the equivalent OMIB rotor angle δ^{OMIB} for the transient following a three-phase fault at bus 4 for the IEEE 14-bus system

```
from cvxopt.umfpack import solve as umfsolve

def timedomain():

  # check settings
  iter_max = system.Settings.dynmit
  tol = system.Settings.tol
  nx = system.DAE.nx
  ny = system.DAE.ny
  In = spmatrix(1, range(nx), range(nx), (nx, nx), 'd')

  # initializations
  t = system.Settings.t0
  step = 0
  h = first_time_step()
  inc = matrix(0, (nx + ny, 1), 'd')
  inc = matrix(0, (nx + ny, 1), 'd')
  system.DAE.factorize = True
  system.DAE.mu = 1.0
  system.DAE.kg = 0.0
  switch = False
  nextpc = 0.1

  # time vector faults and breaker events
```

```
fixed_times = system.Device.get_times()

# compute max rotor angle difference
diff_max = anglediff()

# Main loop
while t <= system.Settings.tf and t + h > t and not diff_max:

    if t + h > system.Settings.tf: h = system.Settings.tf - t
    actual_time = t + h

    # check for the occurrence of a disturbance
    for item in fixed_times:
        if item > t and item < t + h:
            actual_time = item
            h = actual_time - t
            switch = True
            break

    # set global time
    system.DAE.t = actual_time

    # backup of actual variables
    xa = matrix(system.DAE.x)
    ya = matrix(system.DAE.y)

    # initialize NR loop
    iteration = 0
    fn = matrix(system.DAE.f)

    # applying faults, breaker interventions and disturbances
    if switch:
        system.Fault.intervention(actual_time)
        system.Breaker.intervention(actual_time)
        switch = False

    if system.Settings.disturbance: system.File.disturbance(actual_time)

    # main loop of the Newton's method
    system.Settings.error = tol + 1 # force at least one iteration
    while system.Settings.error > tol and iteration < iter_max:

      # DAE equations
      exec system.Device.call_int

      # complete Jacobian matrix DAE.Ac
      if system.Settings.method == 'euler':
          system.DAE.Ac = sparse([[In - h*system.DAE.Fx, system.DAE.Gx], \
                                  [-h*system.DAE.Fy, system.DAE.Gy]])
          system.DAE.q = system.DAE.x - xa - h*system.DAE.f
      else: # default is implicit trapezoidal method
          system.DAE.Ac = sparse([[In - h*0.5*system.DAE.Fx, system.DAE.Gx],\
                                  [-h*0.5*system.DAE.Fy, system.DAE.Gy]])
          system.DAE.q = system.DAE.x - xa - h*0.5*(system.DAE.f + fn)
```

```
        # anti-windup limiters
        exec system.Device.call_windup
        if system.DAE.factorize:
            F = symbolic(system.DAE.Ac)
            system.DAE.factorize = False
        inc = -matrix([system.DAE.q, system.DAE.g])
        try:
            umfsolve(system.DAE.Ac, numeric(system.DAE.Ac, F), inc)
        except ArithmeticError:
            print 'Singular matrix'
            iteration = iter_max + 1
        except ValueError:
            # Unexpected refactorization of the power flow Jacobian matrix
            F = symbolic(system.DAE.Ac)
            try:
                umfsolve(system.DAE.Ac, numeric(system.DAE.Ac, F), inc)
            except ArithmeticError:
                print 'Singular matrix'
                iteration = iter_max + 1
        system.DAE.x += inc[:nx]
        system.DAE.y += inc[nx:nx + ny]
        system.Settings.error = max(abs(inc))
        iteration += 1

    if iteration >= iter_max:
        h = time_step(False, iteration, t)
        print 'Reducing time step (delta t = %.5f s)' % h
        system.DAE.x = matrix(xa)
        system.DAE.y = matrix(ya)
        system.DAE.f = matrix(fn)
        continue

    # update output variables and time step
    t = actual_time
    step += 1
    system.Varout.store(t, step)
    # avoid freezing at t == system.Settings.tf
    if h == 0: break
    h = time_step(True, iteration, t)

    # plot variables and display iteration status
    perc=(t - system.Settings.t0)/(system.Settings.tf - system.Settings.t0)
    if perc > nextpc:
      print ' # Simulation time = %.4f s (%.1f%%)' % (system.DAE.t, perc*100)
      nextpc += 0.1

    # compute max rotor angle difference
    diff_max = anglediff()
```

8.5 Electro-magnetic Transients

In [78], Dommel proposed a compensation method based on nodal analysis for a systematic study of electro-magnetic transients of electrical circuits. The Dommel's method is a numerical integration substitution that transforms the basic electrical elements, namely resistances, inductances and capacitances, into an equivalent parallel circuit composed of a resistance and a current generator. A fixed-step trapezoidal method allows obtaining this result.

Inductance

For an inductance, one has:

$$v_h - v_k = L\frac{di_{hk}}{dt} \tag{8.58}$$

where h and k are the terminal nodes of the inductance. At a generic time step t, the current variations for $t + \Delta t$ is:

$$i_{hk}(t + \Delta t) = i_{hk}(t) + \frac{1}{L}\int_t^{t+\Delta t}(v_h - v_k)dt \tag{8.59}$$

Applying the trapezoidal rule:

$$i_{hk}(t + \Delta t) = \hat{i}_{hk}(t) + \frac{\Delta t}{2L}(v_h(t + \Delta t) - v_k(t + \Delta t)) \tag{8.60}$$

where the equivalent current $\hat{i}_{hk}(t)$ depends on the system state at t:

$$\hat{i}_{hk}(t) = i_{hk}(t) + \frac{\Delta t}{2L}(v_h(t) - v_k(t)) \tag{8.61}$$

Capacitance

For a capacitance, one has:

$$v_h(t + \Delta t) - v_k(t + \Delta t) = \frac{1}{C}\int_t^{t+\Delta t} i_{hk}dt + v_h(t) - v_k(t) \tag{8.62}$$

Applying the trapezoidal rule:

$$i_{hk}(t + \Delta t) = \hat{i}_{hk}(t) + \frac{2C}{\Delta t}(v_h(t + \Delta t) - v_k(t + \Delta t)) \tag{8.63}$$

where the equivalent current $\hat{\imath}_{hk}(t)$ depends on the system state at t:

$$\hat{\imath}_{hk}(t) = -\imath_{hk}(t) - \frac{2C}{\Delta t}(v_h(t) - v_k(t)) \tag{8.64}$$

Resistance

The resistance is a memory-less element. Thus, one has:

$$\imath_{hk}(t + \Delta t) = \frac{1}{R}(v_h(t + \Delta t) - v_k(t + \Delta t)) \tag{8.65}$$

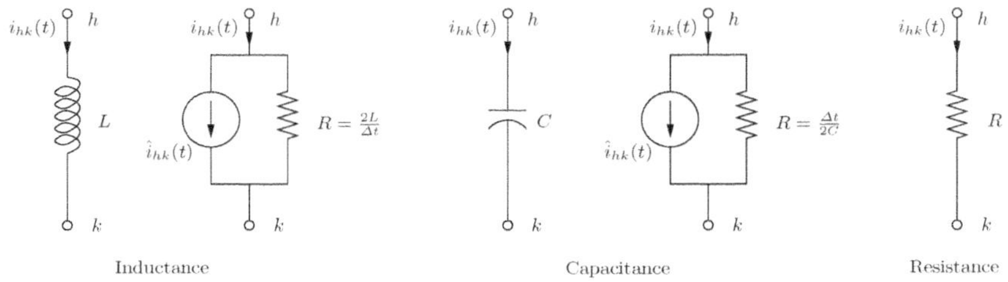

Fig. 8.12 Dommel's equivalents

Figure 8.12 summarizes the compensated models introduced by the Dommel's method. In conclusion, an electrical network composed of linear devices can be represented as:

$$\boldsymbol{G}\boldsymbol{v}(t + \Delta t) = \boldsymbol{i}(t + \Delta t) + \hat{\boldsymbol{i}}(t) \tag{8.66}$$

where $\boldsymbol{G}$ is the nodal conductance matrix, $\boldsymbol{v}(t + \Delta t)$ is the vector of node voltages at time $t + \Delta t$, $\boldsymbol{i}(t + \Delta t)$ is the vector of currents injected at nodes at time $t + \Delta t$, and $\hat{\boldsymbol{i}}(t)$ is the vector of known current sources at time t.

The main advantage of (8.66) is that the conductance matrix $\boldsymbol{G}$ is constant (as long as the network topology does not change) and thus it can be factorized only once at the beginning of the time integration. The efficiency of the Dommel's method has made it a standard *de facto* in electro-magnetic transients and has been adopted by most popular EMT software tools (e.g., PSCAD and ATP).

However, the proposed method is not flawless, at least if one needs to integrate it into a transient stability analysis tool. Main drawbacks are as follows [154]:

1. If the network topology changes several times (e.g., due to converter switching), the method is not so effective since the matrix $\boldsymbol{G}$ has to be recomputed and re-factorized several times.

2. Matrix $\boldsymbol{G}$ is constant only if the system element are linear. At most, non-linear current sources can be used [79]. In case of nonlinear elements, $\boldsymbol{G}$ has to be re-factorized at each time step.
3. Since nodal equations are reformulated as integral-differential algebraic equations (IDAE), the network nodal conductance matrix may result poorly conditioned and may show numerical instabilities [109].

Although in EMT programs the issues above can be overcome, the integration of electro-magnetic transient analysis into a transient stability program is still an open task [151]. The integration is more and more a necessity due to the large penetration of non-conventional energy sources, most of which are dc systems (e.g., photo-voltaic and fuel cells). Thus, there is the need of modelling both static converters and dc circuits. Chapters 17 and 18 describe in details dc models and ac/dc converters, respectively. In those chapters, a general nonlinear current-injection model (8.13) is used.

8.6 Quasi-static Analysis

In some applications, the variations of the inputs are relatively slow with respect to transient dynamics. A relevant example is the study of the effect of long term voltage stability phenomena, such as the daily load ramp or voltage collapse [336]. In this case load powers are modelled as time dependent controllable parameters $\boldsymbol{\eta}(t)$. Since load variations take from tens of minutes to some hours, any transient dynamic can be considered steady-state. The resulting system equations are obtained by imposing $\dot{\boldsymbol{x}} = \boldsymbol{0}$ in (8.12):

$$\begin{aligned} \boldsymbol{0} &= \boldsymbol{f}(\boldsymbol{x}, \boldsymbol{y}, \boldsymbol{\eta}(t)) \\ \boldsymbol{0} &= \boldsymbol{g}(\boldsymbol{x}, \boldsymbol{y}, \boldsymbol{\eta}(t)) \end{aligned} \tag{8.67}$$

which is generally referred to as *quasi-static* or *quasi-steady-state* model. For example, a quasi-steady-state simulator called WPSTAB has been developed by the National Technical University of Athens [340].

Loads can be modeled as linear time ramps:

$$\boldsymbol{p}_L(t) = \boldsymbol{p}_L^0 + \boldsymbol{r}_L t \tag{8.68}$$

In practice, the load ramp can be assumed as:

$$\boldsymbol{r}_L \equiv k_t \boldsymbol{p}_L^0 \tag{8.69}$$

where k_t is a coefficient that imposes the time rate. For example $k_t = 1/3600$ s^{-1} means the that load powers double in one hour.

Solving the original system (8.12) with a very small time rate k_t provides practically same results as (8.67). However, (8.67) allows drastically reducing the computing time. In fact, the maximum time step that can be used for integrating (8.12) depends on the system time constants. On the other hand, the solution of (8.67) is a sequence of equilibrium points and thus the step length size is limited only by the convergence properties of the method used for solving (8.67) and by the initial guess, which is generally set as the solution of the previous point.

It is relevant to note the similarity between (8.68) and the load direction model (5.13) given in Chapter 5. Section 5.4.4 suggests that the homotopy (5.59) can be interpreted as a differential equation where the continuation parameter μ is used as an independent variable (e.g., Davidenko's method). Analogously, in (8.67), the independent variable t can viewed as a continuation parameter. The point is that any homotopy method described in Chapter 5 can be used for studying (8.67) and for determining how much the system can be loaded until a bifurcation occurs.

The quasi-static system (8.67) can show three kinds of bifurcations. Besides saddle-node and limit-induced bifurcations that were introduced in Chapter 5, synchronous machines and primary voltage regulation (e.g., AVRs) can lead to Hopf bifurcations. This kind of bifurcations is characterized by a pair of complex eigenvalues that crosses the imaginary axis as the load increases [45, 115, 337]. After the occurrence of an Hopf bifurcation the system shows undamped oscillations that can lead to a collapse, to a limit cycle or to more complex phenomena such as period doubling or chaos [6]. In any case, the occurrence of the Hopf bifurcation has to be considered a maximum loading condition similar to a saddle-node or a critical limit-induced bifurcation.

It is important to note that there is a conceptual difference between the solution of (8.67) through an homotopy method and the solution of a continuation power flow using the static model (5.10) with the initialization of dynamic device state variables for each point of the nose curve. In fact, although both analyses compute a series of equilibrium points, the continuation power flow approach imposes constant bus voltage magnitudes at generator buses, while the quasi-steady-state analysis imposes constant regulator reference signals. This difference may lead to quite different results in terms of small-signal stability analysis as it is qualitatively discussed in the following example.

Example 8.9 Quasi-static Integration for the IEEE 14-Bus System

This example shows the results of the quasi-static time domain analysis for the IEEE 14-bus system. The dynamic models include synchronous machines and AVRs as described in Appendix D. The perturbation consists in increasing the load using a time ramp (8.68). Figure 8.13 shows voltages at load buses 12, 13 and 14. Only the upper part of the curve has a physical meaning, i.e., up to the maximum loading condition at which a saddle-node bifurcation occurs. In this case, the loading level $\mu \equiv k_t t$.

When considering synchronous machines and AVRs, limit-induced bifurcations are implicitly taken into account by field voltage and AVRs hard limits.

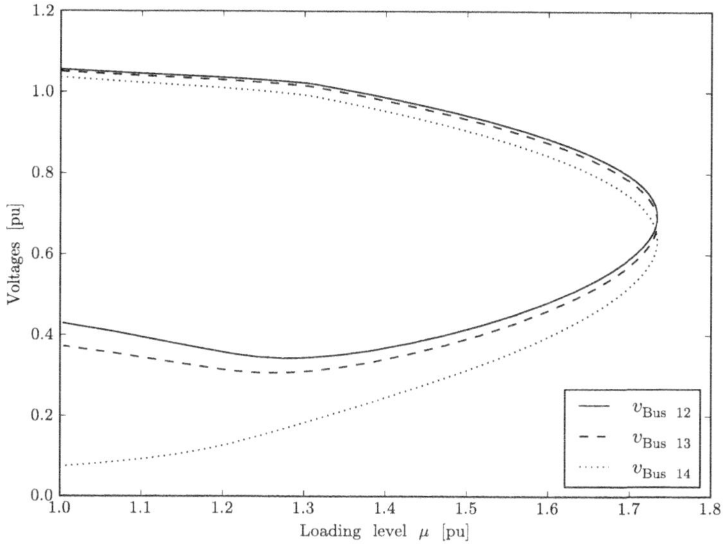

Fig. 8.13 Quasi-static time domain analysis through homotopy method for the IEEE 14-bus system with generator field voltage limits

Figure 8.14 shows the saturation of field voltage of synchronous machine 2 to 5 (which is a consequence of the saturation of AVR state variable v_{r1}) and the generated reactive powers for the IEEE 14-bus system. As long as field voltages increase, also the reactive powers injected into the network increase. As field voltages saturate, synchronous machines limit the reactive power output. The saturation of v_{r1} and thus of the field voltage is a limit-induced bifurcation since, once saturated, the AVR control loop opens and system equations change.

The results of the quasi-static analysis can be compared with the static CPF analysis shown in Example 5.4 of Chapter 5. Figure 8.15 compares such simulations in a synoptic plot. The CPF analysis and the homotopy method applied to the quasi-static system provides similar nose curves. Actually, the classical approximation of synchronous machines with primary voltage regulation as constant PV generators is quite reasonable. If the reactive power limits of the PV generators are properly chosen, the two models show a LIB for similar values of the loading level. In quasi-static analysis, the bus voltage is not perfectly constant since the AVR control is not integral (see Section 16.2.1 of Chapter 16). As a consequence, the machine bus voltage decreases even though the field voltage is not saturated.

Figure 8.15 also shows the loading level at which a Hopf bifurcation occurs for the two approaches. In this case, the two approaches provide comparatively different results. For the static CPF analysis, state matrix eigenvalues

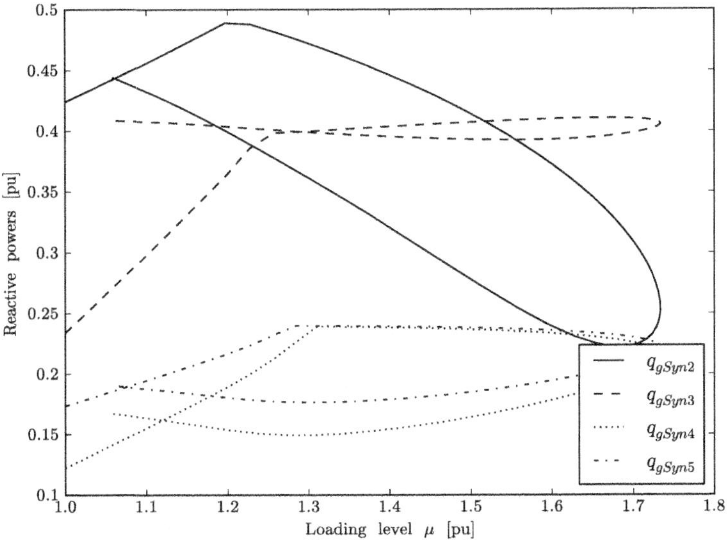

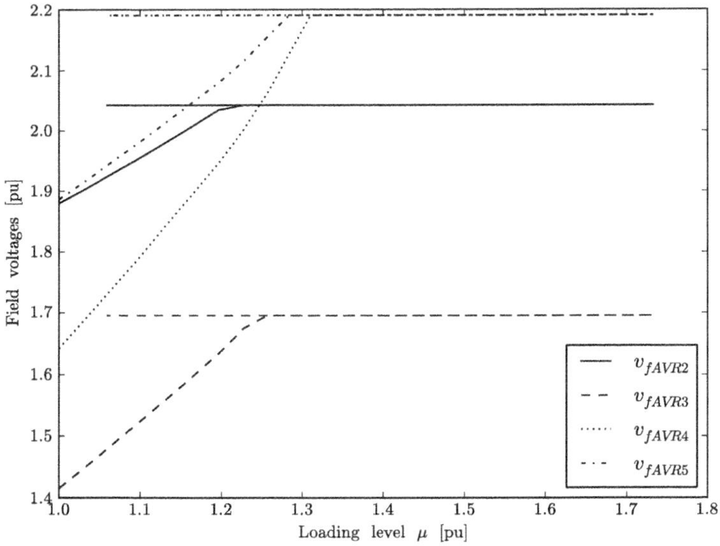

Fig. 8.14 Synchronous machine field voltages and reactive powers for the IEEE 14-bus system

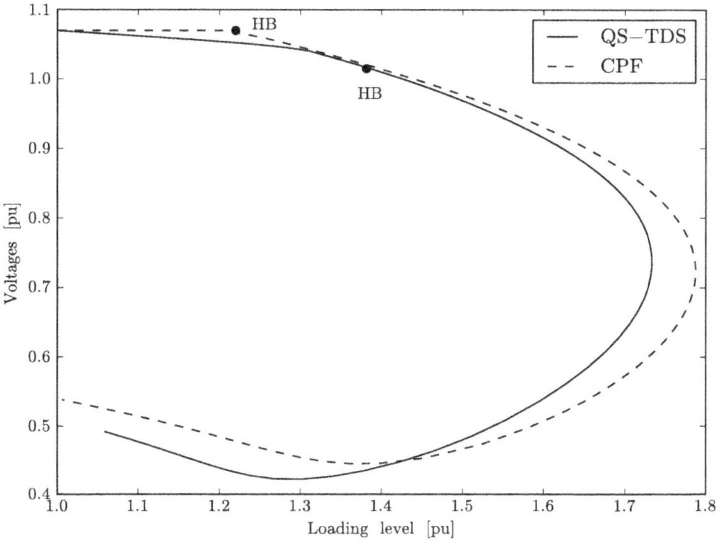

Fig. 8.15 Comparison between the quasi-static time domain simulation and the CPF analysis for the IEEE 14-bus system. The plot shows the voltage magnitude of bus 6

are obtained after initializing state variables using the procedure described in Section 9.1.1 of Chapter 9. For the quasi-static analysis, eigenvalues are computed by forming the state matrix for each point of the curve. In this case, the static CPF analysis is conservative with respect to the quasi-static simulation.

8.7 Summary

This section summarizes most relevant concepts related to time-domain analysis.

Time Scale: It is important to clearly define the time scale. Depending on the time scale, the full set of state variables can be divided into *fast*, *normal* and *slow*. Time constants associated with fast dynamics can be considered zero and the associated variables are modelled as algebraic. State variables associated with state variables can be considered "frozen" and do not move during the transients. In transient stability analysis the time scale is between 10^{-2} and 10^{1} s. This means that electro-magnetic transients can be considered fast (i.e., bus voltages are algebraic variables) and long-term dynamics are slow (e.g., daily load ramp).

Integration Method: There is huge variety of numerical integration methods. Explicit methods such as Runge-Kutta's formulæ are well suited for non-stiff ODE systems. However, transient stability analysis concerns the integration of a nonlinear and generally stiff DAE systems. In this case, the most adequate methods are A-stable implicit and semi-implicit methods such as the implicit trapezoidal method or Rosenbrock's formulæ.

Step Length: The step length Δt can be fixed or variable. Implicit A-stable numerical methods allow relatively large step length without loosing accuracy. The step length has to take into account the time constants of the system. With this aim, an eigenvalue analysis for the initial value can provide information about the time scales of the system.

Disturbance: Typical disturbances for transient stability analysis are short-circuit and device outages. However a general routine for time domain analysis has to be able to support any user-defined disturbance. The latter can be easily obtained if using an open-source tool based on a scripting language.

Stop Criterion: When dealing with contingency analysis, it is important to save time by stopping the numerical integration routine if the trajectories are surely stable or unstable. Monitoring the maximum synchronous machine rotor angle difference provides such information. The SIME method provides sufficient instability conditions and necessary stability ones based on the computation of an equivalent OMIB system at each time step.

Part III
Device Models

Chapter 9
Device Generalities

This chapter is about modelling and scripting of general purpose devices. Section 9.1 defines the mathematical model of an as general as possible device, while Section 9.2 provides the conceptual basis for implementing a device as a class. Subsection 9.2.1 presents a Python example of a base device class, while Subsection 9.2.3 provides an example of specific device methods, namely the two-axis model of the synchronous machine.

9.1 General Device Model

Chapter 8 defines a general power injection model, as follows:

$$\dot{\boldsymbol{x}} = \boldsymbol{f}(\boldsymbol{x}, \hat{\boldsymbol{y}}, \bar{\boldsymbol{v}}, \boldsymbol{\eta}) \tag{9.1}$$

$$\boldsymbol{0} = \hat{\boldsymbol{g}}(\boldsymbol{x}, \hat{\boldsymbol{y}}, \bar{\boldsymbol{v}}, \boldsymbol{\eta}) \tag{9.2}$$

$$\boldsymbol{0} = \bar{\boldsymbol{s}}(\boldsymbol{x}, \hat{\boldsymbol{y}}, \bar{\boldsymbol{v}}, \boldsymbol{\eta}) - \bar{\boldsymbol{V}}\bar{\boldsymbol{Y}}^*(\boldsymbol{x}, \hat{\boldsymbol{y}}, \boldsymbol{\eta})\bar{\boldsymbol{v}}^* \tag{9.3}$$

Each device is a subset of equations (9.1)-(9.3). In particular, for each device, one can define a certain set of differential equations $\boldsymbol{f}_i \subset \boldsymbol{f}$ and algebraic equations $\hat{\boldsymbol{g}}_i \subset \hat{\boldsymbol{g}}$. Moreover, if the device is connected to a network bus, say bus h, the element h of the vector $\bar{\boldsymbol{s}}$ contains the power injection of the considered device. In general, there can be more than one device connected to the same bus. Thus, the element h of the vector $\bar{\boldsymbol{s}}$ is obtained as the sum of the contributions of each device connected to the bus h. Similarly, the element h of the vector $\bar{\boldsymbol{V}}\bar{\boldsymbol{Y}}^*(\boldsymbol{x}, \boldsymbol{\eta})\bar{\boldsymbol{v}}^*$ can be also viewed as the sum of all powers injected to branches (e.g., transmission lines and transformers) connected to bus h.

According to the discussion above, equation (9.3) can be rewritten as:

$$0 = \sum_{i \in \hat{\Omega}_h} \bar{s}_{h,i}(\boldsymbol{x}, \hat{\boldsymbol{y}}, \bar{\boldsymbol{v}}, \boldsymbol{\eta}) - \bar{v}_h \sum_{k \in \mathcal{C}_h} \bar{y}^*_{hk}(\boldsymbol{x}, \hat{\boldsymbol{y}}, \boldsymbol{\eta})\bar{v}^*_k, \quad h \in \mathcal{B} \tag{9.4}$$

where $\hat{\Omega}_h$ and $\mathcal{C}_h$ are the sets of all devices and all branches, respectively, connected to bus h, and $\mathcal{B}$ is the set of buses. Further generalizing (9.4),

F. Milano: Power System Modelling and Scripting, Power Systems, pp. 221–246.
springerlink.com

i.e., considering also transmission lines and transformers as devices that inject powers at buses, leads to:

$$0 = \sum_{i \in \Omega_h} \bar{s}_{h,i}(\boldsymbol{x}, \hat{\boldsymbol{y}}, \bar{\boldsymbol{v}}), \quad h \in \mathcal{B} \tag{9.5}$$

where $\Omega_h = \hat{\Omega}_h \cup \mathcal{C}_h$.

A similar equation holds for node current balance equations of dc devices:

$$0 = \sum_{i \in \Omega_{\mathrm{dc},h}} i_{h,i}(\boldsymbol{x}, \hat{\boldsymbol{y}}, \boldsymbol{v}), \quad h \in \mathcal{N} \tag{9.6}$$

where $i_{h,i}$ is the current injected at node h by device i, $\Omega_{\mathrm{dc},h}$ is the set of all dc devices connected to node h, and $\mathcal{N}$ is the set of nodes of the dc network.

In conclusion, the most general model of an ac device is as follows:

$$\begin{aligned} \dot{\boldsymbol{x}}_i &= \boldsymbol{f}_i(\boldsymbol{x}_i, \boldsymbol{x}_e, \hat{\boldsymbol{y}}_i, \hat{\boldsymbol{y}}_e, \boldsymbol{v}, \boldsymbol{\theta}, \boldsymbol{\eta}) \\ \boldsymbol{0} &= \hat{\boldsymbol{g}}_i(\boldsymbol{x}_i, \boldsymbol{x}_e, \hat{\boldsymbol{y}}_i, \hat{\boldsymbol{y}}_e, \boldsymbol{v}, \boldsymbol{\theta}, \boldsymbol{\eta}) \\ \boldsymbol{p}_{h,i} &= \boldsymbol{g}_{p,hi}(\boldsymbol{x}_i, \boldsymbol{x}_e, \hat{\boldsymbol{y}}_i, \hat{\boldsymbol{y}}_e, \boldsymbol{v}, \boldsymbol{\theta}, \boldsymbol{\eta}) \\ \boldsymbol{q}_{h,i} &= \boldsymbol{g}_{q,hi}(\boldsymbol{x}_i, \boldsymbol{x}_e, \hat{\boldsymbol{y}}_i, \hat{\boldsymbol{y}}_e, \boldsymbol{v}, \boldsymbol{\theta}, \boldsymbol{\eta}) \end{aligned} \tag{9.7}$$

where the sub-index i indicates internal device variables and equations and the sub-index e indicates external variables shared with other devices. Hence, $\boldsymbol{x} = [\boldsymbol{x}_i^T, \boldsymbol{x}_e^T]^T$ and $\hat{\boldsymbol{y}} = [\hat{\boldsymbol{y}}_i^T, \hat{\boldsymbol{y}}_e^T]^T$. In (9.7), complex voltage phasors $\bar{\boldsymbol{v}}$ are split into magnitudes $\boldsymbol{v}$ and phase angles $\boldsymbol{\theta}$. A similar general model can be defined for dc devices.

Typical device models satisfy the following conditions:

1. The vector of state variables $\boldsymbol{x}$ can be divided into internal device variables, say $\boldsymbol{x}_i$, and external ones, say $\boldsymbol{x}_e$. If $\boldsymbol{x}_i \in \mathbb{R}^{n_{xi}}$, then $\boldsymbol{f}_i$ has to map to $\mathbb{R}^{n_{xi}}$.
2. The vector of algebraic variables $\hat{\boldsymbol{y}}$ can be divided into internal device variables, say $\hat{\boldsymbol{y}}_i$, and external ones, say $\hat{\boldsymbol{y}}_e$. If $\hat{\boldsymbol{y}}_i \in \mathbb{R}^{n_{yi}}$, then $\hat{\boldsymbol{g}}_i$ has to map to $\mathbb{R}^{n_{yi}}$.
3. If the device is connected to n_i buses, then both $\boldsymbol{g}_{p,hi}$ and $\boldsymbol{g}_{q,hi}$ have to map to $\mathbb{R}^{n_i}$.

If all devices satisfy these rules, then the complete system (9.1)-(9.3) is well defined, i.e., the number of state variables is equal to the number of differential equations and the number of algebraic variables is equal to the number of algebraic equations. However, the rules above are only sufficient conditions for obtaining a consistent DAE system. As long as $\boldsymbol{f} : \mathbb{R}^{n_x} \times \mathbb{R}^{n_y} \times \mathbb{R}^{n_\eta} \mapsto \mathbb{R}^{n_x}$ and $\boldsymbol{g} : \mathbb{R}^{n_x} \times \mathbb{R}^{n_y} \times \mathbb{R}^{n_\eta} \mapsto \mathbb{R}^{n_y}$, it is not necessary that each device strictly follows these rules.

Finally, in (9.7), it is implicitly assumed that the device depends on a set of parameters, say $\boldsymbol{a}$, which characterize the behavior of the device itself. By default, parameters $\boldsymbol{a}$ are constant.

Example 9.1 Two-Axis Synchronous Machine Model

Consider the two-axis synchronous machine model that is formally deduced in Chapter 15. According to the notation introduced in (9.7), the machine DAE can be subdivided as follows:

$$\boldsymbol{f}_i \Rightarrow \begin{cases} \dot{\delta} = \Omega_b(\omega - \omega_s) \\ \dot{\omega} = (\tau_m - \tau_e - D(\omega - \omega_s))/2H \\ \dot{e}'_q = (-e'_q - (x_d - x'_d)i_d + v_f)/T'_{d0} \\ \dot{e}'_d = (-e'_d + (x_q - x'_q)i_q)/T'_{q0} \end{cases} \tag{9.8}$$

$$\hat{\boldsymbol{g}}_i \Rightarrow \begin{cases} 0 = (v_d + r_a i_d)i_d + (v_q + r_a i_q)i_q - \tau_e \\ 0 = v_q + r_a i_q - e'_q + x'_d i_d \\ 0 = v_d + r_a i_d - e'_d - x'_q i_q \\ 0 = v_h \sin(\delta - \theta_h) - v_d \\ 0 = v_h \cos(\delta - \theta_h) - v_q \\ 0 = \tau_{m0} - \tau_m \\ 0 = v_{f0} - v_f \end{cases}$$

$$\boldsymbol{g}_{p,hi} \Rightarrow \begin{cases} p_{h,i} = v_d i_d + v_q i_q \end{cases}$$

$$\boldsymbol{g}_{q,hi} \Rightarrow \begin{cases} q_{h,i} = v_q i_d - v_d i_q \end{cases}$$

where:

1. The state variables are $\boldsymbol{x} = \boldsymbol{x}_i = [e'_d, e'_q, \delta, \omega]^T$ and $\boldsymbol{x}_e = []$.
2. The algebraic variables are $\hat{\boldsymbol{y}} = \hat{\boldsymbol{y}}_i = [i_d, i_q, v_d, v_f, v_q, \tau_e, \tau_m]^T$ and $\hat{\boldsymbol{y}}_e = []$.
3. The bus voltage is $\bar{v}_h = v_h \angle \theta_h$.
4. The controllable variables are $\boldsymbol{\eta} = [v_{f0}, \tau_{m0}]^T$.
5. The constant parameters are $\boldsymbol{a} = [D, H, r_a, T'_{d0}, T'_{q0}, x_d, x'_d, x_q, x'_q, \Omega_b, \omega_s]^T$.

Observe the use of the trivial equations for assigning the mechanical torque τ_m and the field voltage v_f. Using such dummy algebraic equations is a common practice that allows easily interfacing devices. For example, the mechanical torque τ_{m0} can be set as the output of a turbine governor and the field voltage v_{f0} can be set as the output of an automatic voltage regulator (see also Chapter 16).

9.1.1 Initialization of Device Internal Variables

As discussed in Chapter 4, the solution of the power flow problem provides an operating point, i.e., voltage phasors and complex power injections at each node of the network. Since the standard power flow analysis does not contain dynamic models, the calculation of the equilibrium point $(\boldsymbol{x}_0, \boldsymbol{y}_0)$ requires a specific routine for dynamic devices.

If a devices is initialized after the power flow analysis, bus voltages ($\boldsymbol{v}_0$ and $\boldsymbol{\theta}_0$) and power injections ($\boldsymbol{p}_{h,i0}$ and $\boldsymbol{q}_{h,i0}$) are used as input parameters for computing the initial state variable vector $\boldsymbol{x}_{i0}$, internal algebraic variables $\hat{\boldsymbol{y}}_{i0}$, and/or controllable parameters $\boldsymbol{\eta}_0$, as depicted in Figure 9.1.

The following remarks about the device variable initialization step are relevant:

1. Figure 9.1 only considers the case of a device directly connected to the network. This is not the case of most controllers. For example, AVRs are connected to synchronous machines and PSSs are connected to AVRs. In this case, the initialization has to proceed from device to device, starting for the device connected to the ac grid (see Figure 9.2).
2. Since during the initialization step input and output variables are swapped, the initialization step requires an *ad hoc* procedure for each device.
3. The solution of the system:

$$\begin{aligned} \mathbf{0} &= \boldsymbol{f}_i(\boldsymbol{x}_i, \hat{\boldsymbol{y}}_i, \boldsymbol{v}_0, \boldsymbol{\theta}_0, \boldsymbol{\eta}) \\ \mathbf{0} &= \hat{\boldsymbol{g}}_i(\boldsymbol{x}_i, \hat{\boldsymbol{y}}_i, \boldsymbol{v}_0, \boldsymbol{\theta}_0, \boldsymbol{\eta}) \end{aligned} \tag{9.9}$$

   can be a non-trivial task depending on equation nonlinearity. Typical control schemes are linear and allow defining an explicit solution. In other cases, nonlinearity forces to use an iterative method, such as the Newton's method. See for example the TCSC and the wind turbine models described in Sections 19.2 and 20.2, respectively.

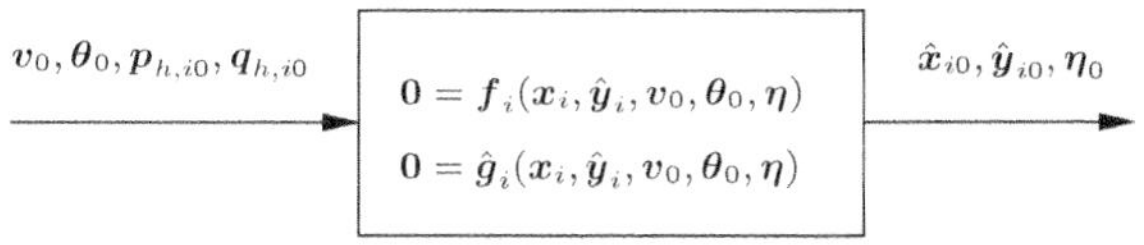

Fig. 9.1 Initialization of a typical dynamic device connected to the ac grid

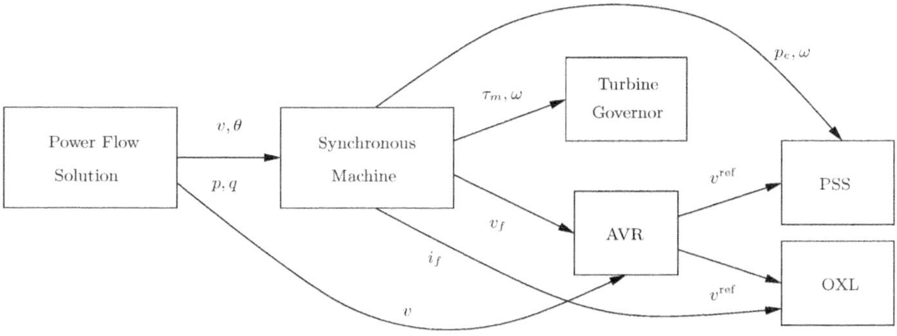

Fig. 9.2 Initialization chain of the synchronous machines and its regulators

Example 9.2 Initialization of the Synchronous Machine Two-Axis Model

To clarify the procedure for initializing state variables, consider the synchronous machine two-axis model of the previous Example 9.1. The set of equations for initializing the two-axis model of the synchronous machine are (see also Section 15.1):

$$\begin{aligned}
0 &= \Omega_b(\omega - \omega_s) \\
0 &= (\tau_m - \tau_e - D(\omega - \omega_s))/2H \\
0 &= (v_q + r_a i_q)i_q + (v_d + r_a i_d)i_d - \tau_e \\
0 &= v_d i_d + v_q i_q - p_0 \\
0 &= v_q i_d - v_d i_q - q_0 \\
0 &= -e'_q - (x_d - x'_d)i_d + v_f \\
0 &= -e'_d + (x_q - x'_q)i_q) \\
0 &= (v_d + r_a i_d)i_d + (v_q + r_a i_q)i_q - \tau_e \\
0 &= v_q + r_a i_q - e'_q + x'_d i_d \\
0 &= v_d + r_a i_d - e'_d - x'_q i_q \\
0 &= v_0 \sin(\delta - \theta_0) - v_d \\
0 &= v_0 \cos(\delta - \theta_0) - v_q
\end{aligned} \tag{9.10}$$

where v_0, θ_0, p_0 and q_0 are the input variables as obtained from the power flow solution and e'_d, e'_q, δ, ω, i_d, i_q, v_d, v_f, v_q, τ_e, τ_m, τ_{m0}, v_{f0} are the quantities to be initialized. In this case, (9.10) have an explicit solution, as follows. The rotor angle δ can be obtained as:

$$\delta_0 = \angle(\bar{v} + (r_a + jx_q)\bar{\imath}) \tag{9.11}$$

where $\bar{v}_0 = v_0 e^{j\theta_0}$ and $\bar{\imath} = (p_0 - jq_0)/\bar{v}_0^*$. Then, it follows that:

$$\begin{aligned}
v_{d0} + jv_{q0} &= \bar{v}e^{-j(\delta_0 - \pi/2)} \\
i_{d0} + ji_{q0} &= \bar{\imath}e^{-j(\delta_0 - \pi/2)} \\
p_{e0} &= (v_{q0} + r_a i_{q0})i_{q0} + (v_{d0} + r_a i_{d0})i_{d0} \\
e'_{q0} &= v_{q0} + r_a i_{q0} + x'_d i_{d0} \\
e'_{d0} &= v_{d0} + r_a i_{d0} - x'_q i_{q0} \\
v_{f0} &= e'_{q0} + (x_d - x'_d)i_{d0}
\end{aligned} \tag{9.12}$$

Finally, in order to impose $\dot{\delta} = 0$, it must be:

$$\omega_0 = \omega_s = 1.0 \text{ pu} \tag{9.13}$$

i.e., the machine speed is synchronous at the initial equilibrium point. This is a consequence of using a fictitious synchronous reference for $\dot{\delta}$. The latter condition leads to

$$\tau_{m0} = \tau_{e0} \tag{9.14}$$

that completes the initialization of the synchronous machine two-axis model.

9.2 Devices as Classes

In computer programming, a *class* is a complex type that contains a variety of properties, namely *attributes* (e.g., data and settings) and *methods* (e.g., functions and procedures). Class attributes are, in some aspects, similar to Leibniz's *monads*, since attributes are entities of irreducible simplicity.[1] Class methods operates on attributes and allows devices interacting with the rest of the program. Thus, unlike monads, class methods do communicate with the rest of the world.

When designing a project based on classes one has to have clear that there is a distinction between the code that implements a certain class and the instances of that class. The code that implements the class is an abstract structure that only defines a new type and states how that new type behaves. On the other hand, an instance of a class is a specific variable whose behavior is defined by the class definition. While the class is unique, there can be any number of class instances. Using a human language analogy, a class is similar to a grammatical object, such as the noun, the verb, the adjective, etc. Words are similar to instances of some grammatical object. For example, *script* is an instance of the class *noun*.

Given this premise, we are interested in defining power system devices as classes. Before proceeding with the definition of classes, there is an important decision that has to be taken. There are two possibilities for implementing such device classes (see Figure 9.3).

1. To create an instance for each device defined in the system. In this case, the interface class that coordinates the functioning of all devices[2] will handle series (or arrays) of class instances.
2. To create an unique instance for each group of devices of the same kind. In this case, the interface class that coordinates the functioning of all devices will handle series of class methods of each device. Each method will internally work on an array of elements of the same device.

At a first glance, these two approaches above may seem equivalent, although, probably, the most intuitive is the first one. In fact, considering again the analogy with human languages, the nouns that are contained in a dictionary

[1] Actually, class attributes may be instances of other classes, but this possibility is not used in this chapter.

[2] For a definition of this class see Script 3.2 of Chapter 3.

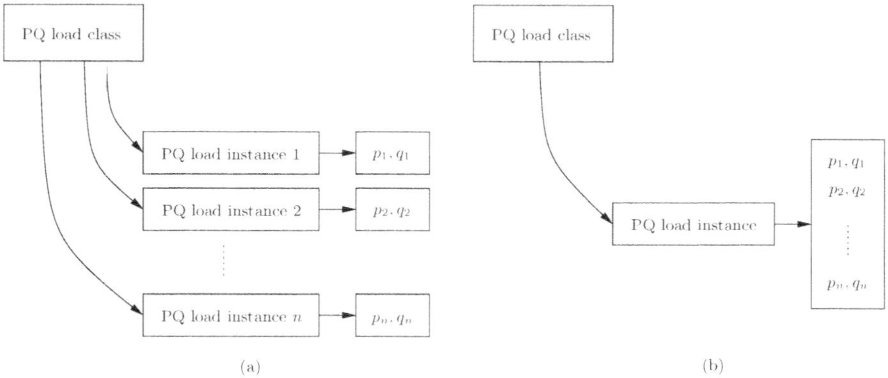

Fig. 9.3 Instancing approaches for device classes: (a) one instance per each physical device, and (b) unique instance per device type

are each one an instance of the grammatical object *noun*. This corresponds to the creation of an instance for each device of the class, say, *PQ load*. According to this approach, whenever the bus power balance (9.5) is computed, one has to look for each instance of the PQ load class and the method that computes the bus power balance has to be called for each one of these instances. The second approach declares only one instance of the PQ load class. Thus, it suffices to call the method for computing the bus power balance only once if PQ loads are defined in the system. The PQ load method internally iterates over each element and computes (9.5).

There is an important difference between iterating over class instances or over elements of a class. In the latter case, operation can be made vectorially since the elements can be grouped into matrices or vectors. When using scripting languages, avoiding explicit for-loops always allows saving CPU time. Thus, for scripting languages, the second approach has to be preferred. For example, in Python, built-in vectorial operations of NumPy arrays or of CVXOPT matrices are by far more efficient than operating over lists or tuples using for loops [164]. On the other hand, system programming languages are not affected by explicit for-loops. Thus, the first approach is also a valid alternative [287]. However, the risk of the first approach is to create a high fragmentation of the information. In fact, for a system with several thousands of devices, there are several thousand of class instances on which operate. In the second approach, since the number of device types is never really high, the program has to handle only a reduced number of class instances. Thus, the second approach should be preferred independently from the programming language adopted.

Assuming that each class has only one instance, the second question that might arise is whether it is really necessary to define a class for each device or if it is better to use some other programming technique (e.g., modules or simple functions). At this regard, it suffices to observe that some properties are

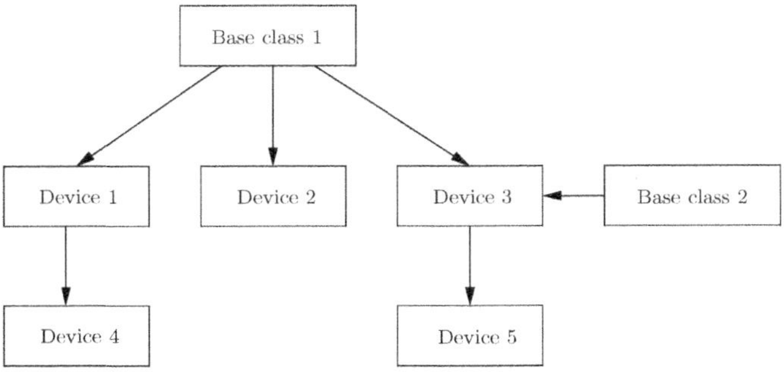

Fig. 9.4 Qualitative representation of class inheritance

common to all devices (as it will be described later on in this section), while other properties are specific of a certain device and make it different from all other devices. Taking advantage of the inheritance property of classes, common methods can be defined in a base class that will be imported by all other specific classes. The inheritance can span more than only one "generation". Furthermore, one device can inherit methods from more than one class. Figure 9.4 illustrates qualitatively the concept of class inheritance.

9.2.1 Base Device Class

The purposes of a base class are twofold: (i) to define common attributes and (ii) to provide common methods for basic device operations. The main difficulty is to be able to build methods sufficiently general to be used smoothly by most devices without the need of write further specific code for each device. Thus, the methods of a base class have to provide a high abstraction level (e.g., *meta-methods*). At this regard, also the attributes of the base class have to be abstract (e.g., *meta-attributes*), otherwise they would not be of general use. A brilliant solution to this problem is provided by the Python language in the form of dictionaries [164].

Let consider an example. Consider the following simple Python class:

```
class simple:

    def __init__(self):

        self.attr1 = 1.0
        self.attr2 = 2.0

    def method1(self):

        print 'method1'

    def method2(self):

        print 'method2'
```

The function `__init__` is the default *constructor* method. This simple class has two attributes (`attr1` and `attr2`) and two methods (`method1` and `method2`). Instantiating this class creates an object with two attributes and two methods, plus some built-in attributes and methods provided by Python. In particular, all attribute names of a class are internally organized in a dictionary type, namely `__dict__`. Thus an instance of the class `simple`, say `instance1` will have an attribute `__dict__` as follows:

```
>>> instance1 = simple()
>>> print instance1.__dict__

{'attr2': 2.0, 'attr1': 1.0}
```

One can thus access any attribute or directly using the attribute name or through the dictionary `__dict__`:

```
>>> print instance1.attr1

1.0

>>> print instance1.__dict__['attr1']

1.0
```

The second syntax may seem only a involved programming style. However, it is the key feature for writing general methods. This concept is exemplified in the following scripts.

Script 9.1 Conversion of Parameter Bases

Let us assume the we want to define a base class for converting the parameters of a generic device to system basis. For the sake of example, let us consider that the possible parameter kinds are only reactances, susceptances and powers. The base class has to operate properly on the parameters depending on their physical functioning.

The base class can be defined as follows.

```
class base:

    def __init__(self):

        self.reactances = []
        self.admittances = []
        self.powers = []

    def setup(self):

        for item in self.reactances:
            self.__dict__[item] = 0
        for item in self.admittances:
            self.__dict__[item] = 0
        for item in self.powers:
            self.__dict__[item] = 0
```

```
    def assign(self, z = [], y = [], p = []):

        # [z] are in Ohm, [y] in Siemens and [p] in MVA

        for item, val in zip(self.reactances, z):
            self.__dict__[item] = val
        for item, val in zip(self.admittances, y):
            self.__dict__[item] = 0
        for item, val in zip(self.powers, p):
            self.__dict__[item] = 0

    def convert(self, Sn, Vn):

        # [Sn] is in MVA and [Vn] in kV

        Zn = Vn*Vn/Sn

        for item in self.reactances:
            self.__dict__[item] /= Zn
        for item in self.admittances:
            self.__dict__[item] *= Zn
        for item in self.powers:
            self.__dict__[item] /= Sn
```

The constructor method only defines three void lists that will be used by device classes for defining the names of reactance, susceptance and power attributes. The three methods of class `base` have the following meaning:

1. Method `setup` initializes the value of each parameters to 0. In this simple example, attributes are assumed to be scalar. A similar although more complex procedure applies for vectorial attributes.
2. Method `assign` assigns custom values to each parameter. Default values for each input argument let that the user is not forced to pass all arguments if these are not used by the device.
3. Method `convert` computes the base conversion using assigned system power and voltage bases, `Sn` and `Vn`, respectively.

An example of device built using class `base` is as follows:

```
class shunt(base):

    def __init__(self):

        base.__init__(self)

        self.admittances = ['b', 'g']
        self.setup()
```

The class above defines a constant impedance device. The constructor of the base class is called by the statement `base.__init__(self)` and declares the three general lists `reactances`, `admittances` and `powers`. This is necessary since the methods of the class `base` expect that all three lists exist. However, the class `shunt` only requires admittances, namely the susceptance `b` and the conductance `g`.

At this point, an instance of the class `shunt` can be created and the system base conversion can be performed.

```
>>> shunt1 = shunt()
>>> shunt1.assign(y = [0.01, 0.02])
>>> shunt1.convert(100.0, 220.0)
>>> print shunt1.b

4.84

>>> print shunt1.g

9.68
```

The class `base` does not "know" anything about the specific parameters of the class `shunt`. However, if the parameters are properly categorized in the meta-attributes `reactances`, `admittances` and `powers` everything works smoothly.

Using meta-methods and meta-attributes also allows easily building up new more general classes. For example, let us assume that we want to define a new model whose parameters include also currents, say a ZIP load. Then, the device code can be as follows:

```
class zip(base):

    def __init__(self):

        base.__init__(self)

        self.impedance = ['r', 'x']
        self.currents = ['ip', 'iq']
        self.powers = ['p', 'q']
        self.setup()

    def setup(self):

        base.setup(self)

        for item in self.currents:
            self.__dict__[item] = 0

    def assign(self, z = [], i = [], p = []):

        base.assign(self, z = z, y = [], p = p)

        for item, val in zip(self.currents, i):
            self.__dict__[item] = val

    def convert(self, Sn, Vn):

        base.convert(self, Sn, Vn)

        # assuming a three-phase power and a phase-to-phase voltage
```

```
        In = Sn/Vn/(3**0.5)

        for item in self.currents:
            self.__dict__[item] /= In
```

All original methods of the class `base` are overloaded to take into account the new parameter type `currents`.

Script 9.2 Meta-attributes of a Base Device Class

A base class for power system devices requires necessarily much more meta-methods and meta-attributes than the ones discussed in the previous example. However, such base class is not conceptually more complex than the class `base`. The following code shows a possible constructor method for a base device class on top of which any other device can be built similarly to the previous example.

```
class device:

    def __init__(self):

        self.n = 0        # number of devices
        self.u = []       # device status
        self.name = []    # device names
        self.int =    # index dictionary

        self.properties = {'gcall':False,   'gycall':False,
                           'fcall':False,   'fxcall':False,
                           'windup':False,  'pflow':False,
                           'xinit':False,   'shunt':False,
                           'series':False,  'flows':False,
                           'connection':False, 'times':False,
                           'stagen':False,  'dyngen':False,
                           'gmcall':False,  'fmcall':False,
                           'dcseries':False, 'opf':False,
                           'obj':False}

        # meta-attributes

        self._type = None   # device type
        self._name = None   # device name
        self._bus = {}      # ac bus indexes
        self._node = {}     # dc node indexes
        self._states = []   # list of state variables
        self._algebs = []   # list of algebraic variables

        self._ymarket = [] # OPF variable indexes
        self._cmarket = []
        self._opf = {}

        # data dictionary
        self._data = {'u':1, 'Sn':100.0, 'Vn':220.0}
```

```
        # parameters
        self._params = ['u', 'Sn', 'Vn']

        # parameters that cannot be zero
        self._zeros = ['Sn', 'Vn']

        # parameter units
        self._units = {'Sn':'MVA', 'Vn':'kV', 'u':'boolean'}

        # parameter descriptions
        self._descr = {'Sn':'rated power', 'Vn':'rated voltage',
                       'u':'connection status'}

        self._powers = []   # powers, inertiae and dampings
        self._voltages = [] # voltages
        self._currents = [] # currents
        self._z = []        # impedances
        self._y = []        # admittances

        self._dccurrents = [] # dc currents
        self._dcvoltages = [] # dc voltages
        self._r = []          # resistances (for dc circuits)
        self._g = []          # susceptances (for dc circuits)

        self._times = []     # time constants
        self._service = []   # auxiliary variables
        self._mandatory = [] # list of mandatory parameters
```

In the previous code, the first four attributes, namely `n`, `u`, `name` and `int`, are common to all system devices. The attribute `n` is an integer that indicates the number of devices of the same kind, while `u` and `name` are lists containing the statuses (e.g., on-line or off-line) and the names of the device elements. Finally, the dictionary `int` contains the pairs of identification codes (or, simply, *id*) and indexes of the device.[3] Since the built-in Python type dictionary is quite efficient,[4] using the dictionary `int` provides both flexibility and speed.

The attribute `properties` is a dictionary that defines the functioning of the device. The base class initializes all properties as `False`. Then the device that inherits the base class switches to `True` the properties that are pertinent. For example, if we want to define a static shunt device to be used in power flow analysis, one can set:

```
self.properties.update({'gcall':True, 'gycall':True, 'pflow':True})
```

[3] An "id" is a number or a string that is assigned by the user to a device item. On the other hand, an index is an integer that is internally assigned and used by the program for locating or pointing to an item of device arrays. To separate the ids from the indexes increases the flexibility of the code and is common practice in computer programming.

[4] For example, if `listvar` and `dictvar` are a list and a dictionary containing the same data, the operation `item in listvar` is much slower than `dictvar.has_key(item)`

The attribute properties is used by the interface class device for properly handling the bridge between devices and routines (for further details see Script 3.2 of Chapter 3).

Meta-attributes are indicated using the Python convention of preceding the name of private attributes with an underscore "_".[5] The proposed list of meta-parameters is commented in the script above. Of course, this is just a possible implementation. Any other meta-parameter can be added directly in the class device or locally in the child class.

The dictionary _data provides the list of data required by the device. By default, three data are defined for any devices, namely the status u, the nominal power Sn and the nominal voltage Vn. Associated to each datum, there is also a default value. The attribute list _params is a subset of the dictionary _data. The attribute names included in _params are those that have to be declared as CVXOPT arrays and that are used in vectorial operations.

To complete this example and clarify the usage of the general class device discussed above, the following script provides the implementation of the constructor method of the synchronous machine two-axis model, whose equations are given in Example 9.1.

```
class synchronous(device):

    def __init__(self):

        device.__init__(self)

        # private data
        self._type = 'Synchronous machine'
        self._name = 'Synchronous machine two-axis model'
        self._data.update({'fn':50, 'bus':None, 'H':3,
                           'ra':0.0, 'xd':1.9, 'xd1':0.302,
                           'xq':1.7, 'xq1':0.5,
                           'Td10':8.0, 'Tq10':0.8,
                           'D':0.0, 'gen':None,
                           'gammap':1.0, 'gammaq':1.0})
        self._params.extend(['D', 'H', 'ra', 'xd', 'xd1',
                             'xq', 'xq1', 'Td10', 'Tq10',
                             'gammap', 'gammaq'])
        self._bus = {'bus':['a', 'v']}
        self._z = ['ra', 'xd', 'xd1', 'xq', 'xq1']
        self._name = 'Syn'
        self._powers= ['H', 'D']
        self._states = ['delta', 'omega', 'e1d', 'e1q']
        self._algebs = ['tm', 'te', 'Id', 'Iq', 'vd', 'vq']
        self._service = ['tm0', 'vf0']
        self._times = ['Td10', 'Tq10']
```

[5] According to official Python conventions, using a single underscore indicates that the attribute is private but is inherited by subclasses. A double underscore indicates a private attribute that is not inherited by subclasses. However, in the current Python release, any parameter is inherited by subclasses, regardless the number of preceding underscores.

```
        self._zeros = ['H', 'Td10', 'Tq10']
        self._mandatory = ['bus', 'gen']
        self._descr.update({'H':'inertia constant',
                            'D':'rotor damping',
                            'fn':'rated frequency',
                            'bus':'bus id',
                            'gen':'static generator id',
                            'ra':'armature resistance',
                            'Td10':'d-axis transient time constant',
                            'Tq10':'q-axis transient time constant',
                            'xd':'d-axis synchronous reactance',
                            'xd1':'d-axis transient reactance',
                            'xd':'d-axis synchronous reactance',
                            'xq1':'q-axis transient reactance',
                            'gammap':'active power ratio',
                            'gammaq':'reactive power ratio'})
        self._units.update({'H':'MWs/MVA', 'D':'pu', 'fn':'Hz',
                            'ra':'pu', 'xd':'pu', 'xd1':'pu', 'xq':'pu',
                            'xq1':'pu', 'Td10':'pu', 'Tq10':'pu',
                            'gammaq':'pu', 'gammap':'pu'})
        self.properties.update({'gcall':True, 'gycall':True,
                                'fcall':True, 'fxcall':True,
                                'xinit':True, 'dyngen':True})
```

In the previous code, the meaning of most variables is self-explicative or is defined in the dictionary `_descr`. The use of the parameter `gen` that indicates the id of the static generator (e.g., PV bus) used for initializing the synchronous machine state and algebraic variables. Parameters `gammap` and `gammaq` that indicate the ratio in per unit of the static generator active and reactive power, respectively, produced by the synchronous machine. These factors allows connecting more than one machine to the same bus and properly initialize all machines (for the initialization process see also Example 9.2). Parameters `ra`, `xd`, `xd1`, `xq` and `xq1` are included in the list `_z`, whereas parameters `H` and `D` in the list `_powers`. In this way, the meta-method that converts device parameters to system bases will operate correctly. It is necessary to include the damping parameter `D` in the list `_powers` since in the differential equation that returns $\dot{\omega}$, the damping is divided by the inertia constant `H`. So, since `H` has to be converted to the system power base, the ratio `D/H`, is not affected by this base conversion. Controllable parameters variables τ_{m0} (`tm0`) and v_{f0} (`vf0`) are included in the attribute `service`. These values can remain constant or be modified by other devices (e.g., a turbine governor). Finally, the modifications to the attribute `properties` indicate that the device contains algebraic and differential equations and Jacobians, that has to be initialized (`xinit`) and that is a dynamic generator (`dyngen`).

9.2.2 *Methods of the Base Class*

The methods (or meta-methods) of the class `device` are basically the same as the ones defined for the class `base` introduced in Script 9.1. A rough list of these methods is as follows.[6]

1. Initialization of all data and indexes defined in the constructor `__init__`. By default, all attributes are initialized as empty lists, as follows.

```
def _init_data(self):

    for arg in self._data:
        self.__dict__[arg] = []

    for key in self._bus:
        for item in self._bus[key]:
            self.__dict__[item] = []

    for arg in self._node.values():
        self.__dict__[arg] = []

    for arg in self._states:
        self.__dict__[arg] = []

    for arg in self._algebs:
        self.__dict__[arg] = []

    for arg in self._service:
        self.__dict__[arg] = []

    if self._name is None: self._name = self._type
```

2. Conversion of parameter values to system bases. In the following code only ac base conversion is considered.

```
def base(self, Sb=100.0, Vb=None):

    for var in self._voltages:
        self.__dict__[var] = mul(self.__dict__[var], self.Vn)
        self.__dict__[var] = div(self.__dict__[var], Vb)

    for var in self._powers:
        self.__dict__[var] = mul(self.__dict__[var], self.Sn)
        self.__dict__[var] /= Sb

    for var in self._currents:
        self.__dict__[var] = mul(self.__dict__[var], self.Sn)
        self.__dict__[var] = div(self.__dict__[var], self.Vn)
        self.__dict__[var] = mul(self.__dict__[var], Vb)
        self.__dict__[var] /= Sb
```

[6] The script examples that are provided in this section refer only to ac devices. For dc devices, similar methods can be defined.

```
    if len(self._z) or len(self._y):
        Zn = div(self.Vn**2, self.Sn)
        Zb = (Vb**2)/Sb

        for var in self._z:
            self.__dict__[var] = mul(self.__dict__[var], Zn)
            self.__dict__[var] = div(self.__dict__[var], Zb)

        for var in self._y:
            if self.__dict__[var].typecode == 'd':
                self.__dict__[var] = div(self.__dict__[var], Zn)
                self.__dict__[var] = mul(self.__dict__[var], Zb)
            elif self.__dict__[var].typecode == 'z':
                self.__dict__[var] = div(self.__dict__[var], Zn + 0j)
                self.__dict__[var] = mul(self.__dict__[var], Zb + 0j)
```

The base conversion of dc parameters undergoes a similar procedure. The method arguments `Sb` and `Vb` are the system power base in MVA and a vector of voltage bases in kV, respectively.

3. Inclusion of a new element to the device instance. This method is used by routines that parse input data. The code below accepts as input arguments the index and the name of the new element, plus a variable-length dictionary of the input data. If no index or name is provided, a default index and name are assigned to the new element. Then the attribute `self.n` is incremented by 1. The following step is to check whether the input data contains mandatory parameters. If not, a warning message is displayed. Then, the default values are assigned to all data. In this way, the user does not need to pass all data when calling the method `add` but only those that are known or relevant. The data provided by the user overwrite default data. While updating data, some consistency check are performed. For example, if a data is not part of the keys of the dictionary `_data` a warning message is displayed. A check of data that cannot be zero is also performed.

```
def add(self, idx=None, name=None, **kwargs):

    if idx is None: idx = self._type + '_' + str(self.n + 1)

    self.int[idx] = self.n
    self.n += 1

    if name is None:
        self.name.append(self._type + ' ' + str(self.n))
    else:
        self.name.append(name)

    # check whether mandatory parameters have been set up
    for key in self._mandatory:
        if not kwargs.has_key(key):
            print 'Mandatory parameter <%s> has not been set up' % key
```

```
    # set default values
    for key, value in self._data.iteritems():
        self.__dict__[key].append(value)

    # overwrite custom values
    for key, value in kwargs.iteritems():

        if not self._data.has_key(key):
            print 'This device has no parameter called <%s>.' % key
            continue

        self.__dict__[key][-1] = value

        # check data consistency
        if not value and key in self._zeros:
            if key == 'Sn':
                default = system.Settings.mva
            elif key == 'Vn':
                default = self._data[key]
            elif key == 'fn':
                default = system.Settings.freq
            else:
                default = self._data[key]
            self.__dict__[key][-1] = default
```

4. Post-parsing operations. These operations consist in setting up the indexes of device algebraic and state variables. These indexes allows locating the elements of the vectors $\boldsymbol{g}$, $\boldsymbol{f}$, $\boldsymbol{y}$ and $\boldsymbol{x}$ and of the Jacobian matrices $\boldsymbol{g_y}$, $\boldsymbol{g_x}$, $\boldsymbol{f_y}$ and $\boldsymbol{f_x}$ associated with the each device element. Assigning indexes is a key operation since it makes each device to properly interact with the the system. For example, in order to update differential equations, a device has to know the position (indexes) of its differential equations $\boldsymbol{f}_i$ in the system vector of differential equations $\boldsymbol{f}$. The following method implements a possible way of assigning algebraic and state variables indexes. It is assumed that if a state variable has a position k in the vector $\boldsymbol{x}$, then the index of the differential equation that computes $\dot{x}_k$ is also k. A similar convention is assumed for the indexes of algebraic variables and equations $\hat{\boldsymbol{y}}_i$ and $\hat{\boldsymbol{g}}_i$.

```
def _xy_index(self, dae):

    zeros = [0]*self.n
    for item in self._states:
        self.__dict__[item] = zeros[:]
    for item in self._algebs:
        self.__dict__[item] = zeros[:]

    for var in range(self.n):

        for item in self._states:
            self.__dict__[item][var] = system.DAE.nx
```

```
            system.DAE.nx += 1

        for item in self._algebs:
            self.__dict__[item][var] = system.DAE.ny
            system.DAE.ny += 1
```

Where the scalars `system.DAE.ny` and `system.DAE.nx` indicate the number of total algebraic and state variables, respectively. A slightly different approach is required for assigning the indexes of ac voltage magnitudes v_h and phases θ_h as well as of equations $\boldsymbol{g}_{p,hi}$ and $\boldsymbol{g}_{q,hi}$. The indexes of these variables and equations are assigned by the device `system.Bus`. Thus, a device connected to a certain bus has to retrieve the indexes of v_h and phases θ_h from `system.Bus` itself. This is easily obtained by providing to the device the indexes of the buses to which are connected.

```
def _bus_index(self):

    for index in self._bus.keys():
        for item in self.__dict__[index]:
            if not system.Bus.int(has_key(idx):
                self.message('Bus index <%s> does not exist',
                             data_tuple=item, level=self.ERROR)
            else:
                idx = system.Bus.int[item]
                self.__dict__[self._bus[index][0]].append(system.Bus.a[idx])
                self.__dict__[self._bus[index][1]].append(system.Bus.v[idx])
```

For example, if `self._bus = {'bus':['a', 'v']}`, the previous code assigns to the attributes `self.a` and `self.v` the elements `self.bus` of `system.Bus.a` and `system.Bus.v`, respectively. Another post-parsing operation consists in converting to CVXOPT arrays all parameters of the list `_param`, as follows:

```
def _list2matrix(self):

    for item in self._params:
        self.__dict__[item] = matrix(self.__dict__[item])
```

5. Deletion of an element from the device instance. Removing an element can be useful for internal operations. For example, after solving the power flow analysis and before running a time domain simulation, it could be necessary to remove static generators at the buses where there are synchronous machines (i.e., synchronous machines substitute static generators in dynamic analysis). The code below accepts as input the index of the element that has to be removed. If the index is not defined, then the procedure exits. Otherwise, all parameter arrays and lists are processed and the element at the position `item` is popped out.

```
def remove(self, idx=None):

    if idx != None:
        if not self.int.has_key(idx):
```

```
        item = self.idx.index[idx]
        key = idx
    else:
        print 'The item <%s> does not exist.' % idx
        return None
else:
    # nothing to remove
    return None

convert = False

if isinstance(self.__dict__[self._params[0]], matrix):
    self._matrix2list()
    convert = True

self.n -= 1
self.int.pop(key, '')
self.idx.pop(item)

for x, y in self.int.iteritems():
    if y > item: self.int[x] = y - 1

for param in self._data:
    self.__dict__[param].pop(item)

for param in self._service:
    if len(self.__dict__[param]) == (self.n + 1):
        if isinstance(self.__dict__[param], list):
            self.__dict__[param].pop(item)
        elif isinstance(self.__dict__[param], matrix):
            service = list(self.__dict__[param])
            service.pop(item)
            self.__dict__[param] = matrix(service)

for x in self._states:
    if len(self.__dict__[x]): self.__dict__[x].pop(item)

for y in self._algebs:
    if self.__dict__[y]: self.__dict__[y].pop(item)

for key, param in self._bus.iteritems():
    if isinstance(param, list):
        for subparam in param:
            if len(self.__dict__[subparam]):
                self.__dict__[subparam].pop(item)
    else:
        self.__dict__[param].pop(item)

for key, param in self._node.iteritems():
    self.__dict__[param].pop(item)

self.name.pop(item)
if convert and self.n: self._list2matrix()
```

The built-in method `pop` is available for both lists and dictionaries and is used in the script for removing the assigned element. Since CVXOPT array dimension cannot be modified and do not provide the method `pop`, CVXOPT arrays are firstly converted into lists through the method `_matrix2list` which implements the opposite conversion than the method `_list2matrix`.

6. Handling windup and anti-windup limiters. A description of windup and anti-windup limiters as well as the implementation of the correspondent meta-methods is provided in Appendix C and Script C.1, respectively.

9.2.3 Specific Device Methods

A description and an example of specific device methods conclude this chapter. Specific methods define the mathematical model of the device, as follows.

1. Initialization of state and algebraic variables (for devices initialized after the power flow analysis).
2. Algebraic equations $\boldsymbol{g}$.
3. Differential equations $\boldsymbol{f}$.
4. Jacobian matrix $\boldsymbol{g}_{\boldsymbol{y}}$.
5. Jacobian matrices $\boldsymbol{g}_{\boldsymbol{x}}$, $\boldsymbol{f}_{\boldsymbol{y}}$ and $\boldsymbol{f}_{\boldsymbol{x}}$.[7]
6. Objective function, inequality constraints and Hessian matrix (for devices that are included in the OPF problem).
7. Windup and anti-windup limiters (for devices with variables than can saturate).

Script 9.3 Methods of the Synchronous Machine Two-Axis Model

The following methods implement the algebraic differential equations and Jacobian matrices as well as the initialization function for the two-axis model of the synchronous machine. It is assumed that the constructor method is the one presented in the previous Script 9.2 and that all parameters are vectors initialized by the function `matrix` of the module CVXOPT. For all functions, the following header is assumed:

```
import system
from cvxopt.base import spmatrix, matrix
from cvxopt.base import mul, div, exp, log, sin, cos
```

1. Initialization of algebraic and state variables. Since the synchronous machine model is not used in power flow analysis, this device has to be initialized after the solution of the power flow problem. The equations

[7] The Jacobian matrix $\boldsymbol{g}_{\boldsymbol{y}}$ is taken apart from other Jacobian matrices to allows implementing efficiently explicit numerical methods (see Chapter 8).

used for initializing the synchronous machine two-axis model are given in Example 9.2.

```
def xinit(self, dae):

    p0 = mul(mul(self.u, system.Bus.pg[self.a]), self.gammap)
    q0 = mul(mul(self.u, system.Bus.qg[self.a]), self.gammaq)
    v0 = mul(self.u, dae.y[self.v])
    theta0 = dae.y[self.a]

    V = mul(v0 + 0j, exp(theta0*1j))
    S = p0 - q0*1j
    I = div(S, V.H.T)
    E = V + mul(self.ra + self.xq*1j, I)
    delta = log(div(E, abs(E) + 0j))
    dae.x[self.delta] = mul(self.u, delta.imag())
    dae.x[self.omega] = matrix(1.0, (self.n, 1), 'd')

    # d- and q-axis voltages and currents
    jpi2 = 1.5707963267948966j
    vdq = mul(self.u + 0j, mul(V, exp(jpi2 - delta)))
    idq = mul(self.u + 0j, mul(I, exp(jpi2 - delta)))

    vd = dae.y[self.vd] = vdq.real()
    vq = dae.y[self.vq] = vdq.imag()
    Id = dae.y[self.Id] = idq.real()
    Iq = dae.y[self.Iq] = idq.imag()

    self.tm0 = mul(vq + mul(self.ra, Iq), Iq) + \
               mul(vd + mul(self.ra, Id), Id)
    dae.y[self.tm] = self.tm0

    dae.x[self.e1q] = vq + mul(self.ra, Iq) + mul(self.xd1, Id)
    dae.x[self.e1d] = vd + mul(self.ra, Id) - mul(self.xq1, Iq)
    self.vf0 = dae.x[self.e1q] + mul(self.xd - self.xd1, Id)
    dae.y[self.vf] = self.vf0

    system.Device.remove_gen(self.gen)
```

In this example, it is assumed that generated active and reactive powers are stored in the vectors `system.Bus.Pg` and `system.Bus.Qg`, while the bus voltage is contained in the variable `dae` (which is passed as an argument of the method and coincides with `system.DAE`). In the last line, the interface class `system.Device` is called for removing, if required, the static generator connected at the synchronous machine buses. A possible implementation of the method `system.Device.remove_gen` is as follows:

```
def remove_gen(self, idx):

    for item, stagen in zip(self.devices, self.stagen):
        if stagen:
            indexes = system.__dict__[item].int.keys()
            for key in idx:
```

```
            if key in indexes:
                system.__dict__[item].remove(key)
```

where `self.devices` is is the list of devices currently in use and the attribute `self.stagen` is a list of the same length as `self.devices` and whose elements are `True` if the correspondent device is a static generator, `False` otherwise.

2. Algebraic equations.

```
def gcall(self, dae):

    delta = dae.x[delta]
    e1d = dae.x[e1d]
    e1q = dae.x[e1q]

    v = mul(self.u, dae.y[self.v])
    theta = dae.y[self.a]

    tm = dae.y[self.tm]
    vf = dae.y[self.vf]
    te = dae.y[self.te]
    vq = mul(self.u, dae.y[self.vq])
    vd = mul(self.u, dae.y[self.vd])
    Iq = mul(self.u, dae.y[self.Iq])
    Id = mul(self.u, dae.y[self.Id])

    # internal algebraic equations

    dae.g[self.te] = mul(vq + mul(self.ra, Iq), Iq) + \
                     mul(vd + mul(self.ra, Id), Id) - te
    dae.g[self.Id] = vq + mul(self.ra, Iq) - e1q + mul(self.xd1, Id)
    dae.g[self.Iq] = vd + mul(self.ra, Id) - e1d - mul(self.xq1, Iq)
    dae.g[self.vd] = mul(v, sin(delta - theta)) - vd
    dae.g[self.vq] = mul(v, cos(delta - theta)) - vq
    dae.g[self.tm] = mul(self.u, self.tm0) - tm
    dae.g[self.vf] = mul(self.u, self.vf0) - vf

    # network interface equations

    # active power
    dae.g[self.a] += spmatrix(mul(vd, Id) + mul(vq, Iq), \
                     self.a, [0]*self.n, (dae.ny, 1), 'd')
    # reactive power
    dae.g[self.v] += spmatrix(mul(vq, Id) - mul(vd, Iq), \
                     self.v, [0]*self.n, (dae.ny, 1), 'd')
```

There is a conceptual difference between internal algebraic equations $\hat{\boldsymbol{g}}_i$ and $g_{p,hi}$ and $g_{q,hi}$. Equations $\hat{\boldsymbol{g}}_i$ are specific of the synchronous machine model and are not shared with other devices. Thus a direct indexation works fine. On the other hand, $g_{p,hi}$ and $g_{q,hi}$ are the synchronous machine contribution to the power balance at the bus h. Since other devices (e.g., transmission lines) sum their power injections to the same bus, the function `gcall` has to update and not to overwrite the current value of $g_{p,hi}$ and $g_{q,hi}$.

The status vector `self.u` systematically multiplies parameters and/or variables to impose $\hat{\boldsymbol{g}}_i = \mathbf{0}$, $g_{p,hi} = 0$ and $g_{q,hi} = 0$ in case some synchronous machine element is off-line.

3. Differential equations.

```
def fcall(self, dae):

    # differential equations

    omega = dae.x[self.omega]
    tm = dae.y[self.tm]
    te = dae.y[self.te]
    vf = dae.y[self.vf]

    iTd = div(self.u, self.Td10)
    xTd = mul(iTd, self.xd - self.xd1)

    iTq = div(self.u, self.Tq10)
    xTq = mul(iTq, self.xq - self.xq1)

    dae.f[self.delta] = system.Settings.rad * \
                        mul(self.u, omega - 1)
    dae.f[self.omega] = mul(self.u, div(tm - te - \
                        mul(self.D, omega - 1), 2*self.H))
    dae.f[self.e1q] = mul(iTd, vf)  - mul(xTd, dae.y[self.Id]) - \
                      mul(iTd, dae.x[self.e1q])
    dae.f[self.e1d] = mul(xTq, dae.y[self.Iq]) - mul(iTq, dae.x[self.e1d])
```

The variable `system.Settings.rad` contains the base synchronous speed in rad/s. The status vector `self.u` imposes $\boldsymbol{f}_i = \mathbf{0}$ if some synchronous machine element is switched off.

4. Jacobian matrices.

```
def gycall(self, dae):

    delta = dae.x[delta]

    v = mul(self.u, dae.y[self.v])
    theta = dae.y[self.a]

    vq = mul(self.u, dae.y[self.vq])
    vd = mul(self.u, dae.y[self.vd])
    Iq = mul(self.u, dae.y[self.Iq])
    Id = mul(self.u, dae.y[self.Id])
    ny_x_ny = (dae.ny, dae.ny)

    # internal Jacobians

    dae.Gy -= spmatrix(1, self.te, self.te, ny_x_ny, 'd')
    dae.Gy += spmatrix(Iq, self.te, self.vq, ny_x_ny, 'd')
    dae.Gy += spmatrix(Id, self.te, self.vd, ny_x_ny, 'd')
```

```
        dae.Gy += spmatrix(vq + 2*mul(self.ra, Iq), self.te, \
                           self.Iq, ny_x_ny, 'd')
        dae.Gy += spmatrix(vd + 2*mul(self.ra, Id), self.te, \
                           self.Id, ny_x_ny, 'd')

        dae.Gy += spmatrix(self.u, self.Id, self.vq, ny_x_ny, 'd')
        dae.Gy += spmatrix(mul(self.u, self.ra), self.Id, self.Iq, \
                           ny_x_ny, 'd')
        dae.Gy += spmatrix(self.xd1, self.Id, self.Id, ny_x_ny, 'd')

        dae.Gy += spmatrix(self.u, self.Iq, self.vd, ny_x_ny, 'd')
        dae.Gy += spmatrix(mul(self.u, self.ra), self.Iq, self.Id, \
                           ny_x_ny, 'd')
        dae.Gy -= spmatrix(self.xq1, self.Iq, self.Iq, ny_x_ny, 'd')

        dae.Gy -= spmatrix(1, self.vd, self.vd, ny_x_ny, 'd')
        dae.Gy += spmatrix(mul(self.u, sin(delta - theta)), \
                           self.vd, self.v, ny_x_ny, 'd')
        dae.Gy -= spmatrix(mul(v, cos(delta - theta)), \
                           self.vd, self.a, ny_x_ny, 'd')

        dae.Gy -= spmatrix(1, self.vq, self.vq, ny_x_ny, 'd')
        dae.Gy += spmatrix(mul(self.u, cos(delta - theta)), \
                           self.vq, self.v, ny_x_ny, 'd')
        dae.Gy += spmatrix(mul(v, sin(delta - theta)), \
                           self.vq, self.a, ny_x_ny, 'd')

        dae.Gy -= spmatrix(1, self.tm, self.tm, ny_x_ny, 'd')
        dae.Gy -= spmatrix(1, self.vf, self.vf, ny_x_ny, 'd')

        # network interface Jacobians

        dae.Gy += spmatrix(Id, self.a, self.vd, ny_x_ny, 'd')
        dae.Gy += spmatrix(vd, self.a, self.Id, ny_x_ny, 'd')
        dae.Gy += spmatrix(Iq, self.a, self.vq, ny_x_ny, 'd')
        dae.Gy += spmatrix(vq, self.a, self.Iq, ny_x_ny, 'd')

        dae.Gy += spmatrix(Id, self.v, self.vq, ny_x_ny, 'd')
        dae.Gy += spmatrix(vq, self.v, self.Id, ny_x_ny, 'd')
        dae.Gy -= spmatrix(Iq, self.v, self.vd, ny_x_ny, 'd')
        dae.Gy -= spmatrix(vd, self.v, self.Iq, ny_x_ny, 'd')

    def fxcall(self, dae):

        delta = dae.x[delta]
        theta = dae.y[self.a]
        v = mul(self.u, dae.y[self.v])

        # Jacobian matrix g_x

        dae.Gx += spmatrix(mul(v, cos(delta - theta)), self.vd, \
                           self.delta, (dae.ny, dae.nx), 'd')
        dae.Gx -= spmatrix(mul(v, sin(delta - theta)), self.vq, \
                           self.delta, (dae.ny, dae.nx), 'd')
```

```
dae.Gx -= spmatrix(self.u, self.Id, self.e1q, (dae.ny, dae.nx), 'd')
dae.Gx -= spmatrix(self.u, self.Iq, self.e1d, (dae.ny, dae.nx), 'd')

# Jacobian matrix f_y

dae.Fy += spmatrix(div(self.u, 2*self.H), self.omega, \
                   self.tm, (dae.nx, dae.ny), 'd')
dae.Fy -= spmatrix(div(self.u, 2*self.H), self.omega, \
                   self.te, (dae.nx, dae.ny), 'd')
dae.Fy += spmatrix(div(self.u, self.Td10), self.e1q, \
                   self.vf, (dae.nx, dae.ny), 'd')
dae.Fy -= spmatrix(div(mul(self.u, self.xd - self.xd1), self.Td10), \
                   self.e1q, self.Id, (dae.nx, dae.ny), 'd')
dae.Fy += spmatrix(div(mul(self.u, self.xq - self.xq1), self.Tq10), \
                   self.e1d, self.Iq, (dae.nx, dae.ny), 'd')

# Jacobian matrix f_x

dae.Fx += spmatrix(1 - self.u, self.delta, \
                   self.delta, (dae.nx, dae.nx), 'd')
dae.Fx += spmatrix(system.Settings.rad*self.u, self.delta, \
                   self.omega, (dae.nx, dae.nx), 'd')
dae.Fx -= spmatrix(div(self.D, 2*self.H), self.omega, \
                   self.omega, (dae.nx, dae.nx), 'd')
dae.Fx -= spmatrix(div(1.0, self.Td10), self.e1q, \
                   self.e1q, (dae.nx, dae.nx), 'd')
dae.Fx -= spmatrix(div(1.0, self.Tq10), self.e1d, \
                   self.e1d, (dae.nx, dae.nx), 'd')
```

In case an element is off-line, all Jacobians matrices are null except for the diagonal elements, which cannot be zero to avoid singularities.

Chapter 10
Power Flow Devices

This chapter describes topological elements as well as standard shunt (i.e., connected to a single bus) devices for power flow analysis. The most important topological element is the bus, while standard devices are constant $v\theta$ generators, PV generators, PQ generators, PQ loads and constant and switched shunt admittances.

10.1 Topological Elements

This section describes the main topological elements used in power system analysis, namely buses, zones, areas, regions and systems. The main data and constraints associated with topological elements are also discussed.

10.1.1 Bus

The minimal network unit is the bus. Shunt devices are connected to buses, while series devices connect two or more buses together.

The name *bus* is an abbreviation of the Latin word *omnibus*, which means *for anybody*. This is actually what a bus should be: a connection point for any device. Unfortunately, the original meaning of the word bus is often misused in power flow analysis. Several software packages and books refer to "PV bus", "slack bus" or "PQ bus". These expressions mix together two different concept: (i) the concept of bus, which provides a topological information and (ii) the concept of generators or loads, which are devices connected to a bus. Actually, there is no real matter in connecting a PQ load and a PV generator at the same bus. Thus, strictly speaking, expressions as "PV bus" or "PQ bus" are misleading and should be avoided.

In power flow analysis, buses are intended as topological nodes, not as physical connections. This may prevent modelling physical bus-bars. For example it can be necessary to model buses composed by different bars or divided into

F. Milano: Power System Modelling and Scripting, Power Systems, pp. 247–261.
springerlink.com

sections. These features can be easily modelled by means of *coupling* devices (see Section 11.1.5 of Chapter 11).

Typical bus parameters are depicted in Table 10.1. Bus numbers (or names) are used by all other devices for identifying the bus to which are connected. Thus, each bus has to be identified by a unique number, code or name. In some old formats (e.g., IEEE common data format [350]), bus identification codes are integers. This was mainly due to old-style system programming languages such as FORTRAN that hardly handle hash types (e.g., lists of key-value pairs). Modern script languages provide a simple manner to handle hashes. The Python implementation of the hash is the built-in type *dictionary*. A key can be any number or string, the only requisite is that keys have to be unique. Then, the value associated to each key is the bus index h. In general, hashes provide a great programming flexibility and should be preferred to other indexing methods.

Table 10.1 Bus parameters

Variable	Description	Unit
-	Bus code	-
-	Bus name	-
-	Area code	-
-	Zone code	-
-	Region code	-
-	System code	-
$v^{(0)}$	Voltage amplitude initial guess	pu
V_b	Voltage rating	kV
$v^{\max}$	Maximum admissible voltage	pu
$v^{\min}$	Minimum admissible voltage	pu
$\theta^{(0)}$	Voltage phase initial guess	rad

The voltage rating V_b is generally used in power flow analysis for fixing the voltage bases for per unit analysis. Voltage magnitudes $v^{(0)}$ and phases $\theta^{(0)}$ can be optionally set if the power flow solution is known or if a custom initial guess is needed. If voltages are not specified, a flat start is used (e.g., $v^{(0)} = 1$ at all buses except for buses where a PV or slack generator is present, and $\theta^{(0)} = 0$). Finally, area, zone, region and system codes are generally used for evaluating inter-area power flows or simply to assign an owner to the bus.

Since buses contain only topological information, no equation is required for defining the bus model. The only issue that can arise is in case a bus is islanded. In that case, the system admittance matrix may become singular if it is not properly conditioned. The easiest solution is to find islanded buses and to impose that the power balance at those buses is zero. This also implies that, the Jacobian matrix has to be properly conditioned, as follows. If a given bus h is islanded, the columns associated with the derivatives with respect to

θ_h and v_h and the rows associated with the derivatives of p_h, q_h have to be set to zero except for diagonal elements that have to set to 1. A simple script that implements this operation is presented in Script C.1 of Appendix C. Further discussion about islanded buses as well as the more general concept of network connectivity is given in Subsection 11.3.3 of Chapter 11.

10.1.2 *Areas, Zones, Regions and Systems*

Apart from buses, most power flow analysis tools allow defining other classes of topological devices such as areas, zones, regions and/or systems. For example, the IEEE common data format defines *interchange area data* [350]. These are generally used for evaluating power transfers between areas. For example:

$$p_{\rm ex} = \sum_{h \in \mathcal{B}_A} p_{G,h} - p_{L,h} \tag{10.1}$$

where $p_{\rm ex}$ is the neat power exchange, $\mathcal{B}_A$ is the set of buses that belong to the area and $p_{G,h}$ and $p_{L,h}$ the generated and consumed powers within the area. If the area is exporting power $p_{\rm ex} > 0$, while $p_{\rm ex} < 0$ otherwise. In power flow analysis, one cannot impose a fixed power exchange, say $p_{\rm ex} = p_{\rm ex0}$, unless some generator active power is left undetermined. Thus, in general, power exchange limits are used only in OPF analysis, where market rules or agreements among area operators impose limits to the import/export of active power between areas.

Typical data for such areas are depicted in Table 10.2. In some cases, an area slack bus can be defined. This slack bus is generally just a "suggestion" and does not affect or redefine $v\theta$ generators. Only in case the area separates from the remaining system as a consequence of line outages, the slack bus is used as reference bus. Similar data can be defined for zones, regions and systems.

Another use of areas is the possibility of grouping variables to be plotted or visualized in the power flow report file. This feature is particularly useful when dealing with networks containing thousands of buses since the complete power flow report would result too large to be understandable. Depicting only a reduced number of buses, for example those pertaining to a given area can simplify the interpretation of the results.

Table 10.2 Area parameters

Variable	Description	Unit
$p_{\rm ex}^{\rm max}$	Maximum interchange export ($> 0 =$ out)	pu
$p_{\rm ex}^{\rm min}$	Minimum interchange export ($< 0 =$ in)	pu
$p_{\rm tol}$	Interchange tolerance	pu
$\Delta p_{\%}$	Annual growth rate	%

10.2 Static Generators

In the classical power flow analysis, generators are only PV and slack ones. This section revises and generalizes the concepts and the models of static generators from the viewpoint of the implementation in a general power system analysis tool.

10.2.1 PV Generator

PV generators impose the voltage magnitude and the power injected at the buses where they are connected, as follows:

$$\begin{aligned} p_h &= p_{G0} \\ v_h &= v_{G0} \end{aligned} \tag{10.2}$$

There are three ways of implementing PV generators for power flow analysis.

1. The classical model considers the voltage v_h as a constant and thus, only imposes one equations for the active power p_h. This model allows reducing the number of power flow equations as well as the size of the Jacobian matrix. On the other hand, handling reactive power limits is complex because the number of variables and equations changes whenever a reactive power limit becomes binding. However, if generator reactive power limits are not considered, this is the most efficient model.
2. To avoid the issue above, one can use two equations, one for the active power p_h and one for the reactive power q_h. If $q_G^{\max} < q_h < q_G^{\min}$, the reactive power equation has to impose that the reactive power balance at node h is satisfied and that the row corresponding to q_h and the column corresponding to v_h in the Jacobian matrix $\boldsymbol{g_y}$ are zero. Only the diagonal element has to be set to 1 to avoid singularity. In this way, the voltage v_h is not varied in the iterations of the Newton's method. If $q_h \geq q_G^{\max}$ or $q_h \leq q_G^{\min}$, then the reactive power is set to $q_h = q_G^{\max}$ or $q_h = q_G^{\min}$ and the Jacobian matrix $\boldsymbol{g_y}$ is not modified so that the voltage v_h can vary. This approach has the advantage of maintaining constant dimensions of the variables, the equations and the Jacobian matrix. However, the number of zeros of the Jacobian matrix changes whenever a reactive power limit becomes binding. This can be an issue if symbolic factorization is used.
3. Using three equations solves the issues of the previous model. Two equations are for the active and reactive power injections p_h and q_h, while the third one imposes the voltage value or the reactive power value as follows:

$$
\begin{aligned}
p_h &= p_{G0} \\
q_h &= q_G \\
&\begin{cases} q_G = q_G^{\max} & \text{if } q_h \geq q_G^{\max} \\ v_h = v_{G0} & \text{if } q_G^{\max} < q_h < q_G^{\min} \\ q_G = q_G^{\min} & \text{if } q_h \leq q_G^{\min} \end{cases}
\end{aligned}
\tag{10.3}
$$

In this model, the reactive power q_G is an internal variable of the PV generator. The drawback of this model is that each PV generator introduces an additional variable. Another drawback is that only one PV generator can be defined at each bus. In fact equation $v_h = v_{G0}$ would be duplicated in case of defining two PV generators at bus h. This is generally not a real issue, since it is not a good practice defining more than one PV generator at the same bus.

In case of using the distributed slack bus model, the active power equation becomes:

$$p_h = (1 + \gamma k_G) p_{G0} \tag{10.4}$$

where k_G is the distributed slack bus variable and γ is the loss participation factor. Table 10.3 depicts PV generator parameters, which include reactive power and voltage limits needed for optimal power flow and continuation load flow analysis. Refer to Chapters 5 and 6 for details.

Table 10.3 PV generator parameters

Variable	Description	Unit
p_{G0}	Active Power	pu
$q_G^{\max}$	Maximum reactive power	pu
$q_G^{\min}$	Minimum reactive power	pu
v_{G0}	Voltage magnitude	pu
$v_G^{\max}$	Maximum voltage	pu
$v_G^{\min}$	Minimum voltage	pu
γ	Loss participation factor	-

Example 10.1 Enforcing Generator Reactive Power Limits

This example focuses on the handling of PV generator reactive power limits. With this aim, consider the IEEE 14-bus system with the following modifications with respect to the base case:

1. The shunt capacity at bus 9 is removed.
2. The load at bus 4 is inductive instead of capacitive and is consuming 0.04 pu of reactive power.

Table 10.4 Power flow results for the IEEE 14-bus system with generator reactive power limit violations

Bus h	v [pu]	θ [rad]	p_G [pu]	q_G [pu]	p_L [pu]	q_L [pu]
1	1.06	0	2.3258	−0.1498	0	0
2	1.045	−0.0871	0.4	0.4882	0.217	0.127
3	1.01	−0.2226	0	0.2737	0.942	0.19
4	1.012	−0.1785	0	0	0.478	0.04
5	1.016	−0.1527	0	0	0.076	0.016
6	1.07	−0.2516	0	0.2251	0.112	0.075
7	1.0493	−0.2309	0	0	0	0
8	1.09	−0.2309	0	**0.2516**	0	0
9	1.0328	−0.2585	0	0	0.295	0.166
10	1.0318	−0.2622	0	0	0.09	0.058
11	1.0471	−0.2590	0	0	0.035	0.018
12	1.0534	−0.2665	0	0	0.061	0.016
13	1.047	−0.2671	0	0	0.135	0.058
14	1.0207	−0.2802	0	0	0.149	0.05
Totals			2.7258	1.0889	2.59	0.814

The power flow results, without enforcing generator reactive power limits are shown in Table 10.4. The PV generator at bus 8 is producing more reactive power than the maximum limit 0.24 pu (see Appendix D for the complete data of the IEEE 14-bus system). Thus, this solution is not acceptable.

Enforcing reactive power limits is a delicate task since it requires to switch the PV generator model to a constant PQ generator, fixing the generated reactive power to $q_G = q_G^{\max}$ or to $q_G = q_G^{\min}$, depending on the limit that is binding. The principal difficulty is that limits should be modelled as inequalities:

$$q_G^{\min} \leq q_G \leq q_G^{\max}$$

Unfortunately, no power flow method discussed in Chapter 4 allows directly modelling and handling inequalities.

A possible approach is to solve a preliminary power flow without enforcing reactive power limit, to check the solution and, if there are reactive power limit violations, to re-run the power flow analysis changing binding PV generators to PQ ones. With this aim, it is not advisable to switch all critical PV generators at a time since doing so could lead to switch more generators than strictly necessary. In fact, switching a PV to a PQ generator leads to a redistribution of all power flows in the network and, as a consequence of this redistribution, some reactive power limit may not be binding anymore. The most secure strategy is to switch one PV generator per iteration, for example starting from the one that exceeds most its reactive power limit. Then, the power flow problem is solved again, and if some reactive power limit is violated, the procedure is repeated.

The method described above is not efficient, especially if the network contains thousands of buses and hundreds of PV generators. In order to save time, a common strategy is to check PV generator reactive powers *on the fly*, i.e., while executing the iterative method used for solving the power flow problem. The idea is to check the reactive power production of the PV generators at each iteration and if there is some limit violation, switch the PV generator to a PQ one. The main issue is to decide not only *how many* generators have to be switched per iteration, but also *when* it is convenient to apply the model switch. In fact, some limit violation can be due to a temporary power mismatch that disappear in the following iterations. On the other hand, if one waits too much, the efficiency can be low.

Table 10.5 shows the result of the power flow analysis for the IEEE 14-bus system allowing switching PV generators since the first iteration of the Newton's method. The reactive power limits of three generators are binding. However, this does not seem a reasonable result since, in Table 10.5, only one PV generator is exceeding its reactive power limit by a relatively small amount.

Table 10.5 Base case power flow results for the IEEE 14-bus system enforcing reactive power limits since the first iteration

Bus h	v [pu]	θ [rad]	p_G [pu]	q_G [pu]	p_L [pu]	q_L [pu]
1	1.06	0	2.331	0.139	0	0
2	1.033	−0.0845	0.4	**0.5**	0.217	0.127
3	0.9666	−0.2182	0	0	0.942	0.19
4	0.9916	−0.1777	0	0	0.478	0.04
5	0.9996	−0.1519	0	0	0.076	0.016
6	1.053	−0.2545	0	**0.24**	0.112	0.075
7	1.028	−0.2321	0	0	0	0
8	1.068	−0.2321	0	**0.24**	0	0
9	1.012	−0.2608	0	0	0.295	0.166
10	1.012	−0.2647	0	0	0.09	0.058
11	1.028	−0.2617	0	0	0.035	0.018
12	1.036	−0.2698	0	0	0.061	0.016
13	1.029	−0.2704	0	0	0.135	0.058
14	1.001	−0.2836	0	0	0.149	0.05
Totals			2.7312	1.119	2.59	0.814

Solving once again the power flow problem and enabling the check of PV generator reactive powers only after the second iteration provide the results that are shown in Table 10.6. As expected, only the generator at bus 8 is switched to a constant PQ model.

As a final remark, consider the following question: which is the better solution between the two depicted in Tables 10.5 and 10.6? From the mathematical point of view, both solve the power flow problem and provide a solution

Table 10.6 Base case power flow results for the IEEE 14-bus system enforcing generator reactive power limits

Bus h	v [pu]	θ [rad]	p_G [pu]	q_G [pu]	p_L [pu]	q_L [pu]
1	1.06	0	2.326	−0.1488	0	0
2	1.045	−0.0871	0.4	0.4916	0.217	0.127
3	1.01	−0.2227	0	0.2758	0.942	0.19
4	1.012	−0.1784	0	0	0.478	0.04
5	1.016	−0.1527	0	0	0.076	0.016
6	1.07	−0.2518	0	0.2298	0.112	0.075
7	1.048	−0.2308	0	0	0	0
8	1.087	−0.2308	0	**0.24**	0	0
9	1.032	−0.2585	0	0	0.295	0.166
10	1.031	−0.2622	0	0	0.09	0.058
11	1.047	−0.259	0	0	0.035	0.018
12	1.053	−0.2666	0	0	0.061	0.016
13	1.047	−0.2672	0	0	0.135	0.058
14	1.02	−0.2802	0	0	0.149	0.05
Totals			2.726	1.0883	2.59	0.814

within technical limits. Thus, if one looks at each solution separately, both are acceptable. Actually, the solution shown in Table 10.5 is characterized by 0.1412 pu of active losses and by 0.3050 pu of reactive losses while the solution shown in Table 10.5 is characterized by 0.1360 pu of active losses and by 0.2743 pu of reactive losses. Thus, the latter solution is preferable if the goal is to minimize losses.

In conclusion, enforcing generator reactive power limit cannot be conveniently solved using a simple power flow problem. As shown in this example, using the Newton's method or similar iterative techniques, one can obtain a feasible solution, but there is no guarantee that there not exist a *better* solution. Only formulating the power flow problem as a nonlinear programming optimization problem as described in Chapter 6 can provide, under certain hypotheses, the *best* solution with respect to a given objective function.

10.2.2 Constant Voltage Phasor Generator

Constant voltage phasor generators are modelled as follows:

$$\begin{aligned} v_h &= v_{G0} \\ \theta_h &= \theta_{G0} \end{aligned} \tag{10.5}$$

In principle, any number of constant $v\theta$ generators can be included in a network. In fact, consider the results shown in Table 10.6. The power flow solution would not change if one assumes that the system has two $v\theta$ generators, say at bus 1 and bus 2, where:

$$
\begin{aligned}
v_1 &= 1.060 \qquad & \theta_1 &= 0 \\
v_2 &= 1.045 \qquad & \theta_2 &= -0.0871 \text{ rad}
\end{aligned}
$$

However, since bus voltage phase angles are generally not known *a priori*, it is quite uncommon to define more than one $v\theta$ generator per interconnected ac network. The unique $v\theta$ generator is generally called *slack bus*. As discussed in Chapter 4, a slack bus is not fully justified unless it is an equivalent of a *strong* network with unlimited active and reactive power capacity. In general, a distributed slack bus model should be preferred.

From the implementation viewpoint, a $v\theta$ generator can inherit from the PV generators the voltage/reactive power model. In other words, the $v\theta$ generator can be a subclass of the PV one. Then, similarly to the PV generator, the angle/active power model can be defined in three ways (see also the previous section).

1. To assume that θ_h is a constant parameter. Thus no equations for the active power injection is needed. This model allows reducing the number of equations and the size of the Jacobian matrix. If an active power limit becomes binding, both equations and Jacobian matrix have to be re-sized. If no active power limits are considered, which is the standard case, this is the most efficient model.
2. To impose that the active power balance at bus h is always satisfied. In this case, one has to impose that the active power mismatch at bus h is zero and that the row corresponding to p_h and the column corresponding to θ_h in the Jacobian matrix $\boldsymbol{g_y}$ are zero. Only diagonal elements have to be set to 1 to avoid singularity. In this way, the voltage θ_h is not varied during the Newton's method. This approach has the advantage of maintaining constant dimensions of the variables, the equations and the Jacobian matrix.
3. To add auxiliary variables and equations for the active and reactive powers produced by the generator. This model includes two additional variables, k_G and q_G, for the active and reactive powers, respectively:

$$
\begin{aligned}
p_h &= k_G \\
q_h &= q_G \\
&\begin{cases}
k_G = p_G^{\max} & \text{if } p_h \geq p_G^{\max} \\
\theta_h = \theta_{G0} & \text{if } p_G^{\max} < p_h < p_G^{\min} \\
k_G = p_G^{\min} & \text{if } p_h \leq p_G^{\min}
\end{cases} \\
&\begin{cases}
q_G = q_G^{\max} & \text{if } q_h \geq q_G^{\max} \\
v_h = v_{G0} & \text{if } q_G^{\max} < q_h < q_G^{\min} \\
q_G = q_G^{\min} & \text{if } q_h \leq q_G^{\min}
\end{cases}
\end{aligned}
\tag{10.6}
$$

As remarked above, it is unusual to consider active power limits for the slack bus. In fact, if the slack cannot provide the required active power, the power flow problem has no solution. However, this model has the advantage of providing a unique formulation for the single and the distributed slack bus model. In case of distributed slack bus model, the active power injection becomes:

$$p_h = (1 + \gamma k_G) p_{G0} \tag{10.7}$$

where p_{G0} is the scheduled active power production for the $v\theta$ generator and all other equation are unchanged. Introducing the variable k_G allows writing an unique code for the single and the slack bus model. In fact the number of variables and equations is always the same.

In case of the single slack bus model, k_G is the active power production of the single slack bus.

Table 10.7 depicts the constant $v\theta$ generator parameters, which also contains data used in optimal power flow and continuation power flow analysis. In case of distributed slack bus model, the parameters p_{G0} and γ are required.

Table 10.7 Slack generator parameters

Variable	Description	Unit
p_{G0}	Scheduled active power	pu
$q_G^{\max}$	Maximum reactive power	pu
$q_G^{\min}$	Minimum reactive power	pu
v_{G0}	Voltage magnitude	pu
$v_G^{\max}$	Maximum voltage	pu
$v_G^{\min}$	Minimum voltage	pu
γ	Loss participation coefficient	-
θ_{G0}	Reference angle	pu

10.2.3 PQ Generator

PQ generators are modeled as constant active and reactive powers:

$$\begin{aligned} p_h &= p_{G0} \\ q_h &= q_{G0} \end{aligned} \tag{10.8}$$

as long as voltages are within the specified limits. If a voltage limit is violated, PQ generators are converted into constant impedances, as follows:

$$\begin{aligned} p_h &= p_{G0} v^2 / (v_G^{\lim})^2 \\ q_h &= q_{G0} v^2 / (v_G^{\lim})^2 \end{aligned} \tag{10.9}$$

where $v_G^{\lim}$ is $v_G^{\max}$ or $v_G^{\min}$ depending on the case. For example, maximum and minimum voltage limits can be assumed 1.1 and 0.9 pu, respectively.

Table 10.8 depicts PQ generator parameters. The maximum and minimum reactive powers $q_G^{\max}$ and $q_G^{\min}$ can be defined in analogy with PV and $v\theta$ generators and can be used in CPF and OPF analyses.

From the implementation viewpoint, PQ generators have the same model as a PQ loads, which are described in the next section. Thus, PQ generators can be implemented as a subclass of the PQ load class. The only difference is in the sign of active and reactive powers:

$$\begin{aligned} p_{L0} &= -p_{G0} \\ q_{L0} &= -q_{G0} \end{aligned} \tag{10.10}$$

Alternatively, one can define a PQ generator using a PQ load and declaring negative power consumptions. However, a specific class for PQ generators is useful for separating power productions and power consumptions in the power flow report.

Table 10.8 PQ generator parameters

Variable	Description	Unit
p_{G0}	Active Power	pu
q_{G0}	Reactive Power	pu
$q_G^{\max}$	Maximum reactive power	pu
$q_G^{\min}$	Minimum reactive power	pu
$v_G^{\max}$	Maximum voltage	pu
$v_G^{\min}$	Minimum voltage	pu

10.3 Static Loads

In the classical power flow analysis, loads are constant PQ or shunt admittances. In the following, static load models are revised and generalized from the viewpoint of the implementation in a general power system analysis tool.

10.3.1 PQ Load

PQ loads are modelled as constant active and reactive powers:

$$\begin{aligned} p_h &= -p_{L0} \\ q_h &= -q_{L0} \end{aligned} \tag{10.11}$$

as long as voltages are within the specified limits. If a voltage limit is violated, PQ loads are converted into constant impedances,[1] as follows:

$$
\begin{aligned}
p_h &= -p_{L0}v^2/(v_L^{\lim})^2 \\
q_h &= -q_{L0}v^2/(v_L^{\lim})^2
\end{aligned}
\tag{10.12}
$$

where $v_L^{\lim}$ is $v_L^{\max}$ or $v_L^{\min}$ depending on the case. For example, maximum and minimum voltage limits can be assumed 1.1 and 0.9 pu, respectively. Table 10.9 depicts PQ load parameters.

Table 10.9 PQ load parameters

Variable	Description	Unit
p_{L0}	Active Power	pu
q_{L0}	Reactive Power	pu
$v_L^{\max}$	Maximum voltage	pu
$v_L^{\min}$	Minimum voltage	pu

In the standard transient stability analysis, PQ loads are converted to constant impedances after the power flow solution (see Section 8.2 of Chapter 8). In this case, PQ loads are forced to switch to constant admittances, as follows:

$$
\begin{aligned}
p_h &= -p_{L0}v^2/v_0^2 \\
q_h &= -q_{L0}v^2/v_0^2
\end{aligned}
\tag{10.13}
$$

where v_0 is the voltage value obtained through the power flow analysis. However, the adequacy of constant admittance or other load models depends on the simulation time frame as it is discussed in Chapter 14.

Example 10.2 Constant Power vs. Constant Impedance Load Models in Transient Stability Analysis for the IEEE 14-Bus System

Using a constant impedance or a constant power load model can drastically modify simulation results. Figure 10.1 shows the results for the IEEE 14-bus

[1] Some software package such as Eurostag allows defining the exponent of the voltage, as follows:

$$
\begin{aligned}
p_h &= -p_{L0}v^2/(v_L^{\lim})^{\alpha_p} \\
q_h &= -q_{L0}v^2/(v_L^{\lim})^{\alpha_q}
\end{aligned}
$$

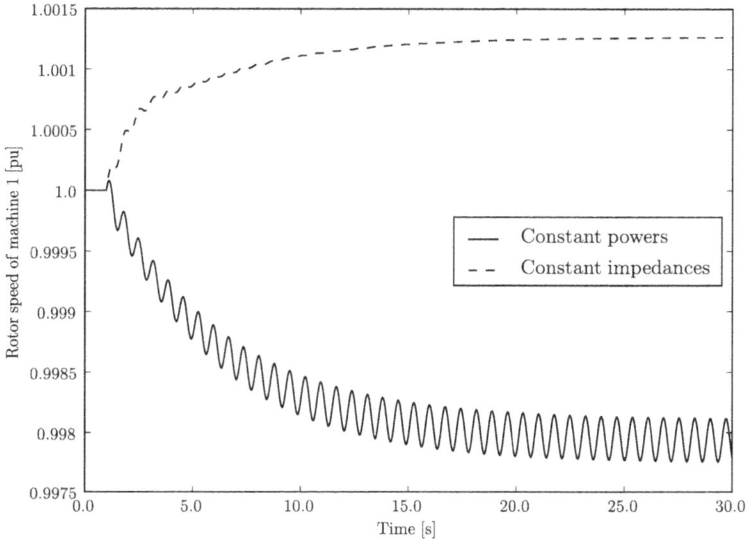

Fig. 10.1 Comparison of the transient analysis using constant impedance and constant power load models for the IEEE 14-bus system

system of the transient following the line 2-4 outage at $t = 1$ s. In order to dramatize the effect of load models, the load power consumption is increased by 20% with respect to the base case. As shown in Example 16.2 of Chapter 16, the IEEE 14-bus system is unstable for such loading level and for line 2-4 outage due to the occurrence of a Hopf bifurcation. The instability drives the system trajectory to fall into a limit cycle. However, the Hopf bifurcation only occurs if using constant power load models. For constant impedance load models, the Hopf bifurcation and the consequent limit cycle disappear.

10.3.2 Constant Power Factor Load

In some industrial applications, loads can be known in terms of active power consumption p_{L0} and power factor $\cos\phi_{L0}$. The conversion to constant PQ load is readily obtained as:

$$\begin{aligned} p_h &= -p_{L0} \\ q_h &= -p_{L0} \tan\phi_{L0} \end{aligned} \tag{10.14}$$

where:

$$\tan\phi_{L0} = \frac{\sin\phi_{L0}}{\cos\phi_{L0}} = \frac{\sqrt{1-(\cos\phi_{L0})^2}}{\cos\phi_{L0}} \tag{10.15}$$

The only drawback of this kind of input data is that purely reactive loads cannot be defined. P$\cos\phi$ loads can be defined as a subclass of the PQ load class.

10.3.3 Shunt Admittance

Constant shunt admittances are described by the following equations:

$$\begin{aligned} p_h &= -g v_h^2 \\ q_h &= b v_h^2 \end{aligned} \tag{10.16}$$

where g and b are the shunt conductance and susceptance, respectively. In (10.16), the susceptance b is negative for inductive charges, positive for capacitive ones. Despite the simplicity of this model, shunt admittances can be used as base class for several other devices. For example, simplified SVC models can be based on (10.16) (see Chapter 19).

In most software packages, constant admittances are included in the admittance matrix $\boldsymbol{Y}$ built for transmission lines and transformers. This practice has the advantage of reducing the computational effort of the power flow analysis. However, there are two relevant drawbacks:

1. If shunt admittances are merged into the transmission line admittance matrix, any connection and disconnection of shunt admittances requires re-building the admittance matrix.
2. In the power flow report, shunt admittances are not considered loads but, rather, included in the transmission system losses. This may be reasonable only in case the shunt admittance is purely reactive. On the other hand, in case of modelling some active loads as constant shunt admittances, merging these loads into the admittance matrix leads to an inconsistent power flow report. In fact, active losses are computed including shunt conductance consumptions.

10.3.4 Switched Shunt Admittances

A simple way to regulate the bus voltage can be obtained using a set of shunt admittances that can be switched on or off depending on the value of the voltage of the bus at which the admittances are connected. Strictly speaking, switched shunt admittances do not provide a voltage control, since admittance variations are not continuous. The base model of switched shunts is given by (10.16) but the values of g and b are not fixed and are defined using a switching logic. A possible switching rule is as follows:

$$b = \begin{cases} b + b_i, \text{if } v_h - v^{\text{ref}} < \Delta v \text{ and } b < b^{\max} \\ b, \text{if } |v_h - v^{\text{ref}}| < \Delta v \\ b - b_i, \text{if } v_h - v^{\text{ref}} > \Delta v \text{ and } b > 0 \end{cases} \tag{10.17}$$

where v^{ref} is the assigned reference voltage, Δv is the voltage error tolerance and b is the current susceptance value and b_i the susceptance value of the next element of the admittance array. The value of b_i can vary if there are admittance blocks of different sizes. The admittance is not varied if all admittances are switched on (i.e., $b = b^{\max}$) or off (e.g., $b = 0$). Assuming that there are s blocks and that each block has n_i elements, each of which characterized by a susceptance b_i, then:

$$b^{\max} = \sum_{i=1}^{s} n_i b_i \tag{10.18}$$

The conductances g_i are generally not physical resistors, but model the losses of the capacitors or reactors. In this case, the conductances g_i are switched on or off if the correspondent susceptance b_i is switched on or off. Finally, during time domain simulations, the switch of shunt elements are delayed by a time Δt_s to avoid activating too frequently admittance breakers. Table 10.10 summarizes the parameters required for defining switched shunt admittances. Other more sophisticate models of shunt devices that are able to regulate the bus voltage are described in Chapter 19.

Table 10.10 Switched shunt parameters

Variable	Description	Unit
$[b_1, b_2, \ldots, b_s]$	Susceptance of each element of each block	pu
$[g_1, g_2, \ldots, g_s]$	Conductance of each element of each block	pu
$[n_1, n_2, \ldots, n_s]$	Number of elements of each block	int
v^{ref}	Reference voltage	pu
Δt_s	Time delay for discrete model	s
Δv	Voltage error tolerance	pu

Chapter 11
Transmission Devices

This chapter describes the models of standard series devices, namely transmission lines (Section 11.1) and transformers (Section 11.2). In particular, Subsection 11.2.1 presents static two-winding transformers, Subsections 11.2.2 and 11.2.3 present under load tap changers and phase shifters, respectively, and Subsection 11.2.4 describes three-winding transformers. Finally, Section 11.3 discusses efficient vectorial computation for static series devices.

11.1 Transmission Line

Several power system books provide a rigorous description of the determination of transmission line parameters [84, 114, 294]. In this section, it is assumed that a short line can be represented as π lumped model as shown in Figure 11.1. *Short* means that

$$\ell \ll \lambda \tag{11.1}$$

where ℓ is the line length and λ is the wave length defined as:

$$\lambda = \frac{1}{f_n \sqrt{L_\ell C_\ell}} \tag{11.2}$$

where f_n is the rated frequency of the ac system, and L_ℓ and C_ℓ are the per-unit length inductance and capacity respectively, of the transmission line. Assuming typical values of L_ℓ and C_ℓ (that depend on the line geometry) and considering $f_n = 50$ Hz, high voltage overhead transmission lines are characterized by $\lambda \approx 6000$ km, while cables by $\lambda \in (2000, 2800)$ km. In practice, transmission line length is never $\ell > \lambda/4$ and the vast majority satisfies the condition $\ell < \lambda/8$ (e.g., ≈ 750 km at 50 Hz). For $\ell \leq \lambda/30$, the error introduced using lumped parameters is $\leq 1\%$. At 50 Hz, this condition leads to $\ell \leq 200$ km. The hypothesis of short line is assumed for the remainder of this section.

The equivalent circuit of Figure 11.1 includes a series resistance, a series reactance and four shunt elements, namely sending-end conductance and

F. Milano: Power System Modelling and Scripting, Power Systems, pp. 263–289.
springerlink.com

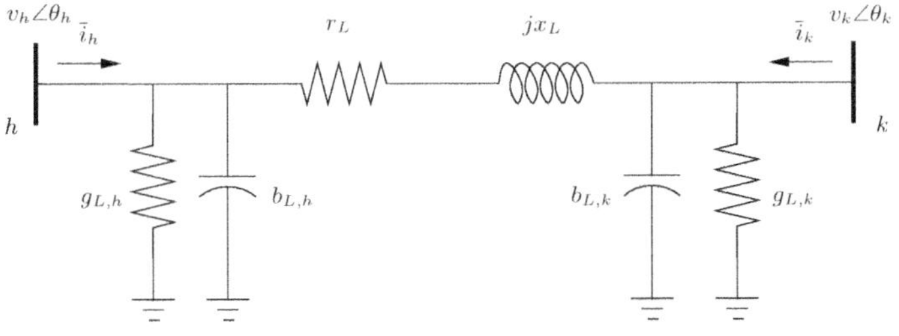

Fig. 11.1 Transmission line lumped π-circuit

susceptance and receiving-end conductance and susceptance. The complex powers injected at each node are:

$$\begin{aligned} \bar{s}_h &= \bar{v}_h \bar{i}_h^* \\ \bar{s}_k &= \bar{v}_k \bar{i}_k^* \end{aligned} \tag{11.3}$$

According to the π model of Figure 11.1, the injected currents $\bar{i}_h$ and $\bar{i}_k$ can be written as:

$$\begin{bmatrix} \bar{i}_h \\ \bar{i}_k \end{bmatrix} = \begin{bmatrix} \bar{y}_L + \bar{y}_{L,h} & -\bar{y}_L \\ -\bar{y}_L & \bar{y}_L + \bar{y}_{L,k} \end{bmatrix} \begin{bmatrix} \bar{v}_h \\ \bar{v}_k \end{bmatrix} \tag{11.4}$$

where

$$\begin{aligned} \bar{y}_L &= g_L + jb_L = (r_L + jx_L)^{-1} \\ \bar{y}_{L,h} &= g_{L,h} + jb_{L,h} \\ \bar{y}_{L,k} &= g_{L,k} + jb_{L,k} \end{aligned} \tag{11.5}$$

Hence, (11.3) can be rewritten as:

$$\begin{aligned} p_h &= v_h^2(g_L + g_{L,h}) - v_h v_k (g_L \cos\theta_{hk} + b_L \sin\theta_{hk}) \\ q_h &= -v_h^2(b_L + b_{L,h}) - v_h v_k (g_L \sin\theta_{hk} - b_L \cos\theta_{hk}) \\ p_k &= v_k^2(g_L + g_{L,k}) - v_h v_k (g_L \cos\theta_{hk} - b_L \sin\theta_{hk}) \\ q_k &= -v_k^2(b_L + b_{L,k}) + v_h v_k (g_L \sin\theta_{hk} + b_L \cos\theta_{hk}) \end{aligned} \tag{11.6}$$

where $\theta_{hk} = \theta_h - \theta_k$.

For short lines, $g_{L,h} \approx g_{L,k} \approx 0$. This assumption also derives from the difficulty of evaluating shunt parasite conductances, which are generally neglected. The lumped series resistance and reactance can be computed as:

$$r_L = R_\ell \ell_t / Z_b \tag{11.7}$$

$$x_L \approx \omega_s L_\ell \ell_t / Z_b \tag{11.8}$$

where $\omega_s = 2\pi f_n$ is the synchronous pulsation in rad/s, Z_b is the base impedance in Ω and other parameters are defined in Table 11.1. The per-unit length resistance R_ℓ depends on the temperature, on the section and on resistivity of the conductor. The lumped shunt susceptances $b_{L,h}$ and $b_{L,k}$ can be approximated as:

$$b_{L,h} \approx b_{L,k} \approx \frac{1}{2}\omega_s C_\ell \ell_t Z_b \tag{11.9}$$

Even though for standard transmission lines $b_{L,h} \approx b_{L,k}$, it is better to maintain separated these parameters for the sake of generality. In this way, line sectioning as well as transformers can share the same programming code as transmission lines.

Table 11.1 defines all transmission line parameters. $I^{\max}$, $P^{\max}$ and $S^{\max}$ indicate the limits for currents, active power flows and apparent power flows. These limits are generally not required in power flow analysis, but can be used for CPF and OPF analyses (see Chapters 5 and 6 for details). Currents or apparent power limits are thermal limits. In some applications it may be required to define two or three values of such limits, depending on the emergency level and or the time that can be waited before taking a corrective action (e.g., disconnecting the line). On the other hand, active power limits generally approximate stability limits and are computed off-line base on some N-1 contingency criterion (see Subsection 5.4.5 of Chapter 5).

Table 11.1 Transmission line parameters

Variable	Description	Unit
$b_{L,h}, b_{L,k}$	Shunt susceptances	pu
C_ℓ	Per-unit length line capacity	F/km
$g_{L,h}, g_{L,k}$	Shunt conductances	pu
$I^{\max}$	Current limit	kA
L_ℓ	Per-unit length line inductance	H/km
ℓ_t	Total line length	km
$P^{\max}$	Active power limit	MW
R_ℓ	Per-unit length line resistance	Ω/km
r_L	Resistance	pu
$S^{\max}$	Apparent power limit	MVA
x_L	Reactance	pu

11.1.1 Line Sections

In some cases, it may be required to define a transmission line as a series of line sections, each of which is characterized by a lumped π-circuit model as in (11.6). Sections originate from transposed lines or from connections that are composed of sections of different materials or characteristic parameters (e.g., a branch composed of the series of overhead lines and underground cables). It is certainly possible to model each section as a transmission line and including fictitious buses to connect sections. However, including fictitious buses generally results in an unnecessary complication.

The equivalent model of a transmission line composed of various sections can be obtained by alternatively using the star-delta (Y-Δ) and the delta-star (Δ-Y) transformations (see Figure 11.2). This can lead to the definition of an equivalent line whose sending and receiving-end shunt admittances are not equal.

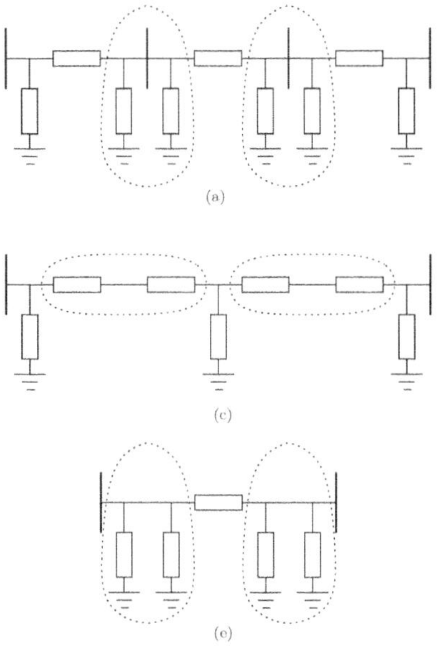

Fig. 11.2 Equivalencing procedure for line sections: (a) original line sections; (b)-(e) intermediate equivalents; (f) final equivalent line

For completeness, the general Y-Δ and Δ-Y transformation formulæ are given below (see also Figure 11.3).

1. Y-Δ transformation:

$$\begin{aligned}\bar{z}_1 &= \frac{\bar{z}_a \bar{z}_b}{\bar{z}_a + \bar{z}_b + \bar{z}_c} \\ \bar{z}_2 &= \frac{\bar{z}_b \bar{z}_c}{\bar{z}_a + \bar{z}_b + \bar{z}_c} \\ \bar{z}_3 &= \frac{\bar{z}_a \bar{z}_c}{\bar{z}_a + \bar{z}_b + \bar{z}_c}\end{aligned} \tag{11.10}$$

2. Δ-Y transformation:

$$\begin{aligned} \bar{z}_a &= \frac{\bar{z}_1\bar{z}_2 + \bar{z}_2\bar{z}_3 + \bar{z}_3\bar{z}_1}{\bar{z}_2} \\ \bar{z}_b &= \frac{\bar{z}_1\bar{z}_2 + \bar{z}_2\bar{z}_3 + \bar{z}_3\bar{z}_1}{\bar{z}_3} \\ \bar{z}_c &= \frac{\bar{z}_1\bar{z}_2 + \bar{z}_2\bar{z}_3 + \bar{z}_3\bar{z}_1}{\bar{z}_1} \end{aligned} \tag{11.11}$$

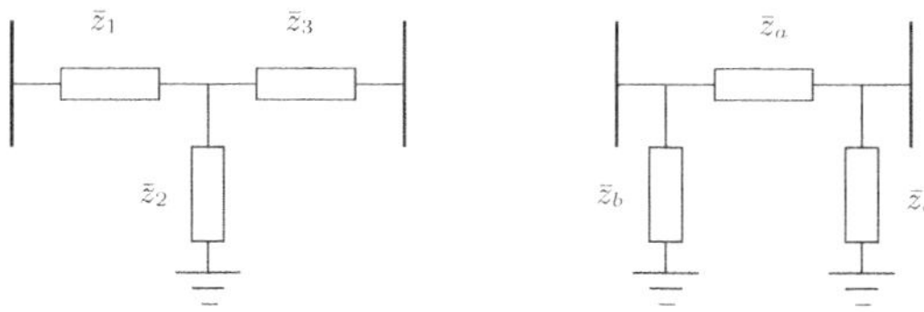

Fig. 11.3 Star and delta circuits used in formulæ (11.10) and (11.11)

11.1.2 Tie Line

Tie lines are branches on which the active power flow is set to a given value or is bounded within assigned limits. Assuming that the active power flow is set to a constant value, the resulting model is (11.6) plus the additional equation:

$$0 = p_h - p^{\text{ref}} \tag{11.12}$$

or the inequality constraints:

$$p^{\min} \leq p_h \leq p^{\max} \tag{11.13}$$

Previous constraints can be easily included in an OPF problem, since generator power productions and load consumptions in case of elastic demand are not assigned. On the other hand, including (11.12) or (11.13) in the power flow problem requires that some parameter becomes a variable. At this regard, there are various possibilities.

A model consists in considering one parameter of the tie-line as a variable, for example, the series reactance x_L. A very similar solution is to define a *compensation* variable c_L, that multiplies a given series reactance, say $c_L x_L$. Considering the tie line series reactance as a variable is not an arbitrary choice. Series FACTS devices (e.g., the TCSC discussed in Chapter 19) vary the series reactance of a transmission line to impose a given active power flow.

Clearly, it is not mandatory that the new variable belongs to the tie line. For example one can use the active power of some generator (thus leading to a constant v generator model). However, in the latter case, it is more reasonable to formulate the problem as an OPF rather than as a power flow.

Example 11.1 Tie Line for the IEEE 14-Bus System

According to the base case power flow solution of the IEEE 14-bus system shown in Table 4.3 of Chapter 4, the active power flow injected at bus 4 by transmission line 2-4 is $p_5 = 0.5445$ pu. Assume that we want to impose an active power flow injection of $p_5^{\rm ref} = 0.60$ pu. Using a tie line that allows varying the series reactance can do the job. In order to obtain this result, the series reactance of line 2-4 must be $x_{24} = 0.14624$ pu. The solution of this power flow problem requires 6 iterations using a standard Newton's method and $x_{24}^{(0)} = 0.17632$ as initial guess for the tie line series reactance. Imposing $p_5^{\rm ref} = 0.64$ requires 9 iterations and provides $x_{24} = 0.13402$ pu, while for $p_5^{\rm ref} = 0.65$, the power flow problem diverges. The critical point is the initial guess of the series reactance x_{24}.

This simple example allows drawing a general conclusion. Series devices always create more convergence issues than shunt devices. Thus, tie line models such the one discussed in this example have to be used with caution.

11.1.3 Distributed Transmission Line Models

As previously discussed, long transmission lines have to be modelled using a distributed parameter model. This leads to the well-known partial-derivative transmission line equations:

$$\begin{aligned} \frac{\partial v(\ell,t)}{\partial \ell} &= R_\ell i(\ell,t) + L_\ell \frac{\partial i(\ell,t)}{\partial t} \\ \frac{\partial i(\ell,t)}{\partial \ell} &= G_\ell v(\ell,t) + C_\ell \frac{\partial v(\ell,t)}{\partial t} \end{aligned} \tag{11.14}$$

where R_ℓ, L_ℓ and C_ℓ are defined in Table 11.1 and G_ℓ is the line conductance per unit length.

Equations (11.14) and the boundary conditions:

$$\begin{aligned} v(0,t) &= v_h(t), & v(\ell_t,t) &= v_k(t) \\ i(0,t) &= i_h(t), & i(\ell_t,t) &= i_k(t) = -i_h(t) \end{aligned} \tag{11.15}$$

define a boundary value problem whose general solution is too complex to be used for systems with hundreds of lines.[1] Thus, some simplifications are required.

The first commonly accepted assumption is to use fast balanced time-varying phasors. The boundary value problem becomes:

---

[1] A very interesting although little exploited continuum model of power systems has been proposed in [274] and recently extended in [233].

$$\begin{aligned}
\frac{\partial \bar{v}(\ell,t)}{\partial \ell} &= R_\ell \bar{i}(\ell,t) + L_\ell \frac{\partial \bar{i}(\ell,t)}{\partial t} + j\omega_s L_\ell \bar{i}(\ell,t) \\
\frac{\partial \bar{i}(\ell,t)}{\partial \ell} &= G_\ell \bar{v}(\ell,t) + C_\ell \frac{\partial \bar{v}(\ell,t)}{\partial t} + j\omega_s C_\ell \bar{v}(\ell,t) \\
\bar{v}(0,t) = \bar{v}_h(t), &\qquad \bar{v}(\ell_t,t) = \bar{v}_k(t) \\
\bar{i}(0,t) = \bar{i}_h(t), &\qquad \bar{i}(\ell_t,t) = -\bar{i}_h(t)
\end{aligned} \tag{11.16}$$

where ω_s is the synchronous pulsation.

Assuming $G_\ell \approx 0$, the boundary value problem (11.16) has an explicit solution [147]. Let define the following quantities:

- Characteristic line admittance $Y_\ell = \sqrt{C_\ell / L_\ell}$.
- Line time delay (or *travelling time*) $\tau_\ell = \ell_t / s$, which is the time required by a wave to pass through the line at the wave speed $s = 1/\sqrt{L_\ell C_\ell}$. Typically $s \approx c$, where $c = 299\,792\,458$ m/s is the light speed.
- Line phase shift $\alpha_\ell = \omega_s \tau_\ell$.
- Attenuation factor $\zeta_\ell = \frac{R_\ell \ell_t}{2} Y_\ell$.

Then, (11.16) has the solution:

$$\begin{aligned}
\bar{i}_h(t) &= \bar{i}_h(t - 2\tau_\ell) e^{-2(\zeta_\ell + j\alpha_\ell)} - Y_\ell \bar{w}_h(t) \\
&\quad + 2Y_\ell \bar{w}_k(t - \tau_\ell) e^{-(\zeta_\ell + j\alpha_\ell)} - Y_\ell \bar{w}_h(t - 2\tau_\ell) e^{-(\zeta_\ell + j\alpha_\ell)} \\
\bar{i}_k(t) &= \bar{i}_k(t - 2\tau_\ell) e^{-2(\zeta_\ell + j\alpha_\ell)} - Y_\ell \bar{w}_k(t) \\
&\quad + 2Y_\ell \bar{w}_h(t - \tau_\ell) e^{-(\zeta_\ell + j\alpha_\ell)} - Y_\ell \bar{w}_k(t - 2\tau_\ell) e^{-2(\zeta_\ell + j\alpha_\ell)}
\end{aligned} \tag{11.17}$$

where $\bar{w}_h$ and $\bar{w}_k$ satisfy the following set of complex differential equations:

$$\begin{aligned}
\dot{\bar{w}}_h &= -j\omega_s \bar{w}_h - \frac{R_\ell}{2L_\ell} \bar{w}_h + \dot{\bar{v}}_h + j\omega_s \bar{v}_h \\
\dot{\bar{w}}_k &= -j\omega_s \bar{w}_k - \frac{R_\ell}{2L_\ell} \bar{w}_k + \dot{\bar{v}}_k + j\omega_s \bar{v}_k
\end{aligned} \tag{11.18}$$

If $R_\ell \approx 0$ (e.g., loss-less line), equations (11.17) become:

$$\begin{aligned}
\bar{i}_h(t) &= \bar{i}_h(t - 2\tau_\ell) e^{-j2\alpha_\ell} - Y_\ell \bar{v}_h(t) \\
&\quad + 2Y_\ell \bar{v}_k(t - \tau_\ell) e^{-j\alpha_\ell} - Y_\ell \bar{v}_h(t - 2\tau_\ell) e^{-j2\alpha_\ell} \\
\bar{i}_k(t) &= \bar{i}_k(t - 2\tau_\ell) e^{-j2\alpha_\ell} - Y_\ell \bar{v}_k(t) \\
&\quad + 2Y_\ell \bar{v}_h(t - \tau_\ell) e^{-j\alpha_\ell} - Y_\ell \bar{v}_k(t - 2\tau_\ell) e^{-j2\alpha_\ell}
\end{aligned} \tag{11.19}$$

Equations (11.17) and (11.18) or (11.19) are a set of functional differential equations with constant delay τ_ℓ. As discussed in Example 1.2 of Chapter 1, the solution of such systems is rather complex. However, in principle, the expressions of the line currents defined in (11.17) or (11.19) can be used for recomputing (11.3).[2]

11.1.4 Effect of Frequency Variation

One of the most well accepted hypothesis of transient stability analysis is that transmission system parameters do not depend on the frequency. The assumption is that, although synchronous machine rotor speeds vary, the frequency deviations in transmission line parameters is negligible:

$$\begin{aligned} x_L &= \frac{\omega L_\ell \ell_t}{Z_b} \approx \frac{\omega_s L_\ell \ell_t}{Z_b} \\ b_{L,h} = b_{L,k} &= \frac{1}{2}\omega C_\ell \ell_t Z_b \approx \frac{1}{2}\omega_s C_\ell \ell_t Z_b \end{aligned} \tag{11.20}$$

where Z_b is the impedance base and ω_s is the synchronous speed in pu (e.g., $\omega_s = 1$ pu).

Removing this assumption leads to the following equations:

$$\begin{aligned} p_h &= v_h^2(g_{L,hk}(\omega) + g_{L,h}) - v_h v_k(g_L(\omega)\cos\theta_{hk} + b_L(\omega)\sin\theta_{hk}) \\ q_h &= -v_h^2(b_{L,hk}(\omega) + b_{L,h}(\omega)) - v_h v_k(g_L(\omega)\sin\theta_{hk} - b_L(\omega)\cos\theta_{hk}) \\ p_k &= v_k^2(g_L(\omega) + g_{L,k}) - v_h v_k(g_{L,hk}(\omega)\cos\theta_{hk} - b_L(\omega)\sin\theta_{hk}) \\ q_k &= -v_k^2(b_L(\omega) + b_{L,k}(\omega)) + v_h v_k(g_L(\omega)\sin\theta_{hk} + b_L(\omega)\cos\theta_{hk}) \end{aligned} \tag{11.21}$$

The value of ω can be defined in various ways. Two common choices are:

1. Computing a local frequency as the derivative of the bus phase angle. This technique is described in Section 13.4 of Chapter 13.
2. Using a system-wide frequency such as the center of inertia (COI), which is described in Subsection 15.1.9 of Chapter 15.

Example 11.2 Effect of Frequency on Line Parameters

Figure 11.4 compares the numerical integration for the IEEE 14-bus system using three different models of transmission lines, namely (i) constant parameters as in (11.6), (ii) COI dependent parameters, and (iii) local bus frequency dependent parameters. The disturbance consists in line 2-4 outage for $t = 1$ s. For clarity, Figure 11.4 only shows a window from 10 to 15 s. As it can be noted, the classical approximation of using constant parameters does not introduce a significant error, at least for small variations of machine rotor speeds. Furthermore, COI and local bus approximations provides practically identical results.

[2] Equations (11.14) to (11.19) are in absolute values. Thus, before substituting (11.17) or (11.19) in (11.3) a proper per unit conversion has to be carried out.

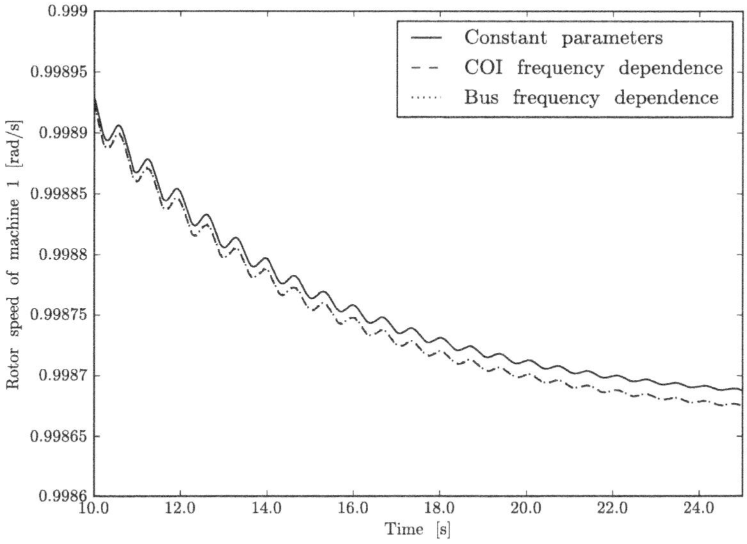

Fig. 11.4 Comparison of the transient behavior of transmission lines with constant and frequency-dependent parameters for the IEEE 14-bus system.

11.1.5 Coupling Device and Zero-Impedance Line

Coupling devices such as disconnecting switches or very short lines can be modelled as zero-impedance lines. Unfortunately, equations (11.6) are not adequate for modelling such coupling devices. In fact, using a very small series impedance (e.g., $x_L < 10^{-6}$ pu) may lead to numerical issues.

Thus, coupling devices require an *ad hoc* model. For example, the following model was proposed in [210]:

$$\begin{aligned} 0 &= \theta_h - \theta_k \\ 0 &= v_h - v_k \\ p_h &= p_c \\ q_h &= q_c \\ p_k &= -p_c \\ q_k &= -q_c \end{aligned} \tag{11.22}$$

where the internal variables p_c and q_c are the active and reactive power flowing in the coupling device from bus h to bus k. The only drawback of the model (11.22) is that it does not allows putting two or more coupling devices in parallel since the powers p_c and q_c in each device would result indeterminate.

11.2 Transformer

This section describes fixed tap and regulating two- and three-winding transformers.

11.2.1 *Two-Winding Transformer*

Two-winding transformers can be modelled as a transmission line with a series impedances $\bar{z}_T = r_T + jx_T$ and a shunt admittance at the sending-end bus, which models iron losses $g_{\rm Fe}$ and the magnetizing susceptance b_μ.[3] Thus, substituting transformer parameters in (11.6), the following correspondences hold:

$$\begin{aligned} r_L &= r_T & \qquad x_L &= x_T \\ b_{L,h} &= b_\mu & \qquad g_{L,h} &= g_{\rm Fe} \\ b_{L,k} &= 0 & \qquad g_{L,k} &= 0 \end{aligned}$$

All transformer parameters are defined in Table 11.2.

Table 11.2 Transformer parameters

Variable	Description	Unit
b_μ	Magnetizing susceptance	pu
$k_T = V_{n,h}/V_{n,k}$	nominal voltage ratio	kV/kV
$g_{\rm Fe}$	Iron losses	pu
$I^{\max}$	Current limit	kA
m	Fixed tap ratio	pu/pu
$P^{\max}$	Active power limit	MW
r_T	Resistance	pu
$S^{\max}$	Apparent power limit	MVA
$V_{n,h}$	Primary voltage rating	kV
$V_{n,k}$	Secondary voltage rating	kV
x_T	Reactance	pu
ϕ	Fixed phase shift	rad

From the modelling viewpoint, the main difference between transformers and transmission lines is that transformers can introduce a complex off-nominal tap ratio $me^{j\phi}$ that allows modifying the magnitude and the phase angle of the receiving or sending-end bus voltage. For static transformers,

[3] Strictly speaking, a transformer should be modelled using a T model. However the error introduced by the approximated circuit is acceptable. Furthermore, in power flow analysis, transformer shunt admittances are generally neglected.

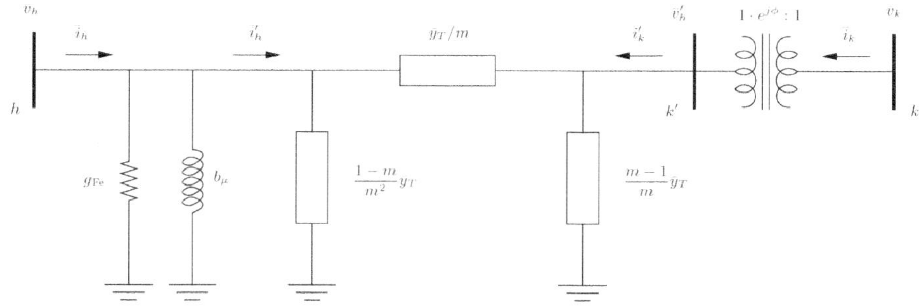

Fig. 11.5 Transformer equivalent circuit

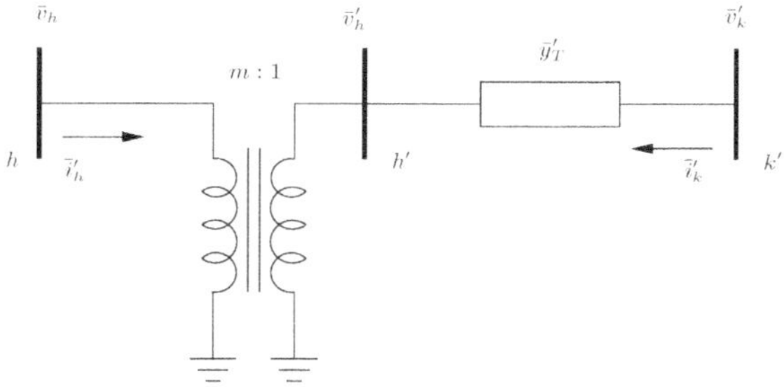

Fig. 11.6 Equivalent circuit of the tap ratio module and series impedance

both m and ϕ are constant. Regulating transformers with under load variable tap ratio are described in Subsection 11.2.2. Assuming that the tap is on the primary side, the complete equivalent circuit of a generic two-winding transformer is depicted in Figure 11.5.

The π-circuit that depends on the series admittance $\bar{y}_T$ and on the off-nominal tap ratio m can be obtained from the circuit depicted in Figure 11.6 [163]. The currents $\bar{i}'_h$ and $\bar{i}'_k$ can be written as:

$$
\begin{aligned}
\bar{i}'_h &= \frac{1}{m}\bar{y}'_T(\bar{v}'_h - \bar{v}'_k) = \frac{1}{m}\bar{y}'_T(\bar{v}_h/m - \bar{v}'_k) \\
\bar{i}'_k &= \bar{y}'_T(\bar{v}'_k - \bar{v}'_h) = \bar{y}'_T(\bar{v}'_k - \bar{v}_h/m)
\end{aligned}
\tag{11.23}
$$

where $\bar{y}'_T = \bar{y}_T = (r_T + jx_T)^{-1}$, $\bar{v}'_h = \bar{v}_h/m$ and $\bar{v}'_k = \bar{v}_k e^{j\phi} = v_k e^{j(\theta_k + \phi)}$. Equations (11.23) in vectorial form become:

$$
\begin{bmatrix} \bar{i}'_h \\ \bar{i}'_k \end{bmatrix} = \bar{y}'_T \begin{bmatrix} \frac{1}{m^2} & -\frac{1}{m} \\ -\frac{1}{m} & 1 \end{bmatrix} \begin{bmatrix} \bar{v}_h \\ \bar{v}'_k \end{bmatrix}
\tag{11.24}
$$

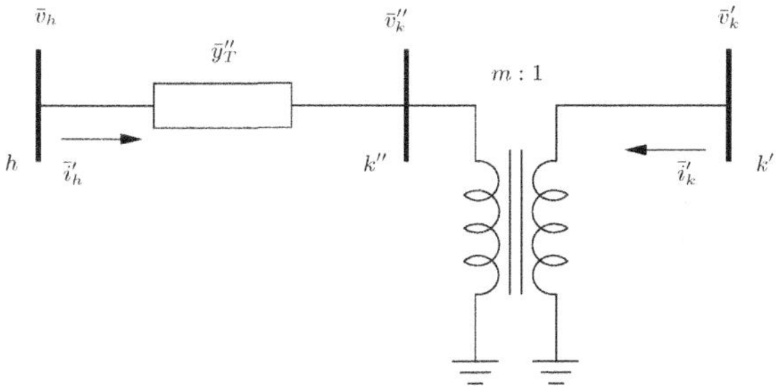

Fig. 11.7 Alternative equivalent circuit of the tap ratio module and series impedance

The same result can be obtained using the equivalent circuit shown in Figure 11.7 [2, 184]. In this case, the currents $\bar{i}'_h$ and $\bar{i}'_k$ are:

$$\begin{aligned} \bar{i}'_h &= \bar{y}''_T(\bar{v}_h - \bar{v}''_k) = \bar{y}''_T(\bar{v}_h - m\bar{v}'_k) \\ \bar{i}'_k &= m\bar{y}''_T(\bar{v}''_k - \bar{v}_h) = m\bar{y}''_T(m\bar{v}'_k - \bar{v}_h) \end{aligned} \tag{11.25}$$

where $\bar{y}''_T = \bar{y}_T/m^2$ and $\bar{v}''_h = m\bar{v}'_k$. Equations (11.25) in vectorial form become:

$$\begin{bmatrix} \bar{i}'_h \\ \bar{i}'_k \end{bmatrix} = \bar{y}''_T \begin{bmatrix} 1 & -m \\ -m & m^2 \end{bmatrix} \begin{bmatrix} \bar{v}_h \\ \bar{v}'_k \end{bmatrix} \tag{11.26}$$

Equations (11.23) and (11.25) are obtained assuming that the tap is on the transformer primary side. Nevertheless, if the tap is on the secondary winding and the off-nominal tap ratio is $\tilde{m} = 1/m$, then the admittances in (11.23) and (11.25) have to be redefined as $\bar{y}'_T = \bar{y}_T/\tilde{m}^2$ and $\bar{y}''_T = \bar{y}_T$, respectively. Thus, it is important to check on which transformer side the tap changer is installed.

In conclusion, the algebraic equations of the power injections are as follows:

$$\begin{aligned} p_h &= v_h^2(g_{\rm Fe} + g_T/m^2) \\ &\quad -v_h v_k(g_T\cos(\theta_{hk} - \phi) + b_T\sin(\theta_{hk} - \phi))/m \\ q_h &= -v_h^2(b_\mu + b_T/m^2) \\ &\quad -v_h v_k(g_T\sin(\theta_{hk} - \phi) - b_T\cos(\theta_{hk} - \phi))/m \\ p_k &= v_k^2 g_T - v_h v_k(g_T\cos(\theta_{hk} - \phi) - b_T\sin(\theta_{hk} - \phi))/m \\ q_k &= -v_k^2 b_T + v_h v_k(g_T\sin(\theta_{hk} - \phi) + b_T\cos(\theta_{hk} - \phi))/m \end{aligned} \tag{11.27}$$

where $g_T + jb_T = \bar{y}_T$.

The fixed tap ratio unit is pu/pu since it represents the ratio of the primary voltage in pu by the secondary voltage in pu. For example, if the nominal voltages are $V_{n,h} = 220$ kV and $V_{n,k} = 128$ kV, and the actual tap positions of the transformer are $V_h = 231$ kV and $V_k = 128$ kV, the tap ratio m is:

$$m = \frac{V_h}{V_{n,h}} \cdot \frac{V_{n,k}}{V_k} = \frac{231}{220} \cdot \frac{128}{128} = 1.05 \, \frac{\text{pu}}{\text{pu}}$$

11.2.2 Under Load Tap Changer

Under Load Tap Changer (ULTC) transformers control the voltage or the reactive power varying the tap ratio. There are two models of ULTC transformers: a discrete model and a continuous one [189, 238, 241].

1. The discrete model consists in modelling the step ratio as a discrete variable, which can vary between the minimum and the maximum tap values $m^{\max}$ and $m^{\min}$ by a fixed step Δm. The regulator model simply switches up or down by one step Δm the tap ratio if the deviation of the regulated quantity v_k (e.g., the voltage on the secondary winding) with respect to the reference v^{ref} exceeds a given tolerance Δv, which works similarly to a dead zone (see Figure 11.8.a). The switching logic is as follows:

$$m = \begin{cases} m + \Delta m, \text{if } v_k - v^{\text{ref}} < \Delta v \text{ and } m < m^{\max} \\ m, \text{if } |v_k - v^{\text{ref}}| < \Delta v \\ m - \Delta m, \text{if } v_k - v^{\text{ref}} > \Delta v \text{ and } m > m^{\min} \end{cases} \tag{11.28}$$

   Each tap switch is a delicate process, since requires moving physically the tap position. In order to avoid unnecessary switching operations, the regulator is delayed so that a tap switch can occur only if a minimum time Δt_s has passed since the last switch. For this reason, ULTC controllers are relatively slow.[4]

2. The continuous model assumes that the tap ratio step Δm is small so that discrete switches can be approximated with a continuous variation of the tap ratio m. The time delay is approximated as a lag transfer function (see Figure 11.8.b). Hence, the tap ratio differential equation is:

$$\dot{m} = -K_d m + K_i (v_k - v^{\text{ref}}) \tag{11.29}$$

   where all parameters are defined in Table 11.3 and the tap m undergoes an anti-windup limiter and the sign of the error $\epsilon_v = v_k - v^{\text{ref}}$ is due to the stability characteristic of the nonlinear control loop. In fact, as shown in Figure 11.10, the regulator stable equilibrium point occurs for a negative tangent slope of the ULTC-load characteristic. Example 11.3 provides a graphical proof of this statement. A similar control can be obtained by regulating the voltage of a remote bus or regulating the reactive power output of the transformer, e.g.:

$$\dot{m} = -K_d m + K_i (q^{\text{ref}} + q_k) \tag{11.30}$$

---

[4] New generations of ULTC are equipped with thyristor-switched controllers, which are characterized by a fast time response.

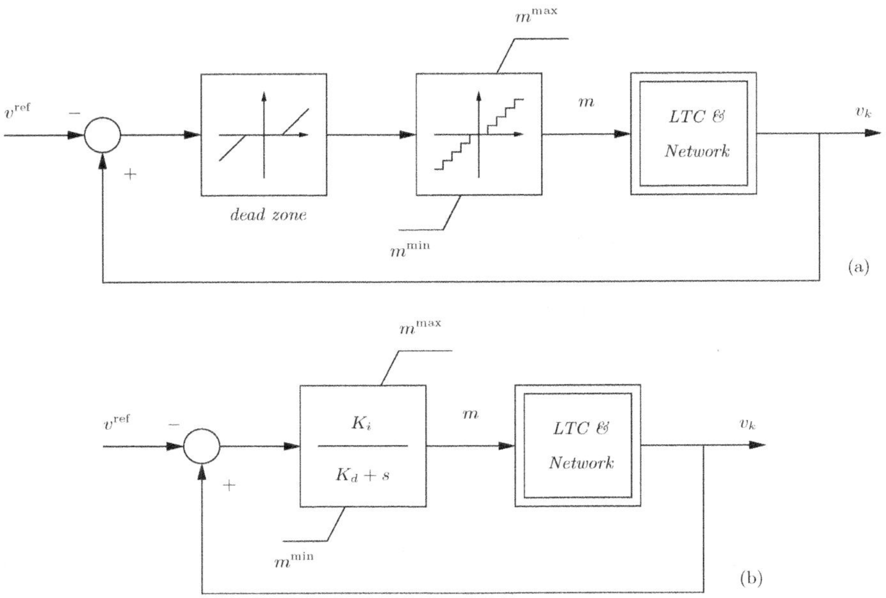

Fig. 11.8 Voltage control diagram of the ULTC transformer: (a) discrete control and (b) continuous control

The discrete model reproduces precisely the physical behavior of the ULTC regulator. However, as discussed in Chapter 1, discrete variables complicate the analysis of DAE systems. For this reason, the continuous model is preferred for stability analysis [55]. An interesting stability study that consists in bounding the discrete behavior through an upper and a lower continuous models is proposed in [339].

Table 11.3 Under load tap changer control parameters

Variable	Description	Unit
K_d	Integral deviation	1/s
K_i	Integral gain	1/s/pu
$m^{\max}$	Maximum tap ratio	pu/pu
$m^{\min}$	Minimum tap ratio	pu/pu
v^{ref} or q^{ref}	Reference voltage or reactive power	pu
Δm	Tap ratio step	pu/pu
Δt_s	Time delay for discrete model	s
Δv	Voltage dead zone	pu

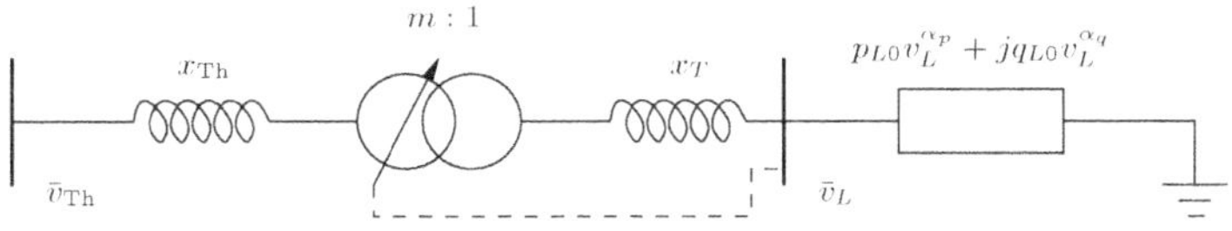

Fig. 11.9 2-bus system with tap changer and voltage dependent load

Example 11.3 Voltage-Tap Ratio Characteristic of Loads Fed by an ULTC

Figure 11.9 shows a simple system composed of a Thevenin equivalent modelling the external transmission system and an under load tap changer feeding a load. The load is a monomial function of the voltage v_L (see also Sections 14.1 and 14.4 of Chapter 14). If $\alpha_p = \alpha_q = 2$, the load is a constant impedance, say $r_L + jx_L$. Assuming that the Thevenin equivalent is a constant voltage v_{Th} behind a reactance x_{Th} and that the transformer reactance is x_T, one has:

$$\bar{v}_L = \bar{v}_{\mathrm{Th}} \frac{m(r_L + jx_L)}{m^2 r_L + j(x_{\mathrm{Th}} + m^2 x'_L)} \tag{11.31}$$

where $x'_L = x_T + x_L$. The maximum voltage is obtained for:

$$m^* = \sqrt{\frac{x_{\mathrm{Th}}}{\sqrt{r_L^2 + {x'_L}^2}}} \tag{11.32}$$

And the maximum voltage value is:

$$v_L^{\max} = \sqrt{v_{\mathrm{Th}}^2 a \frac{r_L}{2x_{\mathrm{Th}}}} \tag{11.33}$$

where

$$a = 1 + b^2(\sqrt{1 + c^2} - c), \quad b = \frac{x_L}{r_L}, \quad c = \frac{x'_L}{r_L} \tag{11.34}$$

Similar expressions can be obtained for constant current ($\alpha_p = \alpha_q = 1$) and constant power ($\alpha_p = \alpha_q = 0$) loads. The characteristics of the load voltage v_L as a function of m and of the load voltage dependence is shown in Figure 11.10. There are two possible equilibrium points, that correspond to the intersections with the regulator reference voltage v^{ref}. The only feasible equilibrium point (i.e., the one for which $m^{\min} \le m \le m^{\max}$) is characterized by a negative tangent slope of the curve (v_L, m).

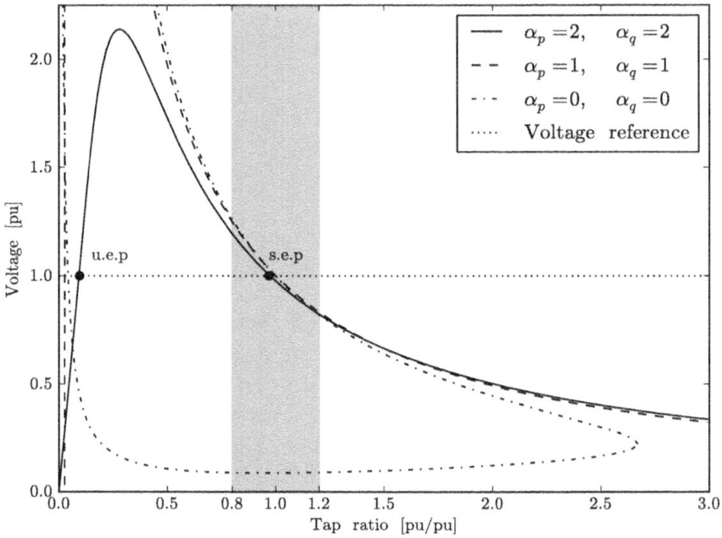

Fig. 11.10 Characteristic of the load with embedded tap changer. The curve is obtained assuming $v_{\mathrm{Th}} = 1.05$ pu, $x_{\mathrm{Th}} = x_T = 0.05$ pu, $p_{L0} = 0.7$ pu and $q_{L0} = 0.5$ pu

Example 11.4 Comparison of ULTC Discrete and Continuous Models

Figure 11.11 shows a comparison of the transient behavior of the discrete and continuous ULTC control models. The plot is obtained substituting in the IEEE 14-bus system the fixed transformer connecting buses 4 and 9 with a voltage regulating transformer. The disturbance is line 2-4 outage at $t = 1$ s. For the continuous model, $K_d = 0$ and $K_i = 0.1$ 1/s/pu, while for the discrete model $\Delta v = 5\%$, $\Delta m = 0.02$ pu/pu and $\Delta t_s = 5$ s. For both models, the regulated voltage is that of bus 9, $v^{\mathrm{ref}} = 1.0563$ pu, $m^{\max} = 1.2$ pu/pu and $m^{\min} = 0.8$ pu/pu. As expected, the behavior of the two models is similar. The discrete model introduces discontinuities in the algebraic variables (e.g., bus voltages).

11.2.3 Phase Shifting Transformer

Phase Shifting Transformers (PhSTs) are able to vary the phase shifting angle ϕ to control the active power flow. These devices are used in meshed networks for reducing the congestion on some transmission lines and/or properly redistributing active power flows in transmission lines. A fairly complete review of PhST technologies can be found in [333]. However, regardless the PhST

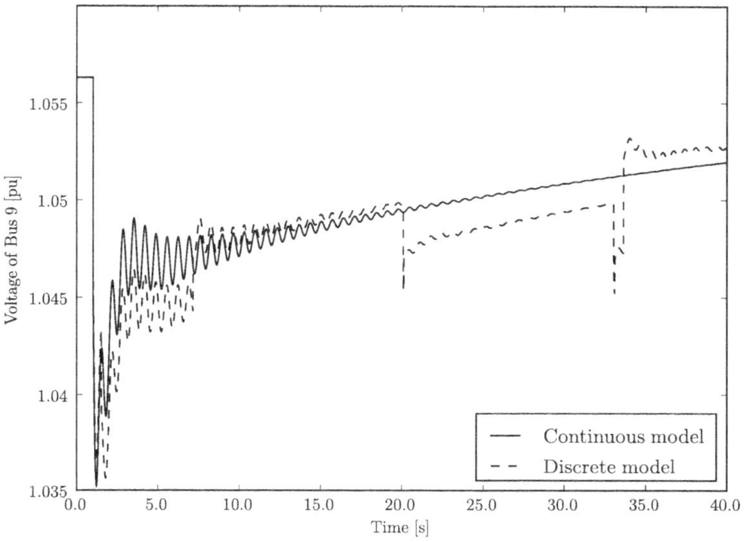

Fig. 11.11 Comparison of ULTC discrete and continuous models

type, the resulting behavior can be described by (11.27) with a variable phase shifting angle ϕ.

Similarly to the ULTC, both a discrete and a continuous models can be formulated. The discrete model is the same as (11.28) but for the regulated quantity ϕ. Figure 11.12 depicts the continuous control diagrams. The measure $p^{\rm mes}$ of the real power flow p_k is compared with the desired power flow $p^{\rm ref}$ and a PI controller is used for varying the phase angle ϕ. Differential equations are as follows:

$$\begin{aligned} \dot{\phi} &= K_p(p_k - p^{\rm mes})/T_m + K_i(p^{\rm mes} - p^{\rm ref}) \\ \dot{p}^{mes} &= (p_k - p^{\rm mes})/T_m \end{aligned} \tag{11.35}$$

The phase angle ϕ is subjected to an windup limiter. PhST parameters are defined in Table 11.4. It is relevant to note that connecting two areas of a network only by means of PhSTs can lead to unsolvable cases, as PhSTs impose the total real power transfer between the two areas.

11.2.4 Three-Winding Transformer

Three-winding transformers can be modelled as three two-winding transformers in a star connection, as depicted in Figure 11.13. The transformation requires adding one fictitious bus. The data of three-winding transformers

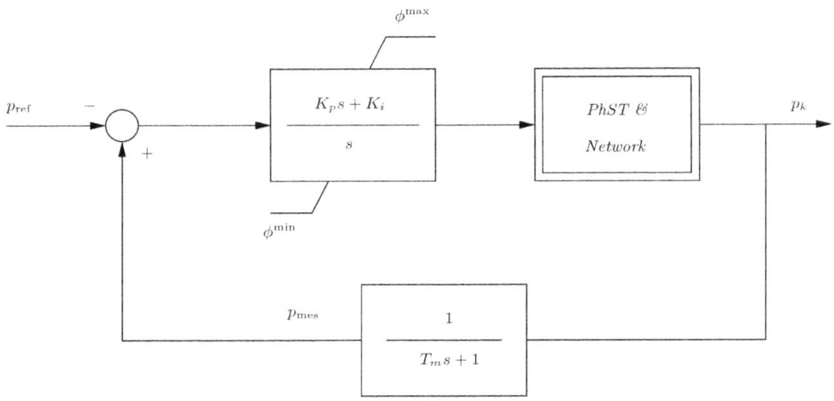

Fig. 11.12 Phase shifting transformer control diagram

Table 11.4 Phase shifting transformer control parameters

Variable	Description	Unit
K_i	Integral gain	rad/s/pu
K_p	Proportional gain	rad/pu
p^{ref}	Reference power	pu
T_m	Measurement time constant	s
$\phi^{\max}$	Maximum phase angle	rad
$\phi^{\min}$	Minimum phase angle	rad

typically are the impedances of the triangle branches, whose relationships with the resulting star impedances are as follows:

$$\begin{aligned}\bar{z}_{12} &= \bar{z}_1 + \bar{z}_2 \\ \bar{z}_{13} &= \bar{z}_1 + \bar{z}_3 \\ \bar{z}_{23} &= \bar{z}_2 + \bar{z}_3\end{aligned} \tag{11.36}$$

Thus, one has:

$$\begin{aligned}\bar{z}_1 &= (\bar{z}_{12} + \bar{z}_{13} - \bar{z}_{23})/2 \\ \bar{z}_2 &= (\bar{z}_{12} + \bar{z}_{23} - \bar{z}_{13})/2 \\ \bar{z}_3 &= (\bar{z}_{13} + \bar{z}_{23} - \bar{z}_{12})/2\end{aligned} \tag{11.37}$$

Table 11.5 defines three-winding transformer parameters. Tap ratios and the phase shifts are with respect to each winding. For example m_1 is the ratio $V_1/V_{n,1}$.

From the implementation viewpoint, three-winding transformer can be converted into a two-winding transformers during the parsing of input data.

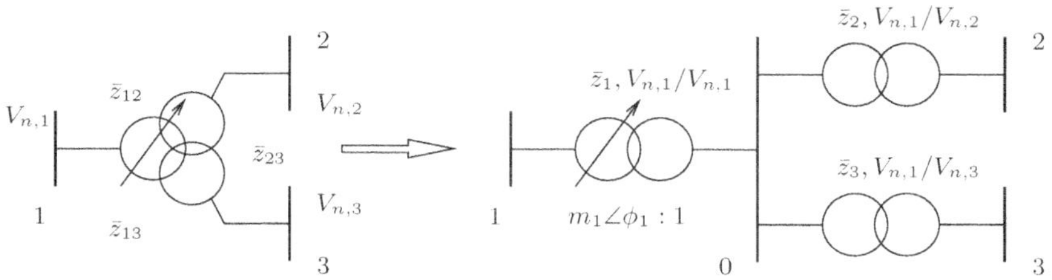

Fig. 11.13 Three-winding transformer equivalent circuit

Table 11.5 Three-winding transformer parameters

Variable	Description	Unit
$I_1^{\max}$, $I_2^{\max}$, $I_3^{\max}$	Winding current limits	kA
m_1, m_2, m_3	Fixed tap ratios	kV/kV
$P_1^{\max}$, $P_2^{\max}$, $P_3^{\max}$	Winding active power limits	kA
r_{12}, r_{23}, r_{13}	Branch resistances	pu
$S_1^{\max}$, $S_2^{\max}$, $S_3^{\max}$	Winding apparent power limits	kA
x_{12}, x_{23}, x_{13}	Branch reactances	pu
$V_{n,1}$, $V_{n,2}$, $V_{n,3}$	Winding voltage ratings	kV
ϕ_1, ϕ_2, ϕ_3	Fixed phase shifts	rad

Thus, the three-winding transformer class requires only a method for creating the set of equivalent two-winding transformers.

Example 11.5 Three-Winding Transformer of the IEEE 14-Bus System

The IEEE 14-bus system contains a three-winding transformer that connects buses 4, 8 and 9. Provided that $x_{47} = 0.20912$ pu, $x_{87} = 0.17615$ pu and $x_{79} = 0.11001$ (see Appendix D), from (11.36) one obtains the original impedances of the three-winding transformer as:

$$x_{48} = 0.38527 \text{ pu}$$
$$x_{89} = 0.28616 \text{ pu}$$
$$x_{49} = 0.31913 \text{ pu}$$

Finally, the winding at bus 4 has the tap ratio $m_4 = 0.978$ kV/kV.

Bus 7 is a fictitious bus introduced by the transformation of the three-winding transformer into three branches. I guess that the IEEE 14-bus system is not known as the $(14 - 1)$-bus system just for superstition.

11.3 Vectorial Implementation

Equations (11.6) and (11.27) are written as power injections at buses h and k. This is the most versatile form of modelling transmission lines. However, in most power system analysis books and software packages, particular relevance is devoted to the construction of the admittance matrix $\bar{\boldsymbol{Y}}$, which allows rewriting (11.3) in vectorial form:

$$\bar{\boldsymbol{s}} = \bar{\boldsymbol{v}} \bar{\boldsymbol{Y}}^* \bar{\boldsymbol{v}}^* \tag{11.38}$$

The admittance matrix $\bar{\boldsymbol{Y}}$ is formed as follows:

1. The diagonal element (h, h) is computed as the sum of all shunt and series admittances of line incident at bus h.
2. The element (h, k) is computed as the sum of series admittances of branches that connect buses h and k changed of sign.

Table 11.6 depicts the full admittance matrix for the IEEE 14-bus system.

Using adequate matrix manipulation libraries (e.g., BLAS), the admittance matrix allows efficiently calculating current and power flow vectors, as well as forming the Jacobian and Hessian matrix of the transmission system. Sample scripts implementing these calculations are provided below. However, the admittance matrix also has important drawbacks, as follows.

1. The information provided by (11.38) is "less" than the information provided by (11.6) and (11.27). In fact, the admittance matrix merges together all parallel lines and shunt devices. In other words, from the knowledge of the admittance matrix, one cannot unequivocally go back to line and transformer parameters.
2. Any change in the system topology (e.g., line outages) as well as in branch parameters requires re-building the whole admittance matrix. Line outages are relatively uncommon events and re-building the admittance matrix can be acceptable. However, regulating transformers that vary continuously or almost continuously the tap ratio and/or the phase shift cannot be efficiently handled through the admittance matrix.
3. As discussed in Subsection 11.1.5, simple coupling devices cannot be included in the admittance matrix. Actually, most series FACTS devices as well as HVDC systems cannot be modelled through the admittance matrix.

The limitations above lead to some issues when it comes to implement a general power system devices model. Moreover, using the admittance matrix model prevents from using the transmission line model as a base class for other devices, such as series devices. Nevertheless, the admittance matrix approach is so widely accepted that it deserves further discussion.

Table 11.6 Admittance matrix of the IEEE 14-bus system

Bus	1	2	3	4
1	$6.025 - 19.447j$	$-4.999 + 15.263j$	0	0
2	$-4.999 + 15.263j$	$9.521 - 30.271j$	$-1.135 + 4.782j$	$-1.686 + 5.116j$
3	0	$-1.135 + 4.782j$	$3.121 - 9.811j$	$-1.986 + 5.069j$
4	0	$-1.686 + 5.116j$	$-1.986 + 5.069j$	$10.513 - 38.635j$
5	$-1.026 + 4.235j$	$-1.701 + 5.194j$	0	$-6.841 + 21.579j$
6	0	0	0	0
7	0	0	0	$-0.000 + 4.890j$
8	0	0	0	0
9	0	0	0	$-0.000 + 1.855j$
10	0	0	0	0
11	0	0	0	0
12	0	0	0	0
13	0	0	0	0
14	0	0	0	0

Bus	5	6	7	8
1	$-1.026 + 4.235j$	0	0	0
2	$-1.701 + 5.194j$	0	0	0
3	0	0	0	0
4	$-6.841 + 21.579j$	0	$-0.000 + 4.890j$	0
5	$9.568 - 35.528j$	$-0.000 + 4.257j$	0	0
6	$-0.000 + 4.257j$	$6.580 - 17.341j$	0	0
7	0	0	$0.000 - 19.549j$	$-0.000 + 5.677j$
8	0	0	$-0.000 + 5.677j$	$0.000 - 5.677j$
9	0	0	$-0.000 + 9.090j$	0
10	0	0	0	0
11	0	$-1.955 + 4.094j$	0	0
12	0	$-1.526 + 3.176j$	0	0
13	0	$-3.099 + 6.103j$	0	0
14	0	0	0	0

Bus	9	10	11	12
1	0	0	0	0
2	0	0	0	0
3	0	0	0	0
4	$-0.000 + 1.855j$	0	0	0
5	0	0	0	0
6	0	0	$-1.955 + 4.094j$	$-1.526 + 3.176j$
7	$-0.000 + 9.090j$	0	0	0
8	0	0	0	0
9	$5.326 - 24.283j$	$-3.902 + 10.365j$	0	0
10	$-3.902 + 10.365j$	$5.783 - 14.768j$	$-1.881 + 4.403j$	0
11	0	$-1.881 + 4.403j$	$3.836 - 8.497j$	0
12	0	0	0	$4.015 - 5.428j$
13	0	0	0	$-2.489 + 2.252j$
14	$-1.424 + 3.029j$	0	0	0

Bus	13	14
1	0	0
2	0	0
3	0	0
4	0	0
5	0	0
6	$-3.099 + 6.103j$	0
7	0	0
8	0	0
9	0	$-1.424 + 3.029j$
10	0	0
11	0	0
12	$-2.489 + 2.252j$	0
13	$6.725 - 10.670j$	$-1.137 + 2.315j$
14	$-1.137 + 2.315j$	$2.561 - 5.344j$

Script 11.1 Sparse Matrix Implementation of the Admittance Matrix

The following script is an efficient sparse matrix-based implementation for the computation of the admittance matrix. The following code can be used as a method of the class that models both transmission lines and two-winding transformers.

```
def build_y(self):

    # process line data and build admittance matrix [Y]
    u = self.u + 0j
    yh0 = mul(u, self.gh0 + self.bh0*1j)
    yk0 = mul(u, self.gk0 + self.bk0*1j)
    yhk = div(u, self.rhk + self.xhk*1j)
    m = mul(self.m + 0j, exp(self.phi*0.017453292519943295*1j))
    m2 = abs(t)**2 + 0j

    self.Y  = spmatrix(div(yhk + yh0, m2), self.afr, \
              self.afr, (self.nb, self.nb), 'z')
    self.Y -= spmatrix(div(yhk, m.H.T), self.afr, \
              self.ato, (self.nb, self.nb), 'z')
    self.Y -= spmatrix(div(yhk, m), self.ato, self.afr, \
              (self.nb, self.nb), 'z')
    self.Y += spmatrix(yhk + yk0, self.ato, self.ato, \
              (self.nb, self.nb), 'z')

    # check for missing connections (0 diagonal elements)
    for item in range(self.nb):
        if abs(self.Y[item, item]) == 0:
            self.Y[item, item] = 1e-6 + 0j
```

In the code above, `self.nb` is the number of network buses, while `self.afr` and `self.ato` are lists of indexes of branch sending and receiving-end buses, respectively. All other parameters are vectors whose length is the number of branch elements. The function `spmatrix` works properly also in case of parallel branches since elements of repeated indexes are summed.

11.3.1 Incidence Matrix

Subsection 4.4.8 of Chapter 4 describes the dc power flow model. The resulting dc power flow equations are synthesized in (4.64) which is repeated below for clarity:

$$p = \boldsymbol{X}^{-1}\boldsymbol{C}\boldsymbol{\theta} + \boldsymbol{C}^T\boldsymbol{X}^{-1}\boldsymbol{\phi} \tag{11.39}$$

where $\boldsymbol{C}$ is the incidence matrix of the transmission system. this matrix is built as follows:

1. If the sending-end of branch h is bus k, set $c_{hk} = 1$.
2. If the receiving-end of branch h is bus k, set $c_{hk} = -1$.
3. Otherwise, set $c_{hk} = 0$.

Table 11.6 depicts the full incidence matrix for the IEEE 14-bus system.

Table 11.7 Incidence matrix of the IEEE 14-bus system

Line	1	2	3	4	5	6	7	8	9	10	11	12	13	14
1	1	-1	0	0	0	0	0	0	0	0	0	0	0	0
2	1	0	0	0	-1	0	0	0	0	0	0	0	0	0
3	0	1	-1	0	0	0	0	0	0	0	0	0	0	0
4	0	1	0	-1	0	0	0	0	0	0	0	0	0	0
5	0	1	0	0	-1	0	0	0	0	0	0	0	0	0
6	0	0	1	-1	0	0	0	0	0	0	0	0	0	0
7	0	0	0	1	-1	0	0	0	0	0	0	0	0	0
8	0	0	0	1	0	0	-1	0	0	0	0	0	0	0
9	0	0	0	1	0	0	0	0	-1	0	0	0	0	0
10	0	0	0	0	1	-1	0	0	0	0	0	0	0	0
11	0	0	0	0	0	1	0	0	0	0	-1	0	0	0
12	0	0	0	0	0	1	0	0	0	0	0	-1	0	0
13	0	0	0	0	0	1	0	0	0	0	0	0	-1	0
14	0	0	0	0	0	0	1	-1	0	0	0	0	0	0
15	0	0	0	0	0	0	1	0	-1	0	0	0	0	0
16	0	0	0	0	0	0	0	0	1	-1	0	0	0	0
17	0	0	0	0	0	0	0	0	1	0	0	0	0	-1
18	0	0	0	0	0	0	0	0	0	1	-1	0	0	0
19	0	0	0	0	0	0	0	0	0	0	0	1	-1	0
20	0	0	0	0	0	0	0	0	0	0	0	0	1	-1

Script 11.2 Incidence Matrix Implementation

The following script implements the incidence matrix $\boldsymbol{C}$ using the CVXOPT function `spmatrix`. the variables `self.n` and `self.nb` are the number of branches and network buses, respectively. The status vector `self.u` is used for removing off-line branches from the incidence matrix. For being used in (11.39), the column correspondent to the slack bus should be removed from the resulting matrix `self.C`.

```
def incidence(self):

    self.C = spmatrix(self.u, range(self.n), self.afr, \
                      (self.n, self.nb), 'd') - \
             spmatrix(self.u, range(self.n), self.ato, \
                      (self.n, self.nb), 'd')
```

11.3.2 Jacobian and Hessian Matrices

The admittance matrix along with efficient routines for manipulating sparse matrices allows a very compact implementation of the Jacobian and Hessian matrices of the transmission system. This compactness allows speeding up power flow, continuation power flow and optimal power flow solvers.

The computation of the Hessian matrix has a further complication due to the size of the Hessian matrix itself. By definition, $\boldsymbol{g_{yy}}$ is a three-dimensional array, which can be difficult to handle and manipulate. Fortunately, the Hessian matrix always appears in the calculations multiplied by a vector. For example, in the IPM-OPF problem, the Hessian matrix of the transmission system is multiplied by dual variables of power flow equations. The product of the Hessian matrix by a vector is a two-dimensional array that can be handled in a conventional way.

Script 11.3 Transmission System Power Flow Jacobian Matrix

The following scripts implements the Jacobian matrix for equations (11.6) and (11.27). The variables `self.a` and `self.v` are the indexes of active and reactive power flow equations, respectively, as well as the indexes of bus voltage angles and magnitudes, respectively. Finally, `self.nb` indicates the number of network buses.

```
def jacobian(self, dae):

    Vn = exp(dae.y[self.a]*1j)
    Vc = mul(dae.y[self.v] + 0j, Vn)
    Ic = self.Y*Vc

    diagVn = spmatrix(Vn, self.a, self.a, (self.nb, self.nb), 'z')
    diagVc = spmatrix(Vc, self.a, self.a, (self.nb, self.nb), 'z')
    diagIc = spmatrix(Ic, self.a, self.a, (self.nb, self.nb), 'z')

    dS = self.Y*diagVn
    dS = diagVc*dS.H.T
    dS += diagIc.H.T*diagVn

    dR = diagIc
    dR -= self.Y*diagVc
    dR = diagVc.H.T*dR

    return sparse([[dR.imag(), dR.real()], \
                   [dS.real(), dS.imag()]])
```

Script 11.4 Transmission System Power Flow Hessian Matrix

The following scripts implements the Hessian matrix for equations (11.6) and (11.27). Variables have the same meanings as in the previous Script 11.3. Furthermore, `dae.rho` indicates the vector of dual variables associated with power flow equations.

```
def hessian(self, dae):

    ang = exp(dae.y[self.a]*1j)
    vol = dae.y[self.v]
    V = matrix(1.0, (1, self.nb), 'd')
    Ma = ang*V
    MW = dae.rho[self.a]*V
    MM = dae.rho[self.v]*V
```

```
S0 = mul(Ma, mul(self.Y, Ma.H))
S1 = mul(Ma, mul(self.Y.H.T, Ma.H))
S2 = S0.H.T

A1 = S1.real()
B1 = S1.imag()
A2 = S2.real()
B2 = S2.imag()

H22 = mul(MW, A1) + mul(MW.T, A2) + \
      mul(MM, B1) + mul(MM.T, B2)
H11 = mul(vol*vol.T, H22)
H11 = H11 - spdiag(V*H11)
H21 = mul(V.T*vol.T, mul(MW, B1) - mul(MW.T, B2) - \
      mul(MM, A1) + mul(MM.T, A2))
H21 = H21 - spdiag(V*H21.T)

return sparse([[H11, H21], [H21.T, H22]])
```

11.3.3 Network Connectivity

The network connectivity provides information about the existence of islanded buses or islanded regions of a given network. This information is important for various reasons.

1. For power flow analysis, islanded buses have to be carefully treated to avoid singularities in the Jacobian matrix. Furthermore, each islanded region requires a reference bus.
2. For time domain simulations, islanded regions may lead to inconsistent calculations of the center of inertia or cause loss of synchronism of generators.

The network connectivity can be determined efficiently using the topological data of series devices. Let us define the connectivity matrix $\boldsymbol{T}$ as a binary matrix built as follows:

1. The element $t_{hh} = 1$ if there exists a branch starting or ending at bus h, $t_{hh} = 0$ otherwise.
2. The element $t_{hk} = 1$ if there exists a branch connecting buses h and k, $t_{hk} = 0$ otherwise.

In practice, $\boldsymbol{T}$ has the same non-zero elements as the admittance matrix $\bar{\boldsymbol{Y}}$.

A straightforward information provided by the connectivity matrix $\boldsymbol{T}$ is that if a diagonal element is $t_{hh} = 0$, then bus h is islanded. Furthermore, each non-zero element t_{hk} indicates the first-level connectivity of buses h and k. Another interesting property of $\boldsymbol{T}$ is that off-diagonal elements of $\boldsymbol{T}^n$ provide the n-level connectivity of buses. For example, if an off-diagonal element (h, k) of $\boldsymbol{T}^2$ is not zero, then there is a path that connects buses h and k through a third bus j. The Goderya's algorithm takes advantage of the properties of $\boldsymbol{T}$ powers to set up an efficient method that assesses the connectivity of the whole network [108]. In fact, by multiplying iteratively $\boldsymbol{T}$ by itself, one can find two situations:

1. For a given iteration n, $\boldsymbol{T}^n$ is full. Thus, the network is fully connected.
2. For a given iteration n, the non-zeros elements of $\boldsymbol{T}^n$ are the same of $\boldsymbol{T}^{n-1}$. Then, the system presents some islands and the non-zeros elements of each row (or column) correspond to the buses that belong to a certain island.

Script 11.5 Network Connectivity

The following script finds islanded buses as well as islanded areas using the Goderya's algorithm. The variables `self.n`, `self.nb`, `self.afr` and `self.ato` have the meaning of the previous Script 11.3.

```
def connectivity(self, bus):

    n = self.nb
    fr = self.afr
    to = self.ato
    os = [0]*self.n

    # find islanded buses
    diag = list(matrix(spmatrix(1.0, to, os, (n, 1), 'd') + \
                       spmatrix(1.0, fr, os, (n, 1), 'd')))
    nib = bus.n_islanded_buses = diag.count(0)
    bus.islanded_buses = []
    for bus in range(n):
        if diag[bus] == 0:
            bus.islanded_buses.append(bus)

    # find islanded areas
    temp = spmatrix(1.0, fr + to + fr + to, \
                    to + fr + fr + to, (n, n), 'd')
    cons = temp[0, :]
    nelm = len(cons.J)
    conn = spmatrix([], [], [], (1, n), 'd')
    bus.island_sets = []
    idx = islands = 0

    while 1:
        while 1:
            cons = cons*temp
            new_nelm = len(cons.J)
            if new_nelm == nelm: break
            nelm = new_nelm
        bus.island_sets.append(list(cons.J))
        conn += cons
        islands += 1
        nconn = len(conn.J)
        if nconn >= (n - nib): break
        for element in conn.J[idx:]:
```

```
            idx += 1
            if not diag[element]:
                # this is an isolated bus
                continue
            if element > (idx - 1): break
        cons = temp[idx, :]
```

It is worth observing that in the discussion of [108], a reviewer argues that a minimum spanning tree algorithm could do the same job as the Goderya's algorithm, but without the need of matrix multiplications. This concern was very adequate having in mind system programming languages available in 1980, which is the date of publication of [108]. However, as discussed in Chapter 3, for scripting languages, for-loops are much less efficient than matrix multiplications obtained using external FORTRAN-based libraries.

The following code implements a recursive minimum spanning tree algorithm and does the same as the double while-loop in the previous script:

```
def find_conn(item, group, fr, to):
    while 1:
        if item in fr:
            idx = fr.index(item)
            fr.pop(idx)
            new_item = to.pop(idx)
        elif item in to:
            idx = to.index(item)
            to.pop(idx)
            new_item = fr.pop(idx)
        else:
            break
        group.add(new_item)
        group, fr, to = find_conn(new_item, group, fr, to)
    return group, fr, to

bus.island_sets = []
islands = 0
while 1:
    if not len(fr): break
    islands += 1
    group = set([])
    group, fr, to = find_conn(fr[0], group, fr, to)
    bus.island_sets.append(group)
```

However, the minimum spanning tree algorithm is much slower than the Goderya's algorithm, at least for a scripting language as Python. For example, for the 11856-bus system discussed in Example 4.3 of Chapter 4, the Goderya's algorithm takes 3.5 s to find the 208 islands, while the minimum spanning tree algorithm takes about 66 s, i.e., it is almost 20 times more time consuming.

Chapter 12
OPF Devices

The object of this chapter are the models that define the objective function and inequality constraints for the optimal power flow analysis discussed in Chapter 6. Sections 12.1 describes typical network security constraints. Section 12.2 describes technical limits and offering functions of generators. Finally Section 12.3 describes technical limits and bidding functions of loads.

12.1 Network Constraints

This section briefly outlines typical network technical limits, namely the bus voltage limits (Subsection 12.1.1) and transmission line flow limits (Subsection 12.1.2).

12.1.1 Bus Voltage Limits

Bus voltage limits are:

$$v^{\min} \leq v_h \leq v^{\max} \tag{12.1}$$

These limits are generally required for all buses, hence also for pure transit nodes (see Table 10.1). However, generator and load devices can impose more restrictive limits.

12.1.2 Transmission Line limits

Branch flow constraints are:

$$\begin{aligned} \phi_{hk} &\leq \phi_{hk}^{\max} \\ \phi_{kh} &\leq \phi_{hk}^{\max} \end{aligned} \tag{12.2}$$

where ϕ_{ij} and ϕ_{ji} are current, active power or apparent power flows (see Tables 11.1, 11.2 and 11.5). Transformer tap ratios and phase shifts can also be subjected to inequalities constraints:

F. Milano: Power System Modelling and Scripting, Power Systems, pp. 291–304.
springerlink.com

$$\begin{aligned} m^{\min} &\leq m \leq m^{\max} \\ \phi^{\min} &\leq \phi \leq \phi^{\max} \end{aligned} \tag{12.3}$$

Finally, as discussed in Subsection 11.1.2 of Chapter 11, tie line constraints can be included in an OPF problem as inequalities:

$$p^{\min} \leq p_h \leq p^{\max} \tag{12.4}$$

12.2 Generator Constraints

This section describes most relevant generator constraints an technical limits. Subsection 12.2.1 defines the static synchronous generator capability curve. Subsection 12.2.2 to 12.2.5 describe generator supply offers, the reactive power payment function, and reserve and ramp limits.

12.2.1 Capability Curve

Generator technical limits form the so-called capability curve and are of particular relevance in OPF analysis. Active power limits are simply:

$$p_G^{\min} \leq p_h \leq p_G^{\max} \tag{12.5}$$

where $p_G^{\max}$ are associated with the generator capacity whereas $p_G^{\min}$ are associated with the minimum technical power output. For thermal plants, $p_G^{\min}$ cannot be zero because the plant requires power for feeding internal compressors and pumps. Hydro plants can show $p_G^{\min} = 0$.

Analogously to (12.5), reactive power limits can be approximated as:

$$q_G^{\min} \leq q_h \leq q_G^{\max} \tag{12.6}$$

where both $q_G^{\min}$ and $q_G^{\max}$ are constant and approximate minimum and maximum field current limits. Constraints (12.5) and (12.6) define the simplest and most commonly used generator capability curve, as shown in Figure 12.1.a.

A more precise curve can be defined considering the physical cause behind reactive power limits. These are thermal constraints, namely (i) stator current limit, (ii) rotor current limit, and (iii) under-excitation limit.

The stator thermal limit or stator current limit can be written as:

$$|\bar{v}_h \bar{\imath}_h^*| = \sqrt{p_h^2 + q_h^2} \leq s_G^{\max} \tag{12.7}$$

where $s_G^{\max}$ is the nominal generator power in pu. The rotor thermal limit is as follows:

$$p_h^2 + \left(q_h + \frac{v_h^2}{x_d}\right)^2 \leq \left(\frac{v_h i_f^{\max}}{x_d}\right)^2 \tag{12.8}$$

where

$$1.7\ \text{pu} \leq i_f^{\max} \leq 1.79\ \text{pu} \tag{12.9}$$

for salient-pole alternators and

$$2.6\ \text{pu} \leq i_f^{\max} \leq 2.73\ \text{pu} \tag{12.10}$$

for turbo-generators. Finally, the under-excitation limit is also a thermal limit. Due to the low field current i_f, the leakage flux re-closes in the end of stator windings. This flux creates eddy currents that overheat the end of stator windings. The under-excitation limit can be approximated as a straight line:

$$q_h \leq -q_0(v_h) + \beta p_G^{\max} \tag{12.11}$$

where typical values of the parameters β and q_0 are $\beta \in (0.1, 0.2)$ pu/pu, and q_0 depends on the bus voltage set point and can be approximated with $q_0 \approx 0.4$ pu. Figure 12.1.b illustrates the detailed generator capability curve while Table 12.1 defines the parameters required for defining generator limits.

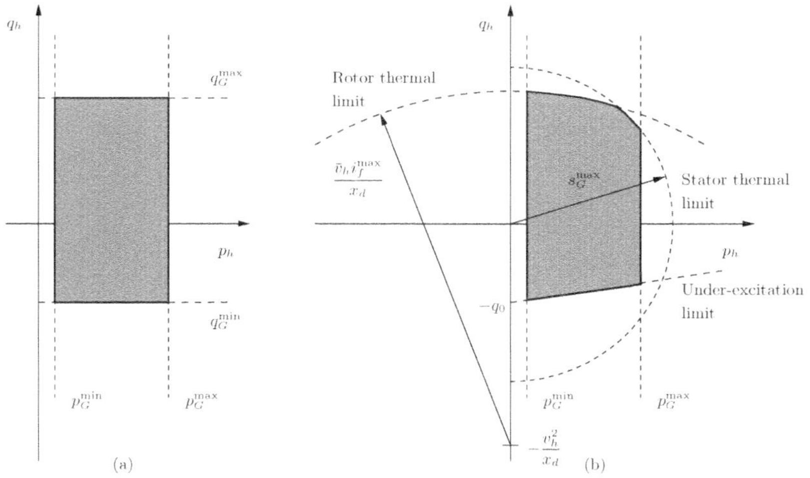

Fig. 12.1 Capability curve: (a) simplified model; (b) detailed model

12.2.2 Supply Offer

Apart from the technical limits discussed in the previous section, electricity markets require the definition of economical data. These are expressed in form of offers, as depicted in Table 12.2. Generator costs have the same mathematical form as price offers. However, in restructured power systems, generator costs are confidential. Since electricity markets are nowadays the common trend, this section refers to offers rather than to costs. Offers can

Table 12.1 Capability curve parameters

Variable	Description	Unit
$i_f^{\max}$	Maximum field current	pu
$p_G^{\max}$	Generator capacity	pu
$p_G^{\min}$	Technical minimum	pu
q_0	Under-excitation limit off-set	pu
$q_G^{\max}$	Maximum reactive power	pu
$q_G^{\min}$	Minimum reactive power	pu
x_d	Synchronous reactance	pu
β	Slope of the under-excitation limit	pu/pu

Table 12.2 Supply offer parameters

Variable	Description	Unit
C_{S0}	Fixed offer price	€/h
C_{S1}	Proportional offer price	€/MWh
C_{S2}	Quadratic offer price	€/MW^{2}h
k_{TB}	Tie breaking cost	€/MW^{2}h
$p_S^{\max}$	Maximum power offer	pu
$p_S^{\min}$	Minimum power offer	pu

be associated with active power are generally modelled using a polynomial model:[1]

$$c_S(p_S) = c_{S0} + c_{S1}p_S + c_{S2}p_S^2 \tag{12.12}$$

Another kind of supply data are *tie breaking* costs k_{TB}. The tie breaking involves a penalty cost k_{TB} prorated by the amount scheduled over the maximum amount that could be scheduled for the generator by means of a quadratic function added to the objective function:

$$c_{\mathrm{TB}} = k_{\mathrm{TB}} \frac{p_S^2}{(p_S^{\max})^2} \tag{12.13}$$

If the generator does not supply power, this cost is zero, whereas if p_S is close to the maximum power the tie breaking cost increases quadratically and penalizes the generator. Thus two otherwise tied energy offers will be

[1] The difference in the notation of upper case prices C_S in Table 12.2 and lower case price c_S in (12.12) is due to quantity units. C_S is used for absolute values, while c_S for pu ones. The relation is as follows:

$$\begin{aligned} c_{S0} &= C_{S0}/S_n \\ c_{S1} &= C_{S1} \\ c_{S2} &= C_{S2}S_n \end{aligned}$$

where S_n is the device nominal base.

scheduled to the point where their modified costs are identical, effectively achieving a prorated result. Generally, the value of $k_{\rm TB}$ is small (e.g., 0.0005).

Supply offer blocks are:

$$p_S^{\min} \leq p_S \leq p_S^{\max} \tag{12.14}$$

Several offer blocks can be defined for each generator (piece-wise bids). The constraints that have to be satisfied are:

$$\begin{aligned} \sum_{j \in \mathcal{S}} p_{S,j} &\leq p_G^{\max} \\ \sum_{j \in \mathcal{S}} p_{S,j}^{\min} &\geq p_G^{\min} \end{aligned} \tag{12.15}$$

where $\mathcal{S}$ is the set of supply blocks of the generator. A more precise supply offer block includes a unit commitment binary variable u_C:

$$u_C p_S^{\min} \leq p_S \leq u_C p_S^{\max} \tag{12.16}$$

The unit commitment discrete variable u_C allows putting off-line generators whose technical minimum $p_S^{\min} \neq 0$. However, including discrete variables in the OPF leads to a MINLP problem whose optimum can be hardly found (see also the discussion provided in Example 1.1 of Chapter 1). Thus, nonlinear OPF problems are generally formulated without unit commitment variables.

Script 12.1 Implementation of Supply Offers

This example provides a possible implementation of the methods for computing the objective function, $\boldsymbol{g}(\boldsymbol{z})$, $\boldsymbol{h}(\boldsymbol{z})$ as well as Jacobian and Hessian matrices associated with the supply offer constraints. These methods are called by the IPM-OPF method described in Script 6.1 of Chapter 6.

```
from cvxopt.base import spmatrix, sparse, matrix, mul, div
from cvxopt.blas import dotu

def gcall_opf(self, dae):

    zeros = [0]*self.n
    dae.h += spmatrix(self.pmin - dae.z[self.p],  \
                      self.pn, zeros, (dae.nh, 1), 'd')
    dae.h += spmatrix(dae.z[self.p] - self.pmax,  \
                      self.px, zeros, (dae.nh, 1), 'd')
    dae.g -= spmatrix(dae.z[self.p], self.a, zeros,  \
                      (dae.ng, 1), 'd')

def gycall_opf(self, dae):

    dae.Hz -= spmatrix(1.0, self.pn, self.p,  \
                       (dae.nh, dae.nz), 'd')
```

```
    dae.Hz += spmatrix(1.0, self.px, self.p,  \
                       (dae.nh, dae.nz), 'd')

    dae.Gz -= spmatrix(1.0, self.a, self.p,  \
                       (dae.ng, dae.nz), 'd')
    dae.Oz[self.p] = self.cp1 + 2*mul(self.cp2 +  \
                                    self.tie, dae.z[self.p])

    dae.Hes += spmatrix(2*(self.cp2 + self.tie),  \
                        self.p, self.p, (dae.nz, dae.nz), 'd')

def call_obj(self, dae):

    p = dae.z[self.p]

    dae.obj += sum(self.cp0)
    dae.obj += dotu(self.cp1, p)
    dae.obj += dotu(self.cp2 + self.tie, p**2)
```

In the code above, `dae.nh` is the number of inequality constraints $\boldsymbol{h}$, `dae.ng` is the number of equality constraints $\boldsymbol{g}$, and `dae.nz` is the number of variables $\boldsymbol{z}$. Furthermore, `self.p` are the indexes of variables p_h in the vector $\boldsymbol{z}$, whereas `self.px` and `self.pn` are the indexes of dual variables associated with the maximum and minimum limits of p_h, respectively.

12.2.3 Reactive Power Payment Function

In restructured power systems, generator reactive powers can be subjected to payment. Reactive power payments are generally associated with the *loss of opportunity* of generating active power. Figure 12.2 shows the payment function proposed in [85, 358]. The production of reactive power can be divided into three regions:

1. Region I: for $q_G^{\rm C} \leq q_h \leq 0$, the generator is under-excited and produces inductive reactive power.
2. Region II: for $0 \leq q_h \leq q_G^{\rm A}$, the generator is over-excited and produces capacitive reactive power without the need of reducing its active power production.
3. Region III: for $q_G^{\rm A} \leq q_h \leq q_G^{\rm B}$, the generator is over-excited and produces capacitive reactive power but has to reduce its active power production to provide the required reactive power. It is assumed that, in this region, the payment increases quadratically with respect to the reactive power.

Regions I and II can be further divided into two regions:

1. Mandatory reactive power availability: for $q_G^{\text{lead}} \leq q_h \leq q_G^{\text{lag}}$, the generator has to provide reactive power at no cost.
2. Payment for the cost of losses: for $q_G^{\text{lag}} \leq q_h \leq q_G^{\text{A}}$ and $q_G^{\text{C}} \leq q_h \leq q_G^{\text{lead}}$, the losses due to the reactive power production justify a payment for generators. In this case, the payment is a linear function of the reactive power.

While q_G^{lead} and q_G^{lag} are assigned constant parameters, q_G^{A}, q_G^{B} and q_G^{C} depend on the generator operating point[2] and can be determined based on the capability curve (see Figure 12.1). In summary, the payment function is:

$$c_Q(q_h) = \begin{cases} 0 \quad \text{if} \quad q_G^{\text{lead}} \leq q_h \leq q_G^{\text{lag}} \\ c_{Q0} - c_{Q1}(q_h - q_G^{\text{lead}}) \quad \text{if} \quad q_G^{C} \leq q_h \leq q_G^{\text{lead}} \\ c_{Q0} + c_{Q2}(q_h - q_G^{\text{lag}}) \quad \text{if} \quad q_G^{\text{lag}} \leq q_h \leq q_G^{\text{A}} \\ c_{Q0} + c_{Q2}(q_G^{\text{A}} - q_G^{\text{lag}}) + 0.5 c_{Q3}(q_h - q_G^{\text{A}})^2 \quad \text{if} \quad q_G^{\text{A}} \leq q_h \leq q_G^{\text{B}} \end{cases} \tag{12.17}$$

Table 12.3 resumes the parameters required for defining the generator reactive power payment function.

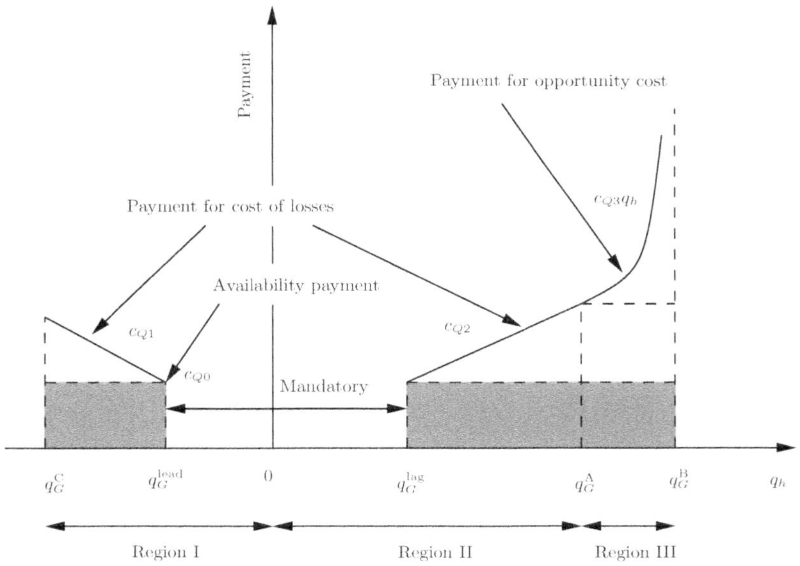

Fig. 12.2 Generator reactive power payment function

[2] In particular, q_G^{A} and q_G^{B} depend on the stator and rotor current limits whereas q_G^{C} depends on the under-excitation limit.

Table 12.3 Generator reactive power payment parameters

Variable	Description	Unit
C_{Q_0}	Availability cost	€/h
C_{Q_1}	Under-excitation loss cost	€/MVArh
C_{Q_2}	Over-excitation loss cost	€/MVArh
C_{Q_3}	Cost of opportunity loss	€/MVAr^{2}h
S_n	Power rating	MVA
q_G^{lead}	Minimum mandatory reactive power	pu
q_G^{lag}	Maximum mandatory reactive power	pu

12.2.4 Generator Power Reserve

The operating reserve of a system is associated with the power that is not directly used by loads but can be requested and generators have to provide quickly. The power reserve has an associated offer:

$$c_R(p_R) = c_{R1} p_R \tag{12.18}$$

and limits:

$$p_R^{\min} \le p_R \le p_R^{\max} \tag{12.19}$$

along with the inequalities that ensure that the sum of the power supply and the power reserve is less than the total available power supply $p_S^{\max}$ and that the total power reserve must be less than the total power demand:

$$\begin{aligned} p_S + p_R &\le p_S^{\max} \\ \sum_{j \in \mathcal{R}} p_{R,j} &\le \sum_{i \in \mathcal{D}} p_{D,i} \end{aligned} \tag{12.20}$$

where $\mathcal{D}$ and $\mathcal{R}$ are the set of demands and reserves, respectively. Reserve parameters are shown in Table 12.4.

Table 12.4 Generator reserve parameters

Variable	Description	Unit
-		-
C_{R1}	Reserve offer price	€/MWh
$p_R^{\max}$	Maximum power reserve	pu
$p_R^{\min}$	Minimum power reserve	pu
S_n	Power rating	MVA

12.2.5 Generator Power Ramp

Generation facilities have limits on their ability to move from one level of production to another, and these limits are generally taken in account by the so-called ramp constraints. Some typical generator ramp parameters are depicted in Table 12.5.

The parameters used in the multi-period OPF problems are the up and down ramp rates, i.e. r^{up} and r^{dw}. These quantities express the amount of power that can be moved each minute up or down by the generator and are associated to technical limits of the generation plants. Ramp constraints are modelled as follows:

$$\begin{aligned} p_S(t) - p_S(t - \Delta t) &\leq p_S^{\max} r^{\mathrm{up}} \Delta t, \quad t \in \mathcal{T} \\ -p_S(t) + p_S(t - \Delta t) &\leq p_S^{\max} r^{\mathrm{dw}} \Delta t, \quad t \in \mathcal{T} \end{aligned} \tag{12.21}$$

where $\mathcal{T}$ is set of periods of the multi-period time horizon, Δt is the time interval in which is divided the complete time horizon spanned by the multi-period OPF.

Along with ramp limits (12.21), one should also define a maximum reserve ramp rate $r_r^{\max}$, that multiplied by the time interval Δt, expresses the maximum amount of power that can be dedicated to the reserve, thus:

$$p_R(t) \leq r_r^{\max} \Delta t, \quad t \in \mathcal{T} \tag{12.22}$$

If the generator output is low, also the operating reserve can decrease, and the operating reserve loading point p_{RLP} allows to reduce the power reserve for low outputs:

$$p_R(t) \leq p_S(t) \frac{r_r^{\max} \Delta t}{p_{\mathrm{RLP}}}, \quad t \in \mathcal{T} \tag{12.23}$$

Thus, the power reserve p_R is the minimum value as obtained from (12.22) and (12.23).

Table 12.5 Generator power ramp parameters

Variable	Description	Unit
DT	Minimum # of period down	h
p_{RLP}	Operating reserve loading point	pu
r^{up}	Ramp rate up	pu/h
r^{dw}	Ramp rate down	pu/h
$r_r^{\max}$	Maximum reserve ramp rate	pu/h
UT	Minimum # of period up	h
α_0	# of periods up for $t = 0$	int
β_0	# of periods down for $t = 0$	int

Minimum on-line and off-line time constraints can be formulated as proposed in [66] and in [65]. These are as follows:

Minimum up time:

$$\sum_{t=1}^{UT_0}(1 - u_{\rm OL}(t)) = 0 \tag{12.24}$$
$$\sum_{\tau=t}^{t+UT-1} u_{\rm OL}(\tau) \geq UT u_{\rm SU}(t), \quad \forall t = UT_0 + 1, \dots, T - UT + 1$$
$$\sum_{\tau=t}^{T}(u_{\rm OL}(\tau) - u_{\rm SU}(t)) \geq 0, \quad \forall t = T - UT + 2, \dots, T$$

Minimum down time:

$$\sum_{t=1}^{DT_0} u_{\rm OL}(t) = 0 \tag{12.25}$$
$$\sum_{\tau=t}^{t+DT-1} (1 - u_i(\tau)) \geq DT u_{\rm SD}(t), \quad \forall t = DT_0 + 1, \dots, T - DT + 1$$
$$\sum_{\tau=t}^{T}(1 - u_{\rm OL}(\tau) - u_{\rm SD}(t)) \geq 0, \quad \forall t = T - DT + 2, \dots, T$$

where $u_{\rm OL}(t)$ is a binary variable that is equal to 1 if the generator is on-line in period t;[3] $u_{\rm SU}(t)$ is a binary variable that is equal to 1 if the generator is started up at the beginning of period t; $u_{\rm SD}(t)$ is a binary variable that is equal to 1 if the generator is shut down at the beginning of period t; T is the scheduled time horizon (e.g. 24 hours); and UT_0 and DT_0 are the number of period units during which the generator must be on-line and off-line at the beginning of the time horizon respectively, as follows:

$$\begin{aligned} UT_0 &= \min\{T, (UT - \alpha_0) u_{\rm OL}(0)\} \\ DT_0 &= \min\{T, (DT - \beta_0)(1 - u_{\rm OL}(0))\} \end{aligned} \tag{12.26}$$

where α_0 is the number of time periods that the generator has been on-line at the beginning of the market horizon; and β_0 is the number of time periods that the generator has been off-line at the beginning of the market horizon. Finally, the start-up and the shut-down status of the units are managed as follows:

[3] Observe that, if considering minimum up and down constraints, (12.21), (12.22) and (12.23) have to be modified by multiplying the supply powers $p_S(t)$ and $p_S(t - \Delta t)$ by the binary variable $u_{\rm OL}(t)$ and $u_{\rm OL}(t - \Delta t)$, respectively.

$$u_{\mathrm{SU}}(t) - u_{\mathrm{SD}}(t) = u_{\mathrm{OL}}(t) - u_{\mathrm{OL}}(t - \Delta t), \quad \forall t \in \mathcal{T} \tag{12.27}$$
$$u_{\mathrm{SU}}(t) + u_{\mathrm{SD}}(t) \leq 1, \quad \forall t \in \mathcal{T}$$

Equations (12.27) are necessary to avoid simultaneous commitment and decommitment of a unit.

The constraints above make clear that multi-period OPF problems that include generator ramp limits, start-up and shut-down constraints, minimum number of up and down periods, etc., are generally modelled as MILP problems. The interested reader can found detailed model in [65, 66, 212]. However, an use of generator ramp limits can be also associated to stability constrained OPF problems. The interested reader can found an example of such use of ramp limits in [356].

12.3 Load Constraints

This section discusses constraints related to loads. Subsection 12.3.1 describes demand bid functions. Subsection 12.3.2 describes a model of load daily profile. Finally, Subsection 12.3.3 describes load ramp limits.

12.3.1 Demand Bid

In some electricity markets, loads are elastic, i.e., participate to the auction providing bid blocks in the same way suppliers provide offer blocks. The basic parameters for elastic demand bids are shown in Table 12.6 and include maximum and minimum power consumption as well as the coefficient of the bid function. The bid function has the same structure as (12.12), as follows:

$$c_D(p_D) = c_{D0} + c_{D1} p_D + c_{D2} p_D^2$$

The reactive power can be a function of the active power demand through a constant power factor $\cos \phi_D$:

$$q_D = p_D \frac{\sqrt{1 - \cos^2 \phi_D}}{\cos \phi_D} = p_D \tan \phi_D \tag{12.28}$$

As for the constraints, one has:

$$p_D^{\min} \leq p_D \leq p_D^{\max} \tag{12.29}$$

Similarly to generator bid blocks, one can define a unit commitment discrete variable u_C that allows modelling non-dispatched demands with $p_D^{\min} \neq 0$:

$$u_C p_D^{\min} \leq p_D \leq u_C p_D^{\max} \tag{12.30}$$

Finally, similarly to generator bids, one can define a tie breaking cost k_{TB}. The tie breaking involves a penalty cost k_{TB} prorated by the amount scheduled over the maximum amount that could be scheduled for the load by means of a quadratic function added to the objective function:

$$c_{TB} = k_{TB} \frac{p_D^2}{p_D^{\max}} \tag{12.31}$$

If the load does not consume power, this cost is zero, whereas if p_D is close to the maximum power the tie breaking cost increases quadratically and penalizes the load. Thus, two otherwise tied energy demands will be scheduled to the point where their modified costs are identical, effectively achieving a prorated result. Generally, the value of k_{TB} is small (e.g., 0.0005).

Table 12.6 Demand bid parameters

Variable	Description	Unit
C_{D0}	Fixed bid price	€/h
C_{D1}	Proportional bid price	€/MWh
C_{D2}	Quadratic bid price	€/MW^{2}h
k_{TB}	Tie breaking cost	€/MWh
$p_D^{\max}$	Maximum power bid	pu
$p_D^{\min}$	Minimum power bid	pu

12.3.2 Demand Daily Profile

Multi-period market clearing procedures require the definition of a daily demand profile. The simplest model is simply a sequence of demand values, one per each period in which the time horizon is subdivided:

$$\boldsymbol{d} = [d_1, d_2, \ldots, d_{n_T}] \tag{12.32}$$

For example, for a $\Delta t = 1$ h, $n_T = 24$. Each element d_t of the demand profile array is a percentage of the load p_{L0} or, in case of elastic demands, of the demand bid limits $p_D^{\max}$ and $p_D^{\min}$. Thus, one has:

$$p_L(t) = \frac{d_t}{100} p_{L0}, \quad \forall t \in \mathcal{T} \tag{12.33}$$

or

$$\begin{aligned} p_D^{\max}(t) &= \frac{d_t}{100} p_D^{\max}, \quad \forall t \in \mathcal{T} \\ p_D^{\min}(t) &= \frac{d_t}{100} p_D^{\min}, \quad \forall t \in \mathcal{T} \end{aligned} \tag{12.34}$$

where $\mathcal{T} = \{1, 2, \ldots, 24\}$. Figure 12.3 shows an example of daily demand profile.

The daily demand profile can be defined depending on kind of the day, the day of the week, and the week of the year (see Table 12.7). For example, the coefficients d_t can be computed as:

$$d_t = \frac{k_\alpha(t,\alpha)}{100} \cdot \frac{k_\beta(\beta)}{100} \cdot \frac{k_\gamma(\gamma)}{100} 100, \quad t \in \mathcal{T} \tag{12.35}$$

where $k_\alpha(t,\alpha)$ (24 elements), k_β (scalar) and k_γ (scalar) are in % and represent the kind of the day, the day of the week and the week of the year, respectively, and the indexes α, β and γ are as follows:

α: index of the kind of the day in the range from 1 to 6, with the following notation:

1: winter working day
2: winter weekend
3: summer working day
4: summer weekend
5: spring/fall working day
6: spring/fall weekend

β: day of the week in the range from 1 (Monday) to 7 (Sunday).
γ: week of the year in the range from 1 to 52.

Table 12.7 Demand profile parameters

Variable	Description	Unit
$k_\alpha(t,1)$	Daily profile for a winter working day	%
$k_\alpha(t,2)$	Daily profile for a winter weekend	%
$k_\alpha(t,3)$	Daily profile for a summer working day	%
$k_\alpha(t,4)$	Daily profile for a summer weekend	%
$k_\alpha(t,5)$	Daily profile for a spring/fall working day	%
$k_\alpha(t,6)$	Daily profile for a spring/fall weekend	%
k_β	Profile for the days of the week	%
k_γ	Profile for the weeks of the year	%
α	Kind of the day	$\{1, \ldots, 6\}$
β	Day of the week	$\{1, \ldots, 7\}$
γ	Week of the year	$\{1, \ldots, 52\}$

12.3.3 Demand Power Ramp

Although less commonly used than generation ramp rates, also demands can undergo ramp constraints. These take in account the ability of demands to move from one level of consumption to another. Demand ramp parameters

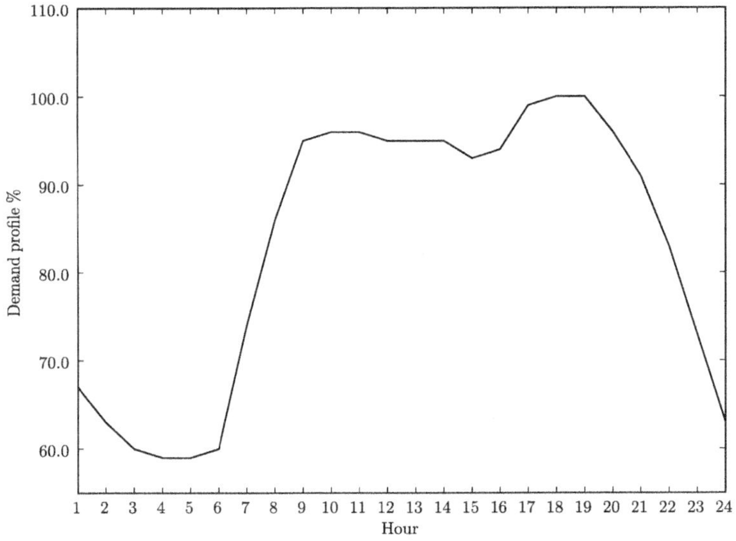

Fig. 12.3 Example of daily demand profile

are basically the same as generator ones (see Table 12.5). For example, up and down ramp constraints express the amount of demand power that can be moved up or down during each period Δt and are associated with technical limits in demand facilities, as follows:

$$\begin{aligned} p_D(t) - p_D(t - \Delta t) &\leq p_D^{\max} r^{\mathrm{up}} \Delta t, \quad t \in \mathcal{T} \\ -p_D(t) + p_D(t - \Delta t) &\leq p_D^{\max} r^{\mathrm{dw}} \Delta t, \quad t \in \mathcal{T} \end{aligned} \tag{12.36}$$

These equations are conceptually similar to (12.21) for the generation ramp rate, and uses the same time interval Δt defined in the OPF time horizon. Finally, if pertinent, minimum up and down demand periods can be modelled by means of constraints similar to (12.24)-(12.27).

Chapter 13
Faults and Protections

This chapter describes symmetrical three phase faults (Section 13.1), breakers (Section 13.2) and relays (Section 13.3). The chapter also describes measurement devices of non-standard quantities during time domain simulations, namely the Phasor Measurement Unit (Section 13.4) and the bus frequency measurement (Section 13.5).

13.1 Fault

As discussed in Chapter 8, transient stability analysis studies the effects of short circuits on the dynamic of synchronous machines. From the modelling viewpoint, three-phase faults can be considered as static shunt impedances that are connected to a bus at a given time t_f (fault time) and cleared at a given time t_c (clearing time). Equations are:

$$\begin{aligned} p_h &= -u(t) g_f v_h^2 \\ q_h &= -u(t) b_f v_h^2 \\ u(t) &= \begin{cases} 1 & \text{if } t_f \leq t \leq t_c \\ 0 & \text{if } t < t_f \text{ or } t > t_c \end{cases} \end{aligned} \tag{13.1}$$

where $g_f + jb_f = 1/(r_f + jx_f)$. The main difference with shunt impedances is the time dependence and the impedance value that, for faults, is typically low. Table 13.1 defines the parameters required for three phase faults.

Script 13.1 Fault Interventions

The following script implements fault interventions. This method is called by the numerical integration routine described in Script 8.2 of Chapter 8. The method turns on or off the status of the fault impedance. To avoid convergence issues when clearing the fault, voltage angles and magnitudes

F. Milano: Power System Modelling and Scripting, Power Systems, pp. 305–312.
springerlink.com

Table 13.1 Fault parameters

Variable	Description	Unit
r_f	Fault resistance	pu
t_c	Clearing time	s
t_f	Fault time	s
x_f	Fault reactance	pu

are recovered to the pre-fault values. The statement `system.DAE.factorize = True` forces the symbolic re-factorization of the system Jacobian matrix for any topological change following a fault status switch. The set `self._t` contains all fault and clearing times.

```
import system

def intervention(self, t):

    if not t in self._t: return False

    for item in range(self.n):

        if t == self.tf[item]: # fault occurrence

            print 'Apply fault at t = %s s' % self.tf[item]

            # enable fault
            self.u[item] = 1.0
            system.DAE.factorize = True

            # store pre-fault bus angles
            self._ang = matrix(system.DAE.y[system.Bus.a])
            self._vol = matrix(system.DAE.y[system.Bus.v])

        if t == self.tc[item]: # fault clearance

            print 'Clear fault at t = %s s' % self.tc[item]

            # disable fault
            self.u[item] = 0.0
            system.DAE.factorize = True

            # recover bus voltages
            system.DAE.y[system.Bus.a] = matrix(self._ang)
            system.DAE.y[system.Bus.v] = matrix(self._vol)
```

13.2 Breaker

In transient stability analysis, a breaker is a simple switch that contains the information on when another device has to be put on- or off-line. Electro-magnetic transients following breaker operations are quite fast with respect to transient stability and can thus be neglected (see Figure 1.6 of Chapter 1).

Since each device has its own status u, the breaker only has to set to 1 (on-line) or to 0 (off-line) the status of other devices at assigned times. In case of switching transmission lines, each switch also requires rebuilding the admittance matrix as well as checking the network connectivity. Scripts 11.1 and 11.5 of Chapter 11 provide further details on the admittance matrix and the assessment of network connectivity, respectively.

13.3 Relay

Relays are devices that, based on certain measures, decide and coordinate the times at which breaker actions occur. The decision logic depends on the relay type. There exist over-current relays, over- and under-voltage relays, distance relays, etc. The decision logic is translated into code by if-then-else logic depending on the relay type. For example, for an instantaneous over-current relay, one has:

$$\text{if } i(t) \geq i_{\text{sp}} \text{ then } t_{\text{op}} = t \tag{13.2}$$

where i_{sp} is the relay current set point and t_{op} is the operating time at which the relay sends to the breaker the switching signal. If the controlled quantity is measured remotely, a measurement delay t_m can be required:

$$\text{if } i(t + t_m) \geq i_{\text{sp}} \text{ then } t_a = t + t_m \tag{13.3}$$

A lag block can also do the job:

$$\begin{aligned} \dot{i}_m &= K_m(i - i_m)/T_m \\ \text{if } i_m(t) &\geq i_{\text{sp}} \text{ then } t_a = t \end{aligned} \tag{13.4}$$

where i_m is the measured current that is delayed with respect to the real current i. Instantaneous relays are affected by a delay t_r which is the time required by the breaker to actually switch after receiving the signal by the relay. Thus the breaker time intervention can be estimated as:

$$t_b = t_a + t_r \tag{13.5}$$

Inverse definite time lag relays have a more interesting model then instantaneous ones. The inverse time function can be defined by a series of $(i_{\text{sp}}, t_{\text{op}})$ pairs. Another possibility is to approximate the operating curve of the relay by means of a function, for example a logarithm [14]:

$$t_{\text{op}} = \begin{cases} 3.0/\log(i), & \text{if } 1.1 \leq i \leq 20 \text{ pu} \\ \infty, & \text{if } i < 1.1 \end{cases} \tag{13.6}$$

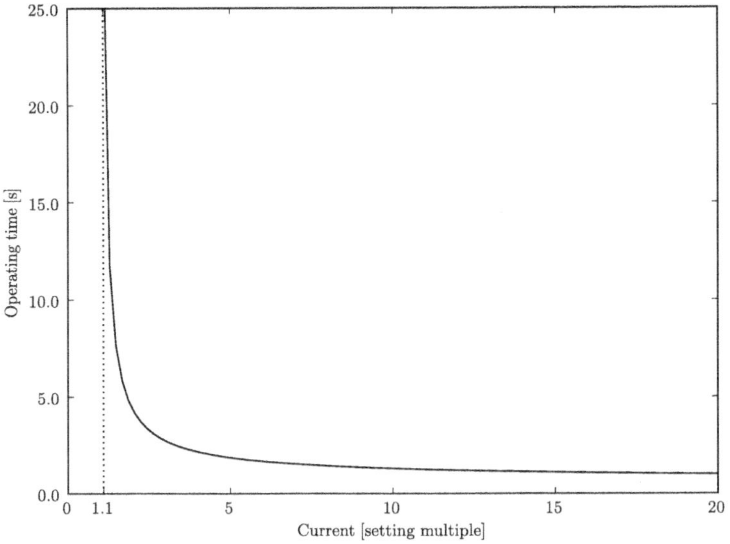

Fig. 13.1 Relay inverse time characteristic curve

Figure 13.1 illustrates (13.6).

For electro-mechanical relays, the dynamic of the disc travel d can be modelled as:

$$\dot{d} = \begin{cases} \log(i/s_{\rm pb})/(3s_{\rm tm}), & \text{if } i \geq 1.1s_{\rm pb} \\ -1/(T_r s_{\rm tm}), & \text{if } i < 1.1s_{\rm pb} \\ 0, & \text{if } d = 0 \text{ and } \dot{d} < 0 \end{cases} \tag{13.7}$$

where $s_{\rm pb}$ and $s_{\rm tm}$ are the plug bridge and the time multiplier settings, respectively, and T_r is the time required by the disc to reset due to spring action. In (13.7), it is assumed that the initial $d(0)$ after a complete resetting is zero. Then the operating time $t_{\rm op}$ is defined as:

$$d(t_{\rm op}) = 1 \text{ pu} \tag{13.8}$$

In numerical integration, the following formula can be used for approximating (13.8) [14]:

$$t_{\rm op} = t - \frac{d(t) - 1}{d(t) - d(t - \Delta t)} \Delta t, \quad \text{for } d(t) \geq 1 \text{ and } d(t - \Delta t) \leq 1 \tag{13.9}$$

where Δt is the step length of the numerical integration. Finally, the breaker intervention time is defined by (13.5). Table 13.2 defines the parameters required for an electro-mechanical inverse time over-current relay.

Table 13.2 Over-current relay parameters

Variable	Description	Unit
K_m	Measurement lag block gain	pu/pu
$i_{\rm sp}$	Current set point	s
$s_{\rm pb}$	Plug bridge setting	-
$s_{\rm tm}$	Time multiplier setting	-
T_m	Measurement lag block time constant	s
t_r	Breaker delay	s
T_r	Spring reset time	s

13.4 Phasor Measurement Unit

A Phasor Measurement Unit (PMU) is a device able to measure the magnitude and the angle of a phasor. Basic theory, definitions and applications about PMUs can be found in [349]. Brief outlines are provided below.

Let us define a sinusoidal quantity:

$$x(t) = X_M \cos(\omega t + \phi) \tag{13.10}$$

its phasor representation is:

$$\bar{X} = \frac{X_M}{\sqrt{2}} e^{j\phi} \tag{13.11}$$

The phasor is defined for a pure constant sinusoid, but it can also be used for transients, assuming that the phasor is the fundamental frequency component of a waveform over a finite interval (observation window).

PMU devices work on sampled measures (see Figure 13.2). In the case of $x(t)$, we can define the samples signal x_k at $t = k\tau_s$, where τ_s is the sampling interval. Using a Discrete Fourier Transform (DFT), the phasor $\bar{X}$ is given by:

$$\bar{X} = \frac{1}{\sqrt{2}}\frac{2}{n}(X_c - jX_s) \tag{13.12}$$

where n is the number of samples in one period of the nominal fundamental frequency f_n, and:

$$X_c = \sum_{k=1}^{n} x_k \cos k\theta \tag{13.13}$$

$$X_s = \sum_{k=1}^{n} x_k \sin k\theta \tag{13.14}$$

and θ is the sampling angle associated with the sampling interval τ_s, as follows:

$$\theta = \frac{2\pi}{n} = 2\pi f_n \tau \tag{13.15}$$

A typical sampling rate used in relaying and measurements functions is 12 times the power system frequency (e.g., 600 Hz for a 50 Hz system or 720 Hz for a 60 Hz system).

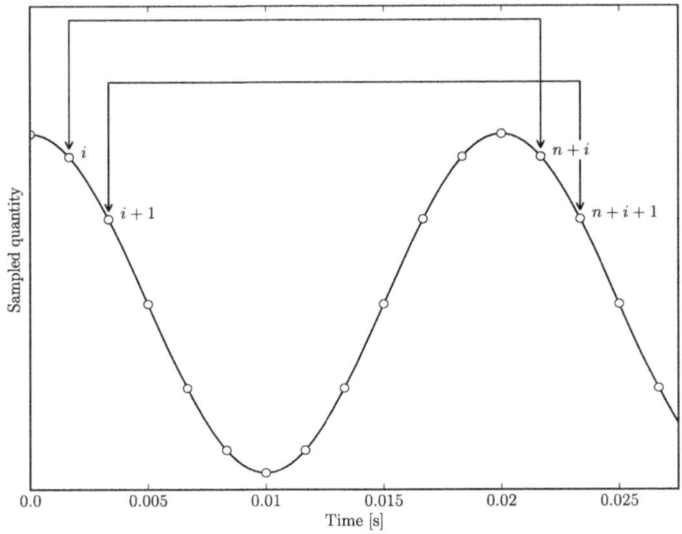

Fig. 13.2 Data sampling windows for phasor measurements

Equation (13.12) represents a non-recursive DFT calculation. A recursive calculation is an efficient method for time varying phasors. Let $\bar{X}^r$ be the phasor corresponding to the data set $x\{k = r, r+1, \ldots, n+r-1\}$, and let a new data sample be obtained to produce a new data set $x\{k = r+1, r+2, \ldots, n+r\}$. The recursive phasor corresponding to the new data window $\bar{X}^{r+1}$ is as follows:

$$\bar{X}^{r+1} = \bar{X}^r + \frac{1}{\sqrt{2}}\frac{2}{n}(x_{n+r} - x_r)e^{-jr\theta} \tag{13.16}$$

A recursive calculation through a moving window data sample is faster than a non-recursive one, needs only two sample data at each calculation (x_{n+r} and x_r) and provides a stationary phasor.

If the quantity $x(t)$ undergoes a transient, the moving window detects the amplitude and angle variations with a delay which depends on the time sample rate τ_s. If the system frequency f_n undergoes a variation Δf, at the

r^{th} time sampling, the positive sequence of the phasor undergoes the following change:

$$\bar{X}^r(f_n + \Delta f) = \bar{X} e^{-j(n-1)\pi \Delta f \Delta t} \frac{\sin(n \Delta f \Delta t)}{n \sin(\Delta f \Delta t)} e^{j2\pi r \Delta f \Delta t} \tag{13.17}$$

Thus, the rate of change of the phasor angle is as follows:

$$\frac{d\phi}{dt} = 2\pi \Delta f \tag{13.18}$$

PMU devices can be used for measuring both phasor magnitudes and phase angles. The detailed functioning of these device is rather complex. Furthermore, different (and not always clearly documented) technologies are used by different manufacturers. However, for transient analysis, the measurement can be approximated as a simple low-pass filter, as follows:

$$\dot{v}_m = (v_h - v_m)/T_v \tag{13.19}$$

$$\dot{\theta}_m = (\theta_h - \theta_m)/T_\theta \tag{13.20}$$

where v_m and θ_m are the measured voltage magnitude and phase, respectively, and T_v and T_θ are the filter time constants.

13.5 Bus Frequency Estimation

The estimation of bus frequency deviation described in this section is based on the bus voltage phase angle time derivative. The frequency estimation is obtained by means of a high-pass and a low-pass filter, as depicted in Figure 13.3. The high-pass filter approximates the derivative of the input signal. Differential equations are as follows:

$$\dot{x}_\theta = \frac{1}{T_f}\left(\frac{1}{2\pi f_n}\frac{1}{T_f}(\theta - \theta_0) - x_\theta\right) \tag{13.21}$$

$$\dot{\omega} = (\Delta\omega + \omega_s - \omega)/T_\omega$$

where θ_0 is the initial phase angle (e.g., the phase angle obtained by the power flow analysis), f_n is the nominal frequency in Hz, ω_s is the synchronous frequency in pu (e.g., $\omega_s = 1$ pu), T_f and T_ω are the time constants of the high-pass and of the low-pass filters, respectively, and $\Delta\omega$ is defined as:

$$\Delta\omega = -x_\theta + \frac{1}{2\pi f_n}\frac{1}{T_f}(\theta - \theta_0) \tag{13.22}$$

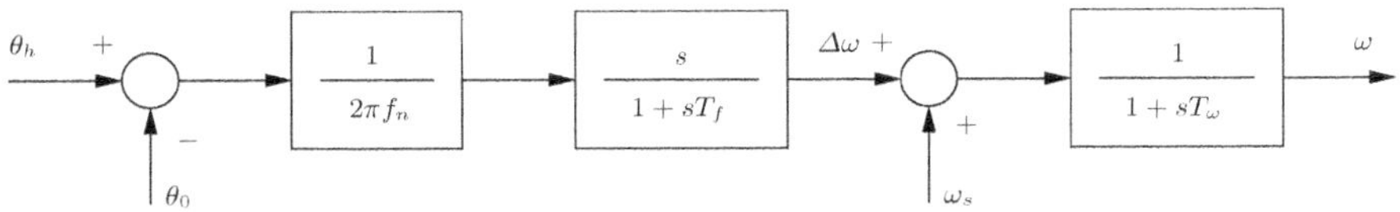

Fig. 13.3 Bus frequency measurement filter

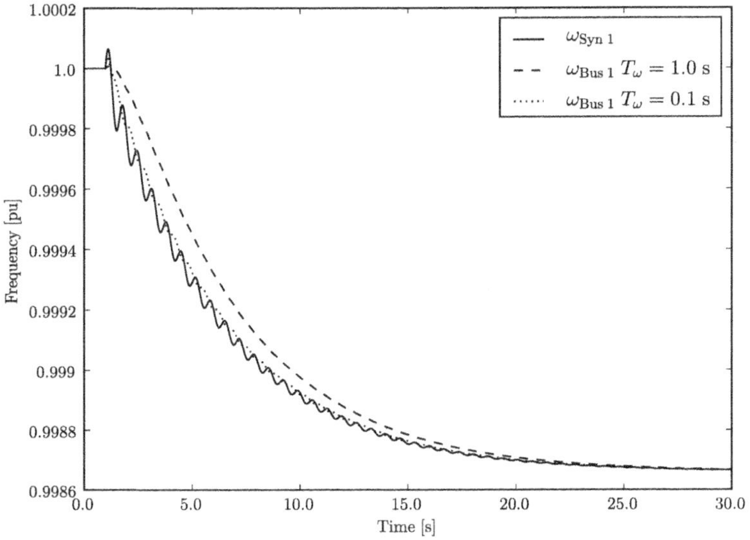

Fig. 13.4 Comparison of rotor speed and bus frequency measurements for the IEEE 14-bus system

Example 13.1 Bus Frequency Measurements for the IEEE 14-Bus System

Figure 13.4 shows the rotor machine speed as well as the bus 1 frequency computed as described above for the IEEE 14-bus system. The disturbance consists in line 2-4 outage for $t = 1$ s. In the simulation, $T_f = 0.1$ s is used. The low-pass filter of the bus measurement allows following the trend of the rotor speed but removes swing oscillations. As expected, the smaller T_ω, the more accurate the measurement output.

Chapter 14
Loads

This chapter describes static and dynamic nonlinear loads. Since traditional loads used in power flow and transient analysis are constant PQ and shunt admittances, the loads described in this chapter are also called *non-conforming* loads. These are the voltage dependent load (Section 14.1), the ZIP load (Section 14.2), the frequency dependent load (Section 14.3), the exponential recovery load (Section 14.5), the thermostatically controlled load (Section 14.6), the Jimma's load (Section 14.7), and the mixed load (Section 14.8).

Non-conforming loads are generally initialized after the power flow analysis. However, there is no particular difficulty in including non-conforming loads in power flow equations. This possibility is taken into account in the following formulation of non-conforming loads. Moreover, according to the notation given in Chapter 9, load powers p_h and q_h are preceded by a minus because these powers are absorbed from the bus.

14.1 Voltage Dependent Load

Voltage Dependent Loads (VDLs) are loads whose powers are monomial functions of the bus voltage magnitude, as follows:

$$\begin{aligned} -p_h &= p_0(v/v_0)^{\alpha_p} \\ -q_h &= q_0(v/v_0)^{\alpha_q} \end{aligned} \tag{14.1}$$

where v_0 is the initial voltage at the load bus as obtained by the power flow solution. Other parameters are defined in Table 14.1. Equations (14.1) can be directly included in the formulation of power flow equations. However, VDLs are generally initialized after the power flow analysis, and p_0 and q_0 are computed based on constant PQ load powers p_{L0} and q_{L0}:

$$\begin{aligned} p_0 &= \frac{k_p}{100} p_{L0} \\ q_0 &= \frac{k_q}{100} q_{L0} \end{aligned} \tag{14.2}$$

F. Milano: Power System Modelling and Scripting, Power Systems, pp. 313–324.
springerlink.com

Clearly, in this case, a PQ load must be connected to the same bus as the VDL. Equations (14.1) are a simplification of the nonlinear general exponential voltage frequency dependent load described in Section 14.3.

Table 14.1 Voltage dependent load parameters

Variable	Description	Unit
k_p or p_0	Active power rating	% or pu
k_q or q_0	Reactive power rating	% or pu
α_p	Active power exponent	-
α_q	Reactive power exponent	-

In [230], it was recognized that the load characteristic is fundamental to define the PV curves of the system. The standard CPF analysis considers only PQ loads. This assumption is justified by the presence of ULTCs that fix the voltage and thus the power consumption at load buses. However, if the load voltage is not regulated, or if ULTCs saturate, then the load voltage dependency has to be considered. Figure 14.1 compares the voltage dependent load curves with the network curves. Figure 14.1 only shows the active power, but similar curves can be drawn for the reactive power. Depending on the value of α_p, the load characteristics may or may not intersect the network PV curves.

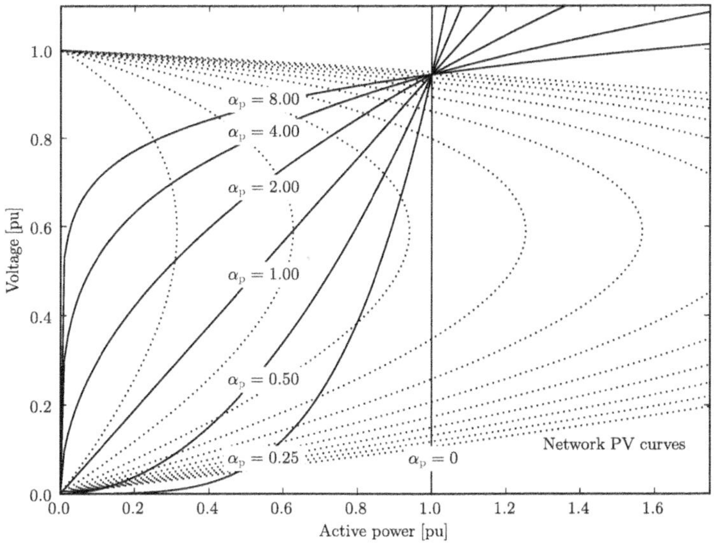

Fig. 14.1 Voltage dependent load characteristics versus network PV curves

Example 14.1 Network PV Curves Considering Load Characteristics

The standard CPF analysis provides an information about the existence of power flow solutions if considering constant PQ load models. Figure 14.2 shows the effect of load characteristics for the determination of PV curves for the IEEE 14-bus system. The higher the exponents α_p and α_q, the higher the maximum loading level.

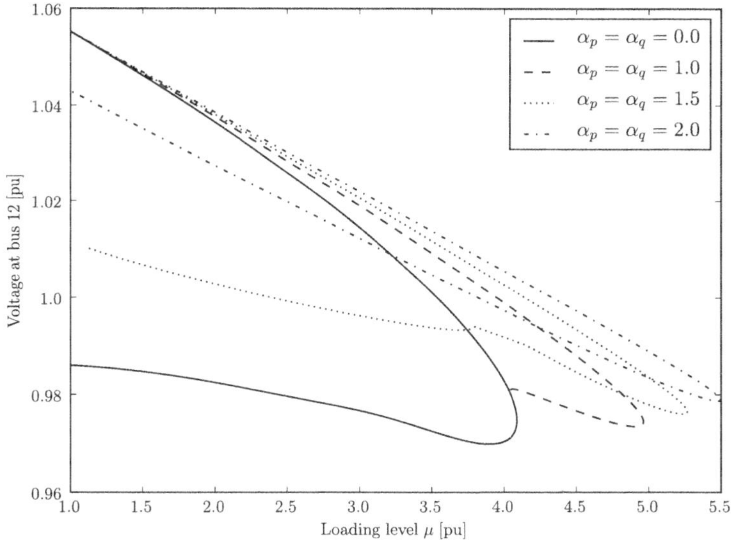

Fig. 14.2 PV curves using difference load characteristics for the IEEE 14-bus system

14.2 ZIP Load

Polynomial or ZIP loads are loads whose powers are a quadratic expression of the bus voltage:

$$\begin{aligned} -p_h &= p_{z0}\left(\frac{v_h}{v_0}\right)^2 + p_{i0}\frac{v_h}{v_0} + p_{p0} \\ -q_h &= q_{z0}\left(\frac{v_h}{v_0}\right)^2 + q_{i0}\frac{v_h}{v_0} + q_{p0} \end{aligned} \tag{14.3}$$

where v_0 is the initial voltage at the load bus as obtained by the power flow solution. Other parameters are defined in Table 14.2. If the ZIP load is initialized after the power flow analysis, the parameters in (14.3) can be defined based on the PQ load powers p_{L0} and q_{L0}:

$$p_{z0} = \frac{k_{pz}}{100}\frac{p_{L0}}{v_0^2}, \qquad p_{i0} = \frac{k_{pi}}{100}\frac{p_{L0}}{v_0}, \qquad p_{p0} = \frac{k_{pp}}{100}p_{L0}; \tag{14.4}$$
$$q_{z0} = \frac{k_{qz}}{100}\frac{q_{L0}}{v_0^2}, \qquad q_{i0} = \frac{k_{qi}}{100}\frac{q_{L0}}{v_0}, \qquad q_{p0} = \frac{k_{qp}}{100}q_{L0}.$$

In this case, a PQ load must be connected at the ZIP load bus.

Table 14.2 ZIP load parameters

Variable	Description	Unit
k_{pi} or p_{i0}	Active current	% or pu
k_{pp} or p_{p0}	Active power	% or pu
k_{pz} or p_{z0}	Conductance	% or pu
k_{qi} or q_{i0}	Reactive current	% or pu
k_{qp} or q_{p0}	Reactive power	% or pu
k_{qz} or q_{z0}	Susceptance	% or pu

14.3 Frequency Dependent Load

A generalized exponential voltage frequency dependent load is modeled by the following set of DAE [130]:

$$\begin{aligned}
\dot{x} &= -\frac{\Delta\omega}{T_f} \\
0 &= x + \frac{1}{2\pi f_n}\frac{1}{T_f}(\theta - \theta_0) - \Delta\omega \\
-p_h &= p_0\left(\frac{v}{v_0}\right)^{\alpha_p}(1 + \Delta\omega)^{\beta_p} \\
-q_h &= q_0\left(\frac{v}{v_0}\right)^{\alpha_q}(1 + \Delta\omega)^{\beta_q}
\end{aligned} \tag{14.5}$$

where the frequency deviation $\Delta\omega$ is approximated by filtering and differentiating the bus voltage phase angle θ (see Figure 14.3). The parameters p_0 and q_0 are the initial active and reactive powers, respectively, computed after the power flow solution and based on the PQ load active and reactive powers p_{L0} and q_{L0} as defined in (14.2), and v_0 and θ_0 are the voltage magnitude and phase angle determined by the power flow analysis.

Table 14.3 defines the parameters of the frequency dependent load whereas Table 14.4 depicts some typical exponent values for characteristic loads [25].

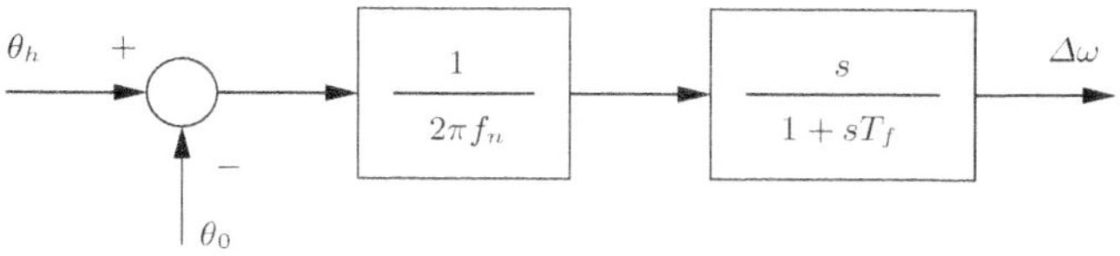

Fig. 14.3 Measure of frequency deviation

Table 14.3 Frequency dependent load parameters

Variable	Description	Unit
k_p	Active power percentage	%
k_q	Reactive power percentage	%
T_f	Filter time constant	s
α_p	Active power voltage exponent	-
α_q	Reactive power voltage exponent	-
β_p	Active power frequency exponent	-
β_q	Reactive power frequency exponent	-

Table 14.4 Typical load exponents [25]

Load	α_p	α_q	β_p	β_q
Filament lamp	1.6	0	0	0
Fluorescent lamp	1.2	3.0	-0.1	2.8
Heater	2.0	0	0	0
Induction motor (half load)	0.2	1.6	1.5	-0.3
Induction motor (full load)	0.1	0.6	2.8	1.8
Reduction furnace	1.9	2.1	-0.5	0
Aluminum plant	1.8	2.2	-0.3	0.6

14.4 Voltage Dependent Load with Dynamic Tap Changer

Figure 14.4 depicts a simplified model of voltage dependent load with embedded dynamic tap changer.[1] The transformer model consists of an ideal circuit with tap ratio m, hence the voltage on the secondary winding is $v_s = v_h/m$. The voltage control is obtained by means of a quasi-integral anti-windup regulator. All constant parameters are defined in Table 14.5.

[1] A more detailed model can be obtained using an ULTC (see Subsection 11.2.2 of Chapter 11) feeding a voltage dependent load (see Section 14.1).

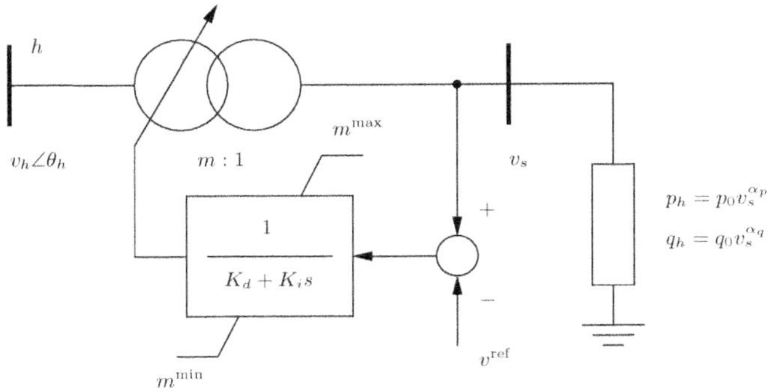

Fig. 14.4 Voltage dependent load with dynamic tap changer

Table 14.5 Load with dynamic tap changer parameters

Variable	Description	Unit
K_d	Integral deviation	1/s
K_i	Integral gain	1/s/pu
$m^{\min}$	Maximum tap ratio	pu/pu
$m^{\max}$	Minimum tap ratio	pu/pu
$v^{\rm ref}$	Reference voltage	pu
p_0	Load active power	pu
q_0	Load reactive power	pu
α_p	Voltage exponent for the active power	-
α_q	Voltage exponent for the reactive power	-

The algebraic equations of the device are:

$$
\begin{aligned}
-p_h &= p_0 \left(\frac{v_h}{m}\right)^{\alpha_p} \\
-q_h &= q_0 \left(\frac{v_h}{m}\right)^{\alpha_q}
\end{aligned}
\tag{14.6}
$$

and the differential equation is:

$$
\dot{m} = -K_d m + K_i \left(\frac{v_h}{m} - v^{\rm ref}\right) \tag{14.7}
$$

The reference voltage sign is negative due to the characteristic of the stable equilibrium point (see Example 11.3).

If voltage dependent loads with embedded dynamic tap changer are initialized after the power flow analysis, the powers p_0 and q_0 are computed based on the PQ load powers as in (14.2), and the state variable m and the voltage reference $v^{\rm ref}$ are initialized as follows:

$$m_0 = v_0 \qquad (14.8)$$
$$v^{\text{ref}} = 1 + \frac{K_d}{K_i} v_0$$

where v_0 is the bus voltage obtained by the power flow solution.

Example 14.2 Effect of Tap Changer Dynamics on Transient Analysis for the IEEE 14-Bus System

Figure 14.5 compares the results of the time domain integration for the IEEE 14-bus system using constant PQ loads, constant impedance loads and voltage dependent loads with embedded dynamic tap changer. Tap changer dependent loads are initialized after the power flow solution and have the following data: $\alpha_p = \alpha_q = 2$, $K_d = 0$ and $K_i = 0.1$ 1/s/pu. When considering tap-changer embedded loads, the response of the system in the first seconds after the disturbance is similar to that of the system with constant impedance load models (see also Example 10.2 of Chapter 10). This is due to the slow response of tap changers. After some seconds, tap changers are able to regulate the voltage magnitude and the bus voltage trajectories approximate those of the system with constant PQ load models.

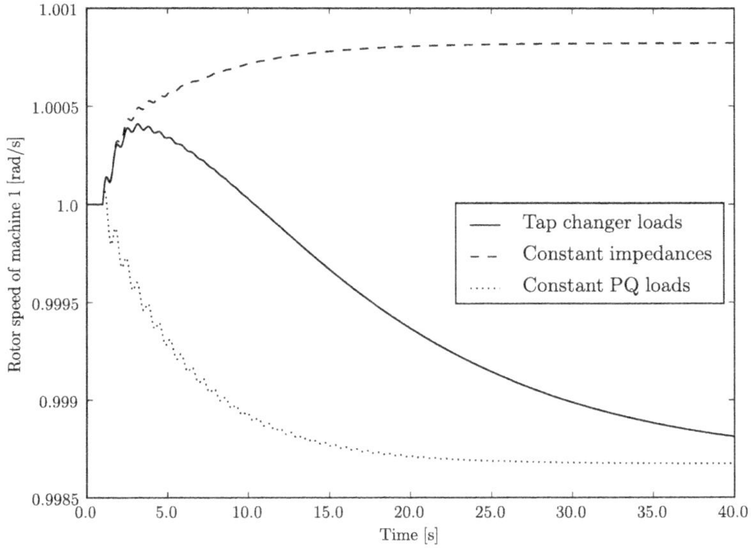

Fig. 14.5 Effect of tap changer dynamics in transient analysis for the IEEE 14-bus system

14.5 Exponential Recovery Load

This section describes an exponential recovery load based on the model proposed in [128, 155]. Equations are:

$$\begin{aligned} \dot{x}_p &= -x_p/T_p + p_s - p_t \\ -p_h &= x_p/T_p + p_t \end{aligned} \tag{14.9}$$

where p_s and p_t are the static and transient real power absorptions, which depend on the load voltage:

$$\begin{aligned} p_s &= p_0(v/v_0)^{\alpha_s} \\ p_t &= p_0(v/v_0)^{\alpha_t} \end{aligned} \tag{14.10}$$

Similar equations hold for the reactive power:

$$\begin{aligned} \dot{x}_q &= -x_q/T_q + q_s - q_t \\ -q_h &= x_q/T_q + q_t \end{aligned} \tag{14.11}$$

and:

$$\begin{aligned} q_s &= q_0(v/v_0)^{\beta_s} \\ q_t &= q_0(v/v_0)^{\beta_t} \end{aligned} \tag{14.12}$$

The power flow solution and the PQ load data are used for determining the values of p_0, q_0 and v_0. In particular, p_0 and q_0 are determined as in (14.2). A PQ load is required to initialize the exponential recovery load bus. All parameters are defined in Table 14.6.

Table 14.6 Exponential recovery load parameters

Variable	Description	Unit
k_p	Active power percentage	%
k_q	Reactive power percentage	%
T_p	Active power time constant	s
T_q	Reactive power time constant	s
α_s	Static active power exponent	-
α_t	Dynamic active power exponent	-
β_s	Static reactive power exponent	-
β_t	Dynamic reactive power exponent	-

14.6 Thermostatically Controlled Load

This section describes a dynamic load with temperature control based on the model given in [130]. This device is initialized after the power flow solution and needs a PQ load connected at the same bus to properly initialize the state variables. The control diagram is depicted in Figure 14.6 that represents the following equations:

$$\begin{aligned}
\dot{\Theta} &= (\Theta_a - \Theta + K_1 p)/T_1 \\
\dot{x} &= K_i(\Theta^{\text{ref}} - \Theta)/T_i \\
g &= K_p(\Theta^{\text{ref}} - \Theta) + x \\
-p_h = p &= g v_h^2 \\
q_h &= 0
\end{aligned} \tag{14.13}$$

where the state variable x undergoes an anti-windup limiter and the algebraic variable G undergoes a windup limiter.

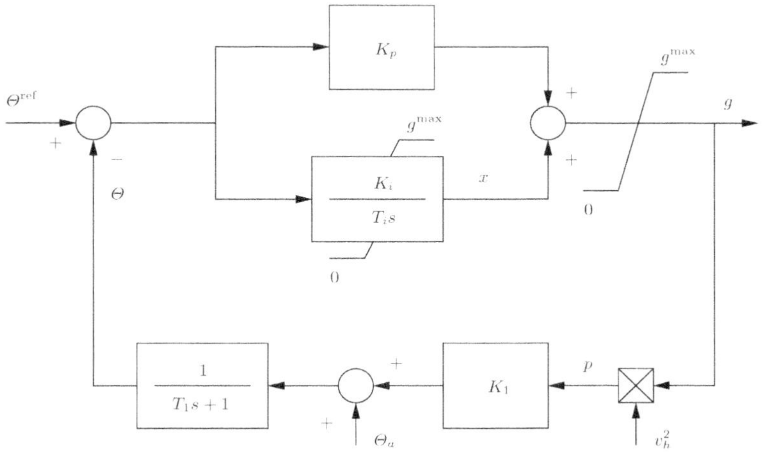

Fig. 14.6 Thermostatically controlled load

The power flow solution provides the initial voltage v_0 and active power p_0 that are used for determining the gain K_1 and the maximum conductance $g^{\max}$, as follows:

$$\begin{aligned}
K_1 &= \frac{\Theta^{\text{ref}} - \Theta_a}{p_0} \\
g^{\max} &= K_L g_0
\end{aligned} \tag{14.14}$$

where $g_0 = p_0/v_0{}^2$ and K_L ($K_L > 1$) is the ceiling conductance output ratio. Finally, the initial load temperature is $\Theta_0 = \Theta^{\text{ref}}$ and Table 14.7 defines all constant parameters required by this device.

Table 14.7 Thermostatically controlled load parameters

Variable	Description	Unit
K_i	Gain of integral controller	pu/K
K_L	Ceiling conductance output	pu/pu
k_p	Percentage of active power	%
K_p	Gain of proportional controller	pu/K
T_1	Time constant of thermal load	s
T_i	Time constant of integral controller	s
Θ_a	Ambient temperature	K
Θ^{ref}	Reference temperature	K

14.7 Jimma's Load

This section describes a load similar to a ZIP model except for the dependence of the reactive power on the time derivative of the bus voltage [152, 341]. This device is not included in the power flow analysis and thus requires a PQ load connected at the same bus to be properly initialized. Since in transient stability analysis bus voltages are not state variables, the time derivative is defined using an auxiliary state variable x_v and a high-pass filter similar to the bus frequency measurement device described in Section 13.5 of Chapter 13 (see Figure 14.7). The differential equation is:

$$\begin{aligned} \dot{x}_v &= (-v_h/T_f - x_v)/T_f \\ \frac{dv_h}{dt} &= x_v + v_h/T_f \end{aligned} \tag{14.15}$$

and the power injections are defined as:

$$-p_h = p_{z0}\left(\frac{v_h}{v_0}\right)^2 + p_{i0}\left(\frac{v_h}{v_0}\right) + p_{p0} \tag{14.16}$$

$$-q_h = q_{z0}\left(\frac{v_h}{v_0}\right)^2 + q_{i0}\left(\frac{v_h}{v_0}\right) + q_{p0} + K_v\frac{dv_h}{dt} \tag{14.17}$$

where the load parameters p_{z0}, p_{i0}, p_{p0}, q_{z0}, q_{i0} and q_{p0} are computed as in (14.4). The power flow analysis provides the initial voltage v_0 that is needed for computing the Jimma's load power injections. Table 14.8 defines the parameters of this device.

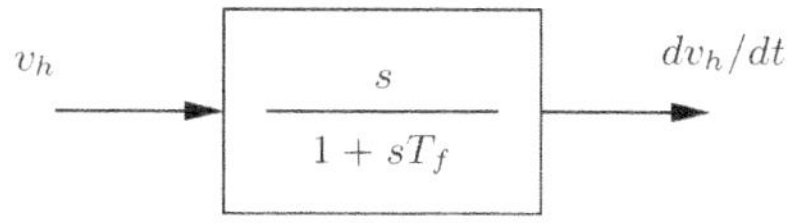

Fig. 14.7 Jimma's load

Table 14.8 Jimma's load parameters

Variable	Description	Unit
k_{pi}	Percentage of active power $\propto v_h$	%
k_{pp}	Percentage of constant active power	%
k_{pz}	Percentage of active power $\propto v_h^2$	%
k_{qi}	Percentage of reactive power $\propto v_h$	%
k_{qp}	Percentage of constant reactive power	%
k_{qz}	Percentage of reactive power $\propto v_h^2$	%
K_v	Coefficient of the voltage time derivative	s pu/pu
T_f	Time constant of the high-pass filter	s

14.8 Mixed Load

This section describes a load similar to a frequency dependent load. In addition, the active and the reactive powers depend on the time derivative of the bus voltage. This device is not included in the power flow analysis and thus requires a PQ load connected at the same bus to be properly initialized. Since in transient stability analysis bus voltage phasors are not state variables, the time derivatives of the voltage magnitude and angle are defined through two auxiliary state variables x_v and x_θ and high-pass filters similar to the bus frequency measurement device described in Section 13.5 of Chapter 13 (see Figures 14.7 and 14.3). The differential equations are:

$$\begin{aligned} \dot{x}_v &= (-v_h/T_{fv} - x_v)/T_{fv} && (14.18)\\ \Rightarrow \quad \frac{dv_h}{dt} &= x_v + v_h/T_{fv} \\ \dot{x}_\theta &= -\frac{1}{T_{ft}}\left(\frac{1}{2\pi f_n}\frac{1}{T_{ft}}(\theta - \theta_0) + x_\theta\right) && (14.19)\\ \Rightarrow \quad \Delta\omega &= x_\theta + \frac{1}{2\pi f_n}\frac{1}{T_{ft}}(\theta - \theta_0) \end{aligned}$$

The bus power injections p_h and q_h are defined as follows:

$$\begin{aligned} -p_h &= K_{pf}\Delta\omega + p_0\left[\left(\frac{v_h}{v_0}\right)^{\alpha_p} + T_{pv}\frac{dv_h}{dt}\right] && (14.20)\\ -q_h &= K_{qf}\Delta\omega + q_0\left[\left(\frac{v_h}{v_0}\right)^{\alpha_q} + T_{qv}\frac{dv_h}{dt}\right] \end{aligned}$$

where p_0 and q_0 are computed based on the PQ load active and reactive powers p_{L0} and q_{L0} as defined in (14.2). The power flow solution provides the initial voltage v_0 that is needed for computing the power injections. Table 14.9 defines all constant parameters of this devices.

Table 14.9 Mixed load parameters

Variable	Description	Unit
k_p	Percentage of active power	%
K_{pf}	Frequency coefficient for the active power	s pu/pu
k_q	Percentage of reactive power	%
K_{qf}	Frequency coefficient for the reactive power	s pu/pu
T_{ft}	Time constant of voltage angle filter	s
T_{fv}	Time constant of voltage magnitude filter	s
T_{pv}	Time constant of dV/dt for the active power	s
T_{qv}	Time constant of dV/dt for the reactive power	s
α_p	Voltage exponent for the active power	-
α_q	Voltage exponent for the reactive power	-

Chapter 15
Alternate-Current Machines

This chapter describes the two most important alternate-current machines used in power systems, namely the synchronous machine and the induction machine. Section 15.1 provides a detailed taxonomy of synchronous machine models, as well as a discussion about saturation models, the center of inertia and the sub-synchronous resonance phenomenon. Section 15.2 describes various induction machine models and provides an example about the induction motor start-up transient.

15.1 Synchronous Machine

Virtually the totality of power system books agree on the Park's "two-reaction theory" of synchronous machine model [234].[1] Thus the basic assumptions about the synchronous machine model are well-known and are not repeated here. Unfortunately, there is not so much agreement on the transfer functions that link stator fluxes with stator currents and the field voltage. Depending on the dynamic order and the detail of these transfer functions, the resulting set of machine DAE changes. This section provides a modular approach to set up a huge variety of synchronous machine models.

The machine scheme considered in this section is shown in Figure 15.1. This machine has a salient-pole rotor with one field (excitation) winding ff' and a three-phase system of stator windings, namely aa', bb' and cc'. The effect of induced currents in the rotor core is modelled as a lumped winding $q_1 q_1'$ in quadrature with the field winding. Finally, damping effects are modelled as two fictitious lumped windings, $d_1 d_1'$ and $q_2 q_2'$, respectively, in the rotor.

As it is well known, the Park's transformation consists in projecting all quantities onto three axes, namely the direct, the quadrature and the homopolar axes, d, q and 0, respectively. Since the direct and the quadrature axes are rotating at the synchronous speed, the Park transformation allows

[1] It is interesting to note that in [330] and in Russian publications in general, the Park's model is called *Park-Gorev's model*.

F. Milano: Power System Modelling and Scripting, Power Systems, pp. 325–353.
springerlink.com

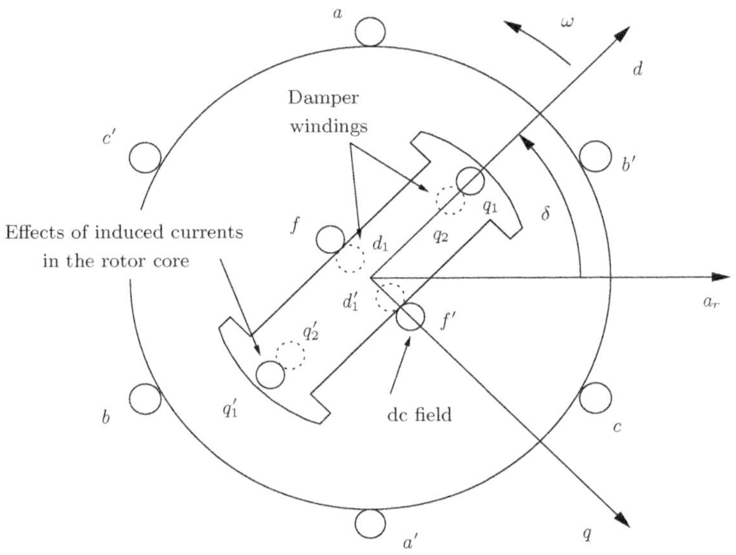

Fig. 15.1 Synchronous machine scheme

reducing machine equations to static phasors if the machine rotor is also rotating at the synchronous speed. Less evident is the advantage of the Park's transformation in case the rotor is not rotating at the synchronous speed. Nevertheless, the Park's model is a standard *de facto* and is so widely used that machine data are generally available according to the axes of the Park's transformation.

Following subsections are organized as follows. The complete set of machine parameters are defined in Subsection 15.1.1. Then the machine variable initialization procedure is outlined in Subsection 15.1.2. The remainder presents the machine equations. The material is organized taking into account implementation issues. Since some equations are common to all models, it is convenient to create a base class that includes such common equations and then to import the base class into specific machine models. No homopolar equations are given since the system and the machine are considered perfectly balanced.

15.1.1 Synchronous Machine Parameters

Table 15.1 defines the complete set of synchronous machine parameters. Factors α_p and α_q are used in case of multiple generators connected to the same bus and indicate the fraction of active and reactive powers that each machine provides with respect to the total power produced by the static generator defined in power flow analysis. The sum of these factors for the machines connected to the same bus has to be 1.

Table 15.1 Synchronous machine parameters

Variable	Description	Unit
D	Damping coefficient	pu
H	Inertia constant	MWs/MVA
r_a	Armature resistance	pu
x_ℓ	Leakage reactance	pu
x_d	d-axis synchronous reactance	pu
x'_d	d-axis transient reactance	pu
x''_d	d-axis sub-transient reactance	pu
x_q	q-axis synchronous reactance	pu
x'_q	q-axis transient reactance	pu
x''_q	q-axis sub-transient reactance	pu
T_{AA}	d-axis additional leakage time constant	s
T'_{d0}	d-axis open circuit transient time constant	s
T''_{d0}	d-axis open circuit sub-transient time constant	s
T'_{q0}	q-axis open circuit transient time constant	s
T''_{q0}	q-axis open circuit sub-transient time constant	s
α_p	Active power ratio at node	[0,1]
α_q	Reactive power ratio at node	[0,1]

15.1.2 Initialization

Dynamic models of synchronous machines are not included in standard power flow analysis. Thus a PQ, PV or a slack generator are required to impose the desired voltage and active power at the synchronous machine bus. Once the power flow solution is determined, v_0, θ_0, p_0 and q_0 at the generator bus are used for initializing the machine state variables, the field voltage v_f and the mechanical torque τ_m. Example 9.2 of Chapter 7 describes the initialization of the two-axis synchronous machine model. The initialization procedure for other machine models is basically the same. The only differences are:

1. Higher order machine models require the initialization of state variables associated with magnetic fluxes and or sub-transient emfs. These can be obtained directly by the algebraic equations (15.11) and any set of magnetic equations (assumed steady-state) provided in Subsection 15.1.5.
2. If more than one machine is connected to the same bus, the total injected powers p_0 and q_0 obtained by the power flow analysis have to be multiplied by the factors α_p and α_q, respectively.
3. In case of the classical machine model (i.e., emf behind the transient reactance), equation (9.11) has to be substituted for:

$$\delta_0 = \angle(\bar{v}_h + (r_a + jx'_d)\bar{\imath}_h) \tag{15.1}$$

In fact, for the classical model, x_q is not defined.

15.1.3 Common Equations

The equations common to all machine models are the interface with the network and mechanical differential equations. The power injection p_h and q_h at bus h are:

$$p_h = v_d i_d + v_q i_q \tag{15.2}$$
$$q_h = v_q i_d - v_d i_q \tag{15.3}$$

whereas the link between the network quasi-static voltage phasor $v_h \angle \theta_h$ and machine voltages v_d and v_q are:

$$\begin{aligned} 0 &= v_h \sin(\delta - \theta_h) - v_d \\ 0 &= v_h \cos(\delta - \theta_h) - v_q \end{aligned} \tag{15.4}$$

where δ is the machine rotor angle. Mechanical differential equations are:

$$\begin{aligned} \dot{\delta} &= \Omega_b(\omega - \omega_s) \\ \dot{\omega} &= \frac{1}{2H}(\tau_m - \tau_e - D(\omega - \omega_s)) \end{aligned} \tag{15.5}$$

where the electro-magnetic torque τ_e is:

$$\tau_e = \psi_d i_q - \psi_q i_d \tag{15.6}$$

where Ω_b is the base synchronous frequency in rad/s (314.16 rad/s at 50 Hz) and ω_s is the reference frequency in pu. If the reference frequency is the synchronous one, then $\omega_s = 1$ pu. Some references define the *machine starting time* as $M = 2H$ (or $T_M = 2H$) and use this quantity in (15.5) instead of the inertia constant H.

Finally, one can define auxiliary equations for the input mechanical torque and field voltage. For the mechanical torque:

$$0 = \tau_{m0} - \tau_m \tag{15.7}$$

and for the field voltage:

$$0 = v_{f0} - v_f \tag{15.8}$$

where τ_{m0} and v_{f0} are the initial values of the mechanical torque and the field voltage, respectively. As discussed in Example 9.1 of Chapter 9, τ_m and v_f are auxiliary algebraic variables that allows easily interfacing the machine model with other devices such as turbine governors and automatic voltage regulators.

15.1.4 Stator Electrical Equations

Stator electrical equations link the voltages to currents and magnetic fluxes, as follows:

$$\begin{aligned} \dot{\psi}_d &= \Omega_b(r_a i_d + \omega\psi_q + v_d) \\ \dot{\psi}_q &= \Omega_b(r_a i_q - \omega\psi_d + v_q) \end{aligned} \tag{15.9}$$

While required for electro-magnetic transients, flux dynamics are relatively fast for transient stability studies. In fact, the inverse of the base frequency $1/\Omega_b \approx 10^{-3}$s for common power systems working at 50 or 60 Hz. Thus, a common simplification is to assume $\dot{\psi}_d \approx \dot{\psi}_q \approx 0$, which leads to:

$$\begin{aligned} 0 &= r_a i_d + \omega\psi_q + v_d \\ 0 &= r_a i_q - \omega\psi_d + v_q \end{aligned} \tag{15.10}$$

Furthermore, considering that rotor speed deviations are small, one can assume $\omega \approx 1$ in (15.10). Hence:

$$\begin{aligned} 0 &= r_a i_d + \psi_q + v_d \\ 0 &= r_a i_q - \psi_d + v_q \end{aligned} \tag{15.11}$$

The three models above, namely (15.9), (15.10) and (15.11), can be used indifferently. Thus, it is convenient to create a class for each electrical equation model and then to form the complete machine model by importing the required electrical equation class. Clearly, using model (15.9) makes sense only in very detailed analyses that require a precise formulation of electro-magnetic dynamics. The most common choice adopted by most power system books is (15.11), which also allows removing the variables ψ_d and ψ_q from the machine model.

15.1.5 Magnetic Equations

As discussed in the introduction of this section, the transfer functions that link stator fluxes with stator currents and the field voltage provide a certain degree of arbitrariness in the synchronous machine model. Most complete models introduce one state variable per rotor winding, real or equivalent. Thus for the machine depicted in Figure 15.1, four state variables and their associated differential equations are required for most detailed models. Finally, two algebraic equations allow defining stator fluxes as functions of the stator currents, the field voltage and the rotor state variables. Simplified models consist in downgrading one or more rotor state variables to algebraic ones.

The dynamic response of damper windings is faster than that of the dc field winding and of the rotor-core induced currents. The standard notation defines *sub-transient* (indicated by a double superscript $''$) the fast dynamics of damper windings and *transient* (indicated by a single superscript $'$) the dynamics of the dc field winding and of the rotor-core induced currents. Finally, steady-state quantities are said *synchronous* and are indicated without superscripts.

Sauer-Pai's Model

The Sauer-Pai's model is as follows [269]:

$$\begin{aligned}
\dot{e}'_q &= (-e'_q - (x_d - x'_d)(i_d + \gamma_{d2}\dot{\psi}''_d) + v_f)/T'_{d0} \\
\dot{e}'_d &= (-e'_d + (x_q - x'_q)(i_q + \gamma_{q2}\dot{\psi}''_q))/T'_{q0} \\
\dot{\psi}''_d &= (-\psi''_d + e'_q - (x'_d - x_\ell)i_d)/T''_{d0} \\
\dot{\psi}''_q &= (-\psi''_q - e'_d - (x'_q - x_\ell)i_q)/T''_{q0}
\end{aligned} \tag{15.12}$$

or using the standard ODE notation:

$$\begin{aligned}
\dot{e}'_q &= (-e'_q - (x_d - x'_d)(i_d - \gamma_{d2}\psi''_d - (1-\gamma_{d1})i_d + \gamma_{d2}e'_q) + v_f)/T'_{d0} \\
\dot{e}'_d &= (-e'_d + (x_q - x'_q)(i_q - \gamma_{q2}\psi''_q - (1-\gamma_{q1})i_q - \gamma_{d2}e'_d))/T'_{q0} \\
\dot{\psi}''_d &= (-\psi''_d + e'_q - (x'_d - x_\ell)i_d)/T''_{d0} \\
\dot{\psi}''_q &= (-\psi''_q - e'_d - (x'_q - x_\ell)i_q)/T''_{q0}
\end{aligned} \tag{15.13}$$

where:

$$\begin{aligned}
\gamma_{d1} &= \frac{x''_d - x_\ell}{x'_d - x_\ell} \\
\gamma_{q1} &= \frac{x''_q - x_\ell}{x'_q - x_\ell} \\
\gamma_{d2} &= \frac{x'_d - x''_d}{(x'_d - x_\ell)^2} = \frac{1-\gamma_{d1}}{x'_d - x_\ell} \\
\gamma_{q2} &= \frac{x'_q - x''_q}{(x'_q - x_\ell)^2} = \frac{1-\gamma_{q1}}{x'_q - x_\ell}
\end{aligned} \tag{15.14}$$

Finally, the following algebraic equations complete the model:

$$\begin{aligned}
0 &= \psi_d + x''_d i_d - \gamma_{d1}e'_q - (1-\gamma_{d1})\psi''_d \\
0 &= \psi_q + x''_q i_q + \gamma_{q1}e'_d - (1-\gamma_{q1})\psi''_q
\end{aligned} \tag{15.15}$$

Marconato's Model

The complete *d*- and *q*-axis diagrams of the Marconato's model are depicted in Figure 15.2. The differential equations are [184]:

$$\begin{aligned}
\dot{e}'_q &= (-e'_q - (x_d - x'_d - \gamma_d)i_d + (1 - \frac{T_{AA}}{T'_{d0}})v_f)/T'_{d0} \\
\dot{e}'_d &= (-e'_d + (x_q - x'_q - \gamma_q)i_q)/T'_{q0} \\
\dot{e}''_q &= (-e''_q + e'_q - (x'_d - x''_d + \gamma_d)i_d + \frac{T_{AA}}{T'_{d0}}v_f)/T''_{d0} \\
\dot{e}''_d &= (-e''_d + e'_d + (x'_q - x''_q + \gamma_q)i_q)/T''_{q0}
\end{aligned} \tag{15.16}$$

where coefficients γ_d and γ_q are defined as follows:

$$\begin{aligned}
\gamma_d &= \frac{T''_{d0}}{T'_{d0}}\frac{x''_d}{x'_d}(x_d - x'_d) \\
\gamma_q &= \frac{T''_{q0}}{T'_{q0}}\frac{x''_q}{x'_q}(x_q - x'_q)
\end{aligned} \tag{15.17}$$

The following algebraic equations complete the model:

$$\begin{aligned}
0 &= \psi_d + x''_d i_d - e''_q \\
0 &= \psi_q + x''_q i_q + e''_d
\end{aligned} \tag{15.18}$$

Anderson-Fouad's Model

The Anderson-Fouad's model, which, apart from [10], is also reported by the majority of books on power systems, e.g., [14] and [179], is as follows:

$$\begin{aligned}
\dot{e}'_q &= (-e'_q - (x_d - x'_d)i_d + v_f)/T'_{d0} \\
\dot{e}'_d &= (-e'_d + (x_q - x'_q)i_q)/T'_{q0} \\
\dot{e}''_q &= (-e''_q + e'_q - (x'_d - x''_d)i_d)/T''_{d0} \\
\dot{e}''_d &= (-e''_d + e'_d + (x'_q - x''_q)i_q)/T''_{q0}
\end{aligned} \tag{15.19}$$

and (15.15). The Anderson-Fouad's model can be considered a simplification of the Sauer-Pai's model. In fact, the Sauer-Pai's model leads to the Anderson-Fouad's one by defining:

$$e''_q = \psi''_d, \qquad e''_d = -\psi''_q \tag{15.20}$$

and assuming:

- $\gamma_{d1} \approx \gamma_{q1} \approx 0$.
- $\gamma_{d2}\dot{\psi}''_d \approx 0$ in the differential equations of $\dot{e}'_q$.
- $\gamma_{q2}\dot{\psi}''_q \approx 0$ in the differential equations of $\dot{e}'_d$.

The Anderson-Fouad's model can be also derived from the Marconato's model by assuming:

$$\gamma_d \approx \gamma_q \approx T_{AA} \approx 0 \tag{15.21}$$

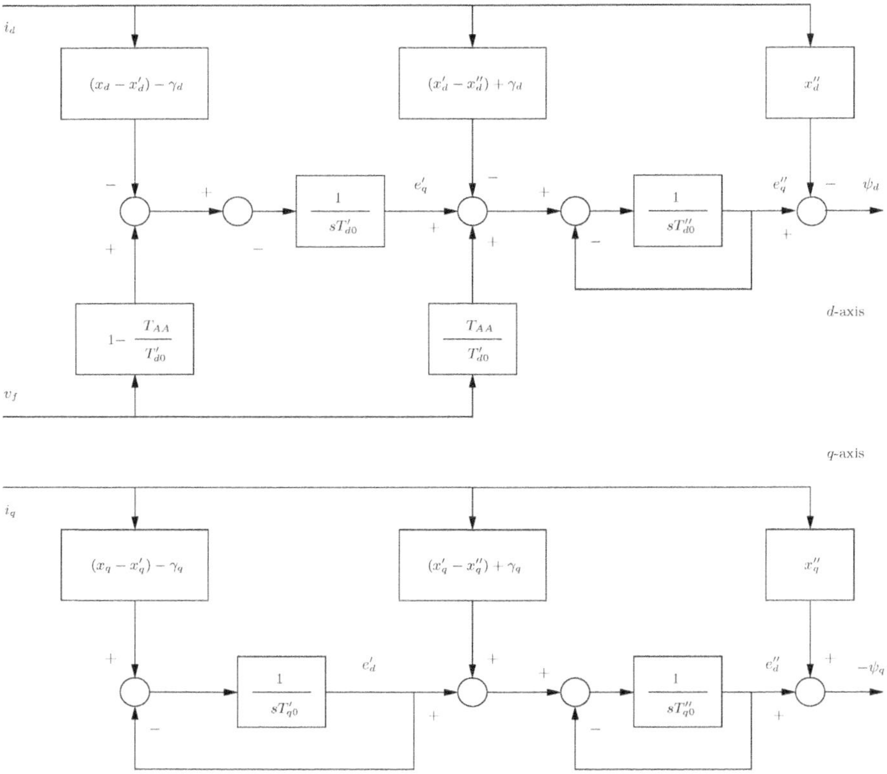

Fig. 15.2 Block diagram of stator fluxes for the Marconato's model of the synchronous machine

15.1.6 *Simplified Magnetic Equations*

There are a variety of possible simplifications for the magnetic equations presented above. Following paragraphs only show some models presented in the literature.

Two d- and One q-Axis Model

Assuming $T'_{q0} \approx 0$ and $x'_q \approx x_q$ in (15.13) leads to $e'_d \approx 0$. Hence, (15.13) can be rewritten as:

$$
\begin{aligned}
\dot{e}'_q &= (-e'_q - (x_d - x'_d)(i_d - \gamma_{d2}\psi''_d - (1 - \gamma_{d1})i_d + \gamma_{d2}e'_q) + v_f)/T'_{d0} \\
\dot{\psi}''_d &= (-\psi''_d + e'_q - (x'_d - x_\ell)i_d)/T''_{d0} \\
\dot{\psi}''_q &= (-\psi''_q - (x'_q - x_\ell)i_q)/T''_{q0}
\end{aligned}
\qquad (15.22)
$$

and algebraic equations (15.15) become:

$$0 = \psi_d + x''_d i_d - \gamma_{d1} e'_q - (1 - \gamma_{d1})\psi''_d \qquad (15.23)$$
$$0 = \psi_q + x''_q i_q - (1 - \gamma_{q1})\psi''_q$$

A second type of two d- and one q-axis model, can be obtained from (15.16) assuming only one additional circuit on the q-axis (e.g., $T'_{q0} \approx 0$) [184]. The resulting model has three magnetic state variables e'_q, e''_q and e''_d as follows:

$$\dot{e}'_q = (-e'_q - (x_d - x'_d - \gamma_d)i_d + (1 - \frac{T_{AA}}{T'_{d0}})v_f)/T'_{d0} \qquad (15.24)$$
$$\dot{e}''_q = (-e''_q + e'_q - (x'_d - x''_d + \gamma_d)i_d + \frac{T_{AA}}{T'_{d0}}v_f)/T''_{d0}$$
$$\dot{e}''_d = (-e''_d + (x_q - x''_q)i_q)/T''_{q0}$$

with the algebraic equations:

$$0 = v_q + r_a i_q - e''_q + x''_d i_d \qquad (15.25)$$
$$0 = v_d + r_a i_d - e''_d - x''_q i_q$$

Similar equations can be obtained using (15.19) or imposing $\gamma_d \approx T_{AA} \approx 0$ in (15.24).

One d- and Two q-Axis Model

Another model presented in [184] assumes:

$$x'_d \approx x''_d \approx x''_q \qquad (15.26)$$

which leads to a single d-axis equation for the variable e'_q. Both q-axis transient and sub-transient dynamics are used. Hence, (15.16) becomes:

$$\dot{e}'_q = (-e'_q - (x_d - x'_d)i_d + v_f)/T'_{d0} \qquad (15.27)$$
$$\dot{e}'_d = (-e'_d + (x_q - x'_q - \gamma_q)i_q)/T'_{q0}$$
$$\dot{e}''_d = (-e''_d + e'_d + (x'_q - x'_d + \gamma_q)i_q)/T''_{q0}$$

with the following algebraic equations:

$$0 = v_q + r_a i_q - e'_q + x'_d i_d \qquad (15.28)$$
$$0 = v_d + r_a i_d - e''_d - x'_q i_q$$

Similar equations can be obtained using (15.19) or imposing $\gamma_q \approx 0$ in (15.27).

One d- and One q-Axis Model

In this model, lead-lag transfer functions are used for modelling the d- and q-axis inductances (e.g., $T''_{d0} \approx T''_{q0} \approx 0$). The resulting magnetic equations have only two state variables, namely e'_q and e'_d, as follows:

$$\begin{aligned} \dot{e}'_q &= (-e'_q - (x_d - x'_d)i_d + v_f)/T'_{d0} \\ \dot{e}'_d &= (-e'_d + (x_q - x'_q)i_q)/T'_{q0} \end{aligned} \tag{15.29}$$

and the following algebraic equations:

$$\begin{aligned} 0 &= v_q + r_a i_q - e'_q + x'_d i_d \\ 0 &= v_d + r_a i_d - e'_d - x'_q i_q \end{aligned} \tag{15.30}$$

This model can be obtained indifferently from any of the complete models (15.13), (15.16) and (15.19). This two-axis model is the highest order model on which there is substantial agreement in the literature. Actually, this model is the most commonly used in power system stability analysis because it provides the right compromise between simplicity and accuracy.

A similar fourth order model can be formulated using the sub-transient d-axis voltage e''_d instead of e'_d (e.g., $T''_{d0} \approx T'_{q0} \approx 0$). The differential equations becomes:

$$\begin{aligned} \dot{e}'_q &= (-e'_q - (x_d - x'_d)i_d + v_f)/T'_{d0} \\ \dot{e}''_d &= (-e''_d + (x_q - x''_q)i_q)/T''_{q0} \end{aligned} \tag{15.31}$$

and the algebraic equations:

$$\begin{aligned} 0 &= v_q + r_a i_q - e'_q + x'_d i_d \\ 0 &= v_d + r_a i_d - e''_d - x''_q i_q \end{aligned} \tag{15.32}$$

One d-Axis Model

A common model used in transient stability consists in neglecting all q-axis electro-magnetic circuits, and using a lead-lag transfer function for the d-axis inductance [325, 330]. The only magnetic state variable is e'_q, with the following differential equation:

$$\dot{e}'_q = (-e'_q - (x_d - x'_d)i_d + v_f)/T'_{d0} \tag{15.33}$$

and with the algebraic equations:

$$\begin{aligned} 0 &= v_q + r_a i_q - e'_q + x'_d i_d \\ 0 &= v_d + r_a i_d - x_q i_q \end{aligned} \tag{15.34}$$

This model is the simplest one to which an automatic voltage regulator can be connected.

Classical Model

The classical electro-mechanical model neglects all electro-magnetic dynamics. As a consequence the field voltage is substituted by a constant e'_q. The electrical equations are (15.11). Since in these equations, $\omega \approx 1$, one can also assume that the electrical power $p_e = \omega\tau_e \approx \tau_e$. Hence, the electrical power p_e can be written as:

$$p_e = (v_q + r_a i_q)i_q + (v_d + r_a i_d)i_d \tag{15.35}$$

If $r_a \approx 0$, then $p_e \approx p_h$. Another assumption is that and assume that $x_q = x'_d$, hence the following relations between voltages and currents hold:

$$\begin{aligned} 0 &= v_q + r_a i_q - e'_q + x'_d i_d \\ 0 &= v_d + r_a i_d - x'_d i_q \end{aligned} \tag{15.36}$$

where e'_q is a *constant emf behind the transient reactance* x'_d. For sake of clarity, the full classical model is given below:

$$\begin{aligned} \dot{\delta} &= \Omega_b(\omega - 1) \\ \dot{\omega} &= (p_m - p_e - D(\omega - 1))/2H \\ 0 &= (v_q + r_a i_q)i_q + (v_d + r_a i_d)i_d - p_e \\ 0 &= v_q + r_a i_q - e'_q + x'_d i_d \\ 0 &= v_d + r_a i_d - x'_d i_q \\ 0 &= v_h \sin(\delta - \theta_h) - v_d \\ 0 &= v_h \cos(\delta - \theta_h) - v_q \\ p_h &= v_d i_d + v_q i_q \\ q_h &= v_q i_d - v_d i_q \end{aligned} \tag{15.37}$$

In most books and software packages, it is also assumed that $r_a \approx D \approx 0$, thus leading to a loss-less model. However, the property of being loss-less is not implicit in the approximation of neglecting all flux dynamics.

It is possible to define a second type of synchronous machine second-order model by assuming constant sub-transient emfs e''_d and e''_q. From (15.11) and (15.18), one obtains:

$$\begin{aligned} v_d &= e''_d - r_a i_d + x''_q i_q \\ v_q &= e''_q - r_a i_q - x''_d i_d \end{aligned} \tag{15.38}$$

This model consists in a *constant emf behind the sub-transient reactance* and is more precise than the classical one in the first instants after a disturbance. Also in this case, it is common practice to assume $r_a \approx D \approx 0$ [184].

15.1.7 Synchronous Machine Model Taxonomy

To complete the machine model taxonomy, Table 15.2 indicates the dynamic order, the equations and the state variables for all models described in previous subsections.

Table 15.3 depicts a quick reference for the usage of time constants and reactances within synchronous machine models. It is assumed that the leakage reactance x_ℓ is used only in models 8.a, 6.a, 6.d and 5.a, whereas the time constant T_{AA} is used only in models 8.b, 6.b, 6.e, 5.b and 5.c.

Example 15.1 Comparison of Synchronous Machine Models of Different Orders

Figure 15.3 shows a comparison of the transient behavior of synchronous machine models 8.a, 6.a and 6.d. The simulation refers to the generator 1 bus voltage of the IEEE 14-bus system. The disturbance is line 2-4 outage at $t = 0.2$ s. In the first few instants after the disturbance, the most detailed model, namely model 8.a, which includes stator fluxes dynamics, shows fast damped oscillations. However, For $t > 0.45$, the trajectory of the three models is practically the same. As expected, flux dynamics are very fast with respect to electro-mechanical time scales. Furthermore, a relatively small step length is required to observe the effect of flux dynamics. Finally, there is practically no difference in the dynamic responses of models 6.a and 6.d. This result confirms the conclusion drawn in Example 11.2 of Chapter 11.

Example 15.2 Comparison of Synchronous Machine Models of Different Types

Figure 15.4 shows a comparison of transient response of Sauer-Pai's, Marconato's and Anderson-Fouad's models. In particular the plots show the generator 1 bus voltage for the IEEE 14-bus system following a line 2-4 outage at $t = 1$ s. The simulation is repeated using three synchronous machines models, namely models 6.a, 6.b and 6.c. The differences in the magnetic equations do not lead to substantial changes in the transient behavior. However, the three models show different oscillation modes and damping. In other words, state matrix eigenvalues change depending on the model used. This eigenvalue uncertainty has to be taken into account when setting up parameters of control systems such as AVRs and PSSs.

Table 15.2 Synchronous machine model taxonomy

Name	Order	Equations	State Variables
Model 8.a	8	(15.2)-(15.8), (15.9), (15.13), (15.15)	δ, ω, e'_q, e'_d, ψ''_d, ψ''_q, ψ_d, ψ_q
Model 8.b	8	(15.2)-(15.8), (15.9), (15.16), (15.18)	δ, ω, e'_q, e'_d, e''_d, e''_q, ψ_d, ψ_q
Model 8.c	8	(15.2)-(15.8), (15.9), (15.19), (15.18)	δ, ω, e'_q, e'_d, e''_d, e''_q, ψ_d, ψ_q
Model 6.a	6	(15.2)-(15.8), (15.11), (15.13), (15.15)	δ, ω, e'_q, e'_d, ψ''_d, ψ''_q
Model 6.b	6	(15.2)-(15.8), (15.11), (15.16), (15.18)	δ, ω, e'_q, e'_d, e''_d, e''_q
Model 6.c	6	(15.2)-(15.8), (15.11), (15.19), (15.18)	δ, ω, e'_q, e'_d, e''_d, e''_q
Model 6.d	6	(15.2)-(15.8), (15.10), (15.13), (15.15)	δ, ω, e'_q, e'_d, ψ''_d, ψ''_q
Model 6.e	6	(15.2)-(15.8), (15.10), (15.16), (15.18)	δ, ω, e'_q, e'_d, e''_d, e''_q
Model 6.f	6	(15.2)-(15.8), (15.10), (15.19), (15.18)	δ, ω, e'_q, e'_d, e''_d, e''_q
Model 5.a	5	(15.2)-(15.8), (15.11), (15.22), (15.23)	δ, ω, e'_q, ψ''_d, ψ''_q
Model 5.b	5	(15.2)-(15.8), (15.11), (15.24), (15.25)	δ, ω, e'_q, e''_d, e''_q
Model 5.c	5	(15.2)-(15.8), (15.11), (15.27), (15.28)	δ, ω, e'_q, e'_d, e''_d
Two-axis model	4	(15.2)-(15.8), (15.11), (15.29), (15.30)	δ, ω, e'_q, e'_d
Model 4.b	4	(15.2)-(15.8), (15.11), (15.31), (15.32)	δ, ω, e'_q, e''_d
One-axis model	3	(15.2)-(15.8), (15.11), (15.33), (15.34)	δ, ω, e'_q
Classical model 2.a	2	(15.2)-(15.8), (15.11), (15.35), (15.36)	δ, ω
Classical model 2.b	2	(15.2)-(15.8), (15.11), (15.35), (15.38)	δ, ω

Table 15.3 Reference table for synchronous machine time constants and reactances

Order	T'_{d0}	T'_{q0}	T''_{d0}	T''_{q0}	x_d	x'_d	x''_d	x_q	x'_q	x''_q
Models 8.x	✓	✓	✓	✓	✓	✓	✓	✓	✓	✓
Models 6.x	✓	✓	✓	✓	✓	✓	✓	✓	✓	✓
Model 5.a	✓		✓	✓	✓	✓	✓	✓		✓
Model 5.b	✓	✓		✓	✓	✓		✓	✓	
Model 5.c	✓		✓	✓	✓	✓	✓	✓		✓
Two-axis model	✓	✓			✓	✓		✓	✓	
Model 4.b	✓			✓	✓	✓		✓		✓
One-axis model	✓				✓	✓		✓		
Classical model 2.a						✓				
Classical model 2.b							✓			✓

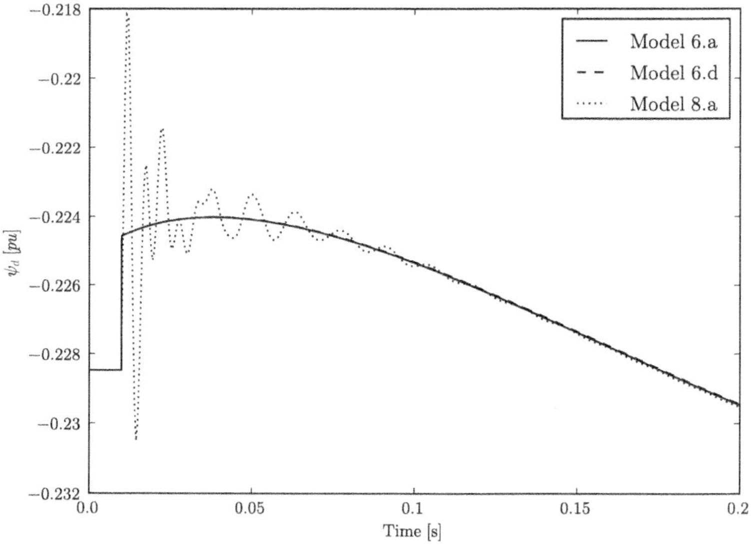

Fig. 15.3 Comparison of synchronous machine models of different orders

Example 15.3 One-Axis Model with Stator Flux Dynamics

For the sake of showing all possible combinations of the sets of equations described above, this example considers an unusual fifth order model described in [263]. This model is formed by equations (15.2)-(15.8), by the dynamical electrical equations (15.9), (15.34), and by the field voltage differential equation:

$$\dot{\psi}_f = (v_f - e'_q)/T'_{d0} \tag{15.39}$$

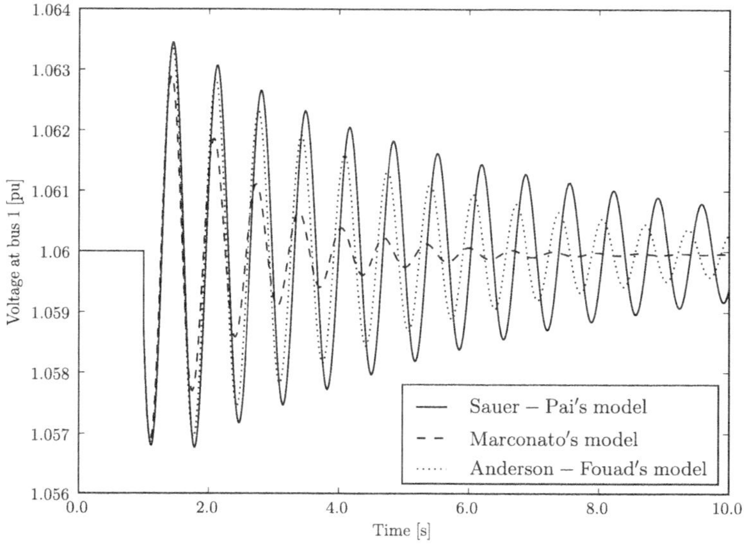

Fig. 15.4 Comparison of synchronous machine models of different types

where the field flux ψ_f is:

$$\psi_f = e'_q - (x_d - x'_d)i_d \tag{15.40}$$

which leads to rewrite (15.39) as:

$$\dot{e}'_q = \frac{x_d}{x'_d}\left(\frac{1}{T'_{d0}}(v_f - e'_q) - \frac{x_d - x'_d}{x_d}\dot{\psi}_d\right) \tag{15.41}$$

Clearly, this model has no practical applications since is too detailed and too simplified at the same time. It is too detailed because it includes stator flux dynamics, thus requiring a comparatively small step length in numerical integration, and it is too simplified because it considers only one dynamic on the d-axis, thus resulting inadequate for detailed transient stability analysis.

15.1.8 Saturation

As discussed above, the variety of magnetic equations allows defining a huge variety of machine models. Taking into account saturation introduces even more arbitrariness in the formulation of the synchronous machine model.

A general model that accounts for saturation is a generalization of the Sauer-Pai's model (15.13) and (15.15) [269].

$$\begin{aligned} \dot{e}'_q &= (-e'_q - (x_d - x'_d)(i_d - \gamma_{d2}\psi''_d - (1-\gamma_{d1})i_d + \gamma_{d2}e'_q \\ &\quad + \gamma_{d2}\varsigma_d(\boldsymbol{z})) - \varsigma_f(\boldsymbol{z}) + v_f)/T'_{d0} \\ \dot{e}'_d &= (-e'_d + (x_q - x'_q)(i_q - \gamma_{q2}\psi''_q - (1-\gamma_{q1})i_q - \gamma_{d2}e'_d \\ &\quad + \varsigma_{q2}(\boldsymbol{z})) + \varsigma_{q1}(\boldsymbol{z}))/T'_{q0} \\ \dot{\psi}''_d &= (-\psi''_d + e'_q - (x'_d - x_\ell)i_d - \varsigma_d(\boldsymbol{z}))/T''_{d0} \\ \dot{\psi}''_q &= (-\psi''_q - e'_d - (x'_q - x_\ell)i_q - \varsigma_{q2}(\boldsymbol{z}))/T''_{q0} \\ 0 &= \psi_d + x''_d i_d - \gamma_{d1}e'_q - (1-\gamma_{d1})\psi''_d - \varsigma_d(\boldsymbol{z}) \\ 0 &= \psi_q + x''_q i_q + \gamma_{q1}e'_d - (1-\gamma_{q1})\psi''_q - \varsigma_q(\boldsymbol{z}) \end{aligned} \tag{15.42}$$

where $\boldsymbol{z} = [e'_q, e'_d, \psi''_d, \psi''_q, i_d, i_q]^T$ and the saturation functions are:

$\varsigma_f(\boldsymbol{z})$: saturation of the dc field winding.
$\varsigma_d(\boldsymbol{z})$ and $\varsigma_q(\boldsymbol{z})$: d- and q-axis saturation of the stator windings.
$\varsigma_{q1}(\boldsymbol{z})$: saturation of current induced in the rotor core.
$\varsigma_{d1}(\boldsymbol{z})$ and $\varsigma_{q2}(\boldsymbol{z})$: saturation of the equivalent damper windings.

Instead of focusing on some particular model that includes saturation, following subsections discuss two examples of saturation functions and the data required for defining such saturation. Once the functions are defined, the inclusion of a saturation function is straightforward using (15.42). For simplicity, in the remainder, it is considered a generic saturation function $\psi = \varsigma(i)$ that links a generic current i to a generic flux ψ.

Piece-Wise Saturation Function

A piece-wise saturation model is discussed in [163] and is shown in Figure 15.5. The saturation function consists of three regions, as follows:

$$\psi = \begin{cases} x_1 i, & \text{if } \psi < \psi_A \\ x_1 i \dfrac{\psi}{\psi + A_s e^{B_s(\psi - \psi_A)}}, & \text{if } \psi_A \leq \psi < \psi_B \\ x_1 i \dfrac{\psi}{\psi_C + x_r(\psi - \psi_B)}, & \text{if } \psi_B \leq \psi \end{cases} \tag{15.43}$$

where $x_r = x_1/x_2$. The curve is completely defined by the parameters ψ_A, ψ_B, ψ_C, x_r, A_s and B_s. The main issue of this model is that for $i_A = \psi_A/x_1$, the function is discontinuous, since for i_A^-, the function returns $x_1 i_A$ and for i_A^+, the function returns $x_1 i_A \psi_A/(\psi_A + A_s)$. This discontinuity is immaterial only if $A_s \ll \psi_A$.

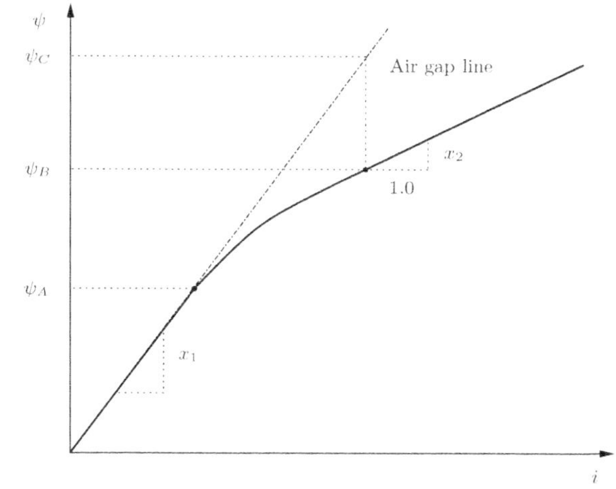

Fig. 15.5 Piece-wise saturation model

Polynomial Interpolation

This model consists in computing the coefficients of the polynomial that best interpolates three points of the saturation curve [272]. The three points are associated with $\psi_{0.8} = 0.8$, $\psi_{1.0} = 1.0$ and $\psi_{1.2} = 1.2$ pu, as shown in Figure 15.6. The saturation curve is assumed to be linear for $\psi < 0.8$. For $\psi \geq 0.8$ one has:

$$\psi = c_2 i^2 + c_1 i + c_0 \tag{15.44}$$

where:

$$\begin{aligned} c_0 &= \boldsymbol{s}\,[15,\ -24,\ 10]^T \\ c_1 &= \boldsymbol{s}\,[-27.5,\ 50,\ -22.5]^T \\ c_2 &= \boldsymbol{s}\,[12.5,\ -25,\ 12.5]^T \end{aligned} \tag{15.45}$$

and

$$\boldsymbol{s} = [0.8,\ 1 - s_1,\ 1.2(1 - s_2)] \tag{15.46}$$

The saturation factors s_1 and s_2 are computed as:

$$\begin{aligned} s_1 &= 1 - \frac{i_{a1}}{i_{b1}} \\ s_2 &= 1 - \frac{i_{a1.2}}{i_{b1.2}} \end{aligned} \tag{15.47}$$

and, along with the slope for $\psi < 0.8$, completely define the saturation curve. The main issues of this model are (i) the point at which the polynomial is computed are fixed, and (ii) high sensitivity with respect to the parameters s_1 and s_2.

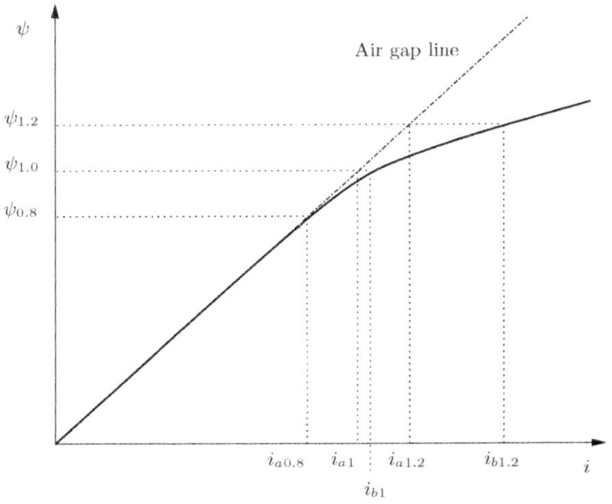

Fig. 15.6 Polynomial interpolation saturation model

15.1.9 *Center of Inertia*

In equation (15.5), the machine rotor angle and speed are assumed relative to the reference angle and speed of a hypothetical machine with constant speed ω_s and constant rotor angle δ_s, where generally $\omega_s = 1$ and $\delta_s = 0$. In some applications, it is useful to refer machine angles and speeds to the *center of inertia* (COI), which is a weighted sum of all machine angles and speeds:

$$\delta_{\mathrm{COI}} = \frac{\sum_{j \in \mathcal{G}} H_j \delta_j}{\sum_{j \in \mathcal{G}} H_j} \tag{15.48}$$

$$\omega_{\mathrm{COI}} = \frac{\sum_{j \in \mathcal{G}} H_j \omega_j}{\sum_{j \in \mathcal{G}} H_j} \tag{15.49}$$

Thus, the first of (15.5) becomes:

$$\dot{\delta} = \Omega_b(\omega - \omega_{\mathrm{COI}}) \tag{15.50}$$

In order to avoid inconsistencies, all speeds used in the computation of the equivalent COI speed have to pertain to an interconnected area. If, as a consequence of line outages, the system separates into two or more areas, a COI speed should be defined for each area.

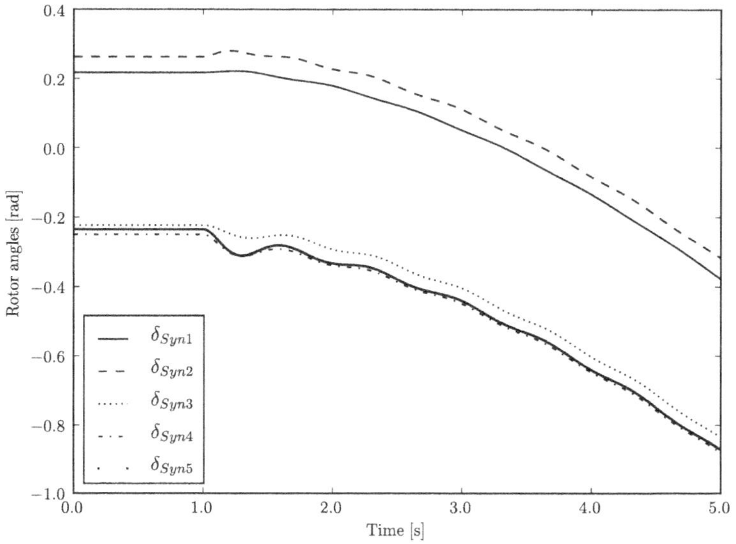

Fig. 15.7 Generator rotor angles using a constant synchronous speed reference

Example 15.4 Effect of Using the Center of Inertia for the IEEE 14-Bus System

Figures 15.7 and 15.8 show the difference between rotor angles referred to a constant synchronous speed and the same angles but using a COI speed reference. These figures represent a line 2-4 outage at $t = 1$ for the IEEE 14-bus system. At a first glance, the angle trajectories of Figure 15.7 could lead to think that the system is losing synchronism. Actually, the relative differences among rotor angles remain bounded, thus the system is stable. This conclusion is straightforward if using the COI speed reference.

15.1.10 Dynamic Shaft

A dynamic mass-spring model is used for defining the shaft of the synchronous machine. Figure 15.9 depicts the shaft scheme (springs are in solid gray). The complete set of differential equations that describe the dynamic shaft is as follows:

$$
\begin{aligned}
\dot{\delta}_{\mathrm{HP}} &= \Omega_b(\omega_{\mathrm{HP}} - \omega_s) \qquad (15.51)\\
\dot{\omega}_{\mathrm{HP}} &= (\tau_m - D_{\mathrm{HP}}(\omega_{\mathrm{HP}} - \omega_s) - D_{12}(\omega_{\mathrm{HP}} - \omega_{\mathrm{IP}})\\
&\quad + K_{\mathrm{HP}}(\delta_{\mathrm{IP}} - \delta_{\mathrm{HP}}))/2H_{\mathrm{HP}}\\
\dot{\delta}_{\mathrm{IP}} &= \Omega_b(\omega_{\mathrm{IP}} - \omega_s)
\end{aligned}
$$

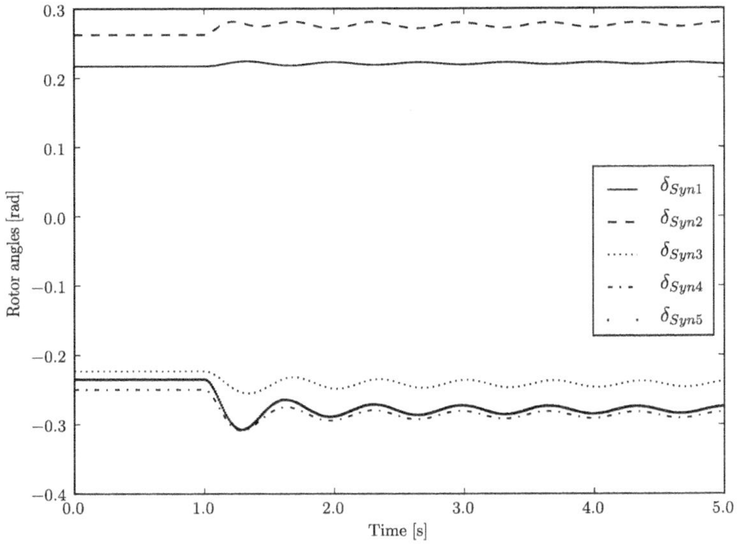

Fig. 15.8 Generator rotor angles using a COI speed reference

$$
\begin{aligned}
\dot{\omega}_{\mathrm{IP}} &= (-D_{\mathrm{IP}}(\omega_{\mathrm{IP}} - \omega_s) - D_{12}(\omega_{\mathrm{IP}} - \omega_{\mathrm{HP}}) - D_{23}(\omega_{\mathrm{IP}} - \omega_{\mathrm{LP}}) \\
&\quad + K_{\mathrm{HP}}(\delta_{\mathrm{HP}} - \delta_{\mathrm{IP}}) + K_{\mathrm{IP}}(\delta_{\mathrm{LP}} - \delta_{\mathrm{IP}}))/2H_{\mathrm{IP}} \\
\dot{\delta}_{\mathrm{LP}} &= \Omega_b(\omega_{\mathrm{LP}} - \omega_s) \\
\dot{\omega}_{\mathrm{LP}} &= (-D_{\mathrm{LP}}(\omega_{\mathrm{LP}} - \omega_s) - D_{23}(\omega_{\mathrm{LP}} - \omega_{\mathrm{IP}}) - D_{34}(\omega_{\mathrm{LP}} - \omega) \\
&\quad + K_{\mathrm{IP}}(\delta_{\mathrm{IP}} - \delta_{\mathrm{LP}}) + K_{\mathrm{LP}}(\delta - \delta_{\mathrm{LP}}))/2H_{\mathrm{LP}} \\
\dot{\delta} &= \Omega_b(\omega - 1) \\
\dot{\omega} &= (-\tau_e - D(\omega - \omega_s) - D_{34}(\omega - \omega_{\mathrm{LP}}) - D_{45}(\omega - \omega_{\mathrm{EX}}) \\
&\quad + K_{\mathrm{LP}}(\delta_{\mathrm{LP}} - \delta) + K_{\mathrm{EX}}(\delta_{\mathrm{EX}} - \delta))/2H \\
\dot{\delta}_{\mathrm{EX}} &= \Omega_b(\omega_{\mathrm{EX}} - \omega_s) \\
\dot{\omega}_{\mathrm{EX}} &= (-D_{\mathrm{EX}}(\omega_{\mathrm{EX}} - \omega_s) - D_{45}(\omega_{\mathrm{EX}} - \omega) \\
&\quad + K_{\mathrm{EX}}(\delta - \delta_{\mathrm{EX}}))/2H_{\mathrm{EX}}
\end{aligned}
$$

Example 15.5 Transient Behavior of Dynamics Shafts

Figure 15.10 shows the transient behavior of typical shaft rotor speed dynamics. The plot is obtained considering a dynamic shaft for the synchronous machine 1 of the IEEE 14-bus system. All shaft data are given in Appendix D.

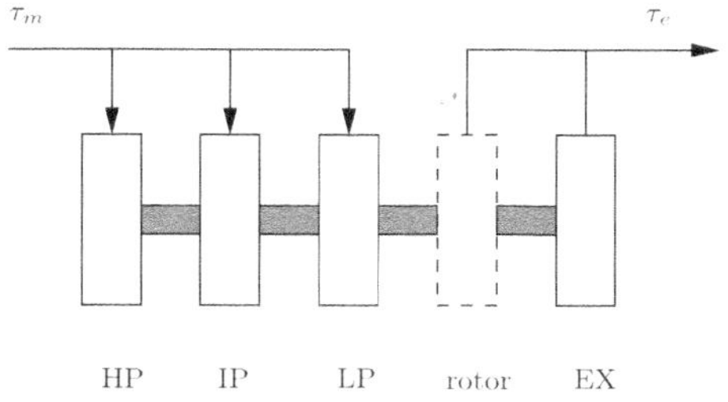

Fig. 15.9 Synchronous machine mass-spring shaft model

Table 15.4 Dynamic Shaft Data

Variable	Description	Unit
-	Synchronous machine code	-
H_{HP}	High pressure turbine inertia constant	MWs/MVA
H_{IP}	Intermediate pressure turbine inertia constant	MWs/MVA
H_{LP}	Low pressure turbine inertia constant	MWs/MVA
H_{EX}	Exciter inertia constant	MWs/MVA
D_{HP}	High pressure turbine damping	pu
D_{IP}	Intermediate pressure turbine damping	pu
D_{LP}	Low pressure turbine damping	pu
D_{EX}	Exciter damping	pu
D_{12}	High-interm. pressure turbine damping	pu
D_{23}	Interm.-low pressure turbine damping	pu
D_{34}	Low pressure turbine-rotor damping	pu
D_{45}	Rotor-exciter damping	pu
K_{HP}	High pressure turbine angle coefficient	pu
K_{IP}	Intermed. pressure turbine angle coefficient	pu
K_{LP}	Low pressure turbine angle coefficient	pu
K_{EX}	Exciter angle coefficient	pu

15.1.11 Sub-synchronous Resonance

Figure 15.11 depicts a generator with shaft dynamics and compensated line, which represents a simple model for studying the sub-synchronous resonance (SSR) problem. Shaft dynamics are the same as those described in previous Section 15.1.10 and are modeled as high, intermediate and low pressure turbine masses, exciter mass and machine rotor. The scheme of Figure 15.11 is one of the simplest models that may show the SSR phenomenon. SSR is a well-known problem of undamped oscillations that may occur when the transmission line to which the machine is connected is compensated by a series capacitor [140, 141, 142, 355].

The dynamics of the RLC circuit cannot be neglected since the line presents two modes whose frequency can be roughly estimated as $\Omega_b(1 \pm \sqrt{x_C/x_L})$. For typical values of the inductive and capacitive reactances, the

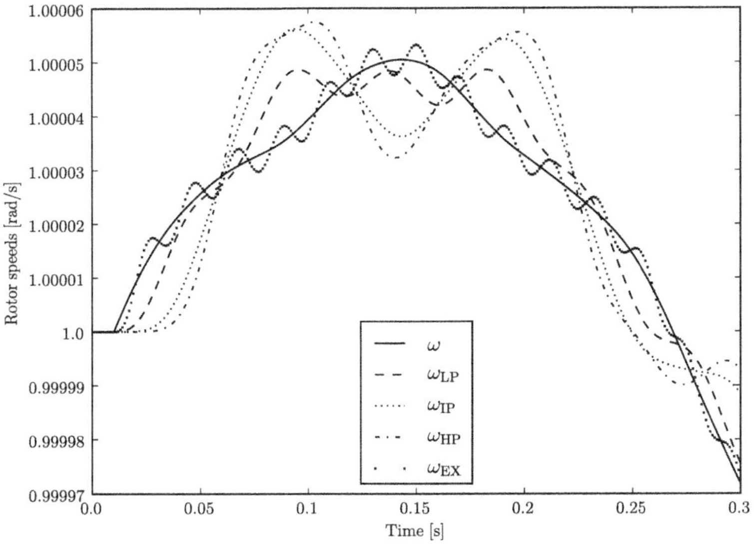

Fig. 15.10 Dynamic shaft rotor speed dynamics

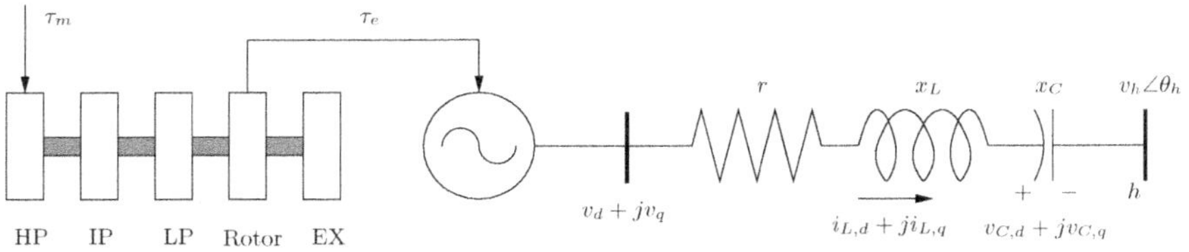

Fig. 15.11 Generator with dynamic shaft and compensated line

lower of these two frequencies can be close to one of the mechanical oscillations of the generator shaft. Thus, beyond a certain value of the compensation level, the machine may experiment a negative damping of one of the mechanical modes that results in dangerous stresses on the shaft. This phenomenon can be also described in terms of Hopf bifurcation [45, 205, 206].

A simple model used for studying SSR is presented in [361]. The machine is modelled through (15.9), (15.51) and

$$\dot{\psi}_f = (v_f - i_f)/T'_{d0} \tag{15.52}$$

The transmission line dynamics are:

$$
\begin{aligned}
\dot{i}_{L,d} &= \Omega_b(i_{L,q} + (v_d - r i_{L,d} - v_{C,d} - V \sin(\delta - \theta))/x_L) \\
\dot{i}_{L,q} &= \Omega_b(-i_{L,d} + (v_q - r i_{L,q} - v_{C,q} - V \cos(\delta - \theta))/x_L) \\
\dot{v}_{C,d} &= \Omega_b(x_C i_{L,d} + v_{C,q}) \\
\dot{v}_{C,q} &= \Omega_b(x_C i_{L,q} - v_{C,d})
\end{aligned}
\tag{15.53}
$$

Finally, the constraints that link the time derivatives of generator fluxes and line currents are:

$$
\begin{aligned}
\dot{\psi}_f &= \dot{i}_f - (x_d - x'_d)\dot{i}_{L,d} \\
\dot{\psi}_d &= \dot{i}_f - x_d \dot{i}_{L,d} \\
\dot{\psi}_q &= -x_q \dot{i}_{L,q}
\end{aligned}
\tag{15.54}
$$

Example 15.6 Sub-synchronous Resonance Transient

Figure 15.12 shows the dynamic behavior of a system that shows sub-synchronous resonance. The plot was obtained using data provided in [206].

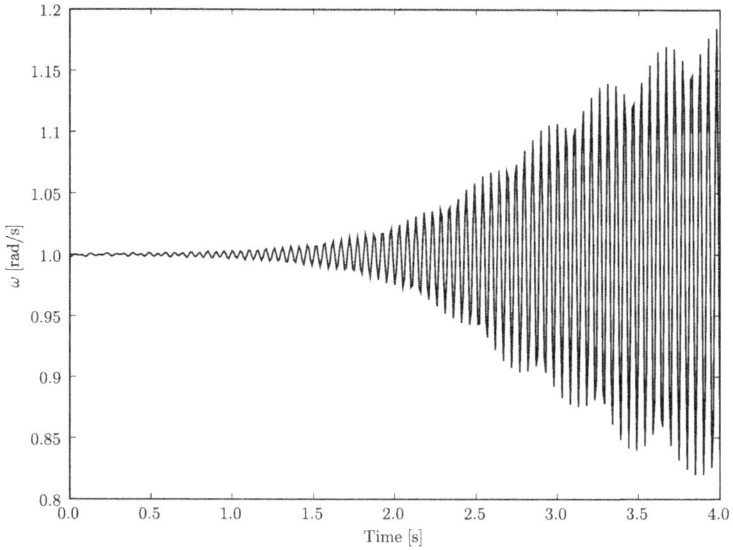

Fig. 15.12 Sub-synchronous resonance transient

15.2 Induction Machine

Induction machine models can be formally formulated using the Park's approach. However, since induction machine rotors have no salient poles and since the rotor angular position is generally irrelevant, the Park's two-reaction approach is not strictly necessary. This section describes three models of increasing complexity. These are pure mechanical model, single-cage rotor model, and double-cage rotor model. Each machine model includes a mechanical torque, which can thus be modeled separately and then included as a common ancestor class.

Table 15.5 defines the parameters of all induction machine models described in this section. Since a typical study related to induction machines is the start-up transient, the parameter list includes start-up parameters that control if and when the machine is started (i.e., s_{up} and t_{up} in Table 15.5). If the machine is marked for start-up, the slip is $\sigma = 1$ (e.g., rotor speed $\omega = 0$) and the machine status is $u = 0$ for $t \leq t_{up}$.

Table 15.5 Induction machine parameters

Variable	Description	Unit
a	1$^{\rm st}$ coeff. of $\tau_m(\omega)$	pu
b	2$^{\rm nd}$ coeff. of $\tau_m(\omega)$	pu
c	3$^{\rm rd}$ coeff. of $\tau_m(\omega)$	pu
H_m	Machine rotor inertia constant	MWs/MVA
r_{R1}	1$^{\rm st}$ cage rotor resistance	pu
r_{R2}	2$^{\rm nd}$ cage rotor resistance	pu
r_S	Stator resistance	pu
s_{up}	Start-up control	$\{0, 1\}$
t_{up}	Start up time	s
x_{R1}	1$^{\rm st}$ cage rotor reactance	pu
x_{R2}	2$^{\rm nd}$ cage rotor reactance	pu
x_S	Stator reactance	pu
x_μ	Magnetization reactance	pu
$\aleph$	Allow working as brake	$\{0, 1\}$

15.2.1 Initialization

As discussed in Chapter 4, the standard power flow problem is formulated describing loads as constant power consumptions. However, induction motors do not behave as constant power consumptions. Thus, if the machine dynamic equations are initialized after the power flow analysis, there will be certainly a data inconsistency. In fact if one uses the bus voltage $\bar{v}_0$ and the active power p_0 to compute the machine slip and mechanical torque, then the reactive power consumed by the machine is assigned. Generally, this reactive power is not equal to the one obtained as the solution of the power flow analysis. To solve this inconsistency, there are two possibilities:

1. Taking into account that induction motors are generally compensated, a shunt capacitor bank can be included to fix the reactive power mismatch at the machine bus.
2. The dynamic machine model can be directly included into the power flow problem. In this case the data to be imposed in the power flow analysis is the mechanical torque at the machine shaft. This method is certainly the most precise. Furthermore, if there is a capacitor bank, this can be modelled using its real capacity.

A detailed discussion on this topic can be found in [262].

15.2.2 Torque Model

A typical model that can be used for the mechanical torque is a quadratic function of the rotor speed:

$$\tau_m = a + b\omega + c\omega^2 \tag{15.55}$$

and given the relationship between the slip σ and the speed ω in pu, e.g. $\sigma = 1 - \omega$, the torque/slip characteristic becomes:

$$\tau_m = \alpha + \beta\sigma + \gamma\sigma^2 \tag{15.56}$$

where

$$\alpha = a + b + c, \quad \beta = -b - 2c, \quad \gamma = c \tag{15.57}$$

In some cases, the previous model is not adequate, for example, it does not allow taking into account the machine duty-cycle. In this case, it is necessary to provide a function of time $\tau_m(t)$. Since accounting for any possible behavior is not possible, the better solution is likely to provide a series of (τ_m, t)-value pairs.

15.2.3 Electromechanical Model

The electrical circuit for the first order induction motor is depicted in Figure 15.13. Only the mechanical state variable is considered, being the circuit in steady-state condition. The differential equation is:

$$\dot{\sigma} = \frac{1}{2H_m}\left(\tau_m(\sigma) - \frac{r_{R1}v_h^2/\sigma}{(r_S + r_{R1}/\sigma)^2 + (x_S + x_{R1})^2}\right) \tag{15.58}$$

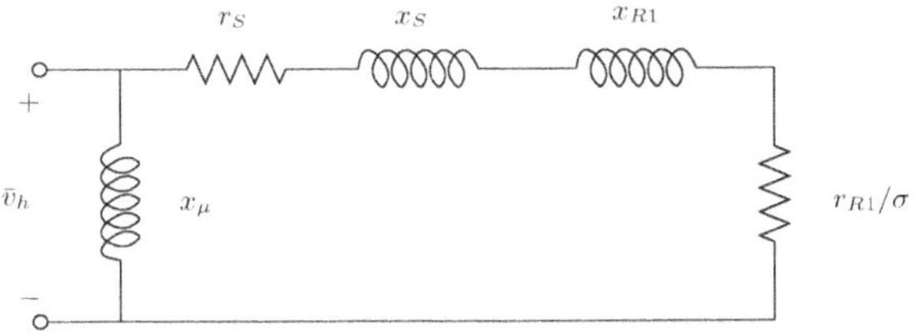

Fig. 15.13 Electrical circuit of the first-order induction machine model

whereas the power injections are:

$$\begin{aligned} p_h &= -\frac{(r_S + r_{R1}/\sigma)v_h^2}{(r_S + r_{R1}/\sigma)^2 + (x_S + x_{R1})^2} \\ q_h &= -\frac{v_h^2}{x_\mu} - \frac{(x_S + x_{R1})v_h^2}{(r_S + r_{R1}/\sigma)^2 + (x_S + x_{R1})^2} \end{aligned} \tag{15.59}$$

The negative sign of the active and reactive power indicates that the machine is working as a motor. If the machine cannot work as a brake, then the differential equation undergoes an anti-windup limiter that activates if $\sigma \leq 0$ and $\dot{\sigma} < 0$ (see Appendix C).

15.2.4 Detailed Single-Cage Model

The simplified electrical circuit for the single-cage induction motor is depicted in Figure 15.14. Equations are formulated in terms of the real (d-) and imaginary (q-) axes, with respect to the network reference angle. In a synchronously rotating reference frame, the link between the network and the stator machine voltages is as follows:

$$\begin{aligned} v_d &= -v_h \sin\theta \\ v_q &= v_h \cos\theta \end{aligned} \tag{15.60}$$

Using the notation of Figure 15.14, the power absorptions are:

$$\begin{aligned} p_h &= -(v_d i_d + v_q i_q) \\ q_h &= -(v_q i_d - v_d i_q) \end{aligned} \tag{15.61}$$

The differential equations in terms of the voltage behind the the stator resistance r_S are:

$$\begin{aligned} \dot{e}'_d &= \Omega_b \sigma e'_q - (e'_d + (x_0 - x')i_q)/T'_0 \\ \dot{e}'_q &= -\Omega_b \sigma e'_d - (e'_q - (x_0 - x')i_d)/T'_0 \end{aligned} \tag{15.62}$$

whereas the link between voltages, currents and state variables is as follows:

$$\begin{aligned} v_d - e'_d &= r_S i_d - x' i_q \\ v_q - e'_q &= r_S i_q + x' i_d \end{aligned} \tag{15.63}$$

where x_0, x' and T_0 can be obtained from the motor parameters:

$$\begin{aligned} x_0 &= x_S + x_\mu \\ x' &= x_S + \frac{x_{R1} x_\mu}{x_{R1} + x_\mu} \\ T'_0 &= \frac{x_{R1} + x_\mu}{\Omega_b r_{R1}} \end{aligned} \tag{15.64}$$

Finally, the mechanical equation is as follows:

$$\dot{\sigma} = (\tau_m(\sigma) - \tau_e)/(2H_m) \tag{15.65}$$

where the electrical torque is:

$$\tau_e \approx e'_d i_d + e'_q i_q \tag{15.66}$$

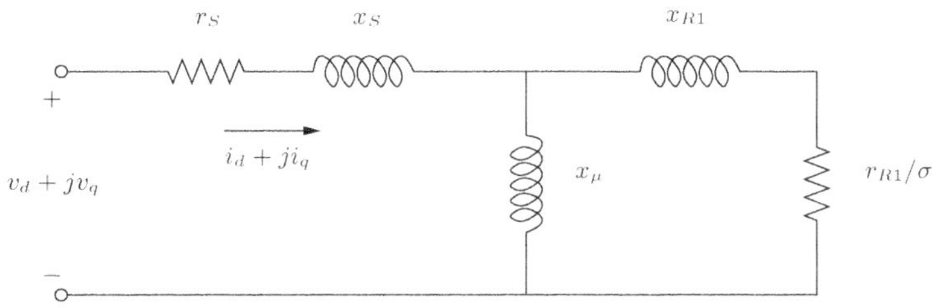

Fig. 15.14 Electrical circuit of the third-order induction machine model

15.2.5 Detailed Double-Cage Model

The electrical circuit for the double-cage induction machine model is depicted in Figure 15.15. As for the single-cage model, real and imaginary axes are defined with respect to the network reference angle, and (15.60) and (15.61) apply. Two voltages behind the stator resistance r_S model the cage dynamics, as follows:

$$\begin{aligned}
\dot{e}'_d &= \Omega_b \sigma e'_q - (e'_d + (x_0 - x')i_q)/T'_0 \\
\dot{e}'_q &= -\Omega_b \sigma e'_d - (e'_q - (x_0 - x')i_d)/T'_0 \\
\dot{e}''_d &= -\Omega_b \sigma (e'_q - e''_q) + \dot{e}'_d - (e'_d - e''_q - (x' - x'')i_q)/T''_0 \\
\dot{e}''_q &= \Omega_b \sigma (e'_d - e''_d) + \dot{e}'_q - (e'_q - e''_d + (x' - x'')i_d)/T''_0
\end{aligned} \tag{15.67}$$

and the links between voltages and currents are:

$$\begin{aligned}
v_d - e''_d &= r_S i_d - x'' i_q \\
v_q - e''_q &= r_S i_q + x'' i_d
\end{aligned} \tag{15.68}$$

where the parameters are determined from the circuit resistances and reactances and are given by equations (15.64) and:

$$\begin{aligned}
x'' &= x_S + \frac{x_{R1} x_{R2} x_\mu}{x_{R1} x_{R2} + x_{R1} x_\mu + x_{R2} x_\mu} \\
T''_0 &= \frac{x_{R2} + x_{R1} x_\mu / (x_{R1} + x_\mu)}{\Omega_b r_{R2}}
\end{aligned} \tag{15.69}$$

The differential equation for the slip is the (15.65), while the electrical torque is defined as follows:

$$\tau_e \approx e''_d i_d + e''_q i_q \tag{15.70}$$

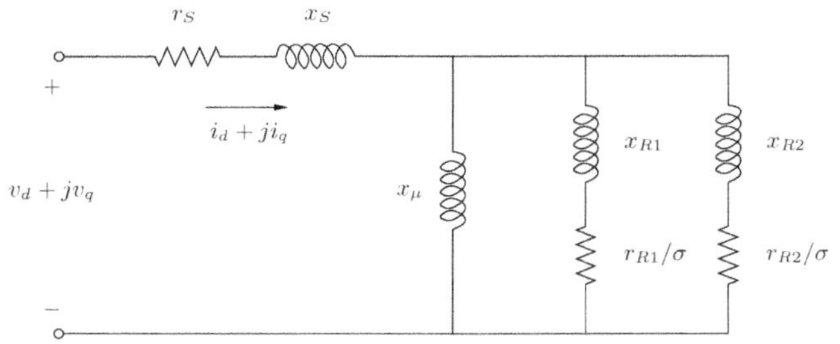

Fig. 15.15 Electrical circuit of the fifth-order induction machine model

Example 15.7 Induction Motor Start-Up

Figure 15.16 shows the start-up of a double-cage induction motor modelled with a fifth order set of DAE as discussed in the previous section. The motor is connected to the network at $t = 1$ s. Fast electrical dynamics damp out in about a second while the mechanical transient takes several seconds. Thus, fast dynamics should be considered in the classical transient stability analysis i.e., for time scale from one to five seconds after the fault clearing. For voltage

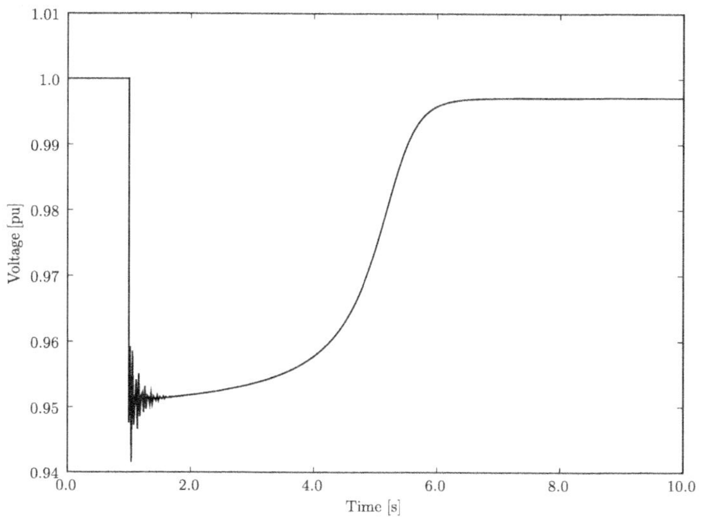

Fig. 15.16 Induction motor start-up transient

or frequency stability analysis (e.g., tens of seconds), a simple mechanical model for the induction motor is adequate.

Chapter 16
Synchronous Machine Regulators

This chapter describes the most relevant synchronous machines primary regulators and limiters. These are the turbine governor, the automatic voltage regulator and the over- and under-excitation limiters. These regulators are the dynamic counterpart of the static capability curve described in Section 12.2.1 of Chapter 12. Furthermore, this chapter also describes the power system stabilizer that allows efficiently damping synchronous machine rotor oscillations. Figure 16.1 provides a synoptic scheme of the synchronous machine regulators described in this chapter.

There is a huge variety of synchronous machine regulator models. For example, the EPRI Extended Transient-Midterm Stability Program (ETMSP) User's Manual contains hundreds of controls schemes for turbine governors, automatic voltage regulators and power system stabilizers [130]. The main object of this chapter is to provide an overview of the basic functioning and some commonly used models. Other models, even very complex, can be implemented starting from the basic schemes given in this chapter. Moreover, providing a complete list of all existing regulator models is likely useless, since it is always possible that someone proves that currently accepted models are actually inadequate [237].

16.1 Turbine Governor

Turbine Governors (TGs) define the primary frequency control of synchronous machines. Figure 16.2 shows the basic functioning of the primary frequency control. In particular, Figure 16.2.a depicts the linearized control loop that includes the turbine governor transfer function $G(s)$ and a simplified machine model. The transfer function $G(s)$ depends on the type of the turbine and on the control (see for example Figures 16.3 and 16.4). In steady-state conditions:

$$\lim_{s \to 0} G(s) = \frac{1}{R} \tag{16.1}$$

F. Milano: Power System Modelling and Scripting, Power Systems, pp. 355–377.
springerlink.com

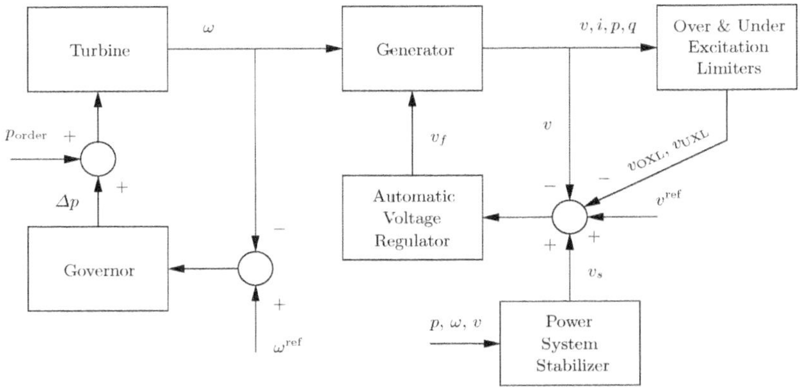

Fig. 16.1 Synoptic scheme of synchronous machine regulators

thus, approximating the mechanical torque with the mechanical power, one has:

$$\Delta p_m = \frac{1}{R}(\Delta\omega^{\rm ref} - \Delta\omega) \tag{16.2}$$

where all quantities are in pu with respect to machine bases. Figure 16.2.b shows the effect of the droop R on the regulation: (i) $R \neq 0$ is the normal situation; (ii) $R = 0$ implies that $G(s)$ contains a pure integrator and, thus, a constant frequency control; and (iii) $R \to \infty$ means constant power control (the primary frequency control loop is open). In general, $R \neq 0$ and $R < \infty$, so that the machine regulates the frequency proportionally to its rated power. Only in islanded systems with one or very few machines, it makes sense to set $R = 0$. Finally, Figure 16.2.c shows the effect of the variation of the system frequency $\Delta\omega$ on a multi-machine system. If $\Delta\omega < 0$ (which implies that the power absorbed by the load has increased), then each machine k increases its power production by a quantity proportional to $1/R_k$.

The droop R is a measure of the participation of each machine to system losses and load power variations. Thus, one can define the loss participation coefficient γ_k in the static generator models (see Sections 10.2.1 and 10.2.2) based on R_k, as follows:

$$\gamma_k = \frac{1/R_k}{\sum_{j\in\mathcal{G}} 1/R_j} \tag{16.3}$$

where $\mathcal{G}$ is the set of synchronous machines. For example, if a system has three machines with $1/R_1 = 3\%$, $1/R_2 = 5\%$, and $1/R_3 = 4\%$, then:

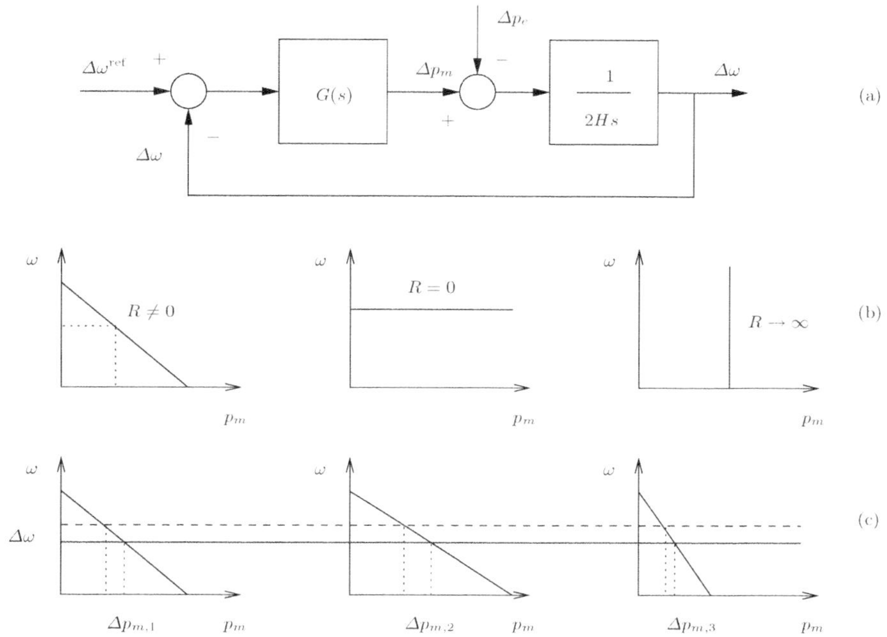

Fig. 16.2 Basic functioning of the primary frequency control: (a) linearized control loop; (b) effect of the droop R on the control loop; and (c) effect of a variation of the rotor speed $\Delta\omega$ on a multi-machine system with $R_j \neq 0$, $\forall j \in \mathcal{G}$

$$\begin{aligned}
\gamma_1 &= \frac{0.03}{0.03 + 0.05 + 0.04} = 0.25 \\
\gamma_2 &= \frac{0.05}{0.03 + 0.05 + 0.04} = 0.42 \\
\gamma_3 &= \frac{0.04}{0.03 + 0.05 + 0.04} = 0.33
\end{aligned}$$

Only the relative values of the coefficients γ_i are relevant, not the absolute ones. Two relevant remarks are as follows:

1. If $R_k \to \infty$, $\gamma_k = 0$ for the machine k.
2. If $R_k = 0$, $\gamma_k = 1$ for the machine k, being all other loss participation coefficients $\gamma_j = 0$, $\forall j \in \mathcal{G}$ and $j \neq i$.

When defining the TG data, the droop R and mechanical power limits are often given in pu with respect to the synchronous machine power rating. If this is the case, during the initialization of turbine governors data, the droops have to be converted to the system power base, as follows:

$$R_{\text{system}} = \frac{S_{\text{system}}}{S_{\text{machine}}} R_{\text{machine}} \tag{16.4}$$

When initializing the turbine governor variable, mechanical power limits have to be checked. If a limit is violated, it means that the turbine governor parameters are not consistent with those of the static generator used in power flow analysis.

Each turbine governor model has two algebraic equations, as follows:

$$0 = \tilde{\tau}_m - \tau_m \tag{16.5}$$

$$0 = \omega_0^{\text{ref}} - \omega^{\text{ref}} \tag{16.6}$$

where (16.5) represents the link between the turbine governor and the synchronous machines, being τ_m the input mechanical power variable used in synchronous machine models, i.e., (16.5) substitutes (15.7); and (16.6) defines the turbine governor reference rotor speed. The reference signal ω_0^{ref} can be modified, for example, by the automatic generation control (e.g., secondary frequency regulation).

Following subsections describe two simple yet commonly used turbine governor models.

16.1.1 Turbine Governor Type I

The TG type I is depicted in Figure 16.3. It includes a governor, a servo and a reheat block. The DAE system that describes this TG model is as follows:

$$\begin{aligned}
\hat{p}_{\text{in}} &= p_{\text{order}} + \frac{1}{R}(\omega^{\text{ref}} - \omega) \\
p_{\text{in}} &= \begin{cases} \hat{p}_{\text{in}} & \text{if } p^{\min} \leq \hat{p}_{\text{in}} \leq p^{\max} \\ p^{\max} & \text{if } \hat{p}_{\text{in}} > p^{\max} \\ p^{\min} & \text{if } \hat{p}_{\text{in}} < p^{\min} \end{cases} \\
\dot{x}_{g1} &= (p_{\text{in}} - x_{g1})/T_s \\
\dot{x}_{g2} &= ((1 - \frac{T_3}{T_c})x_{g1} - x_{g2})/T_c \\
\dot{x}_{g3} &= ((1 - \frac{T_4}{T_5})(x_{g2} + \frac{T_3}{T_c}x_{g1}) - x_{g3})/T_5 \\
\tilde{\tau}_m &= x_{g3} + \frac{T_4}{T_5}(x_{g2} + \frac{T_3}{T_c}x_{g1})
\end{aligned} \tag{16.7}$$

The number of blocks of each part of the turbine can be increased to take into account each stage in detail. However, the structure of the control diagram does not change. Table 16.1 defines the parameters of TG type I.

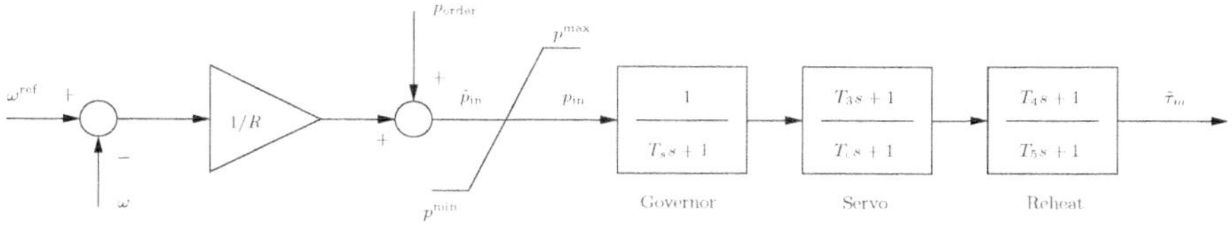

Fig. 16.3 Turbine governor Type I control diagram

Table 16.1 Turbine governor Type I parameters

Variable	Description	Unit
-	Generator code	-
$p^{\max}$	Maximum turbine output	pu
$p^{\min}$	Minimum turbine output	pu
R	Droop	pu
T_3	Transient gain time constant	s
T_4	Power fraction time constant	s
T_5	Reheat time constant	s
T_c	Servo time constant	s
T_s	Governor time constant	s

16.1.2 Turbine Governor Type II

The TG type II is depicted in Figure 16.4 and described by the following equations:

$$\begin{aligned}
\dot{x}_g &= (\frac{1}{R}(1 - \frac{T_1}{T_2})(\omega^{\text{ref}} - \omega) - x_g)/T_2 \\
\hat{\tau}_m &= x_g + \frac{1}{R}\frac{T_1}{T_2}(\omega^{\text{ref}} - \omega) + \tau_{m0} \\
\tilde{\tau}_m &= \begin{cases} \tau^{\max} & \text{if } \hat{\tau}_m > \tau^{\max} \\ \hat{\tau}_m & \text{if } \tau^{\min} \le \hat{\tau}_m \le \tau^{\max} \\ \tau^{\min} & \text{if } \hat{\tau}_m < \tau^{\min} \end{cases}
\end{aligned} \tag{16.8}$$

where τ_{m0} is the initial mechanical torque determined when initializing synchronous machines (see equation (15.7)) and other parameters are defined in Table 16.2. Equations (16.8) simplify the previous model (16.7). The turbine governor type II is typically more than adequate for transient stability analysis.

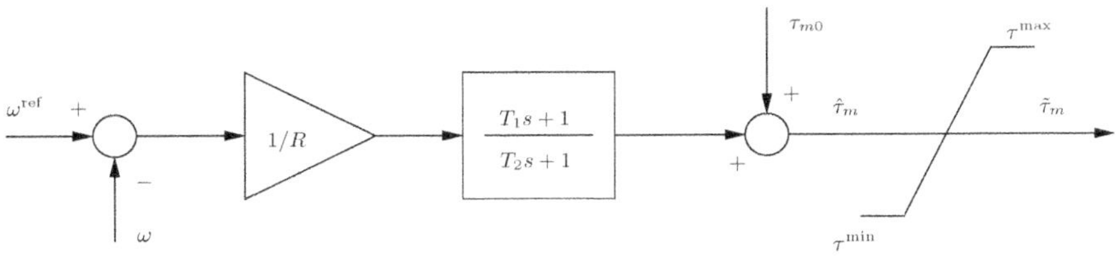

Fig. 16.4 Turbine governor Type II control diagram

Table 16.2 Turbine governor Type II parameters

Variable	Description	Unit
-	Generator code	int
R	Droop	pu
T_1	Transient gain time constant	s
T_2	Governor time constant	s
$\tau^{\max}$	Maximum turbine output	pu
$\tau^{\min}$	Minimum turbine output	pu

Example 16.1 Effect of Turbine Governor on Generator Frequency

Figure 16.5 shows the effect of turbine governors on generator rotor speeds for the IEEE 14-bus system. The plot was obtained including two turbine governors of type II at generators 1 and 2. The complete data are reported in Appendix D. The disturbance is line 2-4 outage at $t = 1$ s. As expected, the turbine governor is able to recover the rotor speed to a value close to the initial synchronous speed. However, since the disturbance considered in this case study implies only a redistribution of losses, the effect of turbine governor is negligible.[1] As discussed above, the primary frequency regulation is never integral, hence the final frequency cannot be equal to the initial one.

Turbine governors regulates the production of synchronous machine active powers. In the case of the IEEE 14-bus system, only two machines produce active power. Since synchronous machine 1 is ten times bigger than machine 2 and the two machines have the same droop, machine 1 takes about the 90% of the active power variation after the disturbance.

[1] In fact, without turbine governors, the frequency error is $< 0.2\%$. This is why turbine governors are not considered in most examples of this book.

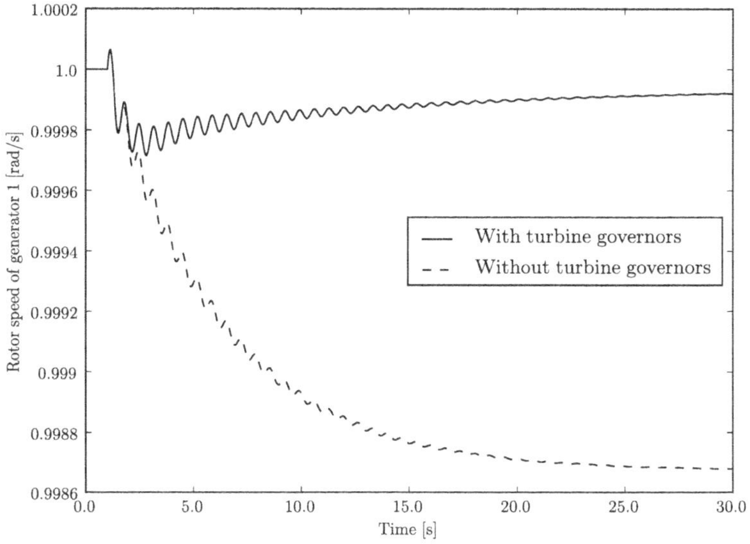

Fig. 16.5 Effect of the turbine governor on the generator frequency for the IEEE 14-bus system

16.2 Automatic Voltage Regulator

Automatic Voltage Regulators (AVRs) define the primary voltage regulation of synchronous machines. Several AVR models have been proposed and realized in practice [89, 145]. In this section, three simple AVR types are described. AVR Type I is a simplified version of the standard dc exciter IEEE type I, whereas AVR Type II is a typical static exciter model. AVR Type III is the simplest AVR model that can be used for rough stability evaluations.

Figure 16.6 depicts a conceptual linearized model of the primary voltage regulation of the synchronous machine. Although the detailed system is much more complicated, the kernel of primary voltage regulation can be reduced to a transfer function composed of the AVR regulator, the exciter and the machine d-axis emf. The system of Figure 16.6 has the closed-loop root loci shown in Figure 16.7. Depending on the AVR gain the system can be stable or unstable (due to the occurrence of a Hopf bifurcation). Clearly, AVR gains are chosen to avoid instability in most operating conditions. However, it is possible that an unusual loading condition and/or line outage causes an increase of the equivalent closed-loop gain and, thus, leads to instability [337]. This phenomenon is illustrated in Example 16.2.

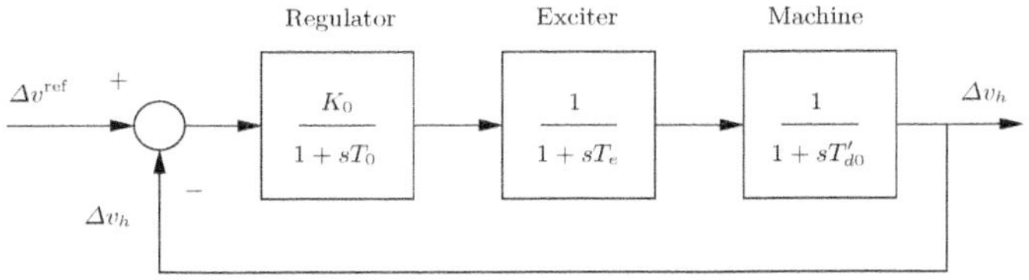

Fig. 16.6 Basic functioning of the primary voltage control

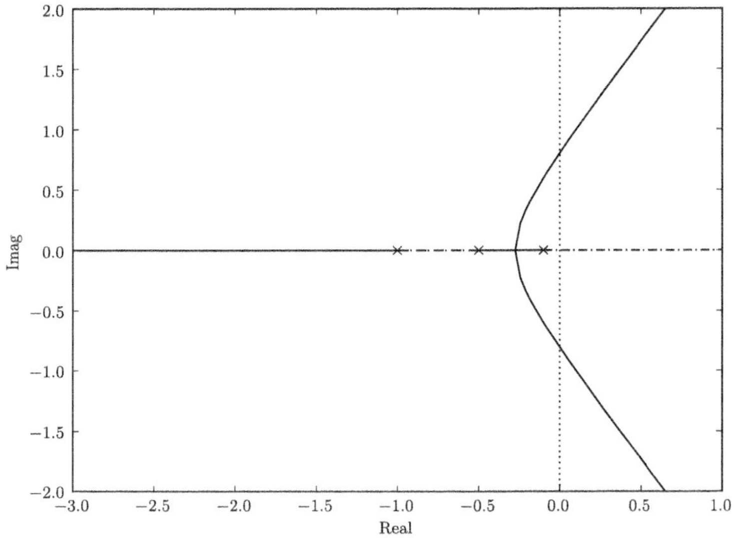

Fig. 16.7 Primary voltage control root loci

Each AVR model has two algebraic equations, as follows:

$$0 = \tilde{v}_f - v_f \tag{16.9}$$

$$0 = v_0^{\mathrm{ref}} - v^{\mathrm{ref}} \tag{16.10}$$

where (16.9) represents the link in between the AVR and the synchronous machines, being v_f the algebraic variable that defines the synchronous machine field voltage, i.e., (16.9) substitutes (15.8). Equation (16.10) defines the AVR reference voltage. It is useful to define the reference voltage as a variable since other devices such as over-excitation limiters or power system stabilizers modify such reference with additional signals. Thus, (16.10) allows interfacing other regulators to the AVR. Furthermore, the bus voltage measure delay is usually modelled as a lag block:

$$\dot{v}_m = (v_h - v_m)/T_r \tag{16.11}$$

where v_h is the generator bus voltage or any bus voltage regulated by the AVR, v_m is the state variable used as voltage signal within the AVR and T_r the measurement block time constant. Equations (16.9)-(16.11) constitute the common equations of AVR models and can be included in a base AVR class.

The reference voltage v_0^{ref} is initialized after the power flow analysis and after the initialization of synchronous machines. In case the violations of AVR limits, the data of the static generator used in power flow analysis are not consistent with the AVR data and, thus, AVR state variables cannot be correctly initialized.

16.2.1 Automatic Voltage Regulator Type I

The AVR Type I depicted in Figure 16.8 represents a typical dc exciter model. The DAE system is (16.9)-(16.11) and:

$$\begin{aligned}
\dot{v}_{r1} &= (K_a(v^{\text{ref}} - v_m - v_{r2} - \frac{K_f}{T_f}\tilde{v}_f) - v_{r1})/T_a \qquad (16.12)\\
\dot{v}_{r2} &= -(\frac{K_f}{T_f}\tilde{v}_f + v_{r2})/T_f \\
\dot{\tilde{v}}_f &= -(\tilde{v}_f(K_e + S_e(\tilde{v}_f)) - v_{r1})/T_e
\end{aligned}$$

where v_h is the generator terminal voltage or a remote-bus regulated voltage and the ceiling function S_e is:[2]

$$S_e(\tilde{v}_f) = A_e e^{B_e|\tilde{v}_f|} \qquad (16.13)$$

This model is a simplified version of the classic IEEE type DC1 [145]. The IEEE DC1 system includes an additional lead-lag block before the amplifier block. However, this lead-lag block is often neglected. The amplifier state variable v_{r1} is subjected to an anti-windup limit. Table 16.3 defines all parameters of AVR Type I.

[2] The coefficients A_e and B_e can be determined by measuring two points of the ceiling function S_e. Typically, one knows the values $S_e^{\max}$ and $S_e^{0.75\cdot\max}$ that correspond to the field voltages $v_f^{\max}$ and $0.75 \cdot v_f^{\max}$, respectively. To compute A_e and B_e, one has to solve the following system:

$$\begin{aligned}
0 &= -(1 + S_e^{\max})v_f^{\max} + v_r^{\max} \\
S_e^{\max} &= A_e e^{B_e v_f^{\max}} \\
S_e^{0.75\cdot\max} &= A_e e^{B_e \cdot 0.75 \cdot v_f^{\max}}
\end{aligned}$$

where $S_e^{\max}$, $S_e^{0.75\cdot\max}$, $v_f^{\max}$, and $v_r^{\max}$ are given values.

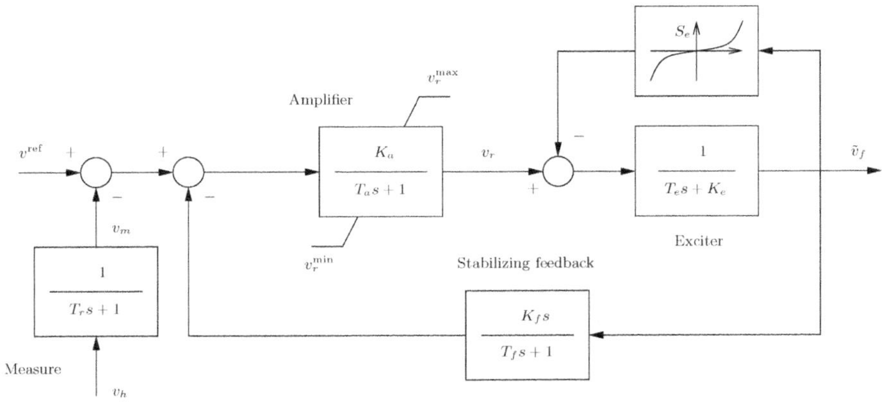

Fig. 16.8 Automatic voltage regulator Type I control diagram

Table 16.3 Automatic voltage regulator Type I parameters

Variable	Description	Unit
-	Generator code	-
-	Regulated voltage code	-
A_e	1st ceiling coefficient	-
B_e	2nd ceiling coefficient	1/pu
K_a	Amplifier gain	pu/pu
K_e	Field circuit integral deviation	-
K_f	Stabilizer gain	s pu/pu
T_a	Amplifier time constant	s
T_f	Stabilizer time constant	s
T_e	Field circuit time constant	s
T_r	Measurement time constant	s
v_r^{max}	Maximum regulator voltage	pu
v_r^{min}	Minimum regulator voltage	pu

16.2.2 Automatic Voltage Regulator Type II

The AVR Type II is shown in Figures 16.9 and 16.10. This AVR models a typical static exciter which is characterized by higher gains and faster response than the previous dc exciter. The DAE system is (16.9)-(16.11) and:

$$\dot{v}_{r1} = (K_0(1 - \frac{T_2}{T_1})(v^{ref} - v_m) - v_{r1})/T_1 \tag{16.14}$$

$$\dot{v}_{r2} = ((1 - \frac{T_4}{T_3})(v_{r1} + K_0\frac{T_2}{T_1}(v^{ref} - v_m)) - v_{r2})/T_3 \tag{16.15}$$

$$\hat{v}_r = v_{r2} + \frac{T_4}{T_3}(v_{r1} + K_0\frac{T_2}{T_1}(v^{ref} - v_m)) \tag{16.16}$$

$$v_r = \begin{cases} v_r^{\max} & \text{if } \hat{v}_r > v_r^{\max}, \\ \hat{v}_r & \text{if } v_r^{\min} \leq \hat{v}_r \leq v_r^{\max}, \\ v_r^{\min} & \text{if } \hat{v}_r < v_r^{\min}. \end{cases} \tag{16.17}$$

$$\dot{\tilde{v}}_f = -(\tilde{v}_f(1 + S_e(\tilde{v}_f)) - v_r)/T_e \tag{16.18}$$

where where v_h is the generator terminal voltage or a remote-bus regulated voltage and S_e is the ceiling function (16.13). For high values of the gain K_0, the state variable v_{r2} takes also high values. To keep the values of v_{r1} and v_{r2} comparable, the following variable change is used:

$$\tilde{v}_{r2} = \frac{v_{r2}}{K_0} \tag{16.19}$$

Hence, (16.15) and (16.16) can be rewritten as follows:

$$\dot{\tilde{v}}_{r2} = ((1 - \frac{T_4}{T_3})(v_{r1} + K_0 \frac{T_2}{T_1}(v^{\text{ref}} - v_m)) - K_0 \tilde{v}_{r2})/(K_0 T_3) \tag{16.20}$$

$$\hat{v}_r = K_0 \tilde{v}_{r2} + \frac{T_4}{T_3}(v_{r1} + K_0 \frac{T_2}{T_1}(v^{\text{ref}} - v_m)) \tag{16.21}$$

Table 16.4 defines all parameters required by the AVR Type II.

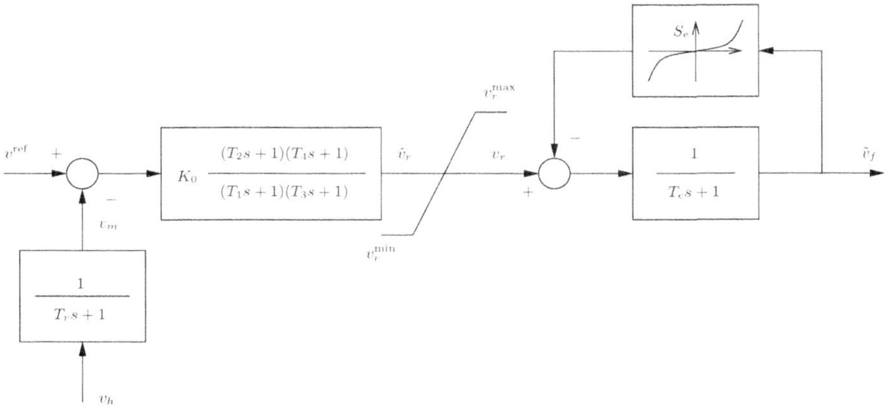

Fig. 16.9 Automatic voltage regulator Type II control diagram

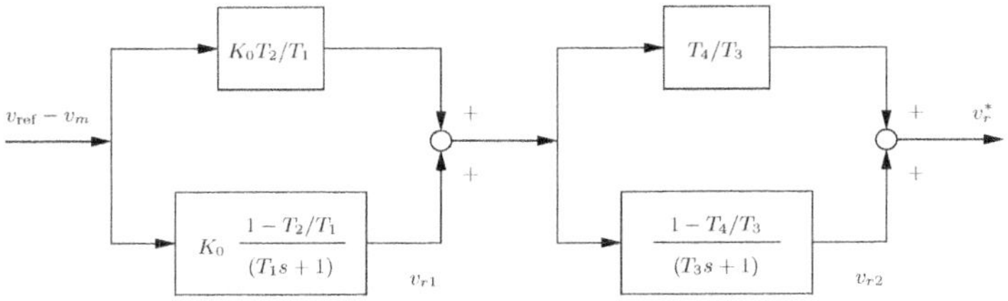

Fig. 16.10 Detail of the double lead-lag block of AVR Type II

Table 16.4 Automatic voltage regulator Type II parameters

Variable	Description	Unit
-	Generator code	-
-	Regulated voltage code	-
A_e	1st ceiling coefficient	-
B_e	2nd ceiling coefficient	1/pu
K_0	Regulator gain	pu/pu
T_1	1st pole	s
T_2	1st zero	s
T_3	2nd pole	s
T_4	2nd zero	s
T_e	Field circuit time constant	s
T_r	Measurement time constant	s
$v_r^{\max}$	Maximum regulator voltage	pu
$v_r^{\min}$	Minimum regulator voltage	pu

16.2.3 Automatic Voltage Regulator Type III

The AVR Type III depicted in Figure 16.11 is a simple model that can be useful for simplified stability studies. The DAE system is (16.9)-(16.11) and:

$$\dot{v}_r = (K_0(1 - \frac{T_1}{T_2})(v^{\rm ref} - v_m) - v_r)/T_2 \tag{16.22}$$
$$\dot{\tilde{v}}_f = ((v_r + K_0\frac{T_1}{T_2}(v^{\rm ref} - v_m) + v_{f0})(1 + s_0(\frac{v_h}{v_0} - 1)) - \tilde{v}_f)/T_e$$

where v_h is the generator terminal voltage or a remote-bus regulated voltage. The initial field voltage v_{f0} and bus voltage v_0 are set during the synchronous machine initialization step. The field voltage $\tilde{v}_f$ is subjected to an anti-windup limiter. Table 16.5 defines all parameters required by the AVR Type III. If s_0 is set to 1, the signal v_h/v_0 is enabled.

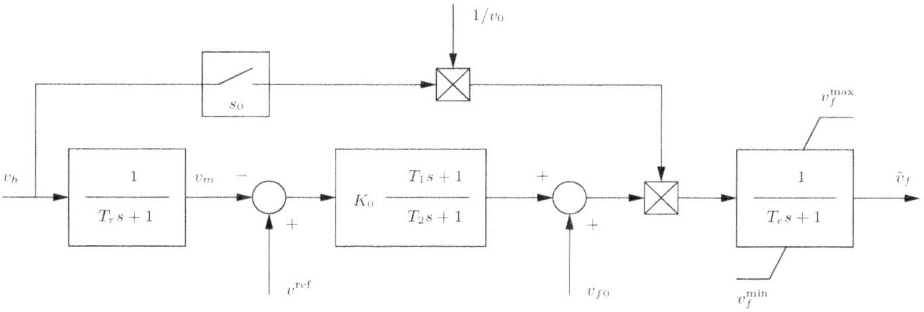

Fig. 16.11 Automatic voltage regulator Type III control diagram

Table 16.5 Automatic voltage regulator Type III parameters

Variable	Description	Unit
-	Generator code	-
-	Regulated voltage code	-
K_0	Regulator gain	pu/pu
s_0	Bus voltage signal	$\{0, 1\}$
T_1	Regulator zero	s
T_2	Regulator pole	s
T_e	Field circuit time constant	s
T_r	Measurement time constant	s
v_f^{max}	Maximum field voltage	pu
v_f^{min}	Minimum field voltage	pu

Example 16.2 Effect of Automatic Voltage Regulation on Synchronous Machine Bus Voltage

Figure 16.12 shows the effect of the automatic voltage regulation on the bus voltage of synchronous machine 1 of the IEEE 14-bus system. The transient refers to line 2-4 outage at $t = 1$ s. As expected, the system without primary voltage regulation does not recover the desired voltage values after the disturbance.

Figure 16.14 shows the transient following the same disturbance but with a 20% load increase.[3] For this loading level, the line outage leads to an unstable equilibrium point, as shown in Figure 16.13. In fact, as discussed in Example 8.9 of Chapter 8, a Hopf bifurcation occurs when increasing the loading level. In summary, the response of the system without AVRs is poor but, depending on the loading level, the inclusion of AVRs can lead to instability. Thus, it is required to include additional controllers to improve the system transient behavior. A solution is provided by power system stabilizers that are described in the following section.

[3] Loads are modelled as constant powers and are switched to constant impedances for low bus voltage magnitude, i.e., $v_h < 0.8$ pu.

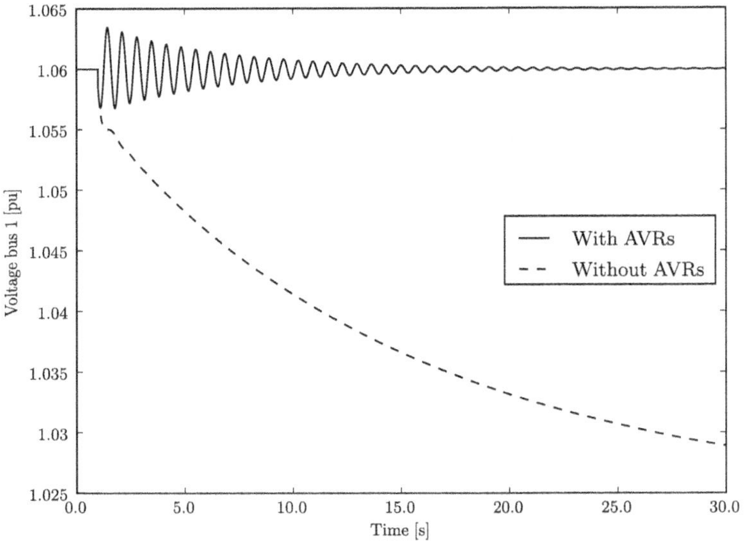

Fig. 16.12 Effect of automatic voltage regulation on synchronous machine bus voltage for the IEEE 14-bus system (100% loading level)

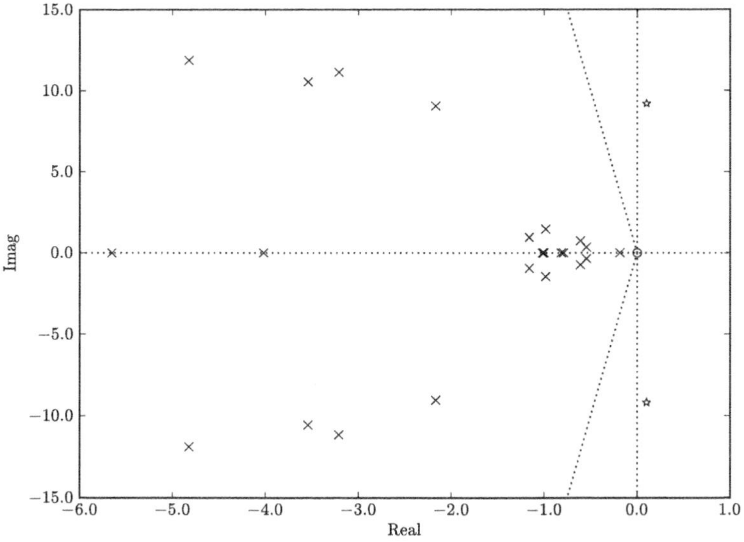

Fig. 16.13 Eigenvalue loci for 120% loading level and line 2-4 outage for the IEEE 14-bus system

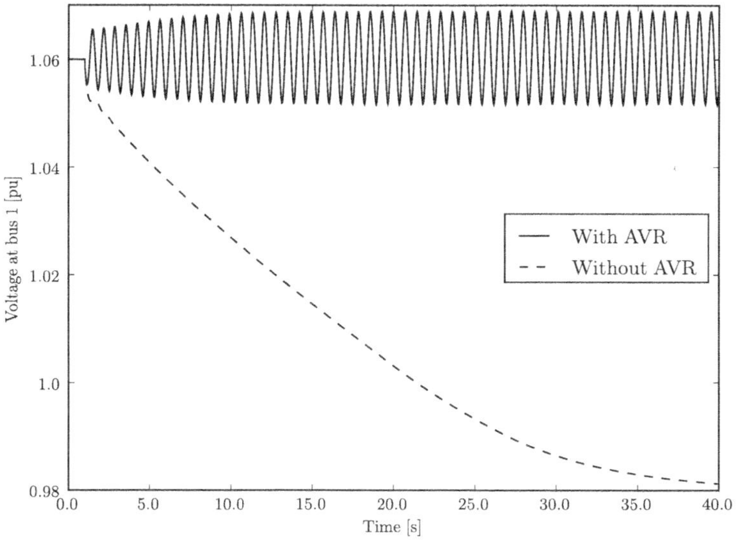

Fig. 16.14 Effect of automatic voltage regulation on synchronous machine bus voltage for the IEEE 14-bus system (120% loading level)

16.3 Power System Stabilizer

Power System Stabilizers (PSSs) are used for damping power system oscillations. Although several PSS models have been proposed in the literature, the rationale behind the PSS functioning is the same for all control schemes. To explain the basic functioning of PSSs, consider a simple electromechanical model of the synchronous machine with no damping:

$$2H\frac{d\omega}{dt} = p_m - p_e(\delta) \tag{16.23}$$

where:

$$p_e(\delta) = \frac{e'_q v_h}{x'_d}\sin(\delta - \theta_h) \tag{16.24}$$

differentiating the above expression and assuming a constant mechanical power p_m leads to:

$$2Hs\Delta\omega = -\frac{\partial p_e}{\partial \delta}\Delta\delta - \frac{\partial p_e}{\partial e'_q}\Delta e'_q - \frac{\partial p_e}{\partial v_h}\Delta v_h \tag{16.25}$$

If e'_q and v_h are constant, one has:

$$2Hs\Delta\omega = -\frac{\partial p_e}{\partial \delta}\Delta\delta = -k\Delta\delta \tag{16.26}$$

where

$$k = \frac{e'_q v_h}{x'_d} \cos(\delta_0 - \theta_0) \tag{16.27}$$

Since $\Delta\omega = s\Delta\delta$:

$$2Hs^2\Delta\delta + k\Delta\delta = 0 \tag{16.28}$$

which has a pair of pure complex eigenvalues (no damping):

$$\lambda_{1,2} = \pm j\sqrt{\frac{k}{2H}} \tag{16.29}$$

The PSS allows imposing

$$\Delta v_h = k_1 \Delta\delta \tag{16.30}$$

Hence, one has:

$$2Hs\Delta\omega = -k\Delta\delta - k_\omega s\Delta\delta \tag{16.31}$$

where

$$k_\omega = k_1 \frac{\partial p_e}{\partial e'_q} \tag{16.32}$$

and, finally:

$$2Hs^2\Delta\delta + k_\omega s\Delta\delta + k\Delta\delta = 0 \tag{16.33}$$

which has a pair of complex eigenvalues with negative real part:

$$\lambda_{1,2} = -\frac{k_\omega}{4H} \pm j\frac{\sqrt{8kH - k_\omega^2}}{4H} \tag{16.34}$$

In practice (16.30) is obtained by introducing a feedback signal proportional to the active power or rotor frequency into the primary voltage control loop. Typical PSS input signals are the rotor speed ω, the active power p_h and also the bus voltage v_h of the generator to which the PSS is connected through the automatic voltage regulator. The PSS output signal is a signal v_s that modifies the reference voltage $v_{\rm ref}$ of the AVR.

In the following subsections, four typical PSS models are described. Except for the simple model described in Subsection 16.3.1, other models have two algebraic equations, as follows:

$$0 = g_s(\boldsymbol{x}_i, \hat{\boldsymbol{y}}_i) - v_s \tag{16.35}$$

$$0 = v_{\rm ref}^0 - v_{\rm ref} + v_s \tag{16.36}$$

where (16.35) defines the PSS signal v_s, and (16.36) sums the signal v_s to the AVR reference voltage (see also (16.10)). All PSS parameters used in the following subsections are defined in Table 16.6.

Table 16.6 Power system stabilizer parameters

Variable	Description	Unit
-	AVR code	-
K_p	Gain for active power	pu/pu
K_v	Gain for bus voltage magnitude	pu/pu
K_w	Stabilizer gain	pu/pu
T_1	First stabilizer time constant	s
T_2	Second stabilizer time constant	s
T_3	Third stabilizer time constant	s
T_4	Fourth stabilizer time constant	s
$v_s^{\max}$	Max stabilizer output signal	pu
$v_s^{\min}$	Min stabilizer output signal	pu
T_w	Wash-out time constant	s

16.3.1 Simplified Power System Stabilizer Model

According to the qualitative discussion above, the simplest PSS model is obtained by introducing in the synchronous machine equations a feedback signal that modifies the field voltage:

$$\tilde{v}_f = v_f + K_\omega(\omega - \omega_s) - K_P(p_h(\boldsymbol{x}_i, v_h, \theta_h) - p_0) \tag{16.37}$$

where p_0 is the initial electric power generated by the machine. The modified field voltage $\tilde{v}_f$ can substitute v_f in equations (15.13), (15.16), (15.19), (15.22), (15.24), (15.27), (15.29), (15.31) or (15.33) defined in Section 15.1 of Chapter 15.

16.3.2 Power System Stabilizer Type I

PSS Type I is depicted in Figure 16.15, and is described by the following differential equation:

$$\begin{aligned} \dot{v}_1 &= -(K_w\omega + K_p p_h + K_v v_h + v_1)/T_w \\ v_s &= K_w\omega + K_p p_h + K_v v_h + v_1 \end{aligned} \tag{16.38}$$

where ω, p_h and v_h are the rotor speed, the active power and the voltage magnitude of the generator to which the PSS is connected through the AVR.

16.3.3 Power System Stabilizer Type II

The PSS Type II is depicted in Figure 16.16, and is described by the equations:

$$\begin{aligned}\dot{v}_1 &= -(K_w v_{SI} + v_1)/T_w \\ \dot{v}_2 &= ((1 - \frac{T_1}{T_2})(K_w v_{SI} + v_1) - v_2)/T_2 \\ \dot{v}_3 &= ((1 - \frac{T_3}{T_4})(v_2 + (\frac{T_1}{T_2}(K_w v_{SI} + v_1))) - v_3)/T_4 \\ v_s &= v_3 + \frac{T_3}{T_4}(v_2 + \frac{T_1}{T_2}(K_w v_{SI} + v_1))\end{aligned} \tag{16.39}$$

Equations (16.39) are an arbitrary choice of DAE that describes the control blocks depicted in Figure 16.16. Another possible formulation is:

$$\begin{aligned}\dot{v}_1 &= (K_w v_{SI} - v_1)/T_w \\ \dot{v}_2 &= (K_w v_{SI} - v_1 - v_2)/T_2 \\ \dot{v}_3 &= ((1 - \frac{T_1}{T_2})v_2 + \frac{T_1}{T_2}(K_w v_{SI} - v_1) - v_3)/T_4 \\ v_s &= (1 - \frac{T_3}{T_4})v_3 + \frac{T_3}{T_4}((1 - \frac{T_1}{T_2})v_2 + \frac{T_1}{T_2}(K_w v_{SI} - v_1))\end{aligned} \tag{16.40}$$

The two sets of DAE (16.39) and (16.40) have same dynamic response and stability properties, but the values of state variables are different.

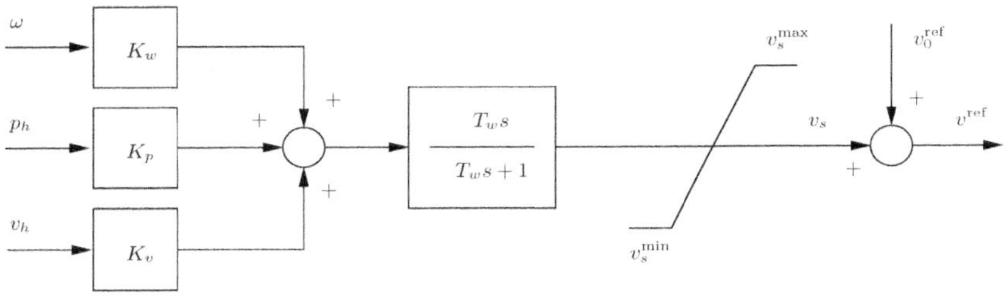

Fig. 16.15 Power system stabilizer Type I control diagram

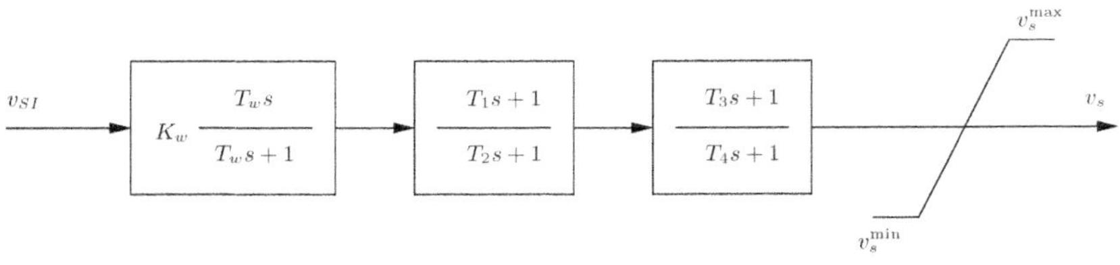

Fig. 16.16 Power system stabilizer Type II control diagram

16.3.4 Power System Stabilizer Type III

The PSS Type III is depicted in Fig. 16.17, and is described by the equations:

$$\begin{aligned}
\dot{v}_1 &= -(K_w v_{SI} + v_1)/T_w \\
\dot{v}_2 &= v_3 \\
\dot{v}_3 &= (K_w v_{SI} + v_1 - T_4 v_3 - v_2)/T_2 \\
v_s &= v_2 + \frac{T_1}{T_2}(K_w v_{SI} + v_1) + (T_3 - \frac{T_1}{T_2}T_4)v_3 + (1 - \frac{T_1}{T_2})v_2
\end{aligned} \tag{16.41}$$

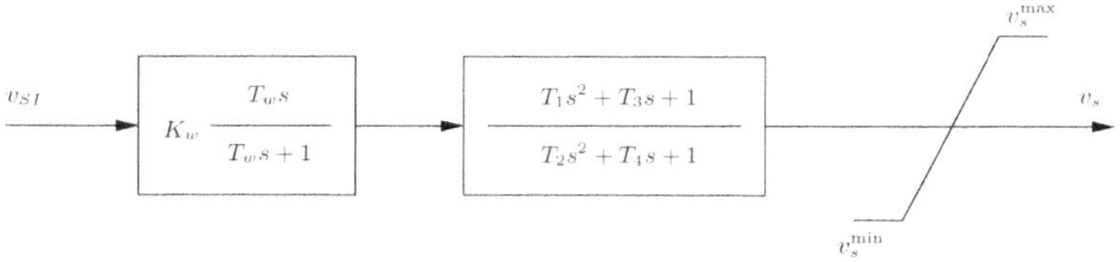

Fig. 16.17 Power system stabilizer Type III control diagram

Example 16.3 Effectiveness of Power System Stabilizers for Removing Hopf Bifurcations

Figure 16.18 shows the eigenvalue loci for the IEEE 14-bus system with a 120% loading level and line 2-4 outage with inclusion of a PSS Type II connected at generator 1. The PSS data are reported in Appendix D. The system is stable and well damped. Thus, in this case, the effect of the PSS is twofold:

1. To remove the Hopf bifurcation.
2. To properly damp the system (compare Figure 16.18 with the eigenvalue loci of Example 7.1 of Chapter 7).

Figure 16.19 compares the transient response of the IEEE 14-bus system with and without PSS at generator 1. As expected from the eigenvalue analysis, the system with PSS is stable and well damped. Furthermore, the PSS allows recovering the generator bus voltage at the desired value. This simulation has to be compared with Figure 16.14 of Example 16.2.

16.4 Over-Excitation Limiter

Over-eXcitation Limiters (OXLs) provide an additional signal v_{OXL} to the reference voltage v_0^{ref} of AVRs [143]. The OXL is modelled as a pure integrator, with anti-windup hard limits (see Figure 16.20). This regulator is generally sleeping, i.e., $v_{\mathrm{OXL}} = 0$, unless the field current is greater than its

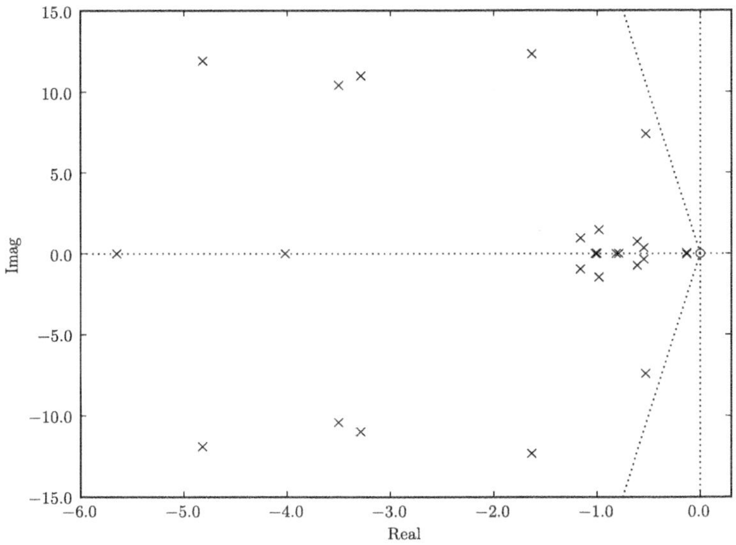

Fig. 16.18 Eigenvalue loci for the IEEE 14-bus system with 120% loading level, line 2-4 outage and a PSS at generator 1

thermal limit ($i_f > i_f^{\rm lim} \approx 2.7$ pu). It is implicitly assumed that at the initial condition given by the power flow solution, all $i_f \leq i_f^{\rm lim}$, thus leading to $v_{\rm OXL} = 0$ at $t = 0$. If the field current exceeds its limits the power flow data are not consistent with dynamic ones.

The output signal $v_{\rm OXL}$ is zero as long as $i_f \leq i_f^{\rm lim}$. If $i_f > i_f^{\rm lim}$, the OXL becomes active and undergoes the following differential equation:

$$\dot{v}_{\rm OXL} = (i_f - i_f^{\rm lim})/T_0 \tag{16.42}$$

In some cases the direct measure of the field current i_f is not available. Thus, the field current has to be estimated using available measures. A possible estimation is:

$$0 = \sqrt{(v_h + \gamma_q)^2 + p_h^2} + \left(\frac{x_d}{x_q} + 1\right) \frac{\gamma_q(v_h + \gamma_q) + \gamma_p^2}{\sqrt{(v_h + \gamma_q)^2 + p_h^2}} - i_f \tag{16.43}$$

where

$$\gamma_p = x_q p_h / v_h$$
$$\gamma_q = x_q q_h / v_h$$

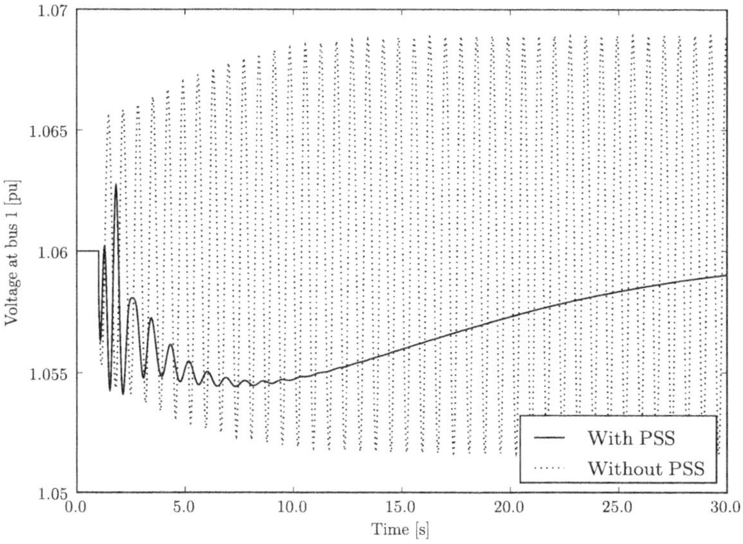

Fig. 16.19 Effect of power system stabilizer on synchronous machine bus voltage for the IEEE 14-bus system (120% loading level)

and v_h is the voltage at the generator bus, and p_h and q_h are the active and the reactive power of the generator, respectively, and all parameters are defined in Table 16.7. Observe that the reactances x_d and x_q of the generator at which the OXL is connected through the AVR are required in (16.43).

The OXL output signal $v_{\rm OXL}$ modifies the reference AVR voltage (see also (16.10)):

$$0 = v_0^{\rm ref} - v^{\rm ref} + v_{\rm OXL} \tag{16.44}$$

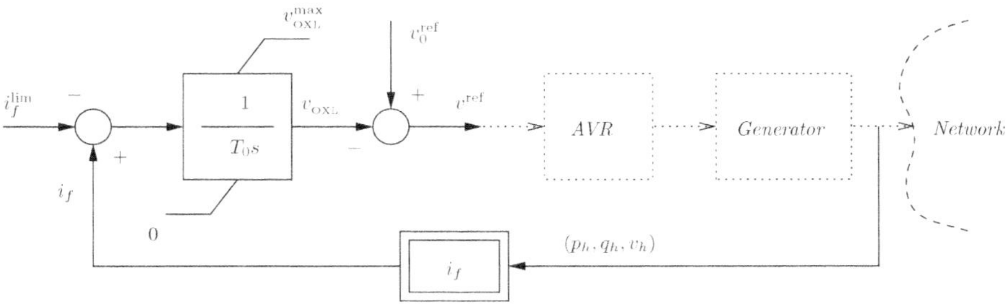

Fig. 16.20 Over-excitation limiter control diagram

Table 16.7 Over-excitation limiter parameters

Variable	Description	Unit
-	AVR code	-
$i_f^{\lim}$	Maximum field current	pu
T_0	Integrator time constant	s
$v_{\mathrm{OXL}}^{\max}$	Maximum output signal	pu
x_d	d-axis estimated generator reactance	pu
x_q	q-axis estimated generator reactance	pu

16.5 Under-Excitation Limiter

Under-eXcitation Limiters (UXL) behave similarly to over-excitation ones, i.e., provide an additional signal to the reference voltage of automatic voltage regulators. Typical UXL schemes take as input signal a measure of generator active and reactive powers and bus voltage as well as the exciter field voltage [144]. A common UXL scheme is shown in Figure 16.21 whereas all parameters are defined in Table 16.8. The complete DAE system is quite lengthy, but it can be easily deduced based on the PSS model Type II described in Section 16.3, which can actually be used as base class for this device.

Similarly to the OXL and the PSS, the UXL output signal v_{UXL} modifies the reference AVR voltage:

$$0 = v_0^{\mathrm{ref}} - v^{\mathrm{ref}} + v_{\mathrm{UXL}} \tag{16.45}$$

The reference value $v_{\mathrm{UXL}0}$ is computed at the initialization step and serves to impose $v_{\mathrm{UXL}} = 0$ at the initial condition.

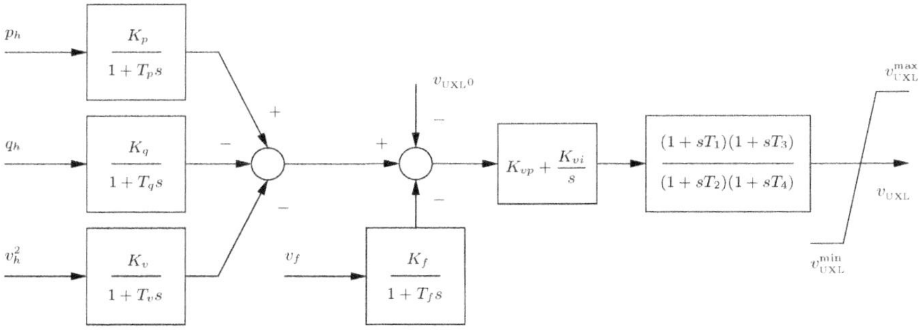

Fig. 16.21 Under-excitation limiter control diagram

Table 16.8 Under-excitation limiter parameters

Variable	Description	Unit
-	AVR code	-
K_f, T_f	Active power measurement lag	pu/pu, s
K_p, T_p	Active power measurement lag	pu/pu, s
K_q, T_q	Reactive power measurement lag	pu/pu, s
K_v, T_v	Field voltage measurement lag	pu/pu, s
K_{vp}, K_{vi}	PI controller gains	pu/pu, pu/pu/s
T_1, T_2, T_3, T_4	Controller time constants	s
$v_{\mathrm{UXL}}^{\max}$	Maximum output signal	pu
$v_{\mathrm{UXL}}^{\min}$	Minimum output signal	pu

Chapter 17
Direct-Current Devices

This chapter describes dc devices used for modelling hybrid electro-magnetic and electro-mechanical power system models. The devices included in this chapter are: dc nodes (Section 17.1), common interface equations for dc devices (Section 17.2), ideal generators (Section 17.3) and basic components such as RLC circuits (Section 17.4), dc machines (Section 17.5), and unconventional dc generators, namely solid oxide fuel cell, solar photovoltaic cell and energy battery (Section 17.6).

17.1 Direct-Current Nodes

Like ac buses, dc nodes has only a topological function. Dc nodes also serve for defining the base dc voltages $V_{\mathrm{dc},b}$ that are used for computing system pu values of dc devices. Table 17.1 defines all dc node parameters.

Table 17.1 DC node parameters

Variable	Description	Unit
-	Node code	-
v_0	Initial voltage guess	pu
$V_{\mathrm{dc},n}$	Dc voltage rating	kV

17.2 Common Interface Equations for Direct-Current Devices

Each dc node introduces a new variable (i.e., the node voltage $v_{\mathrm{dc},h}$) and an equation (i.e., the current balance at that node). The approach used in this section and in the following Chapter 18 for modelling dc devices is the current injection model. As discussed in Chapter 9, the balance equations for the dc system are:

F. Milano: Power System Modelling and Scripting, Power Systems, pp. 379–394.
springerlink.com

$$0 = \sum_{i \in \Omega_h} i_{\mathrm{dc}h,i}(\boldsymbol{x}, \hat{\boldsymbol{y}}, \boldsymbol{v}), \quad h \in \mathcal{N} \tag{17.1}$$

where $i_{\mathrm{dc}h,i}$ is the current injected at node h by device i and $\mathcal{N}$ is the set of nodes of the dc network.

Since all devices described in this section (except for the ground element described in following Section 17.3) connects two nodes, the dc equation interfaces can be standardized in a common class. The general two-node dc device is depicted in Figure 17.1. Using the generator convention, the dc network interface equations are:

$$\begin{aligned} 0 &= v_{\mathrm{dc},h} - v_{\mathrm{dc},k} - v_{\mathrm{dc}}(\boldsymbol{x}_i, \hat{\boldsymbol{y}}_i) \\ i_{\mathrm{dc},h} &= i_{\mathrm{dc}}(\boldsymbol{x}_i, \hat{\boldsymbol{y}}_i) \\ i_{\mathrm{dc},k} &= -i_{\mathrm{dc}}(\boldsymbol{x}_i, \hat{\boldsymbol{y}}_i) \end{aligned} \tag{17.2}$$

With this simple interface, completing a device model only requires defining $v_{\mathrm{dc}}(\boldsymbol{x}_i, \hat{\boldsymbol{y}}_i)$ and $i_{\mathrm{dc}}(\boldsymbol{x}_i, \hat{\boldsymbol{y}}_i)$ along with one differential equation per each element of the internal state variables $\boldsymbol{x}_i$ and one algebraic equation per each element of the internal algebraic variables $\hat{\boldsymbol{y}}_i$.

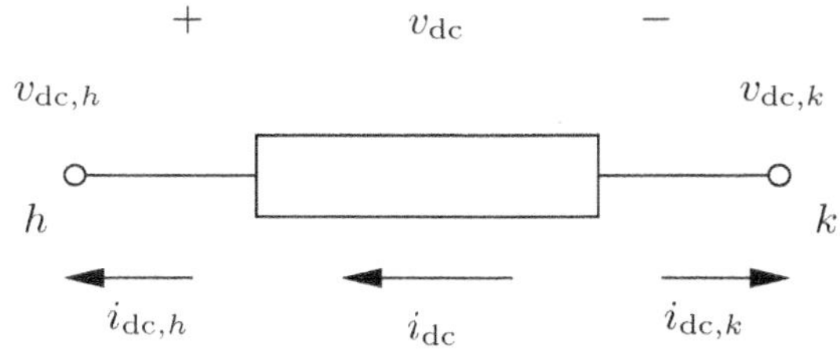

Fig. 17.1 General dc device voltages and currents

This approach does not distinguish between series and shunt devices and allows a high grade of flexibility because the same device (i.e., same programming code) can be used both in series and shunt configurations. This is not the case of ac devices described so far. A pure inductive shunt admittance and a pure reactive transmission line are the same electrical element (i.e., a reactance), but require two different classes for being defined. The key point is that in the ac network the ground bus is implicitly used by all shunt devices. Hence, the proposed dc device modelling approach requires to explicitly define the ground dc node. The different approach used for ac and dc networks has a rationale. For ac networks, defining the ground bus does not provide a significant advantage from the implementation viewpoint since shunt and series devices are conceptually different devices with quite different functioning. On the other hand, in dc networks, most devices can be connected in series or in parallel (e.g., photovoltaic cell grids).

Note on Unit Notation

In this chapter and in the following Chapter 18, the values of dc currents and voltages are considered in pu with respect to the device rated current and voltage, respectively. As discussed in the chapter Notation at the beginning of the book, a dc per-unit system is used for analogy with the ac per-unit system. To maintain consistency, the relative values of resistances, inductances and capacitances are computed as:

$$R_{(\mathrm{pu})} = \frac{R_{(\Omega)}}{R_{\mathrm{dc},n}}, \quad L_{(\mathrm{s})} = \frac{L_{(H)}}{R_{\mathrm{dc},n}}, \quad C_{(\mathrm{s})} = C_{(F)} R_{\mathrm{dc},n} \tag{17.3}$$

where the nominal resistance is computed using the nominal power and voltage device ratings, i.e., $R_{\mathrm{dc},n} = S_n/V_{\mathrm{dc},n}$. In this way, $L_{(\mathrm{s})}/R_{(\mathrm{pu})}$ and $C_{(\mathrm{s})}R_{(\mathrm{pu})}$ are in seconds, as one expects. Whenever absolute values are needed, per-unit quantities are multiplied by the nominal device quantities. For example, to indicate the power in MW, the notation $S_n v_{\mathrm{dc}} i_{\mathrm{dc}}$ will be used.

17.3 Ideal Generators

Ideal generators are modelled as assigned functions of the time and can be voltage or current sources. A special case of ideal generator is the ground device, that fixes the voltage reference for a given dc network.

Ideal Independent Sources

Given the interface (17.2), the equation that defines an ideal independent current source is simply:

$$0 = i_{\mathrm{dc}} - i^{\mathrm{des}} \tag{17.4}$$

where i^{des} is the current imposed by the generator. In this case, since the current injection is imposed, the equation for the voltage v_{dc} (i.e., the first of (17.2)) is not required.

Analogously, the equation that defines an ideal independent voltage source is:

$$0 = v_{\mathrm{dc}} - v^{\mathrm{ref}} \tag{17.5}$$

where v^{des} is the voltage imposed by the generator. The current i_{dc} is needed as an explicit variable due to the current injection model (17.2).

Ground

The ground works for dc networks similarly to the reference angle in ac ones. Since the ground imposes an absolute value to a node voltage, the interface (17.2) is not required. Similarly to the definition of the reference angle discussed in Section 10.2.2 of Chapter 10, there are at least three possible models, as follows.

1. To assign to the ground node a fixed value, thus removing the voltage at that node from the vector of algebraic variables. This solution reduces the size of the resulting DAE system, but complicates the implementation since it cannot be easily generalized.
2. To leave the ground voltage as a variable, but "freezing" it through setting to zero the row and the column of the Jacobian matrix corresponding to that variable. This method can be easily generalized.
3. To leave the ground voltage as a variable and to include an additional algebraic variable for imposing the desired ground voltage value. This method is very general and do not require manipulating the system Jacobian matrix. Equations are:

$$\begin{aligned} i_{\mathrm{dc},h} &= -i_{\mathrm{ground}} \\ 0 &= v_{\mathrm{dc},h} - v_{\mathrm{ground}} \end{aligned} \tag{17.6}$$

where i_{ground} is a dummy variable and v_{ground} is the desired ground voltage. Typically $v_{\mathrm{ground}} = 0$.

17.4 Basic RLC Models

For the sake of example, this section provide the models of the basic circuit elements depicted in Figure 17.2. The resistance, inductance and capacitance models have to be compared with those based on the Dommel's method (see Section 8.5 of Chapter 8). The main differences are that (i) the approach used in this chapter does not depend on the integration step length and (ii) does not require any hypothesis about the linearity of the devices. Table 17.2 summarizes the parameters required for basic *RLC* elements.

Table 17.2 RLC parameters

Variable	Description	Unit
R	Resistance	pu
L	Inductance	s
C	Capacitance	s

- *Resistance* (Figure 17.2.a):

$$0 = v_{\mathrm{dc}}/R + i_{\mathrm{dc}} \tag{17.7}$$

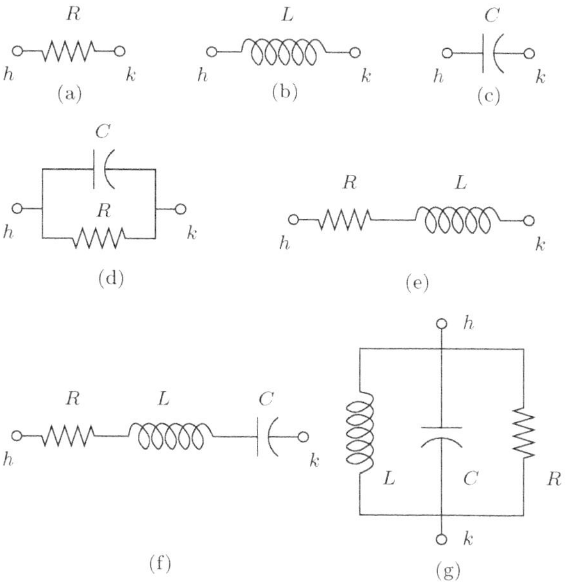

Fig. 17.2 RLC circuits

- *Inductance* (Figure 17.2.b):

$$\begin{aligned} \dot{i}_{\rm L} &= v_{\rm dc}/L \\ 0 &= i_L + i_{\rm dc} \end{aligned} \tag{17.8}$$

- *Capacitance* (Figure 17.2.c):

$$\begin{aligned} \dot{v}_{\rm C} &= -i_{\rm dc}/C \\ 0 &= v_C - v_{\rm dc} \end{aligned} \tag{17.9}$$

- *RC parallel* (Figure 17.2.d):

$$\begin{aligned} \dot{v}_{\rm C} &= -(i_{\rm dc} + v_C/R)/C \\ 0 &= v_C - v_{\rm dc} \end{aligned} \tag{17.10}$$

- *RL series* (Figure 17.2.e):

$$\begin{aligned} \dot{i}_{\rm L} &= (v_{\rm dc} - Ri_L)/L \\ 0 &= i_L + i_{\rm dc} \end{aligned} \tag{17.11}$$

- *RLC series* (Figure 17.2.f):

$$
\begin{aligned}
\dot{i}_L &= (v_{\rm dc} - R i_L - v_C)/L \\
\dot{v}_C &= i_L/C \\
0 &= i_L + i_{\rm dc}
\end{aligned}
\tag{17.12}
$$

- *RLC parallel* (Figure 17.2.g):

$$
\begin{aligned}
\dot{v}_C &= -(i_{\rm dc} + v_C/R + i_L)/C \\
\dot{i}_L &= v_C/L \\
0 &= v_C - v_{\rm dc}
\end{aligned}
\tag{17.13}
$$

Along with the Voltage Source Converter (VSC) described in the following Chapter 18, these basic elements can provide a versatile tool for setting up composite devices. For example, a SMES is an inductance connected to an ac network through a VSC. Furthermore, VSC-based FACTS devices are composed of a VSC or two back-to-back VSCs with a capacitance on the dc side. These devices are described in Chapter 19.

17.5 Direct-Current Machines

A power system book of XXth century would have dedicated an entire chapter to direct-current machines. Unfortunately, these beautiful (from the modelling viewpoint) machines are not very common in these years, although a renewed interest could come from some distributed energy resource applications. However, I believe that a book on power system devices is not complete if it does not include at least a brief outline of basic dc machine models and configurations.

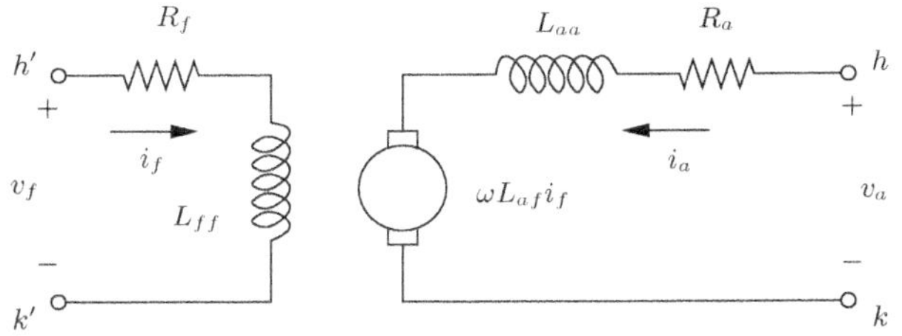

Fig. 17.3 Basic dc machine equivalent circuit

Figure 17.3 shows the basic equivalent circuit of a direct-current machine. The following notation assumes that the machine is working as a motor. The dc machine DAE system is [160]:

$$\begin{aligned}\dot{i}_f &= (v_f - R_f i_f)/L_{ff} \\ \dot{i}_a &= (v_a - R_a i_a - L_{af} i_f \omega)/L_{aa} \\ \dot{\omega} &= (\tau_e - \tau_m - D\omega)/2H \\ 0 &= L_{af} i_f i_a - \tau_e\end{aligned} \tag{17.14}$$

where i_a and i_f are the armature and field currents, respectively, v_a and v_f are the armature and field voltages, respectively, ω is the rotor angular speed, τ_e is the electrical torque, and all other parameters are defined in Table 17.3. The coefficient D models frictions and windage losses, hence $D \ll 1$ pu for $\tau_e \approx 1$ pu.

Table 17.3 Direct-current machine parameters

Variable	Description	Unit
D	Coefficient for frictions and windage losses	pu
H	Machine inertia constant	MWs/MVA
L_{aa}	Armature winding self-inductance	s
L_{af}	Mutual field-armature inductance	s
† L_{as}	Mutual series field-armature inductance	s
L_{ff}	Field winding self-inductance	s
† L_{fs}	Mutual series-shunt field windings inductance	s
† L_{ss}	Series field winding self-inductance	s
R_a	Armature winding resistance	pu
R_f	Field winding resistance	pu
† R_s	Series field winding resistance	pu
τ_m	Mechanical torque	pu

† Parameters required only for the compound-connected dc machine.

The typical configurations of the dc machine described above are: (i) separate winding excitation, (ii) shunt connection, (iii) series connection and (iv) compound connection. The following items provide the interface equations that make each connection type compliant with (17.2).

Separate Winding Connection

The machine is connected to two pairs of dc nodes, i.e. h-k for the armature winding and h'-k' for the field winding (see Figure 17.3). The voltage and current pairs $(v_{\rm dc}, i_{\rm dc})$ and $(v'_{\rm dc}, i'_{\rm dc})$ are associated with the nodes h-k and h'-k', respectively. Thus, for the armature winding:

$$v_{\rm dc} = v_a, \quad i_{\rm dc} = -i_a \tag{17.15}$$

and for the field winding:

$$v'_{\rm dc} = v_f, \quad i'_{\rm dc} = -i_f \tag{17.16}$$

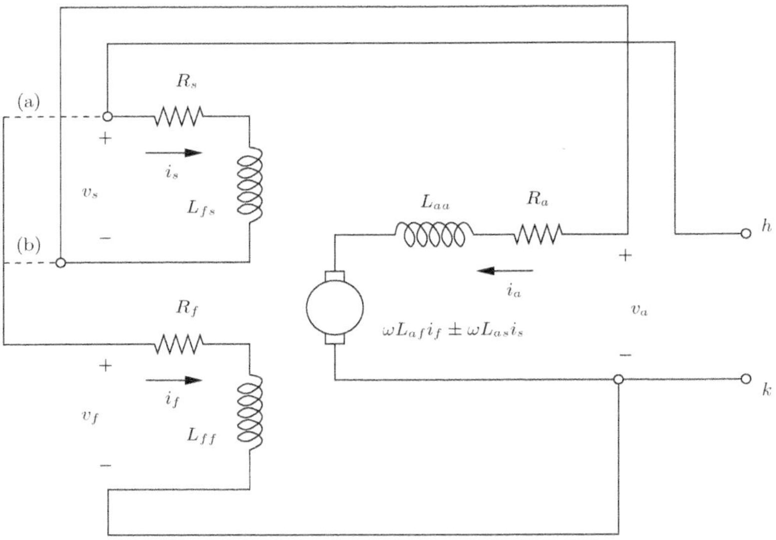

Fig. 17.4 Compound-connected dc machine equivalent circuit: (a) shunt field connected ahead the series field, and (b) shunt field connected behind the series field

Shunt-Connected DC Machine

The armature and the field windings are connected in parallel (i.e., $h \equiv h'$ and $k \equiv k'$). Hence:

$$v_{\rm dc} = v_a = v_f, \quad i_{\rm dc} = -i_a - i_f \tag{17.17}$$

Series-Connected DC Machine

The armature and the field windings are connected in series (i.e., $k \equiv h'$). Hence:

$$v_{\rm dc} = v_a + v_f, \quad i_{\rm dc} = -i_a = -i_f \tag{17.18}$$

Compound-Connected DC Machine

In this case, the machine is equipped with two field windings (see Figure 17.4). The first winding is connected in series, while the second one in parallel with the armature winding. Assuming that v_f and i_f are the voltage and the current, respectively, for the shunt-connected field winding, and v_s and i_s are the voltage and the current, respectively, for the series-connected field winding, the DAE system becomes:

$$
\begin{aligned}
\frac{d}{dt}\begin{bmatrix} i_f \\ i_s \\ i_a \end{bmatrix} &= \begin{bmatrix} L_{ff} & \pm L_{fs} & 0 \\ \pm L_{fs} & L_{ss} & 0 \\ 0 & 0 & L_{aa} \end{bmatrix}^{-1} \\
&\left[\begin{bmatrix} v_f \\ v_s \\ v_a \end{bmatrix} - \begin{bmatrix} R_f & 0 & 0 \\ 0 & R_s & 0 \\ \omega L_{af} & \pm\omega L_{as} & R_a \end{bmatrix}\begin{bmatrix} i_f \\ i_s \\ i_a \end{bmatrix}\right] \\
\dot{\omega} &= (\tau_e - \tau_m - D\omega)/2H \\
0 &= L_{af} i_f i_a \pm L_{as} i_s i_a - \tau_e
\end{aligned}
\tag{17.19}
$$

where the $\pm$ sign indicates either a cumulative or a differential series connection. Finally, the dc-network interface constraints depend on the field circuit connections, as follows.

(a) Shunt field connected ahead the series field:

$$
v_{\mathrm{dc}} = v_s + v_a = v_f, \quad i_{\mathrm{dc}} = -i_s - i_f, \quad i_a = i_s \tag{17.20}
$$

(b) Shunt field connected behind the series field:

$$
v_{\mathrm{dc}} = v_s + v_a = v_s + v_f, \quad i_{\mathrm{dc}} = -i_s, \quad i_a = i_s - i_f \tag{17.21}
$$

17.6 Other Direct-Current Devices

This section describes three models of nonlinear dc devices, namely the solid oxide fuel cell, the solar photovoltaic cell and the energy battery. These devices are of growing interest for distributed and/or renewable resource generation and energy storage. Furthermore, they have interesting nonlinear DAE models, which is enough for being included in this section.

17.6.1 Solid Oxide Fuel Cell

Fuel cells are a promising technology for producing electrical energy. The main issues that complicate the design of efficient and robust fuel cells are related to electrode heating and corrosion. However, fuel cells are expected to play an important role in distributed generation.

A Solid Oxide Fuel Cell (SOFC) model described in this section is based on what was proposed in [124, 159, 226, 273, 362]. Figure 17.5 depicts the fuel cell scheme, which is based on the following equations:

$$
\begin{aligned}
\dot{\Theta} &= \frac{1}{m_g c_p}(Q_e - h_c A_c(\Theta - \Theta_a) - \sigma\varepsilon A_r(\Theta^4 - \Theta_a^4)) \\
\dot{p}_{H_2} &= ((q_{H_2} - 2K_r i_{\mathrm{dc}})/K_{H_2} - p_{H_2})/T_{H_2} \\
\dot{p}_{H_2O} &= (2K_r i_{\mathrm{dc}}/K_{H_2O} - p_{H_2O})/T_{H_2O} \\
\dot{p}_{O_2} &= ((q_{H_2}/r_{HO} - K_r i_{\mathrm{dc}})/k_{O_2} - p_{O_2})/T_{O_2} \\
\dot{q}_{H_2} &= (2K_r i_{\mathrm{dc}}/U_{opt} - q_{H_2})/T_f \\
0 &= -v_{\mathrm{dc}} - R_{\mathrm{dc}}(\Theta) i_{\mathrm{dc}} + \frac{N_0}{V_{\mathrm{dc},n}}(E_0 + \frac{\mathrm{r}\Theta}{2\mathrm{f}} \ln(p_{H_2}\sqrt{p_{O_2}}/p_{H_2O}))
\end{aligned}
\tag{17.22}
$$

Table 17.4 Solid oxide fuel cell parameters

Variable	Description	Unit
A_c	Cell effective convection area	m^2
A_r	Cell effective radiation area	m^2
c_p	Average cell specific heat	J/kg/K
E_0	Ideal standard potential	V
h_c	Convection-cooling coefficient	$W/K/m^2$
K_{H_2}	Valve molar constant for hydrogen	-
K_{H_2O}	Valve molar constant for water	-
K_{O_2}	Valve molar constant for oxygen	-
$\tilde{K}_r$	Mole flow-dc current coefficient	mol/C
m_g	Cell mass	kg
N_0	Number of cells in series in the stack	int.
Q_e	Heat generated by the electrochemical reaction	W
$R^a_{\rm dc}$	Ohmic losses at ambient temperature	pu
r_{HO}	Ratio of hydrogen to oxygen	-
T_f	Fuel processor response time	s
T_{H_2}	Response time for hydrogen flow	s
T_{H_2O}	Response time for water flow	s
T_{O_2}	Response time for oxygen flow	s
U_{opt}	Optimal fuel utilization	-
β_r	Ohmic loss temperature factor	-
ε	Emittance	-
Θ_a	Ambient temperature	K

where $\sigma = 5.670 \cdot 10^{-8}$ W/m^2/K^4 is the Stefan-Boltzmann's constant. The first equation defines the thermodynamic energy balance, while second to fifth equations defines the electrochemical reaction dynamics and the last equation defines the fuel cell voltage. In (17.22), p_{H_2}, p_{O_2} and p_{H_2O} are the hydrogen, oxygen and water mole fractions, respectively, q_{H_2}, q_{O_2} and q_{H_2O} are the hydrogen, oxygen and water flows, respectively, r is the gas constant (r = 8.314 J/mol/K), f is the Faraday constant (f = 96487 C/mol), and the remaining quantities are defined in Table 17.4. The coefficient K_r depends on the number of electrons n_e in the reaction, the Faraday f constant and the current rating $I_{{\rm dc},n} = S_n/V_{{\rm dc},n}$, as follows:

$$K_r = \tilde{K}_r I_{{\rm dc},n} = \frac{n_e I_{{\rm dc},n}}{4{\rm f}} \tag{17.23}$$

The ohmic losses modelled through the resistance $R_{\rm dc}$ are due to the resistance to the flow of ions in the electrolyte and resistance to the flow of electrons through the electrode materials. The resistance depends on the temperature Θ:

$$R_{\rm dc} = R^a_{\rm dc} e^{\beta_r \left(\frac{1}{\Theta_a} - \frac{1}{\Theta}\right)} \tag{17.24}$$

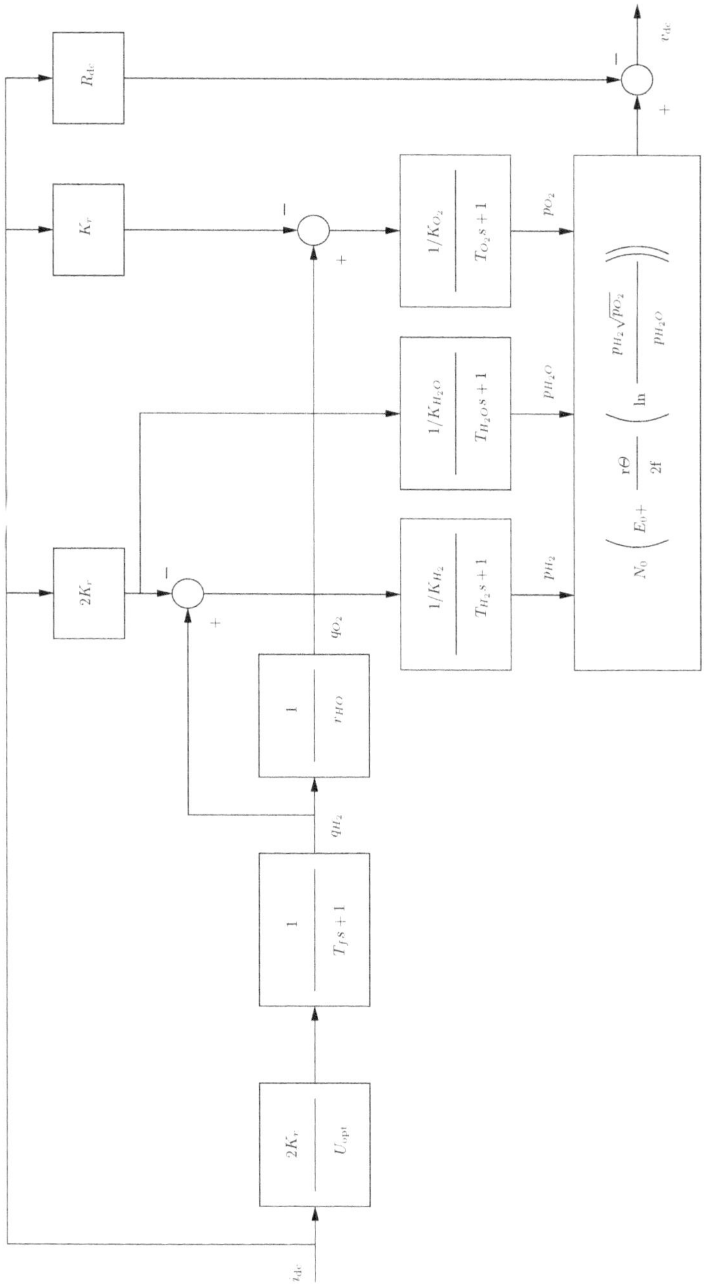

Fig. 17.5 Solid oxide fuel cell scheme

17.6.2 Solar Photovoltaic Cell

Despite their high cost and low efficiency, solar photovoltaic cells are gaining in recent years popularity, thanks to the growing interest in renewable sources and, especially, thanks to public funding. At the moment, the biggest photovoltaic plant capacity is 60 MW,[1] but the vast majority of solar plants have a capacity ≤ 0.2 MW. However, at least from the modelling viewpoint, photovoltaic plants are quite interesting [174, 177].

The electrical circuit is described by the following equations (see Figure 17.6):

$$
\begin{aligned}
i_{\rm dc} &= i_L - i_D - \frac{v_D}{R_{\rm sh}} \\
0 &= v_D - v_{\rm dc} - R_{\rm Se} i_{\rm dc} \\
0 &= i_s(\Theta)(e^{v_D/(\gamma v_\Theta(\Theta))} - 1) - i_D
\end{aligned}
\tag{17.25}
$$

where i_L is the photo-current generated by sunlight, v_D and i_D are the PN junction voltage and current, respectively, v_Θ is the thermal potential, i_s is the reverse saturation current, and remaining parameters are defined in Table 17.5. The variables v_Θ and i_s can be expressed as:

$$
\begin{aligned}
v_\Theta(\Theta) &= \frac{k_B \Theta}{q_\varepsilon} \\
i_s(\Theta) &= i_{s0} \left(\frac{\Theta}{\Theta_a}\right)^3 e^{\varphi(\Theta)}
\end{aligned}
\tag{17.26}
$$

where Θ is the cell temperature and the function $\varphi(\Theta)$ is:

$$
\varphi(\Theta) = \frac{q_\varepsilon}{\gamma k_B} \left(\frac{\mathscr{E}_g(\Theta_a)}{\Theta_a} - \frac{\mathscr{E}_g(\Theta)}{\Theta} \right)
\tag{17.27}
$$

and $k_B = 1.381 \cdot 10^{-23}$ J/K is the Boltzmann's constant, , $q_\varepsilon = 1.602 \cdot 10^{-19}$ C is the electron charge, and $\mathscr{E}_g$ is the energy band gap, which is a function of the temperature:

$$
\mathscr{E}_g(\Theta) = \mathscr{E}_{g0} - \frac{\alpha_g \Theta^2}{\beta_g + \Theta}
\tag{17.28}
$$

The light-generated current can be linearized around a temperature of 298 K and an irradiance of 1000 W/m^2:

$$
0 = (A_a \rho_e G + C_\Theta(\Theta - 298.0)) \frac{G}{1000} - \frac{S_n}{V_{{\rm dc},n}} i_L
\tag{17.29}
$$

where G is the solar irradiance or *insolation* in W/m^2.

---

[1] This plant was completed in 2008 and is located in Olmedilla de Alarcón, Castilla-La Mancha, Spain.

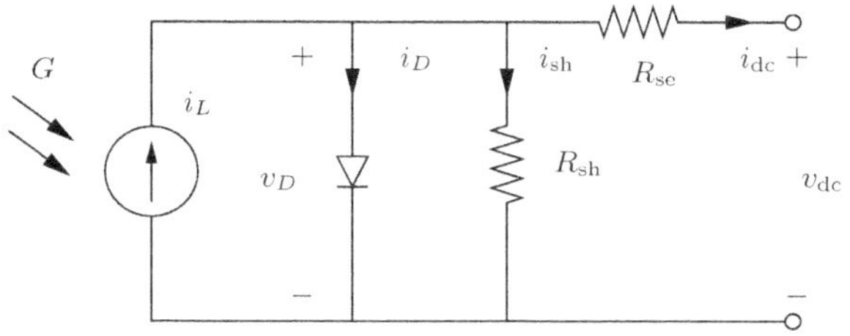

Fig. 17.6 Equivalent circuit of photovoltaic cells

Finally, the model is completed by an energy-balance differential equation that regulates the cell heat transfer with the ambient:

$$\dot{\Theta} = \frac{1}{c_p m_g}\Big(S_n \frac{(v_{dc} - v_D)^2}{R_{se}} + S_n \frac{v_D^2}{R_{sh}} + S_n i_D v_D \\ + (1 - \rho_c - \tau_c - \eta_c) A_a G - h_c A_c (\Theta - \Theta_a) - \sigma \varepsilon A_r (\Theta^4 - \Theta_a^4)\Big) \tag{17.30}$$

where $\sigma = 5.670 \cdot 10^{-8}$ W/m^2/K^4 is the Stefan-Boltzmann's constant and η_c is the cell power conversion efficiency:

$$\eta_c = \frac{S_n v_{dc} i_{dc}}{G A_a} \tag{17.31}$$

The model described above represents a single cell. A panel is composed of a grid of cells connected in series and in parallel. Then a plant is composed of a series of several panels. However, the model above can be also used for representing the entire plant by using proper equivalent parameters. The photovoltaic cell (detailed or equivalent model) has then to be connected to the ac network. This is generally done through a VSC device and proper controllers. An example of controllers are given in Example 18.2 of Chapter 18.

In this model, the solar irradiance G is the input variable since it depends on time t, season, atmospheric conditions, dirtiness of the cell surface, etc. The solar irradiance can be provided as a series of measurements (i.e., (G, t)-value pairs) or approximated using mathematical models [33].

17.6.3 Battery Energy System

Storing electrical energy is a hard task. Since capacitors are far from behaving similarly to an ideal capacitance, the only effective way to store electrical energy is to convert it in another form of energy. For example, batteries are able to store chemical energy, pumping hydro plants store water potential energy, and flywheels store kinetic energy.[2]

[2] Example 18.3 of Chapter 18 provides another example of energy storage device based on a superconducting coil (SMES). In this case the electrical energy is converted into magnetic one.

Table 17.5 Solar photovoltaic cell parameters

Variable	Description	Unit
A_a	Cell active area	m^2
A_c	Cell effective convection area	m^2
A_r	Cell effective radiation area	m^2
C_Θ	temperature coefficient of the photo-current i_L	A/K
c_p	Average cell specific heat	J/kg/K
$\mathscr{E}_{g0}$, α_g, β_g	Energy band gap function parameters	eV, eV/K, K
h_c	Convection-cooling coefficient	$W/K/m^2$
i_{s0}	Saturation current at Θ_a	A
m_g	Cell mass	kg
R_{se}	Cell body series resistance	pu
R_{sh}	Cell body shunt resistance	pu
γ	Diode ideality factor	-
ε	Emittance	-
Θ_a	Ambient temperature	K
ρ_e	Average spectral responsivity	A/W
ρ_c	Cell reflection factor	-
τ_c	Cell transmission factor	-

In recent years, storage devices have gained more and more relevance because they can help systems with stochastic primary energy sources (e.g., wind or solar energy) maintain a smooth power production profile. Among the several existing energy storage devices (e.g., flywheels, electrolyzers, hydrogen storage, etc.), batteries are one of the most promising [235].

A battery is a voltage source that depends on the generated current and on the state of charge (SOC) of the battery itself. There are several battery types, e.g., lead-acid, lithium-ion, lithium-polymer, nickel-cadmium, nickel-metal hydride, zinc-air, etc. A dynamic rechargeable battery model, based on the classical Shepherd's model, is as follows [278, 314]:

$$\begin{aligned} \dot{q}_e &= i_{dc}/3600 \\ \dot{i}_m &= (i_{dc} - i_m)/T_m \\ 0 &= v_{oc} - v_p(q_e, i_m) + v_e e^{-\beta_e q_e} - R_i i_{dc} - v_{dc} \end{aligned} \tag{17.32}$$

where q_e is the per unit extracted capacity normalized with respect to the maximum battery capacity Q_n in Ah, i_m is the battery current i_{dc} passed through a low-pass filter, the polarization voltage $v_p(q_e)$ depends on the sign of i_m, as follows:

$$v_p(q_e, i_m) = \begin{cases} \dfrac{R_p i_m + K_p q_e}{\mathrm{SOC}} & \text{if} \quad i_m > 0 \quad \text{(discharge)} \\ \dfrac{R_p i_m}{q_e + 0.1} + \dfrac{K_p q_e}{\mathrm{SOC}} & \text{if} \quad i_m < 0 \quad \text{(charge)} \end{cases} \tag{17.33}$$

and the SOC is defined as:

$$\mathrm{SOC} = \frac{Q_n - Q_e}{Q_n} = 1 - q_e \tag{17.34}$$

where Q_e is the extracted capacity in Ah and remaining parameters are defined in Table 17.6. Figure 17.7 shows a typical discharge of a battery.

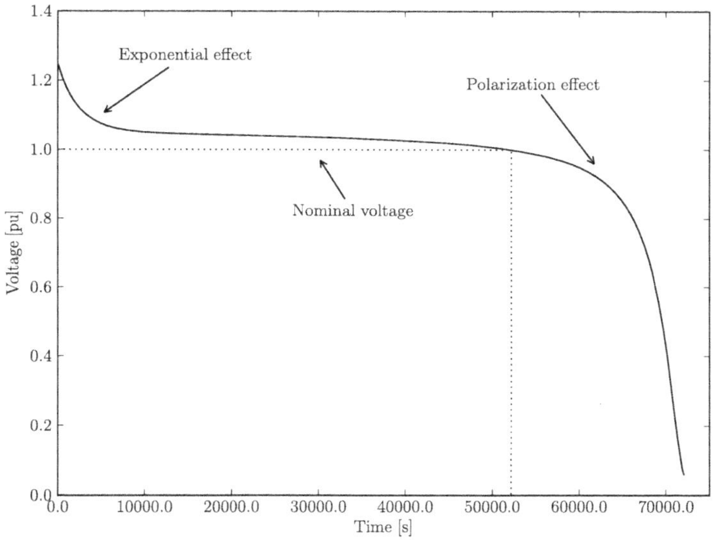

Fig. 17.7 Battery discharge characteristic

In the literature, several other battery models have been proposed. For example, the so-called *universal* model [320] and a Thevenin's equivalent model including resistive and capacitive elements [264]. While the universal model is basically a linearization of the previous model (17.32), the Thevenin's battery equivalent attempts to model the battery capacity through fictitious capacitors. The latter model can be linear (constant parameters) or nonlinear (the parameters depend on the SOC).

Apart from the SOC, another aspect that has to be taken into account in battery models is the parameter dependence on the temperature. In particular, the internal and polarization resistances are a function of the average battery temperature Θ. Figure 17.8 shows some typical empirical curves of the battery equivalent total internal resistance as a function of both Θ and SOC [235]. As a consequence of the internal resistance is that, during the charge and discharge processes, the battery generates heat proportionally to the energy transit in the time interval. To avoid over-heating, the battery has to be cooled down. The battery temperature dynamic can be written as:

$$\dot{\Theta} = \frac{1}{c_p m_g}\Big(S_n v_{\mathrm{dc}} i_{\mathrm{dc}}(1 - \eta_v + \frac{\mathscr{E}_d}{V_{\mathrm{dc},n} v_{\mathrm{oc}}}) \\ -h_c A_c(\Theta - \Theta_a) - \sigma\varepsilon A_r(\Theta^4 - \Theta_a^4)\Big) \tag{17.35}$$

where $\sigma = 5.670 \cdot 10^{-8}$ W/m^2/K^4 is the Stefan-Boltzmann's constant.

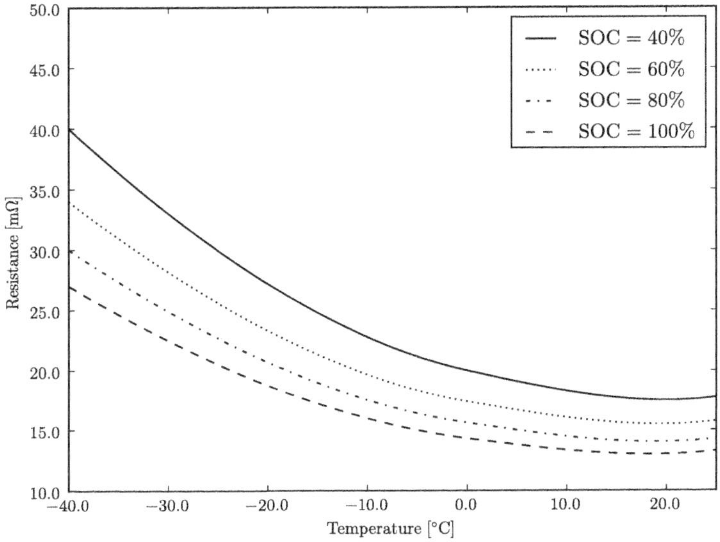

Fig. 17.8 Battery internal resistance as a function of temperature and state of charge

Table 17.6 Energy battery parameters

Variable	Description	Unit
A_c	Cell effective convection area	m^2
A_r	Cell effective radiation area	m^2
c_p	Average battery specific heat	J/kg/K
$\mathscr{E}_d$	Average power loss per ampere of discharge	W/A
h_c	Convection-cooling coefficient	W/K/m^2
K_p	Polarization constant	pu/pu
m_g	Battery mass	kg
Q_n	Rated battery capacity	Ah
R_i	Internal battery resistance	Ω
R_p	Polarization resistance	pu
v_{oc}	Open circuit potential	pu
v_e	Exponential voltage	pu
β_e	Exponential capacity coefficient	1/pu
ε	Emittance	-
η_v	Voltage efficiency factor on discharge	-
Θ_a	Ambient temperature	K

Chapter 18
AC/DC Devices

This chapter describes the two most common ac/dc devices, namely the high-voltage dc transmission system (Section 18.1) and the voltage source converter (Section 18.2). Each section provides the detailed device model and control scheme examples.

18.1 High-Voltage Direct-Current Transmission System

High-Voltage Direct-Current (HVDC) transmission systems consist of a dc line interfaced to the ac network through a rectifier and an inverter. Figure 18.1 shows a typical HVDC scheme. HVDC systems have been intensively studied in the eighties and nineties and raised great expectations [1, 12, 13, 15, 26, 28, 40, 41, 202, 203, 240, 261, 268, 280, 305, 331]. A future of power systems entirely composed of HVDC systems was anticipated. That foreseen scenario has not happened mostly due to rectifier and inverter costs and regulation issues. However, HVDC systems accomplish important niche applications for which there are no other device candidates.

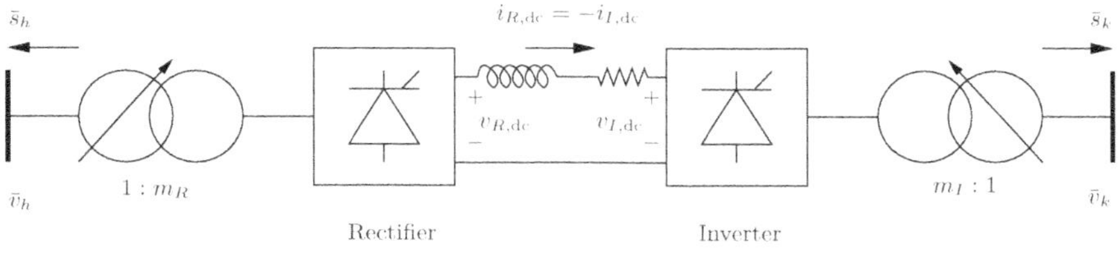

Fig. 18.1 HVDC scheme

HVDC systems are mainly used for long connections that cannot be obtained with standard ac transmission lines due distributed parameter issues,

F. Milano: Power System Modelling and Scripting, Power Systems, pp. 395–412.
springerlink.com

e.g., Ferranti's effect and delays (see Section 11.1 of Chapter 11). HVDC are also the unique choice in case of long submarine connections. Another interesting application is in the back-to-back configuration, which consists in connecting the rectifier and the inverter dc sides without the dc line. The back-to-back configuration is useful to decouple system frequencies. For example a back-to-back HVDC device is necessary to interconnect two ac systems working at different nominal frequencies (e.g., Japanese interconnected system) or if one of the ac sides shows a poor frequency regulation and the other side wants to avoid problems. Finally, a configuration that includes a rectifier and a dc load are used in some industrial applications (e.g., arc furnaces).

The most flexible manner to implement a general HVDC system is to divide the system into its minimal entities:

1. Ac/dc converter (rectifier and inverter).
2. Dc network.
3. Regulators.

In this way, any HVDC configuration (including multi-terminal ones) and control can be implemented with the minimum effort. Dc network elements are described in Chapter 17. Following subsections present the models of converters and HVDC controllers used for a standard two-terminal connection. The interested reader can find a discussion on multi-converter HVDC systems in [16].

18.1.1 Per Unit System for DC Quantities

In [163], the following bases are used for computing the dc per unit quantities:

$$\begin{aligned} V^{\rm dc}_{\rm base} &= \frac{3\sqrt{2}}{\pi} V^{\rm ac}_{\rm base} \\ I^{\rm dc}_{\rm base} &= I^{\rm dc}_{\rm rate} = I^{\rm dc}_{n} \\ S^{\rm dc}_{\rm base} &= V^{\rm dc}_{\rm base} I^{\rm dc}_{\rm base} \\ Z^{\rm dc}_{\rm base} &= V^{\rm dc}_{\rm base} / I^{\rm dc}_{\rm base} \end{aligned} \tag{18.1}$$

Furthermore, $S_n \approx V^{\rm dc}_n I^{\rm dc}_n$ holds to avoid inconsistencies. In the following subsections, the dc power base is assumed equal to the ac one, i.e., $S^{\rm dc}_{\rm base} = S_n$.

18.1.2 Rectifier Model

The rectifier scheme with current and voltage sign notation is shown in Figure 18.2. The dc-side interface equations are (17.2), i.e., the same as any two-terminal dc device described in Chapter 17. According to the notation of Figure 18.2, the active and reactive power injections at the ac-side are:

$$\begin{aligned} p_h &= -v_{R,\mathrm{dc}} i_{R,\mathrm{dc}} \\ q_h &= -k m_R v_h i_{R,\mathrm{dc}} \sin \phi_h \end{aligned} \tag{18.2}$$

where ϕ_h is the angle between the average ac voltage and the ac current of the rectifier, m_R is the transformer tap ratio, and $k = 0.9995 \cdot 3\sqrt{2}/\pi$. The link between the ac and the dc sides is given by:

$$\begin{aligned} 0 &= k m_R v_h \cos \alpha - \frac{3}{\pi} x_{R,c} i_{R,\mathrm{dc}} - v_{R,\mathrm{dc}} \\ 0 &= k m_R v_h \cos \phi_h - v_{R,\mathrm{dc}} \end{aligned} \tag{18.3}$$

where α is the rectifier *firing angle*. Remaining parameters and rectifier variable limits are defined in Table 18.1. Two equations are required to complete the converter model. These equations are used for controlling the transformer tap ratio m_R and the firing angle α (see following Subsection 18.1.4).

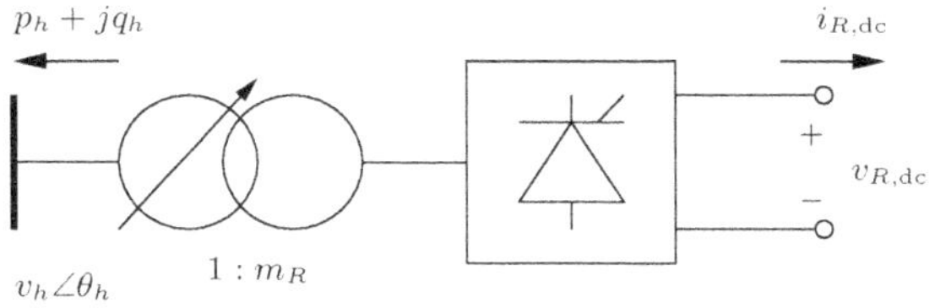

Fig. 18.2 Rectifier scheme

Table 18.1 Rectifier parameters

Variable	Description	Unit
$m_R^{\max}$	Maximum transformer tap ratio	pu/pu
$m_R^{\min}$	Minimum transformer tap ratio	pu/pu
$x_{R,c}$	Commutation reactance	pu
$\alpha^{\max}$	Maximum firing angle α	rad
$\alpha^{\min}$	Minimum firing angle α	rad

18.1.3 Inverter Model

The equations that describe the inverter dc-side interface are (17.2). Furthermore, equations (18.2) and (18.3) have to be rewritten to take into account the *extinction angle* or *commutation margin* γ and the sign of the dc current on the inverter dc side (see Figure 18.3). For the two-terminal scheme of Figure 18.1, $i_{I,\mathrm{dc}} = -i_{R,\mathrm{dc}}$, hence one has:

$$
\begin{aligned}
p_k &= -v_{I,\mathrm{dc}} i_{I,\mathrm{dc}} \\
q_k &= k m_I v_k i_{I,\mathrm{dc}} \sin \phi_k \\
0 &= k m_I v_k \cos(\pi - \gamma) - \frac{3}{\pi} x_{I,c} i_{I,\mathrm{dc}} + v_{I,\mathrm{dc}} \\
0 &= k m_I v_k \cos \phi_k - v_{I,\mathrm{dc}}
\end{aligned}
\tag{18.4}
$$

where ϕ_k is the angle between the ac voltage and the ac current of the inverter and all remaining parameters and inverter variable limits are defined in Table 18.2. Observe that, since $i_{I,} < 0$, the inverter injects active power and consumes reactive power at the ac bus k. As for the rectifier, two equations are required to complete the model. These equations are used for controlling the transformer tap ratio m_I and the extinction angle γ (see Subsection 18.1.4).

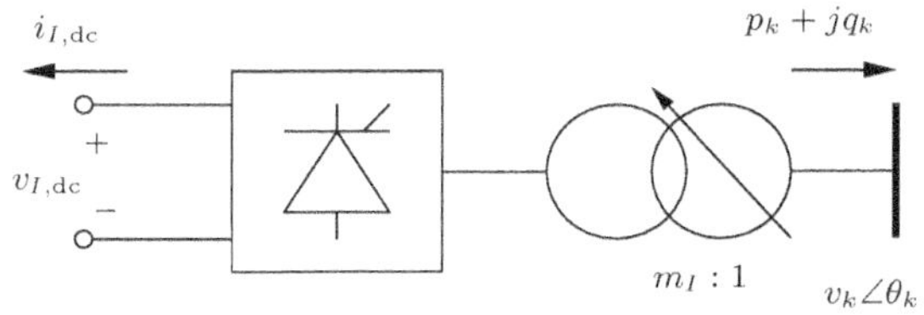

Fig. 18.3 Inverter scheme

Table 18.2 Inverter parameters

Variable	Description	Unit
$m_I^{\max}$	Maximum transformer tap ratio	pu/pu
$m_I^{\min}$	Minimum transformer tap ratio	pu/pu
$x_{I,c}$	Commutation reactance	pu
$\gamma^{\max}$	Maximum extinction angle γ	rad
$\gamma^{\min}$	Minimum extinction angle γ	rad

18.1.4 HVDC Control

HVDC controllers have to coordinate the operations of the rectifier and the inverter devices. The controlling variables are the transformer tap ratio and the firing angle on the rectifier side and the transformer tap ratio and the extinction angle on the inverter side.

Tap ratio controls are necessarily slower than those of firing/extinction angles. The firing and extinction angles, which are characterized by fast dynamics, are considered algebraic variables, while the tap ratio, whose dynamic is relatively slow, are considered state variables.

A common control scheme for the HVDC transmission system scheme of Figure 18.1 is as follows.

Rectifier current control mode (RCCM): Rectifier-side regulators control the dc current i_{dc} through a PI controller that varies the firing angle α, and the ac voltage v_h through the off-nominal tap ratio m_R. Inverter-side regulators maintain constant the extinction angle $\gamma = \gamma^{\mathrm{ref}} \geq \gamma^{\mathrm{min}}$ and control the dc voltage $v_{I,\mathrm{dc}}$ through the tap ratio m_I. The DAE system is as follows:

$$\begin{aligned}
\dot{x}_{R,m} &= (v_h - v_{\mathrm{ac}}^{\mathrm{ref}})/T_R \\
\dot{x}_{I,m} &= (v_{\mathrm{dc}}^{\mathrm{ref}} - v_{I,\mathrm{dc}})/T_I \\
0 &= m_R - x_{R,m} \\
0 &= m_I - x_{I,m} \\
\dot{x}_{R,i} &= K_i(i_{\mathrm{dc}}^{\mathrm{ref}} - i_{\mathrm{dc}}) \\
0 &= x_{R,i} + K_p(i_{\mathrm{dc}}^{\mathrm{ref}} - i_{\mathrm{dc}}) - \alpha \\
0 &= \gamma^{\mathrm{ref}} - \gamma
\end{aligned} \tag{18.5}$$

The PI control undergoes an anti-windup limiter (see Section C.2 of Appendix C) to maintain the firing angle within its limits. The state variable $x_{R,m}$ and $x_{I,m}$ are introduced to allow a general interface with the converter and rectifier models.

Inverter current control mode (ICCM): Inverter-side regulators control the dc current i_{dc} through the extinction angle γ and the the ac voltage v_k through the tap ratio m_I. Rectifier-side regulators maintain constant the firing angle $\alpha = \alpha^{\mathrm{min}}$ and control the dc voltage $v_{R,\mathrm{dc}}$ through the tap ratio m_R. The DAE system is as follows:

$$\begin{aligned}
\dot{x}_{R,m} &= (v_{\mathrm{dc}}^{\mathrm{ref}} - v_{R,\mathrm{dc}})/T_R \\
\dot{x}_{I,m} &= (v_k - v_{ac}^{\mathrm{ref}})/T_I \\
0 &= m_R - x_{R,m} \\
0 &= m_I - x_{I,m} \\
0 &= \alpha^{\mathrm{min}} - \alpha \\
\dot{x}_{I,i} &= K_i(i_{\mathrm{dc}}^{\mathrm{ref}} + i_{\mathrm{dc}} - i_{\mathrm{dc}}^{m}) \\
0 &= x_{I,i} + K_p(i_{\mathrm{dc}}^{\mathrm{ref}} + i_{\mathrm{dc}} - i_{\mathrm{dc}}^{m}) - \gamma
\end{aligned} \tag{18.6}$$

where the *current margin* i_{dc}^{m} avoids that the RCCM and ICCM controls overlap. As for the RCCM, the state variables undergo anti-windup limiters.

Power control: If a power/frequency control is required, the reference current $i_{\mathrm{dc}}^{\mathrm{ref}}$ can be the output of a power (or *master*) control, as follows [184]:

$$i_{\mathrm{dc}}^{\mathrm{ref}} = \min\{i_{\mathrm{dc}}^{\mathrm{des}}, i_{\mathrm{dc}}^{\mathrm{lim}}\} \tag{18.7}$$

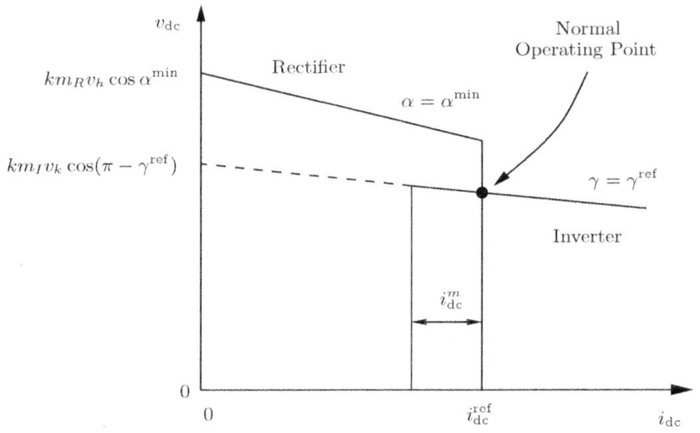

Fig. 18.4 HVDC steady state characteristic for the rectifier current control mode

where $i_{dc}^{des} = p^{ref}/v_{dc}$ is the current that provides the desired reference power p^{ref} and v_{dc} is the dc voltage at one of the two converter terminals, and i_{dc}^{lim} is defined based on the dc voltage value, as follows:

$$i_{dc}^{lim} = \begin{cases} i_{dc}^{min}, & \text{if } v_{dc} < v_{dc}^{min} \\ i_{dc}^{min} + c(v_{dc} - v_{dc}^{min}), & \text{if } v_{dc}^{min} \leq v_{dc} \leq v_{dc}^{max} \\ i_{dc}^{min} + c(v_{dc}^{max} - v_{dc}^{min}), & \text{if } v_{dc} > v_{dc}^{max} \end{cases} \tag{18.8}$$

where, typically, $c = 1$ pu/pu.

Figure 18.4 summarizes the steady-state characteristic of the HVDC control for the RCCM, whereas Table 18.3 summarizes and defines the parameters of the HVDC controllers.

18.2 Voltage Source Converter

The Voltage Source Converter (VSC) device is similar to the rectifier or inverter devices of the HVDC transmission system. Thus, the VSC provides a link between an ac and and dc networks. The main difference with HVDC systems is the technology of power electronic switches. HVDC devices are built using conventional thyristors, which provide the turn-on control, but can be turned off only if the current is zero. The electronic switches used for VSC devices are Gate Turn-Off Thyristor (GTO), MOS Turn-off Thyristor (MTO), Integrated Gate Bipolar Transistor (IGBT) or Integrated Gate-Commutated Thyristor (IGCT). These are called *turn-off* devices since they provide both turn-on and turn-off controls, the latter also for non-zero currents. Turn-off devices are more expensive and have higher losses than conventional thyristors. These drawbacks implies that the typical VSC capacity is smaller than

Table 18.3 HVDC control parameters

Parameter	Description	Units
c	Slope of the voltage/current power control	pu
i_{dc}^{m}	Current margin	pu
i_{dc}^{min}	Minimum dc current	pu
K_i	Integral gain for the current PI control	1/s
K_p	Proportional gain for the current PI control	rad/pu
p^{ref}	Power reference	pu
T_I	Inverter control time constant	s
T_R	Rectifier control time constant	s
v_{dc}^{max}	Maximum dc voltage	pu
v_{dc}^{min}	Minimum dc voltage	pu
v_{dc}^{ref}	Reference dc voltage	pu
v_{ac}^{ref}	Reference ac voltage	pu
α^{ref}	Reference firing angle	rad
γ^{ref}	Reference extinction angle	rad

HVDC ones. However, the turn-off ability allows obtaining controls with higher flexibility than those that can be implemented for HVDC systems. In practice, VSC are used for small to medium power applications such as FACTS devices, wind turbines and other distributed energy resources.

The VSC scheme is shown in Figure 18.5. The typical configuration includes a capacitor, a bi-directional inverter and a transformer. The capacitor allows maintaining constant the dc voltage while the transformer provides a Galvanic insulation. The ac/dc conversion is provided by the VSC, which also includes a series phase reactor and a shunt ac filter.[1] The dc voltage is converted into an ac one whose magnitude depends on the VSC modulating amplitude a_m and whose phase is the inverter firing angle α.

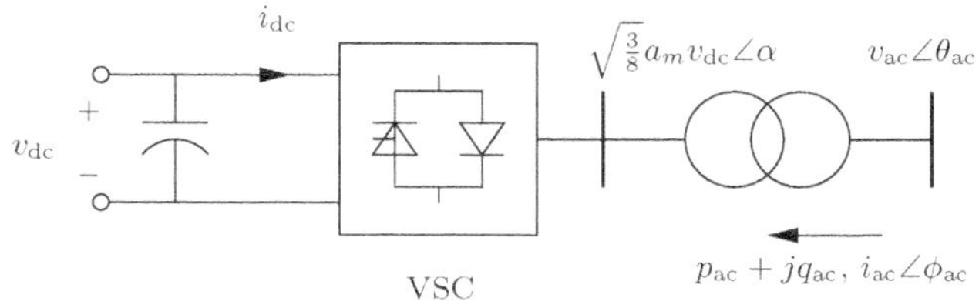

Fig. 18.5 Voltage source converter scheme

The dc-side interface is defined by (17.2). In principle, the elements connected to the dc terminals of the VSC are not part of the VSC model. However, in most cases, the dc side of the VSC contains at least a capacitor used

[1] The phase reactor and the shunt ac filter as well as the phase-locked loop are not modelled in this chapter since their effects are relevant only for electromagnetic transients.

for fixing the dc voltage (see Figure 18.5). The capacitor can be modelled as a parallel RC described by equations (17.10) provided in Chapter 17. To work properly, the dc network must contain two nodes, one of which is connected to the ground. The VSC and the RC element are connected in parallel to the dc nodes. From the implementation viewpoint, it is better that the capacitor is independent from the VSC model. This allows more flexibility in the configuration of the dc side (see for example the different configurations described in [129]). On the other hand, the ac/ac transformer can be embedded in the VSC model. The ac-side interface is:

$$\begin{aligned} p_{\mathrm{ac}} &= g_T v_{\mathrm{ac}}^2 - \sqrt{3/8}\, a_m v_{\mathrm{dc}} v_{\mathrm{ac}} g_T \cos(\theta_{\mathrm{ac}} - \alpha) \\ &\quad - \sqrt{3/8}\, a_m v_{\mathrm{dc}} v_{\mathrm{ac}} b_T \sin(\theta_{\mathrm{ac}} - \alpha) \\ q_{\mathrm{ac}} &= -b_T v_{\mathrm{ac}}^2 + \sqrt{3/8}\, a_m v_{\mathrm{dc}} v_{\mathrm{ac}} b_T \cos(\theta_{\mathrm{ac}} - \alpha) \\ &\quad - \sqrt{3/8}\, a_m v_{\mathrm{dc}} v_{\mathrm{ac}} g_T \sin(\theta_{\mathrm{ac}} - \alpha) \end{aligned} \tag{18.9}$$

where $g_T + jb_T = 1/(r_T + jx_T)$ is the transformer series admittance. The power balance between the dc and the ac sides is:

$$\begin{aligned} 0 &= g_T(\sqrt{3/8}\, a_m v_{\mathrm{dc}})^2 - \sqrt{3/8}\, a_m v_{\mathrm{dc}} v_{\mathrm{ac}} g_T \cos(\theta_{\mathrm{ac}} - \alpha) \\ &\quad + \sqrt{3/8}\, a_m v_{\mathrm{dc}} v_{\mathrm{ac}} b_T \sin(\theta_{\mathrm{ac}} - \alpha) \end{aligned} \tag{18.10}$$

Finally, it is important to monitor the ac current to avoid overloading VSC circuits. The ac current $i_{\mathrm{ac}}\angle\phi_{\mathrm{ac}}$ can be obtained imposing the following constraints:

$$\begin{aligned} 0 &= p_{\mathrm{ac}}^2 + q_{\mathrm{ac}}^2 - v_{\mathrm{ac}} i_{\mathrm{ac}} \\ 0 &= p_{\mathrm{ac}} \sin(\theta_{\mathrm{ac}} - \phi_{\mathrm{ac}}) - q_{\mathrm{ac}} \cos(\theta_{\mathrm{ac}} - \phi_{\mathrm{ac}}) \end{aligned} \tag{18.11}$$

The VSC parameters are defined in Table 18.4 according to the implementation choice of maintaining the capacitor as an independent device.

Table 18.4 Voltage source converter parameters

Variable	Description	Unit
$i_{\mathrm{ac}}^{\max}$	Maximum ac current	pu
$i_{\mathrm{dc}}^{\max}$	Maximum dc current	pu
$a_m^{\max}$	Maximum modulating amplitude	pu
$a_m^{\min}$	Minimum modulating amplitude	pu
r_T	Transformer resistance	pu
x_T	Transformer reactance	pu
$\alpha^{\max}$	Maximum firing angle	rad
$\alpha^{\min}$	Minimum firing angle	rad

VSC models can be connected as shunt or series devices to the ac network. Equations (17.2) and (18.9)-(18.11) can be used as base for both connections, as follows.

VSC Shunt Connection: Assuming that the VSC is connected to the ac bus h, the shunt connection is obtained imposing:

$$p_h = -p_{\mathrm{ac}}, \quad q_h = -q_{\mathrm{ac}}, \quad v_h = v_{\mathrm{ac}}, \quad \theta_h = \theta_{\mathrm{ac}} \tag{18.12}$$

VSC Series Connection: Assuming that the VSC is connected to the ac buses h and k, the series connection is defined by:

$$\begin{aligned} 0 &= p_h + p_k - p_{\mathrm{ac}} \\ 0 &= q_h + q_k - q_{\mathrm{ac}} \\ 0 &= p_h^2 + q_h^2 - (v_h i_{\mathrm{ac}})^2 \\ 0 &= p_h \sin(\theta_h - \phi_{\mathrm{ac}}) - q_h \cos(\theta_h - \phi_{\mathrm{ac}}) \\ 0 &= p_k^2 + q_k^2 - (v_k i_{\mathrm{ac}})^2 \\ 0 &= p_k \sin(\theta_h - \phi_{\mathrm{ac}}) - q_k \cos(\theta_k - \phi_{\mathrm{ac}}) \end{aligned} \tag{18.13}$$

The VSC model is completed by the controls that regulates the modulating amplitude a_m and the firing angle α. These controls depend on the application and should be defined separately to maintain the VSC model as general as possible (similarly to the case of the inverter and rectifier models used in the HVDC link).

A special care has to be devoted to the operating limits of the VSC device. Both the firing angle and the modulating amplitude are limited:

$$\begin{aligned} \alpha^{\max} &\le \alpha \le \alpha^{\min} \\ a_m^{\max} &\le a_m \le a_m^{\min} \end{aligned} \tag{18.14}$$

These limits are used for bounding the controls of the VSC device. Then the ac and dc current limits indicated in Table 18.4 have also to be checked during VSC operation. If a current limit is reached, then some of the VSC controls have to be locked. In other words, current limits impose indirect limits on the firing angle α and the modulating amplitude a_m (e.g., $\alpha^{\max}(i_{\mathrm{dc}})$).

The following examples describe two typical applications of the VSC device for connecting distributed dc resources to the ac network, namely the fuel cell and the photovoltaic cell.

Example 18.1 Solid Oxide Fuel Cell Control

The SOFC model described in Subsection 17.6.1 is linked to ac networks through a shunt-connected VSC device. The two controllers that have to be included to complete the system composed of the SOFC and the VSC are shown in Figure 18.6. The ac voltage magnitude v_h is regulated by means of the VSC inverter modulating amplitude a_m:

$$\dot{a}_m = (K_m(v^{\text{ref}} - v_h) - a_m)/T_m \tag{18.15}$$

The amplitude control undergoes an anti-windup limiter. The fuel cell dc current set point $i_{\text{dc}}^{\text{ref}}$ is defined based on power reference p^{ref}:

$$i_{\text{dc}}^{\text{ref}} = p^{\text{ref}}/v_{\text{dc}} \tag{18.16}$$

The set point $i_{\text{dc}}^{\text{ref}}$ is limited by the dynamic limits proportional to the hydrogen flow:

$$\frac{U^{\min} q_{H_2}}{2K_r} \le i_{\text{dc}}^{\text{ref}} \le \frac{U^{\max} q_{H_2}}{2K_r} \tag{18.17}$$

where q_{H_2} is the hydrogen flow and $U^{\max}$ are $U^{\min}$ are the maximum or the minimum hydrogen utilization, respectively. Then the current i_{dc} is regulated through the VSC firing angle α by means of a PI controller:

$$\begin{aligned} \dot{x}_\alpha &= K_i(i_{\text{dc}}^{\text{ref}} - i_{\text{dc}}) \\ 0 &= K_p(i_{\text{dc}}^{\text{ref}} - i_{\text{dc}}) + x_\alpha - \alpha \end{aligned} \tag{18.18}$$

Table 18.6 defines all parameters of the SOFC controllers.

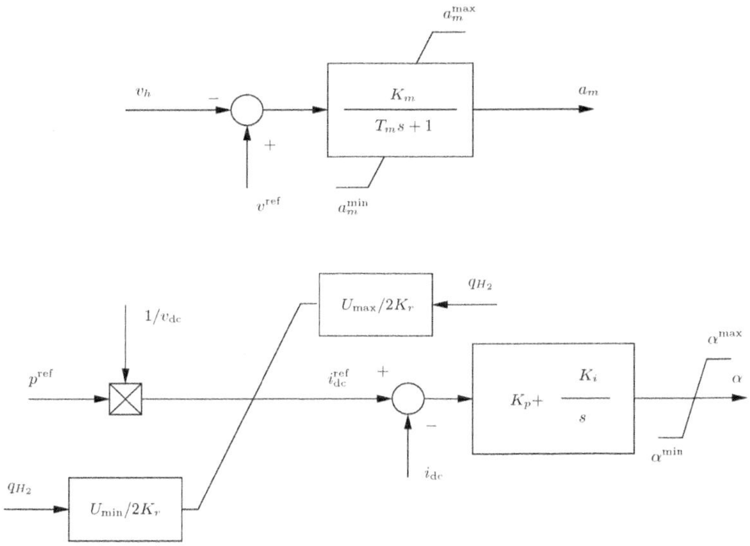

Fig. 18.6 Power and ac voltage controls for the solid oxide fuel cell

Example 18.2 Solar Photovoltaic Cell Control

Similarly to the fuel cell, photovoltaic panels are linked to the ac network through a shunt-connected VSC device. The voltage control on the ac side can

Table 18.5 Solid oxide fuel cell regulator parameters

Variable	Description	Unit
K_i	Integral gain of the current control	rad/s/pu
K_m	Gain of the voltage control loop	pu
K_p	Proportional gain of the current control	rad/pu
p^{ref}	Reference power	pu
T_m	Time constant of the voltage control loop	s
$U^{\max}$	Maximum fuel utilization	-
$U^{\min}$	Minimum fuel utilization	-

be obtained using the VSC inverter modulating amplitude a_m. The voltage control equation is, for example, (18.15).

The other control is aimed to maximize the power production of the photovoltaic cell:

$$\max\{p_{\mathrm{dc}}\} = \max\{v_{\mathrm{dc}} \cdot i_{\mathrm{dc}}(v_{\mathrm{dc}}, \Theta, G)\} \tag{18.19}$$

This control is called Maximum Power Point Tracking (MPPT). As discussed in Section 17.6.2, the output current i_{dc} is a function of the voltage v_{dc}, of the temperature Θ and of the solar irradiance G. The exponential does not allow writing an explicit expression of the current i_{dc} with respect of the other variables. Thus the static pv characteristic of the photovoltaic cell can be found only numerically.

Figure 18.7 shows the pv characteristic of a typical photovoltaic cell. The curve has always a maximum, which depends on the temperature and the solar irradiance values. The power produced by the cell increases as the solar irradiance increases and the temperature decreases.

The maximum power point satisfies the condition:

$$\begin{aligned} 0 &= \frac{dp_{\mathrm{dc}}}{dv_{\mathrm{dc}}} = \frac{d}{dv_{\mathrm{dc}}}(v_{\mathrm{dc}} \cdot i_{\mathrm{dc}}(v_{\mathrm{dc}}, \Theta, G)) \\ &= i_{\mathrm{dc}}(v_{\mathrm{dc}}, \Theta, G) + v_{\mathrm{dc}} \frac{\partial i_{\mathrm{dc}}(v_{\mathrm{dc}}, \Theta, G)}{\partial v_{\mathrm{dc}}} \end{aligned} \tag{18.20}$$

which can be solved numerically. The solution of (18.20) provides the optimal value of the dc voltage reference. Then a PI controller can be used for regulating the VSC firing angle. The resulting equations are:

$$\begin{aligned} 0 &= i_{\mathrm{dc}}(v_{\mathrm{dc}}^{\mathrm{ref}}, \Theta, G) + v_{\mathrm{dc}}^{\mathrm{ref}} \frac{\partial i_{\mathrm{dc}}(v_{\mathrm{dc}}^{\mathrm{ref}}, \Theta, G)}{\partial v_{\mathrm{dc}}^{\mathrm{ref}}} \\ \dot{x} &= K_i(v_{\mathrm{dc}}^{\mathrm{ref}} - v_{\mathrm{dc}}) \\ 0 &= K_p(v_{\mathrm{dc}}^{\mathrm{ref}} - v_{\mathrm{dc}}) + x - \alpha \end{aligned} \tag{18.21}$$

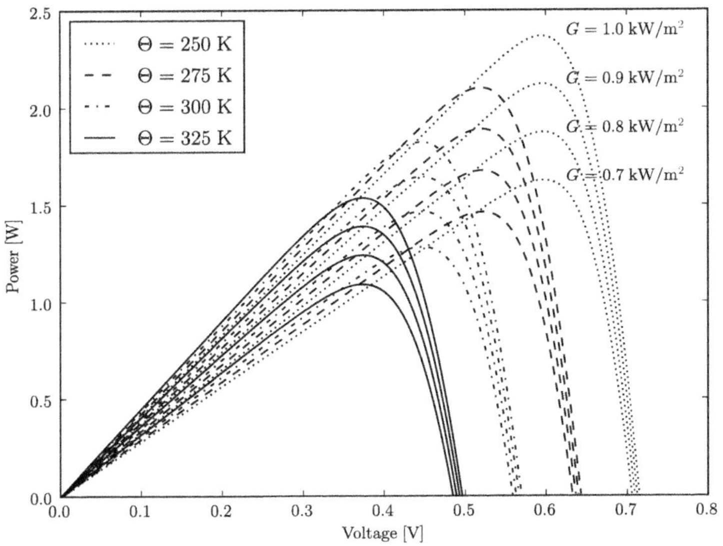

Fig. 18.7 Effect of solar irradiance and temperature on the *pv* characteristic of the photovoltaic cell

In most practical applications, the first equation of (18.21) is often substituted by empirical systems based on measures. Most MPPT tracking systems vary periodically the dc voltage of the photovoltaic cell. The MPPT firstly tries increasing the dc voltage. If the output power increases, then the voltage is further increased, otherwise, the MPPT decreases the dc voltage. This kind of control is easy to implement since it requires only the measures of the electrical dc power (no need of measuring the temperature and the solar irradiance and of knowing cell parameters).

The ac voltage and the power controls of the photovoltaic cell are depicted in Figure 18.8 and all cell regulator parameters are defined in Table 18.6.

Example 18.3 Superconducting Magnetic Energy Storage

The Superconducting Magnetic Energy Storage (SMES) is a shunt-connected VSC-based device where in the dc side a superconducting coil is connected in parallel with the VSC capacitor through a dc/dc system (*chopper*) [123]. The SMES scheme is shown in Figure 18.9. The coil is able to store magnetic energy and to release it to the network depending on the duty cycle of the chopper. Since the resistance of the superconducting coil and of the dc system is very small with respect to the inductance, the time constant of the RL circuit is relatively high with respect to conventional electro-magnetic time scales.

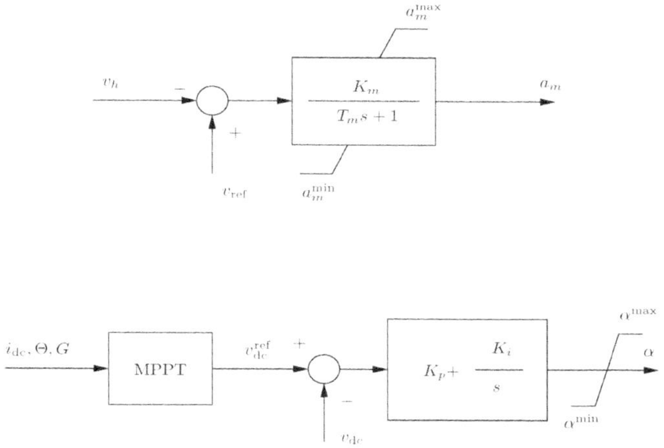

Fig. 18.8 Maximum power point tracking for the photovoltaic cell

Table 18.6 Photovoltaic cell regulator parameters

Variable	Description	Unit
K_i	Integral gain of the current control	rad/s/pu
K_m	Gain of the voltage control loop	pu
K_p	Proportional gain of the current control	rad/pu
T_m	Time constant of the voltage control loop	s

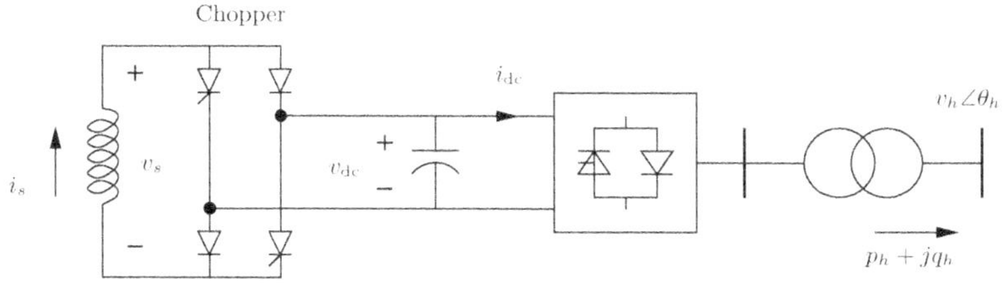

Fig. 18.9 SMES scheme

The equations for the SMES can be divided into three parts: (i) dc network model, (ii) VSC model, and (iii) controllers.

Dc network: The dc network is composed of the parallel the VSC capacitor and the superconducting coil behind the chopper. The capacitor can be considered a parallel RC element described by equations (17.10) and is interfaced to the dc network through (17.2). The coil-chopper block is modelled as:

$$\begin{aligned} \dot{i}_s &= -v_s/L \\ v_s &= (1-2d_c)v_{\rm dc} \\ i_{\rm dc} &= (1-2d_c)i_s \end{aligned} \tag{18.22}$$

where all voltages and currents are average values, L is the superconducting coil inductance and d_c is the chopper duty cycle. The duty cycle can be used a control variable of the SMES, in fact for $d_c = 0.5$ the average coil voltage v_s and the average dc current $i_{\rm dc}$ are zero. For $d_c > 0.5$ the coil is charged, while for $d_c < 0.5$ the coil is discharged.

Finally, the dc network must contain two nodes, one of which is connected to the ground. The VSC and the RL element are connected in parallel to the dc nodes.

Shunt-connected VSC: Equations are (17.2) and (18.9)-(18.11) and (18.12).

Regulators: The regulator are aimed to provide correct handling of the coil charge and discharge. As indicated in (18.22), the coil charge/discharge is defined by the value of the duty cycle. Thus, the active power p_h at the ac side is regulated by means of the duty cycle. Then the VSC firing angle and modulation amplitude can be used for regulating the dc voltage $v_{\rm dc}$ and the ac voltage v_h, respectively. PI or lag controllers similar to the ones defined for the SOFC can be used (see Figure 18.6).

To avoid over- and under-charging, it is necessary to monitor the energy stored in the superconducting coil, for example measuring and integrating the dc voltage and current:

$$\dot{\mathscr{E}} = v_s i_s = v_{\rm dc} i_{\rm dc} \tag{18.23}$$

Then, based on the maximum and minimum coil energy limits, one can define the maximum and minimum duty cycle value and disable the active power control. As discussed in [11], properly handling VSC firing angle and modulating amplitude control limits can also improve the transient behavior of the SMES.

18.2.1 Simplified Dynamic VSC Model

Due to the fast response of the power electronic switches and of the capacitor, in most transient stability applications, the VSC can be modelled taking into account only the power balance and simplified control equations. If the power flow is from the dc side to the ac one, the power balance is:

$$0 = v_{\rm dc} i_{\rm dc} - p_{\rm ac} - p_{\rm loss}(i_{\rm dc}, v_{\rm dc}) \tag{18.24}$$

where $p_{\rm loss}$ is due to commutation and conduction losses of switches and diodes and capacitor losses, and can be evaluated as a function of the rms value of the current circulating in the electronic switches (approximated by the dc current $i_{\rm dc}$) and of the dc voltage $v_{\rm dc}$:

$$p_{\mathrm{loss}} = a i_{\mathrm{dc}}^2 + b i_{\mathrm{dc}} + c + d v_{\mathrm{dc}}^2 \tag{18.25}$$

where the coefficients a, b, c and d are provided by the VSC manufacturer. The simplified control equations do not explicitly include the firing angle α and the modulating amplitude a_m but only consider input and output variables. Hence, to regulate the active and reactive powers on the ac side, the control differential equations can be written as:

$$\begin{aligned} \dot{p}_{\mathrm{ac}} &= (p^{\mathrm{ref}} - p_{\mathrm{ac}})/T_p \\ \dot{q}_{\mathrm{ac}} &= (q^{\mathrm{ref}} - q_{\mathrm{ac}})/T_q \end{aligned} \tag{18.26}$$

Similar equations can be written for controlling the ac voltage, the power factor, the dc current or the dc voltage. VSC current limits can be taken into account by replacing one of the previous equations (18.26) with the current limit that is violated (e.g., $i_{\mathrm{dc}} - i_{\mathrm{dc}}^{\mathrm{max}} = 0$).

18.2.2 Power Flow VSC Model

In power flow analysis, it can be convenient to simplify the VSC model using only static equations. A very simple model is depicted in Figure 18.10, in both versions, i.e., shunt and series connections [2].

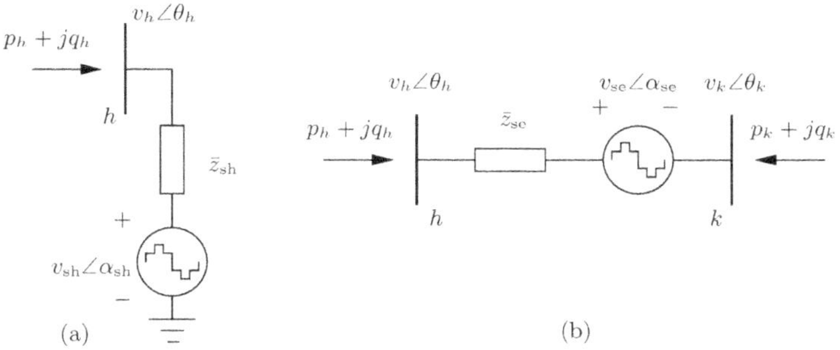

Fig. 18.10 Power flow VSC equivalent circuit: (a) shunt connection and (b) series connection

Shunt-connected VSC: The equations for the shunt-connected VSC are:

$$\begin{aligned} p_h &= v_h^2 g_{\mathrm{sh}} - v_h v_{\mathrm{sh}}(g_{\mathrm{sh}}\cos(\theta_h - \alpha_{\mathrm{sh}}) + b_{\mathrm{sh}}\sin(\theta_h - \alpha_{\mathrm{sh}})) \\ q_h &= -v_h^2 b_{\mathrm{sh}} - v_h v_{\mathrm{sh}}(g_{\mathrm{sh}}\sin(\theta_h - \alpha_{\mathrm{sh}}) - b_{\mathrm{sh}}\cos(\theta_h - \alpha_{\mathrm{sh}})) \\ p_{\mathrm{sh}} &= v_{\mathrm{sh}}^2 g_{\mathrm{sh}} - v_h v_{\mathrm{sh}}(g_{\mathrm{sh}}\cos(\theta_h - \alpha_{\mathrm{sh}}) - b_{\mathrm{sh}}\sin(\theta_h - \alpha_{\mathrm{sh}})) \\ q_{\mathrm{sh}} &= -v_{\mathrm{sh}}^2 b_{\mathrm{sh}} + v_h v_{\mathrm{sh}}(g_{\mathrm{sh}}\sin(\theta_h - \alpha_{\mathrm{sh}}) + b_{\mathrm{sh}}\cos(\theta_h - \alpha_{\mathrm{sh}})) \end{aligned} \tag{18.27}$$

where $g_{\rm sh} + jb_{\rm sh} = 1/(r_{\rm sh} + jx_{\rm sh}) = 1/\bar{z}_{\rm sh}$. These equations approximate the VSC through an independent generator and a connection that models the embedded VSC transformer and internal VSC losses. The variables are $p_{\rm sh}$, $q_{\rm sh}$, $v_{\rm sh}$, $\theta_{\rm sh}$ for the VSC internal bus and v_h, θ_h at the point of connection with the ac network. Since there are six quantities and four equations, two additional equations are needed. These equations strictly depend on the application. In general, one has:

$$
\begin{aligned}
0 &= g_1(p_{\rm sh}, q_{\rm sh}, v_{\rm sh}, \theta_{\rm sh}, p_h, q_h, v_h, \theta_h) \\
0 &= g_2(p_{\rm sh}, q_{\rm sh}, v_{\rm sh}, \theta_{\rm sh}, p_h, q_h, v_h, \theta_h)
\end{aligned}
\tag{18.28}
$$

Section 19.3 of Chapter 19 describes an example of shunt-connected VSC configuration, namely the STATCOM device.

Series-connected VSC: The equations for the series-connected VSC are:

$$
\begin{aligned}
p_h &= v_h^2 g_{\rm se} - v_h v_k (g_{\rm se} \cos(\theta_h - \theta_k) + b_{\rm se} \sin(\theta_h - \theta_k)) \\
&\quad - v_h v_{\rm se} (g_{\rm se} \cos(\theta_h - \alpha_{\rm se}) + b_{\rm se} \sin(\theta_h - \alpha_{\rm se})) \\
q_h &= -v_h^2 b_{\rm se} - v_h v_k (g_{\rm se} \sin(\theta_h - \theta_k) - b_{\rm se} \cos(\theta_h - \theta_k)) \\
&\quad - v_h v_{\rm se} (g_{\rm se} \sin(\theta_h - \alpha_{\rm se}) - b_{\rm se} \cos(\theta_h - \alpha_{\rm se})) \\
p_k &= v_k^2 g_{\rm se} - v_h v_k (g_{\rm se} \cos(\theta_h - \theta_k) - b_{\rm se} \sin(\theta_h - \theta_k)) \\
&\quad - v_k v_{\rm se} (g_{\rm se} \cos(\theta_k - \alpha_{\rm se}) - b_{\rm se} \sin(\theta_k - \alpha_{\rm se})) \\
q_k &= -v_k^2 b_{\rm se} + v_h v_k (g_{\rm se} \sin(\theta_h - \theta_k) + b_{\rm se} \cos(\theta_h - \theta_k)) \\
&\quad + v_k v_{\rm se} (g_{\rm se} \sin(\theta_k - \alpha_{\rm se}) + b_{\rm se} \cos(\theta_k - \alpha_{\rm se})) \\
p_{\rm se} &= v_{\rm se}^2 g_{\rm se} - v_h v_{\rm se} (g_{\rm se} \cos(\theta_h - \alpha_{\rm se}) - b_{\rm se} \sin(\theta_h - \alpha_{\rm se})) \\
&\quad - v_k v_{\rm se} (g_{\rm se} \cos(\theta_k - \alpha_{\rm se}) + b_{\rm se} \sin(\theta_k - \alpha_{\rm se})) \\
q_{\rm se} &= -v_{\rm se}^2 b_{\rm se} + v_h v_{\rm se} (g_{\rm se} \sin(\theta_h - \theta_{\rm se}) + b_{\rm se} \cos(\theta_h - \theta_{\rm se})) \\
&\quad - v_k v_{\rm se} (g_{\rm se} \sin(\theta_h - \alpha_{\rm se}) - b_{\rm se} \cos(\theta_h - \alpha_{\rm se}))
\end{aligned}
\tag{18.29}
$$

where $g_{\rm se} + jb_{\rm se} = 1/(r_{\rm se} + jx_{\rm se}) = 1/\bar{z}_{\rm se}$. These equations approximate the VSC through an independent generator and a series connection that models the embedded VSC transformer and internal VSC losses. The variables are $p_{\rm se}$, $q_{\rm se}$, $v_{\rm se}$, $\theta_{\rm se}$ for the VSC internal bus and v_h, θ_h, v_k and θ_k at the points of connection with the network. Since there are eight quantities and six equations, two additional equations are needed. These equations strictly depend on the application. In general, one has:

$$
\begin{aligned}
0 &= g_1(p_{\rm se}, q_{\rm se}, v_{\rm se}, \theta_{\rm se}, p_h, q_h, v_h, \theta_h, p_k, q_k, v_k, \theta_k) \\
0 &= g_2(p_{\rm se}, q_{\rm se}, v_{\rm se}, \theta_{\rm se}, p_h, q_h, v_h, \theta_h, p_k, q_k, v_k, \theta_k)
\end{aligned}
\tag{18.30}
$$

Section 19.4 of Chapter 19 describes an example of series-connected VSC configuration, namely the SSSC device.

The models (18.27) and (18.29) are nothing more than standard power flow equations of a transmission line between the point-of-connection buses (h and/or k) and fictitious buses of the generator $\bar{v}_{\rm sh}$ or $\bar{v}_{\rm se}$. The detailed dc system has to be replaced using equivalent ac quantities $(p_{\rm sh}, q_{\rm sh}, v_{\rm sh}, \theta_{\rm sh})$ for the shunt connections or $(p_{\rm se}, q_{\rm se}, v_{\rm se}, \theta_{\rm se})$ for the series connection. Another issue is that these models are static and, thus, are not adequate for transient analysis.

As discussed for the detailed model, it is important to impose the adequate limits for both (18.27) and (18.29). These limits depends on the applications but can be resumed as voltage limits and current limits. Hence, for the shunt connected VSC:

$$\begin{aligned} v_{\rm sh}^{\rm min} \leq v_{\rm sh} &\leq v_{\rm sh}^{\rm max} \\ \sqrt{p_h^2 + q_h^2}/v_h &\leq i_{\rm sh}^{\rm max} \end{aligned} \tag{18.31}$$

and for the series-connected VSC:

$$\begin{aligned} v_{\rm se}^{\rm min} \leq v_{\rm se} &\leq v_{\rm se}^{\rm max} \\ \sqrt{p_h^2 + q_h^2}/v_h &\leq i_{\rm se}^{\rm max} \end{aligned} \tag{18.32}$$

Once a limit is violated, the corresponding constraint becomes binding and one of the control equations (18.28) or (18.30) has to be relaxed.

Example 18.4 Power Flow HVDC-VSC Model

This example describes the HVDC-VSC system that can be obtained using two shunt-connected VSC devices as modelled in (18.27). The power flow HVDC-VSC scheme is shown in Figure 18.11 [2]. The main differences with conventional thyristor-based HVDC links are the applications and the controllers of the rectifier and the inverter. The back-to-back configuration is used, for example, in direct-drive synchronous generator for wind turbine applications (see Section 20.2.7 of Chapter 20).

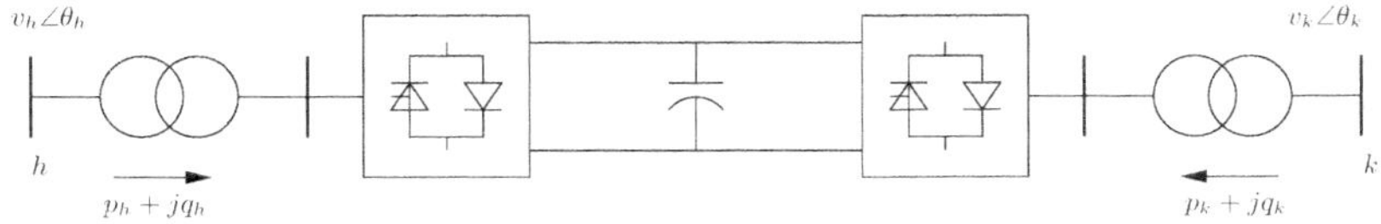

Fig. 18.11 HVDC-VSC scheme

As discussed above, each VSC requires two constraints. One of the four constraints that have to be defined is the power balance of the HVDC-VSC system. In fact, the power flow model (18.27) does not explicitly model the dc systems and, thus the power balance is the only way to link the two VSC devices that form the HVDC system. The power balance can be expressed as (see Figure 18.12):

$$0 = p_{\mathrm{dc},R} + p_{\mathrm{dc},I} + p_{\mathrm{dc\ loss}} \tag{18.33}$$

where back-to-back connections, $p_{\mathrm{dc\ loss}} \approx 0$, while for cable connections $p_{\mathrm{dc\ loss}}$ accounts for the Ohmic losses of the dc line.

It is interesting to note that the detailed hybrid model (17.2), (17.10) and (18.9)-(18.11) takes implicitly into account the power balance since dc node equations impose the Kirchhoff's current law, while ac bus equations impose the active and reactive power balances. This is another advantage of the detailed model over the simplified ones.

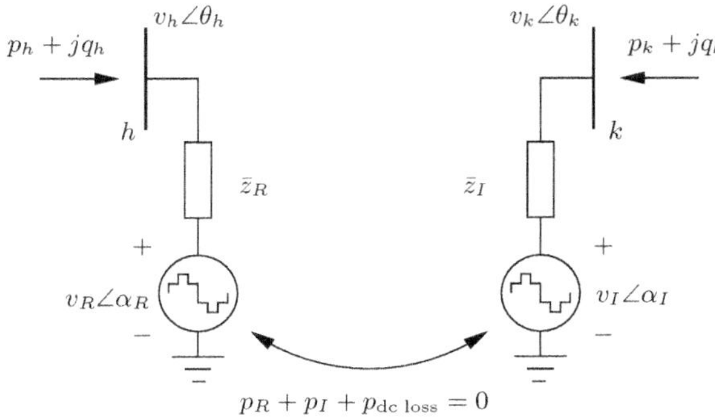

Fig. 18.12 Power flow HVDC-VSC model

Further constraints can be set as follows:

$$\begin{aligned} 0 &= p_h - p_h^{\mathrm{ref}} \\ 0 &= v_h - v_h^{\mathrm{ref}} \\ 0 &= v_k - v_k^{\mathrm{ref}} \end{aligned} \tag{18.34}$$

Chapter 19
FACTS Devices

This chapter describes Thyristor Controlled Reactor (TCR) and Voltage Sourced Converter (VSC) based Flexible AC Transmission System (FACTS) devices. In particular, the considered TCR-based FACTS devices are the static var compensator (Section 19.1) and the thyristor controlled series compensator (Section 19.2), whereas VSI-based FACTS devices are the static synchronous compensator (Section 19.3), the static synchronous series compensator (Section 19.4) and the unified power flow controller (Section 19.5).

19.1 Static Var Compensator

The Static Var Compensator (SVC) is a variable shunt capacitor that is varied to maintain a constant voltage at the bus to which it is connected. The admittance is varied using a thyristor-based switch, as shown in Figure 19.1.a. The firing angle α controls the turn-on period of the thyristor and hence varies the equivalent reactance of the SVC. Assuming a balanced, fundamental frequency operation, the equivalent susceptance of the SVC is a function of the firing angle α [129]:

$$b_{\rm SVC}(\alpha) = \frac{2\alpha - \sin 2\alpha - \pi(2 - x_L/x_C)}{\pi x_L} \tag{19.1}$$

From the numerical viewpoint, using the susceptance as defined in (19.1) is more stable than the reactance, because $1/b_{\rm SVC}(\alpha)$ tends to infinity at the resonant point, defined by:

$$2\alpha_r - \sin 2\alpha_r - \pi(2 - x_L/x_C) = 0 \tag{19.2}$$

19.1.1 SVC Type I

As discussed above, the controlled variable is the firing angle α. Thus, the regulator has to vary α in order to control the bus voltage (see Figure 19.2).

F. Milano: Power System Modelling and Scripting, Power Systems, pp. 413–434.
springerlink.com

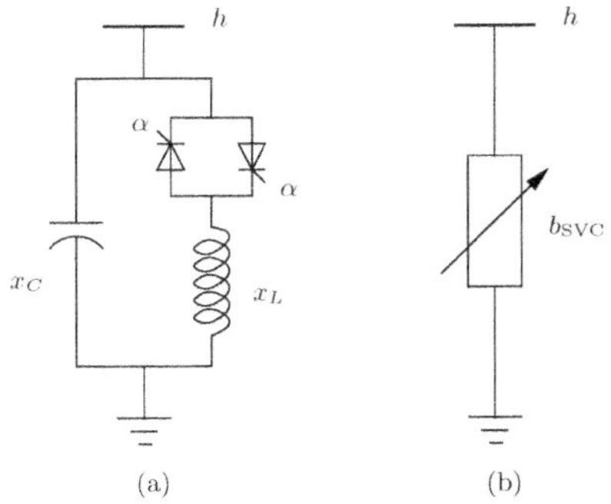

Fig. 19.1 SVC schemes: (a) firing angle model and (b) equivalent susceptance model

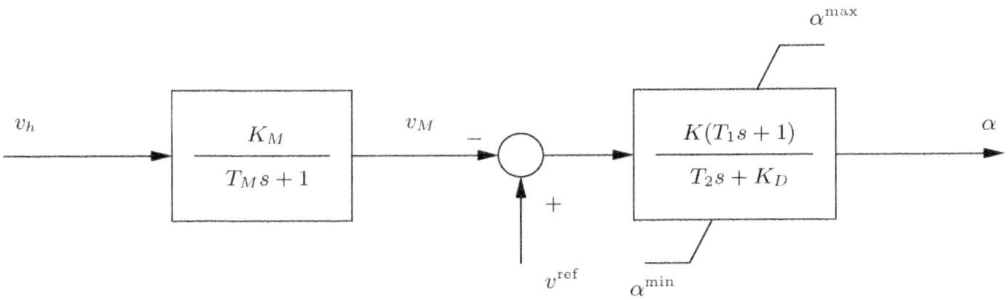

Fig. 19.2 SVC Type I control diagram

The DAE system is follows:

$$\begin{aligned} \dot{v}_M &= (K_M v_h - v_M)/T_M \\ \dot{\alpha} &= (-K_D \alpha + K \frac{T_1}{T_2 T_M}(v_M - K_M v_h) + K(v^{\mathrm{ref}} - v_M))/T_2 \\ q_h &= \frac{2\alpha - \sin 2\alpha - \pi(2 - x_L/x_C)}{\pi x_L} v_h^2 \end{aligned} \tag{19.3}$$

The state variable α undergoes an anti-windup limiter which indirectly allows limiting the SVC current. Table 19.1 defines all parameters of the SVC Type I.

19.1.2 SVC Type II

A common approximation consists in assuming that the controlled variable is b_{SVC} and not the firing angle α (see Figure 19.1.b). The simplified control scheme is depicted in Figure 19.3 and undergoes the following differential equation:

$$\dot{b}_{\mathrm{SVC}} = (K_r(v^{\mathrm{ref}} - v_h) - b_{\mathrm{SVC}})/T_r \tag{19.4}$$

Table 19.1 SVC Type I parameters

Variable	Description	Unit
K	Regulator gain	rad/pu
K_D	Integral deviation	-
K_M	Measure gain	pu/pu
T_1	Transient regulator time constant	s
T_2	Regulator time constant	s
T_M	Measure time delay	s
$v^{\rm ref}$	Reference Voltage	pu
x_L	Reactance (inductive)	pu
x_C	Reactance (capacitive)	pu
$\alpha^{\max}$	Maximum firing angle	rad
$\alpha^{\min}$	Minimum firing angle	rad

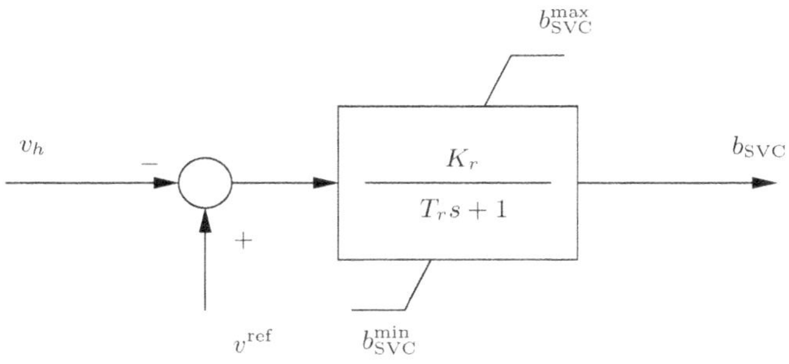

Fig. 19.3 SVC Type II control diagram

The model is completed by the algebraic equation expressing the reactive power injected at the SVC node:

$$q = b_{\rm SVC} v_h^2 \tag{19.5}$$

The regulator has an anti-windup limiter, i.e., the reactance $b_{\rm SVC}$ is locked if one of its limits is reached. Table 19.2 reports data and control parameters for the SVC type 2.

19.1.3 SVC Initialization

Although there is no particular issue in including the detailed SVC model into the power flow analysis, SVC devices are typically initialized after power flow analysis. To impose the voltage regulation a PV generator with $p_G^0 = 0$ can be used. The only difference is that the reactive power limits of the PV generator do not coincide exactly with the susceptance limits of the SVC device, in fact, $q_G^{\max} \neq b_{\rm SVC}^{\max} v_h^2$ only for the nominal value of the bus voltage.

Table 19.2 SVC Type II parameters

Variable	Description	Unit
-	Bus code	-
$b_{\mathrm{SVC}}^{\max}$	Maximum susceptance	pu
$b_{\mathrm{SVC}}^{\min}$	Minimum susceptance	pu
K_r	Regulator gain	pu/pu
T_r	Regulator time constant	s
v^{ref}	Reference Voltage	pu

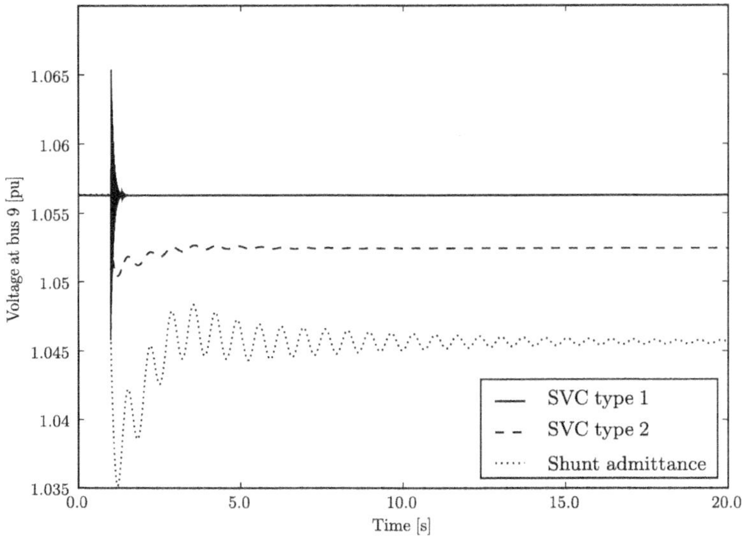

Fig. 19.4 Comparison of SVC models for the IEEE 14-bus system

Example 19.1 Comparison of SVC Models

Figure 19.4 shows a comparison of the transient response of the SVC models Type I and II described above. The plot was obtained substituting the static shunt admittance at bus 9 of the IEEE 14-bus system for an SVC device. All SVC data are given in Appendix D. The reference voltage of SVC regulators is $v^{\mathrm{ref}} = 1.0563$ pu, i.e., the same voltage value as obtained by the power flow analysis using the static shunt admittance. The regulators of both SVC models improve the voltage profile at bus 9. SVC Type II shows a zero static error (due to the integral regulator) but presents high-frequency oscillations.

19.2 Thyristor Controlled Series Compensator

The Thyristor Controlled Series Compensator (TCSC) allows varying the series reactance of a transmission line and, thus, regulating the active flow through the transmission line itself. The functioning of the TCSC is similar to the SVC, but for the fact that the TCSC is a series device, as shown in Figure 19.5.

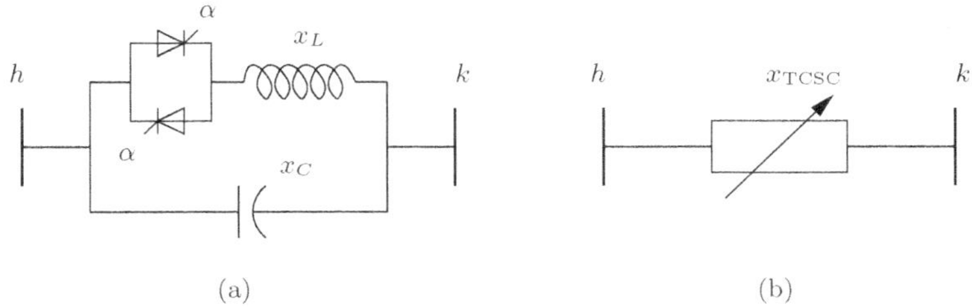

Fig. 19.5 TCSC schemes: (a) firing angle model and (b) equivalent susceptance model

The regulated variable is the firing angle α. The equivalent series reactance x_{TCSC} in balanced, fundamental frequency conditions is [129]:

$$
\begin{aligned}
x_{\mathrm{TCSC}}(\alpha) = & \Big[x_C\Big(\pi k_x^4 \cos k_x(\pi - \alpha) \\
& - \pi \cos k_x(\pi-\alpha) - 2k_x^4\alpha \cos k_x(\pi-\alpha) \\
& + 2\alpha k_x^2 \cos k_x(\pi-\alpha) - k_x^4 \sin 2\alpha \cos k_x(\pi-\alpha) \\
& + k_x^2 \sin 2\alpha \cos k_x(\pi-\alpha) - 4k_x^3 \cos^2\alpha \sin k_x(\pi-\alpha) \\
& - 4k_x^2 \cos\alpha \sin\alpha \cos k_x(\pi-\alpha)\Big)\Big] / [\pi(k_x^4 - 2k_x^2 + 1)\cos k_x(\pi-\alpha)] \\
k_x = & \sqrt{\frac{x_C}{x_L}}
\end{aligned}
\tag{19.6}
$$

The power injections at buses h and k are:

$$
\begin{aligned}
p_h &= v_h v_k b(\alpha) \sin(\theta_h - \theta_k) \\
q_h &= v_h^2 b(\alpha) - v_h v_k b(\alpha) \cos(\theta_h - \theta_k) \\
p_k &= -p_h \quad = \quad -v_h v_k b(\alpha) \sin(\theta_h - \theta_k) \\
q_k &= v_k^2 b(\alpha) - v_h v_k b(\alpha) \cos(\theta_h - \theta_k)
\end{aligned}
\tag{19.7}
$$

where

$$
b(\alpha) = \frac{1}{x_{\mathrm{TCSC}}(\alpha)} \tag{19.8}
$$

The model (19.7) can show numerical issues in case $x_{\mathrm{TCSC}}(\alpha) = 0$.

Since the TCSC device is always connected in series with a transmission line, another common model considers the series of the TCSC and the transmission line as an unique device. Assuming a loss-less line and that the series reactance of the line is x_{hk}, the resulting power flow equations are:

$$
\begin{aligned}
p_{hk} &= v_h v_k b(\alpha, x_{hk}) \sin(\theta_h - \theta_k) \\
p_{kh} &= -p_{hk} \\
q_{hk} &= v_h^2 b(\alpha, x_{hk}) - v_h v_k b(\alpha, x_{hk}) \cos(\theta_h - \theta_k) \\
q_{kh} &= v_k^2 b(\alpha, x_{hk}) - v_h v_k b(\alpha, x_{hk}) \cos(\theta_h - \theta_k)
\end{aligned} \tag{19.9}
$$

where:

$$
b(\alpha, x_{hk}) = \frac{1}{x_{hk} + x_{\text{TCSC}}(\alpha)} \tag{19.10}
$$

This model also allows removing numerically issues of (19.7) of $x_{\text{TCSC}}(\alpha)$ if $|x_{\text{TCSC}}^{\min}| < x_{hk}$.

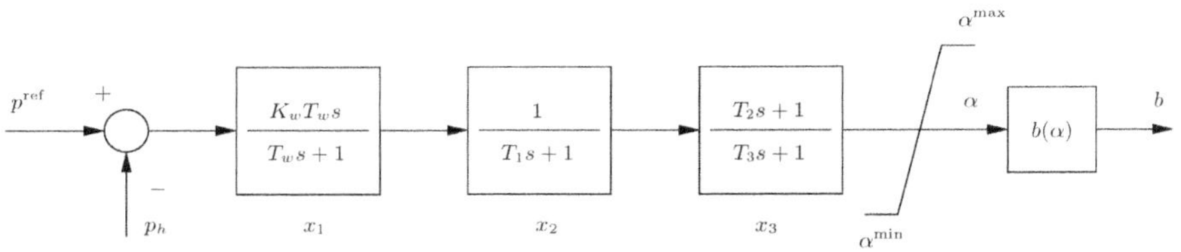

Fig. 19.6 TCSC control diagram

A common control scheme for TCSC devices is shown in Figure 19.6 and works for regulating both the firing angle α and the reactance x_{TCSC}. The DAE system is as follows:

$$
\begin{aligned}
\dot{x}_1 &= -K_w(p^{\text{ref}} - p_h)/T_w - x_1/T_w \\
\dot{x}_2 &= (x_1 - x_2 + (p^{\text{ref}} - p_h))/T_1 \\
\dot{x}_3 &= ((1 - T_2/T_3)x_2 - x_3)/T_3
\end{aligned} \tag{19.11}
$$

where all parameters are defined in Table 19.3. The output signal of the lead-lag block can be either the firing angle α or, using a simplified model, the reactance x_{TCSC}. Considering the firing angle, one has:

$$
\alpha = \frac{T_2}{T_3} x_2 + x_3 \tag{19.12}
$$

If using the reactance x_{TCSC}, equation (19.12) and the control scheme of Figure 19.6 are still valid but substituting α with x_{TCSC}.

Table 19.3 TCSC parameters

Variable	Description	Unit
K_w	Regulator gain	pu/pu
p^{ref}	Reference power	pu
T_1	Low-pass time constant	s
T_2	Lead time constant	s
T_3	Lag time constant	s
T_w	Washout time constant	s
x_C	Reactance (capacitive)	pu
x_L	Reactance (inductive)	pu
$x_{\text{TCSC}}^{\text{max}}$ (α^{max})	Maximum reactance (firing angle)	pu (rad)
$x_{\text{TCSC}}^{\text{min}}$ (α^{min})	Minimum reactance (firing angle)	pu (rad)

19.2.1 TCSC Initialization

There are various methods for initializing a TCSC. A possibility is to include the detailed model of the TCSC in the power flow analysis. However, this is not the common practice. Another strategy is to use a static tie line model such that described in Subsection 11.1.2 of Chapter 11.

19.3 Static Synchronous Compensator

The Static Synchronous Compensator (STATCOM) is a shunt-connected VSC-based FACTS device that regulates the voltage of the ac bus to which it is connected (see Figure 19.7). The effect is similar to the SVC device, but the internal model is rather different, as discussed in the following sections.

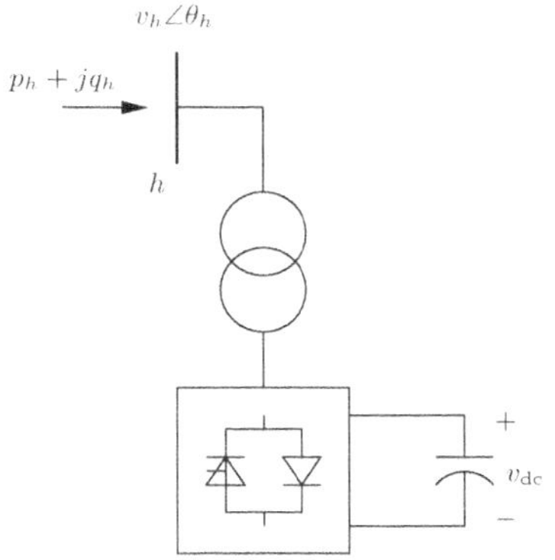

Fig. 19.7 STATCOM scheme

19.3.1 Detailed Model

The detailed model consists of a shunt-connected VSC device with a capacitor in the dc side. The model is composed of three parts, namely the dc network, the VSC and the controllers.

Dc network: The dc side is a parallel RC defined by (17.10) and (17.2). Furthermore, the dc network must contain two nodes, one of which is connected to the ground. The VSC and the RC element are connected in parallel to the dc nodes.

Shunt-connected VSC model: The equations of the VSC are (17.2) and (18.9)-(18.12).

Regulators: The ac voltage control is obtained regulating the modulating amplitude a_m [322]:

$$\dot{a}_m = \frac{1}{T_2}\left[-K_D a_m + K(v_{\mathrm{ac}}^{\mathrm{ref}} - v_{\mathrm{ac}}^m)\right] - \frac{KT_1}{T_2 T_{\mathrm{ac}}}(K_{\mathrm{ac}} v_h - v_{\mathrm{ac}}^m) \tag{19.13}$$

whereas the dc voltage control is regulated through the firing angle α:

$$\dot{\alpha} = \left(\frac{K_P}{T_{\mathrm{dc}}} - K_I\right) v_{\mathrm{dc}}^m + K_I v_{\mathrm{dc}}^{\mathrm{ref}} - \frac{K_P K_{\mathrm{dc}}}{T_{\mathrm{dc}}} v_{\mathrm{dc}} \tag{19.14}$$

Finally, low pass filters are used for modelling both the ac and dc voltage measurements:

$$\begin{aligned} \dot{v}_{\mathrm{ac}}^m &= (-v_{\mathrm{ac}}^m + K_{\mathrm{ac}} v_h)/T_{\mathrm{ac}} \\ \dot{v}_{\mathrm{dc}}^m &= (-v_{\mathrm{dc}}^m + K_{\mathrm{dc}} v_{\mathrm{dc}})/T_{\mathrm{dc}} \end{aligned} \tag{19.15}$$

Voltage and modulating amplitude controls along with the measurement transfer functions are depicted in Figure 19.8. Both the ac and dc voltage controls undergo a windup limiter. For the dc voltage control, the limits on α can be computed imposing the power balance:

$$\begin{aligned} 0 = &\frac{v_{\mathrm{dc}}^2}{r_{\mathrm{dc}}} + r_T i_{\mathrm{ac}}^2 - v_h^2 g_T \\ &+ \sqrt{3/8} v_{\mathrm{dc}} v_h g_T \cos(\alpha) + \sqrt{3/8} v_{\mathrm{dc}} v_h b_T \sin(\alpha) \end{aligned} \tag{19.16}$$

where $g_T + jb_T = 1/(r_T + jx_T)$ is the impedance of the VSC embedded transformer. Imposing $v_{\mathrm{dc}} = v_{\mathrm{dc}}^{\mathrm{ref}}$ and $v_h = v_{\mathrm{ac}}^{\mathrm{ref}}$, from (19.16) it can be obtained:

$$\cos(\alpha) = \frac{bc}{a^2 + b^2} \pm \sqrt{\left(\frac{bc}{a^2 + b^2}\right)^2 - \frac{c^2 - a^2}{a^2 + b^2}} \tag{19.17}$$

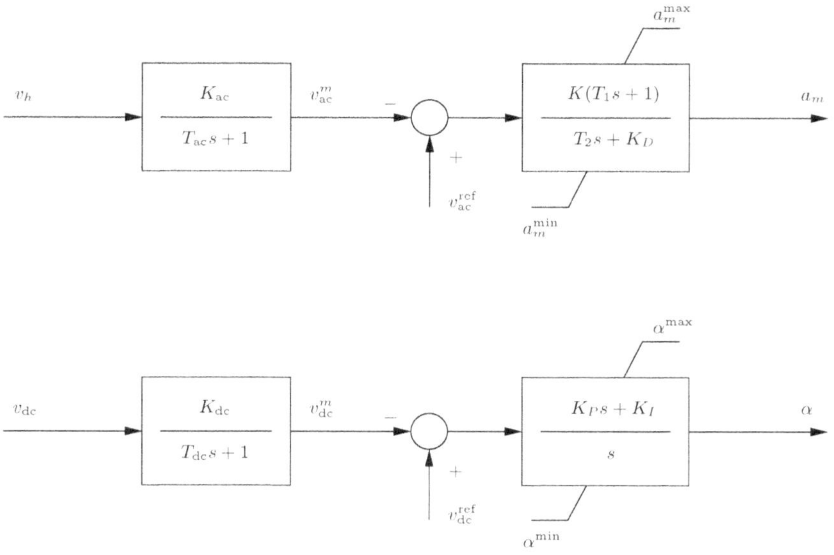

Fig. 19.8 STATCOM ac and dc voltage control diagrams

where

$$
\begin{aligned}
a &= -\sqrt{3/8}v_{dc}^{ref}v_{ac}^{ref}b_T \\
b &= -\sqrt{3/8}v_{dc}^{ref}v_{ac}^{ref}g_T \\
c &= (v_{ac}^{ref})^2 g_T - \frac{(v_{dc}^{ref})^2}{r_{dc}} - r_T i_{ac}^2
\end{aligned}
\tag{19.18}
$$

Finally, the limits for the firing angle α are computed imposing in the equation (19.17) the limits $i^{\max}$ and $i^{\min}$. Table 19.4 defines all parameters required by STATCOM regulators.

19.3.2 Simplified Dynamic Model

A simplified STATCOM current injection model has been proposed in [61, 122, 252]. The STATCOM current i_{sh} is always kept in quadrature in relation to the bus voltage so that only reactive power is exchanged between the ac system and the STATCOM. The equivalent circuit and the control scheme are shown in Figure 19.9. The differential equation and the reactive power injected at the STATCOM node are, respectively:

$$
\begin{aligned}
\dot{i}_{sh} &= (K_r(v^{ref} - v_h) - i_{sh})/T_r \\
q_h &= i_{sh} v_h
\end{aligned}
\tag{19.19}
$$

Table 19.4 STATCOM regulator parameters

Variable	Description	Unit
K	Gain of the ac voltage control	pu/pu
K_{ac}	Gain of the ac measurement	pu/pu
K_D	Integral deviation of the ac voltage control	-
K_{dc}	Gain of the dc measurement	pu/pu
K_I	Integral gain for the dc voltage control	rad/pu/s
K_P	Proportional gain for the dc voltage control	rad/pu
$i^{\max}$	Maximum current	pu
$i^{\min}$	Minimum current	pu
r_{dc}	Resistance of the dc circuit	pu
r_T	Resistance of the ac circuit	pu
T_1	Transient time constant of the ac voltage control	s
T_2	Time constant of the ac voltage control	s
T_{ac}	Time constant of the ac measurement	s
T_{dc}	Time constant of the dc measurement	s
v^{ref}	ac reference voltage	pu
$v_{\mathrm{dc}}^{\mathrm{ref}}$	dc reference voltage	pu
x_T	Reactance of the ac circuit	pu

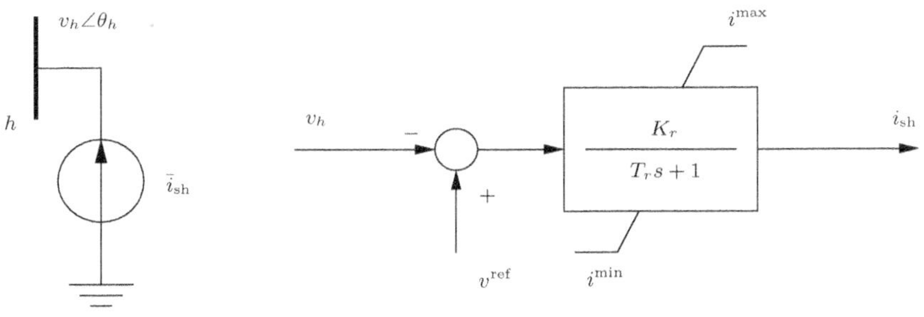

Fig. 19.9 STATCOM circuit and control diagram

where all parameters are defined in Table 19.5. The current i_{sh} undergoes an anti-windup limiter.

19.3.3 Power Flow Model

The STATCOM power flow model is simply obtained based on the power flow model of the VSC device (18.27) and (18.28). In particular, (18.28) are:

$$0 = p_{\mathrm{sh}}, \qquad 0 = v^{\mathrm{ref}} - v_{\mathrm{sh}} \tag{19.20}$$

The latter condition is satisfied only if the current $i_{\mathrm{sh}} \leq i_{\mathrm{sh}}^{\max}$.

Table 19.5 Current-injection STATCOM parameters

Variable	Description	Unit
K_r	Regulator gain	pu/pu
$i^{\max}$	Maximum current	pu
$i^{\min}$	Minimum current	pu
T_r	Regulator time constant	s

19.3.4 STATCOM Initialization

Dynamic STATCOM devices can be initialized using a static PV generator with $p_{G0} = 0$ (similarly to the SVC device), or using the power flow model described in the previous Subsection 19.3.3, which can be viewed as a PV generator behind an impedance.

Example 19.2 Comparison of STATCOM Models

Figure 19.10 shows the transient behavior of the detailed STATCOM model described in Subsection 19.3.1 as well as that of the simplified STATCOM model described in Subsection 19.3.2. The plot was obtained substituting the static shunt admittance at bus 9 of the IEEE 14-bus system for the detailed or the simplified STATCOM models. All data are given in Appendix D. The reference voltage of the STATCOM regulators is $v^{\text{ref}} = 1.0563$ pu, i.e., the same voltage value as obtained by the power flow analysis using the static shunt admittance. The detailed model behaves similarly to the SVC Type I, whereas the simplified model behaves similarly to the SVC Type II (see Figure 19.4). High frequency oscillations shown by the detailed model are due to the interaction between the VSC control and the dc circuit dynamic. Furthermore, due to fast dc dynamics, the numerical integration requires a relatively small step length for the detailed STATCOM model.

19.4 Static Synchronous Series Compensator

The Static Synchronous Series Compensator (SSSC) is a series-connected VSC-based FACTS device that regulates the active power flow between the two ac buses to which it is connected. The effect is similar to the TCSC device or the phase-shifting regulating transformer, but the internal model is rather different, as discussed in the following sections.

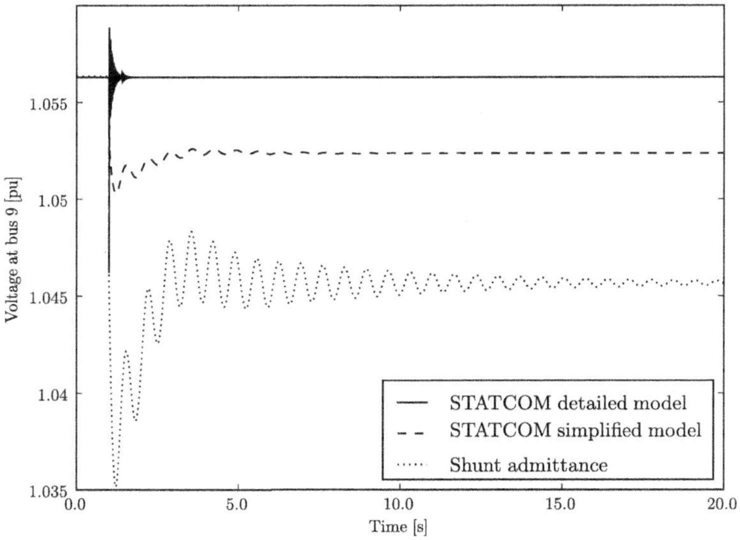

Fig. 19.10 Comparison of STATCOM models for the IEEE 14-bus system

19.4.1 Detailed Model

The detailed model consists of a series-connected VSC device with a capacitor on the dc side (see Figure 19.11). The model is composed of three parts, namely the dc network, the VSC and the controllers.

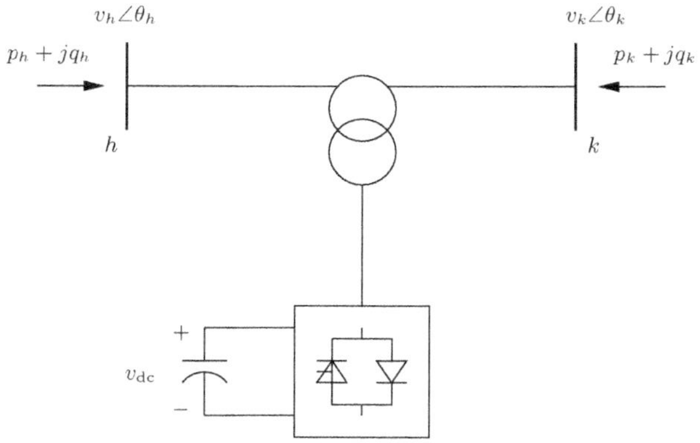

Fig. 19.11 SSSC scheme

Dc network: The dc side is a parallel RC defined by (17.10) and (17.2). Furthermore, the dc network must contain two nodes, one of which is connected to the ground. The VSC and the RC element are connected in parallel to the dc nodes.

Series-connected VSC model: The equations of the VSC are (17.2), (18.9)-(18.11) and (18.13).

Regulators: The control system is then used for controlling the dc voltage $v_{\rm dc}$ and the active power p_h or current flow i_h in the ac side. Figure 19.12 depicts the dc voltage control, the active power control and the measurement transfer functions, which are described by the following DAE system:

$$
\begin{aligned}
\dot{v}_{\rm dc}^m &= (K_{\rm dc} v_{\rm dc} - v_{\rm dc}^m)/T_{\rm dc} \\
\dot{\alpha} &= \left(\frac{K_P}{T_{\rm dc}} - K_I\right) v_{\rm dc}^m + K_I v_{\rm dc}^{\rm ref} - \frac{K_P K_{\rm dc}}{T_{\rm dc}} v_{\rm dc} \\
\dot{p}_{\rm ac}^m &= (K_{\rm ac} p_h - p_{\rm ac}^m)/T_{\rm ac} \\
\dot{a}_m &= \frac{1}{T_2}\left[-K_D a_m + K(p^{\rm ref} - p_{\rm ac}^m)\right] - \frac{K T_1}{T_2 T_{\rm ac}}(K_{\rm ac} p_h - p_{\rm ac}^m)
\end{aligned}
\tag{19.21}
$$

where all parameters are defined in Table 19.12. Both the firing angle α and the modulating amplitude a_m undergo windup limiters. The active power signal in the firing angle control diagram can be substituted by the current i_h.

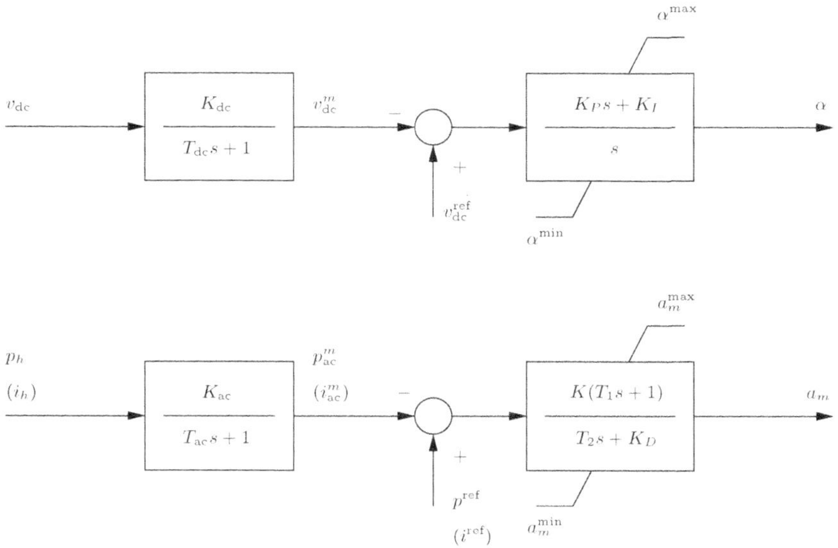

Fig. 19.12 SSSC control diagrams

Table 19.6 SSSC regulator parameters

Variable	Description	Unit
K	Gain of the ac voltage control	pu/pu
K_{ac}	Gain of the ac measurement	pu/pu
K_D	Integral deviation of the ac voltage control	-
K_{dc}	Gain of the ac measurement	pu/pu
K_I	Integral gain for the α control	rad/pu/s
K_P	Proportional gain for the α control	rad/pu
p^{ref} (i^{ref})	Ac reference power (current)	pu
T_1	Transient time constant of the ac control	s
T_2	Time constant of the ac control	s
T_{ac}	Time constant of the ac measurement	s
T_{dc}	Time constant of the dc measurement	s
$v_{\text{dc}}^{\text{ref}}$	Dc reference voltage	pu

19.4.2 Simplified Dynamic Model

Simplified SSSC dynamic models have been proposed in [162, 193, 288] where the SSSC is represented by a series voltage source $\bar{v}_S$, as depicted in Figure 19.13. The voltage $\bar{v}_S$ is in quadrature with line current. Thus, the voltage magnitude v_S is the only controllable variable. The total active and reactive power flows in a transmission line in series with a SSSC device are:

$$\begin{aligned} p_h &= (1+c)\frac{v_h v_k}{x_{hk}}\sin(\theta_h - \theta_k) \\ p_k &= -p_h \\ q_h &= (1+c)\frac{v_h}{x_{hk}}(v_h - v_k\cos(\theta_h - \theta_k) \\ q_k &= (1+c)\frac{v_k}{x_{hk}}(v_k - v_h\cos(\theta_h - \theta_k) \end{aligned} \tag{19.22}$$

where x_{hk} is the series reactance of the transmission line assumed loss-less and:

$$c = \frac{v_S}{\sqrt{v_h^2 + v_k^2 - 2v_h v_k \cos(\theta_h - \theta_k)}} \tag{19.23}$$

The SSSC regulator is shown in Figure 19.14. The regulator differential equations are:

$$\begin{aligned} \dot{x} &= K_I(p^{\text{ref}} - p_h) \\ 0 &= x + K_I(p^{\text{ref}} - p_h) - \tilde{v}_S \\ \dot{v}_S &= (\tilde{v}_S - v_S)/T_r \end{aligned} \tag{19.24}$$

Table 19.7 defines all parameters of the SSSC simplified dynamic model.

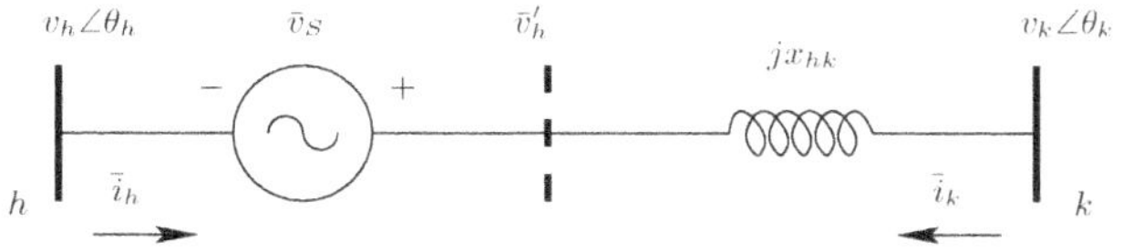

Fig. 19.13 Simplified SSSC circuit

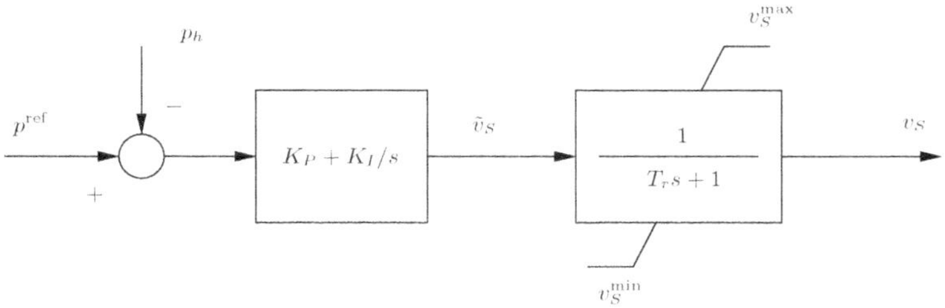

Fig. 19.14 SSSC simplified control diagram

Table 19.7 Simplified SSSC model parameters

Variable	Description	Unit
K_I	Integral gain of PI controller	pu/pu/s
K_P	Proportional gain of PI controller	pu/pu
T_r	Regulator time constant	s
$v_S^{\max}$	Maximum series voltage v_S	pu
$v_S^{\min}$	Minimum series voltage v_S	pu

19.4.3 Power Flow Model

The SSSC power flow model is simply obtained based on the power flow model of the VSC device (18.29) and (18.30). In particular, (18.30) are:

$$0 = p_{\rm se}, \qquad 0 = p_h - p^{\rm ref} \tag{19.25}$$

The latter condition is satisfied only if the current $i_{\rm se} \leq i_{\rm se}^{\max}$.

19.4.4 SSSC Initialization

Dynamic SSSC devices can be initialized using a static tie line, similarly to the TCSC devices. The tie line provides the compensated value $\tilde{x}_{hk}$ of the series reactance that leads to the desired active power flow. Then, the SSSC state variable v_S can be obtained by solving:

$$0 = \frac{x_{hk}}{1 + c(v_S)} - \tilde{x}_{hk} \tag{19.26}$$

Alternatively, one can use the power flow model described in Subsection 19.4.3.

19.5 Unified Power Flow Controller

The Unified Power Flow Controller (UPFC) combines together a shunt and a series VSC with a common capacitor in the dc side. The shunt connection allows regulating the voltage (like a STATCOM) while the series connection allows controlling the power flow (like a SSSC). The UPFC is a very versatile device although it is relatively rare due to its complexity and high cost.

19.5.1 Detailed Model

The detailed model consists in a shunt-connected VSC device and a series-connected VSC device with a common capacitor in the dc side (see Figure 19.15). The model is composed of four parts, namely the dc network, the shunt VSC, the series VSC and the controllers.

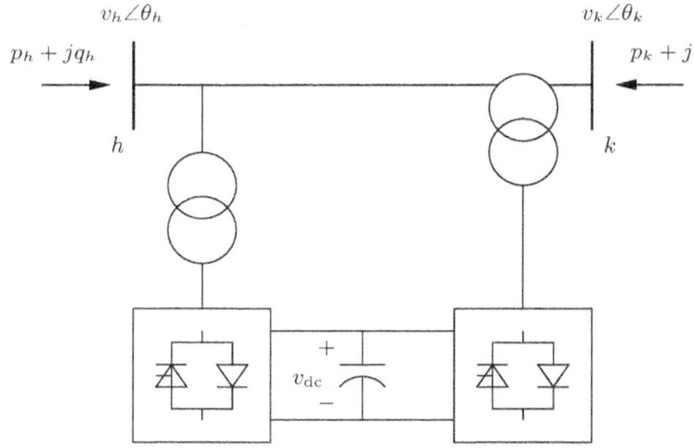

Fig. 19.15 UPFC scheme

Dc network: The dc side is a parallel RC defined by (17.10) and (17.2). Furthermore, the dc network must contain two nodes, one of which is connected to the ground. The shunt VSC, the series VSC and the RC element are connected in parallel.

Shunt-connected VSC model: The equations of the VSC are (17.2) and (18.9)-(18.12).

Series-connected VSC model: The equations of the VSC are (17.2), (18.9)-(18.11) and (18.13).

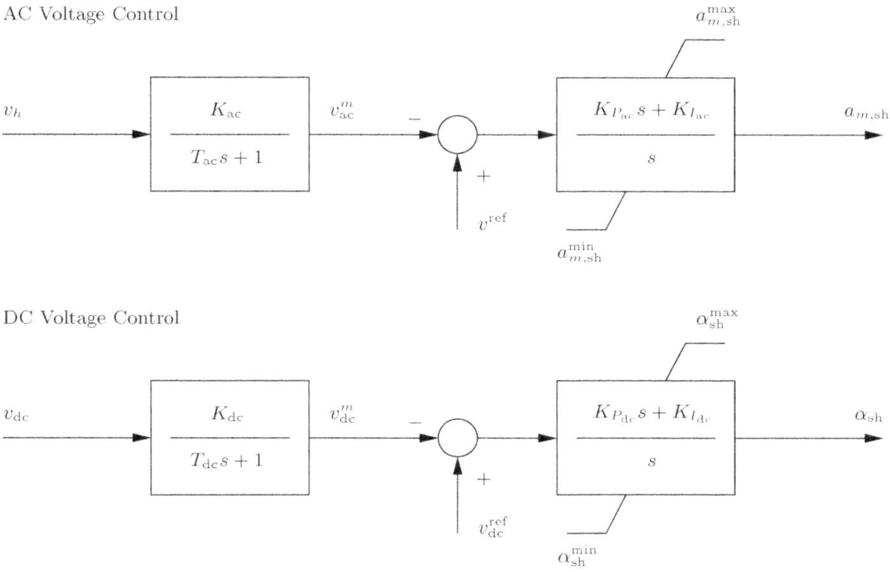

Fig. 19.16 UPFC shunt control diagrams

Regulators: Figure 19.16 depicts the shunt control for the ac voltage v_h and the dc voltage v_{dc}, obtained by means of PI regulators and measurement filters. The equations are:

$$\begin{aligned}
\dot{v}_{\mathrm{ac}}^m &= (-v_{\mathrm{ac}}^m + K_{\mathrm{ac}} v_h)/T_{\mathrm{ac}} \\
\dot{a}_{m,\mathrm{sh}} &= \left(\frac{K_{P_{\mathrm{ac}}}}{T_{\mathrm{ac}}} - K_{I_{\mathrm{ac}}}\right) v_{\mathrm{ac}}^m + K_{I_{\mathrm{ac}}} v^{\mathrm{ref}} - \frac{K_{P_{\mathrm{ac}}} K_{\mathrm{ac}}}{T_{\mathrm{ac}}} v_h \\
\dot{v}_{\mathrm{dc}}^m &= (-v_{\mathrm{dc}}^m + K_{\mathrm{dc}} v_{\mathrm{dc}})/T_{\mathrm{dc}} \\
\dot{\alpha}_{\mathrm{sh}} &= \left(\frac{K_{P_{\mathrm{dc}}}}{T_{\mathrm{dc}}} - K_{I_{\mathrm{dc}}}\right) v_{\mathrm{dc}}^m + K_{I_{\mathrm{dc}}} v_{\mathrm{dc}}^{\mathrm{ref}} - \frac{K_{P_{\mathrm{dc}}} K_{\mathrm{dc}}}{T_{\mathrm{dc}}} v_{\mathrm{dc}}
\end{aligned} \tag{19.27}$$

The power flow regulation is a dq control, as shown in Figure 19.17, which allows decoupling the controls of the active and reactive powers [232, 323].

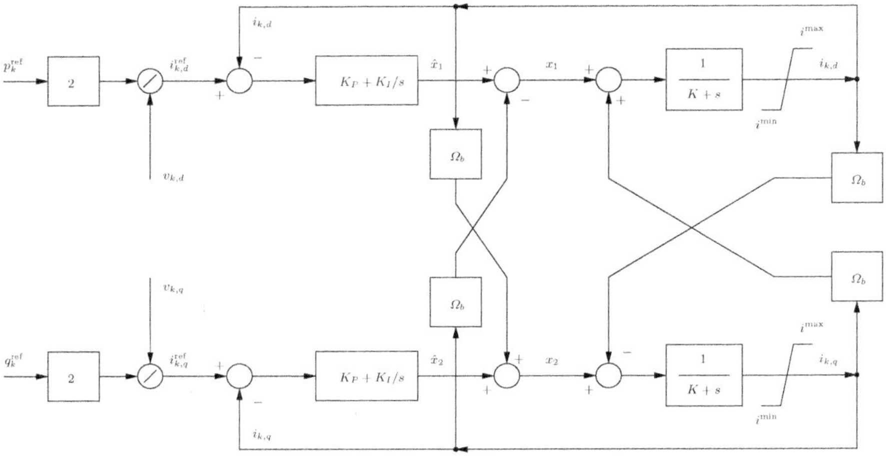

Fig. 19.17 UPFC series dq control diagrams

The DAE system is as follows:

$$
\begin{aligned}
\dot{\hat{x}}_1 &= K_I\left(\frac{2p_k^{\text{ref}}}{v_{k,d}} - i_{k,d}\right) \\
\dot{i}_{k,d} &= \hat{x}_1 - K i_{k,d} + K_P\left(\frac{2p_k^{\text{ref}}}{v_{k,d}} - i_{k,d}\right) \\
\dot{\hat{x}}_2 &= K_I\left(\frac{2q_k^{\text{ref}}}{v_{k,q}} - i_{k,q}\right) \\
\dot{i}_{k,q} &= \hat{x}_2 - K i_{k,q} + K_P\left(\frac{2q_k^{\text{ref}}}{v_{k,q}} - i_{k,q}\right)
\end{aligned}
\tag{19.28}
$$

where the sub-indexes d and q indicate the direct and quadrature components, respectively. For this series branch dq control, the variables $a_{m,\text{se}}$, α_{se} and other control parameters are given by the following relations where Ω_b is the fundamental frequency base in rad/s and other variables are expressed in per unit:

$$
\begin{aligned}
K &= \frac{r_{\rm se}}{\Omega_b x_{\rm se}} \\
v_{k,d} &= \sqrt{2} v_k \\
v_{h,d} &= \sqrt{2} v_h \cos(\theta_k - \theta_h) \\
v_{h,q} &= \sqrt{2} v_h \sin(\theta_k - \theta_h) \\
x_1 &= \hat{x}_1 + K_P \left(\frac{2 p_k^{\rm ref}}{v_{k,d}} - i_{k,d} \right) - \Omega_b i_{k,q} \\
x_2 &= \hat{x}_2 + K_P \left(\frac{2 q_k^{\rm ref}}{v_{k,q}} - i_{k,q} \right) + \Omega_b i_{k,d} \\
v_{{\rm se},d} &= v_{h,d} - v_{k,d} - \frac{x_{\rm se}}{\Omega_b} x_1 \\
v_{{\rm se},q} &= v_{h,q} - \frac{x_{\rm se}}{\Omega_b} x_2 \\
v_{\rm se} &= \frac{1}{\sqrt{2}} \sqrt{v_{{\rm se},d}^2 + v_{{\rm se},q}^2} \\
a_{m,{\rm se}} &= \sqrt{\frac{8}{3}} \frac{v_{\rm se}}{v_{\rm dc}} \\
\alpha_{\rm se} &= \theta_k - \tan^{-1} \left(\frac{v_{{\rm se},q}}{v_{{\rm se},d}} \right)
\end{aligned}
\tag{19.29}
$$

All other parameters are defined in Table 19.8.

19.5.2 Simplified Dynamic Model

Simplified UPFC dynamic models are proposed in [161, 192, 213]. The equivalent circuit is obtained merging together the STATCOM and the SSSC simplified models, i.e., a series voltage source $\bar{v}_S$ and a shunt current source $\bar{\imath}_{\rm sh}$, as depicted in Figure 19.18.

According to the vector diagram of Figure 19.19, the series voltage $\bar{v}_S$ and the shunt current $\bar{v}_{\rm sh}$ sources are defined as:

$$
\begin{aligned}
\bar{v}_S &= v_S e^{j(\gamma + \theta_h)} = (v_p + j v_q) e^{j(\theta_h - \phi)} \\
\bar{\imath}_{\rm sh} &= (i_p + j i_q) e^{j\theta_h}
\end{aligned}
\tag{19.30}
$$

where:

v_p is the component of the voltage $\bar{v}_S$ that is in phase with the line current $\bar{\imath}_h$.

v_q is the component of the voltage $\bar{v}_S$ that is in quadrature with line current $\bar{\imath}_h$.

i_p is the component of the current $\bar{\imath}_{\rm sh}$ in phase with the voltage $\bar{v}_h$. Typically $i_p = 0$.

Table 19.8 UPFC regulator parameters

Variable	Description	Unit
K_{ac}	Gain of the ac measurement	pu/pu
K_{dc}	Gain of the dc measurement	pu/pu
K_I	Integral gain for the dq control	pu/pu/s
$K_{I_{\mathrm{ac}}}$	Integral gain for the ac voltage control	pu/pu/s
$K_{I_{\mathrm{dc}}}$	Integral gain of the v_{dc} control	rad/pu/s
K_P	Proportional gain for the dq control	pu/pu
$K_{P_{\mathrm{dc}}}$	Proportional gain of the v_{dc} control	rad/pu
$K_{P_{\mathrm{ac}}}$	Proportional gain for the ac voltage control	pu/pu
$i^{\max}$	Maximum current	pu
$i^{\min}$	Minimum current	pu
p_k^{ref}	Active reference power	pu
q_k^{ref}	Reactive reference power	pu
r_{se}	Resistance of the series ac circuit	pu
r_{sh}	Resistance of the shunt ac circuit	pu
T_{ac}	Time constant of the ac measurement	s
T_{dc}	Time constant of the dc measurement	s
v^{ref}	ac voltage reference	pu
$v_{\mathrm{dc}}^{\mathrm{ref}}$	dc voltage reference	pu
x_{se}	Reactance of the series ac circuit	pu
x_{sh}	Reactance of the shunt ac circuit	pu

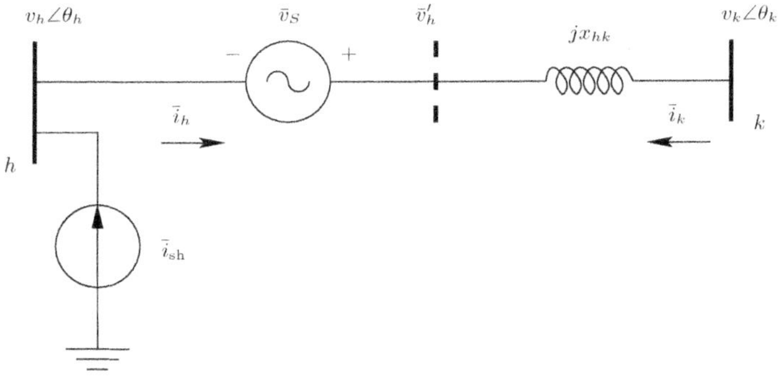

Fig. 19.18 Simplified UPFC circuit

i_q is the component of the current $\bar{i}_{\mathrm{sh}}$ in quadrature with the voltage $\bar{v}_h$. Typically $i_{\mathrm{sh}} = i_q$.

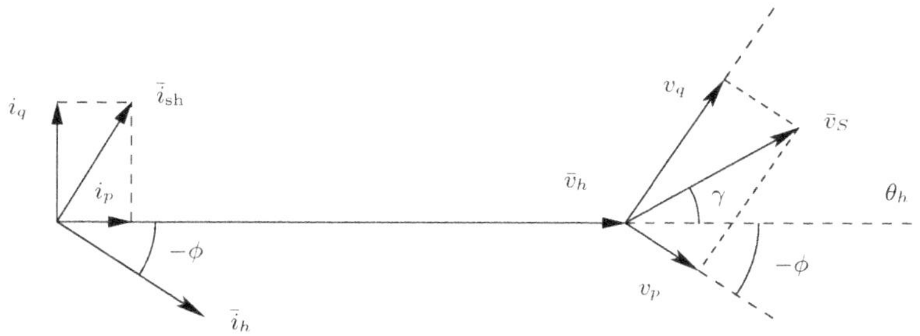

Fig. 19.19 UPFC phasor diagram

The resulting DAE system is as follows. Algebraic equations are:

$$\begin{aligned} p_h &= \frac{1}{x_{hk}}(v_h v_k \sin(\theta_h - \theta_k) + v_h v_S \sin\gamma) \\ p_k &= -p_h \\ q_h &= \frac{1}{x_{hk}}(v_h^2 - v_h v_k \cos(\theta_h - \theta_k) + v_h v_S \cos\gamma) - i_q v_h \\ q_k &= \frac{1}{x_{hk}}(v_k^2 - v_h v_k \cos(\theta_h - \theta_k) - v_k v_S \cos\gamma) \end{aligned} \tag{19.31}$$

where x_{hk} is the series reactance of the loss-less transmission line connected in series to the UPFC device and Table 19.9 defines all other parameters. Differential equations are:

$$\begin{aligned} \dot{v}_p &= \frac{1}{T_r}(v_{p0} - v_p) \\ \dot{v}_q &= \frac{1}{T_r}(v_{q0} - v_q) \\ \dot{i}_{sh} &= \frac{1}{T_r}[K_r(v^{ref} - v_h) - i_{sh}] \end{aligned} \tag{19.32}$$

In general $v_{p0} = 0$ so that the voltage $\bar{v}_S$ is in quadrature with the line current $\bar{i}_h$ (as for the SSSC device). All state variables undergo anti-windup limits.

19.5.3 Power Flow Model

The power flow model is obtained using the series-connected VSC static model (18.27) and (18.28) and the series-connected VSC static model (18.29) and (18.30) as shown in Figure 19.20. Since the UPFC does not produce active power, one has:

$$0 = p_{sh} + p_{se}, \tag{19.33}$$

Table 19.9 Simplified UPFC model parameters

Variable	Description	Unit
K_r	Regulator gain	pu/pu
$i_q^{\max}$	Maximum i_q	pu
$i_q^{\min}$	Minimum i_q	pu
T_r	Regulator time constant	s
$v_p^{\max}$	Maximum v_p	pu
$v_p^{\min}$	Minimum v_p	pu
$v_q^{\max}$	Maximum v_q	pu
$v_q^{\min}$	Minimum v_q	pu

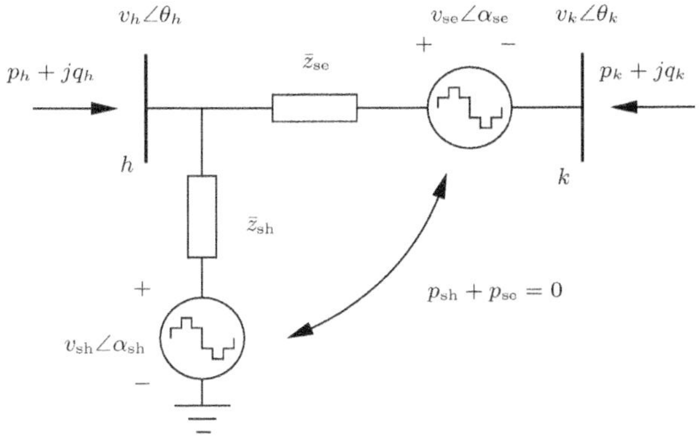

Fig. 19.20 Power flow UPFC equivalent circuit

The other three constraints required can be fixed as follows:

$$0 = v^{\text{ref}} - v_{\text{sh}} \tag{19.34}$$

$$0 = p^{\text{ref}} - p_k \tag{19.35}$$

$$0 = q^{\text{ref}} - q_k \tag{19.36}$$

These constraints can be satisfied as long as $i_{\text{sh}} \leq i_{\text{sh}}^{\max}$ and $i_{\text{se}} \leq i_{\text{se}}^{\max}$.

19.5.4 UPFC Initialization

The initialization of the UPFC device can be obtained using a static PV generator at bus h and a tie line between buses h and k, similarly to what described for the STATCOM and the SSSC devices. Alternatively, the static model described in Subsection 19.5.3 can be used.

Chapter 20
Wind Power Devices

This chapter presents wind speed and wind turbine models. Wind power is particularly interesting from the modelling viewpoint since it combines stochastic models (i.e., wind speed), mechanics (i.e., wind turbine), electrical machines, power electronics (i.e., VSC devices) and controls. For this reason, wind power models conclude this part dedicated to power system modelling.

The chapter is divided into two sections. Section 20.1 describes three wind speed models, namely the Weibull's distribution, a wind model composed of average speed, ramp, gust and turbulence and the normalized Mexican hat wavelet model. Section 20.2 describes three models of wind turbines, namely the constant speed wind turbine with squirrel-cage induction generator, the variable speed wind turbine with doubly-fed (wound rotor) asynchronous generator and the variable speed wind turbine with direct-drive synchronous generator.

20.1 Wind Speed Models

The best way to model the wind speed is by an historical series of measurement data. However, measures are not adequate for comparison and benchmarking. Thus, fictitious mathematical wind speed models are of interest.

The wind speed models described in this section are the Weibull's distribution, a composite model that includes average speed, ramp, gust and turbulence, and the Mexican hat wavelet model.

Table 20.1 defines all parameters used in wind speed models that are described in the following subsections. Air density ρ at 15°C and standard atmospheric pressure is 1.225 kg/m^3, and depends on the altitude (e.g., at 2000 m ρ is 20% lower than at the sea level).

Wind speed time sequences are calculated after solving the power flow and initializing wind turbine variables. To simulate the smoothing of high-frequency wind speed variations over the rotor surface, the actual wind speed

F. Milano: Power System Modelling and Scripting, Power Systems, pp. 435–456.
springerlink.com

values is filtered through low-pass filter before being used for computing the mechanical power of the wind turbine (see Figure 20.1):

$$\dot{v}_m = (\breve{v}_w(t) - v_w)/T_w \tag{20.1}$$

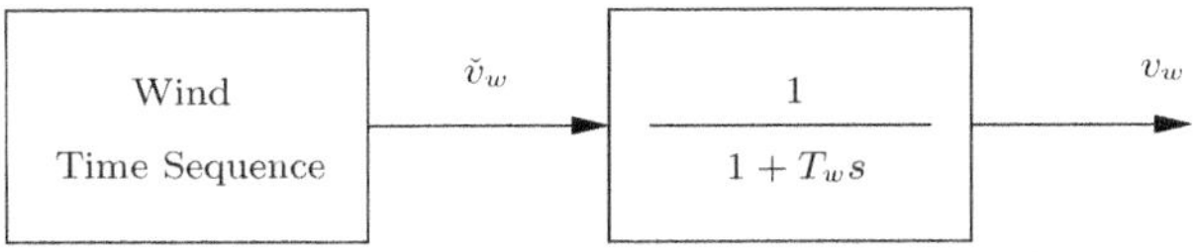

Fig. 20.1 Low-pass filter to smooth wind speed variations

Table 20.1 Wind speed parameters

Variable	Description	Unit
c_w	Scale factor for Weibull's distribution	-
h_w	Height of the wind speed signal	m
k_w	Shape factor for Weibull's distribution	-
n_{har}	Number of harmonics	int
t_0	Centering time of the Mexican hat wavelet	s
t_e^g	Ending gust time	s
t_e^r	Ending ramp time	s
t_s^g	Starting gust time	s
t_s^r	Starting ramp time	s
T_w	Low-pass filter time constant	s
v_w^g	Gust speed magnitude	m/s
v_w^r	Ramp speed magnitude	m/s
v_{wn}	Nominal wind speed	m/s
z_0	Roughness length	m
Δf	Frequency step	Hz
Δt	Sample time for wind measurements	s
ρ	Air density	kg/m^3
σ	Shape factor of the Mexican hat wavelet	-

20.1.1 Weibull's Distribution

A common way to describe the wind speed is by means of the Weibull's distribution, which is as follows:

$$f(v_w, c_w, k_w) = \frac{k_w}{c_w^k} v_w^{k-1} e^{-(\frac{v_w}{c_w})^{k_w}} \tag{20.2}$$

where v_w is the wind speed and c_w and k_w are constants as defined in the wind model data matrix. Time variations $\xi_w(t)$ of the wind speed are then obtained by means of a Weibull's distribution, as follows:

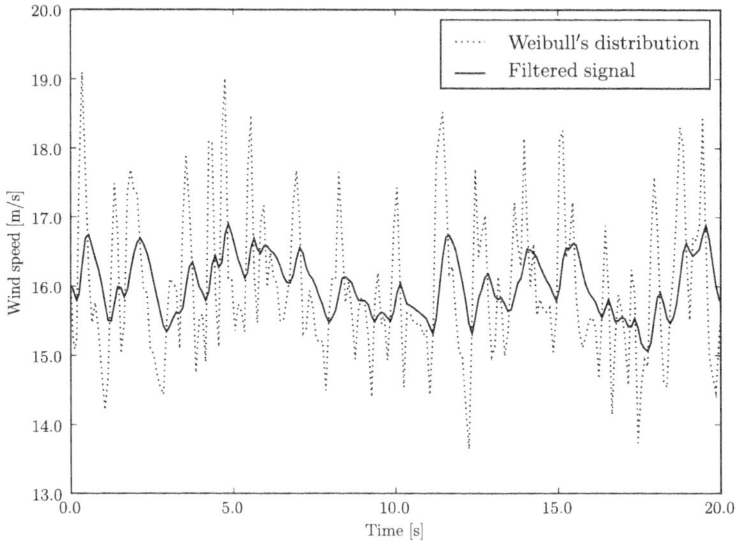

Fig. 20.2 Weibull's distribution model of the wind speed

$$\xi_w(t) = (-\frac{\ln \iota(t)}{c_w})^{\frac{1}{k_w}} \tag{20.3}$$

where $\iota(t)$ is a generator of random numbers ($\iota \in [0,1]$). Usually the shape factor $k_w = 2$, which leads to the Rayleigh's distribution, while $k_w > 3$ approximates the normal distribution and $k_w = 1$ gives the exponential distribution. The scale factor c should be chosen in the range $c_w \in (1, 10)$. Finally, the wind speed is computed setting the initial average speed v_w^a determined at the initialization step as mean speed:

$$\check{v}_w(t) = (1 + \xi_w(t) - \xi_w^a)v_w^a \tag{20.4}$$

where ξ_w^a is the average value of $\xi_w(t)$.

Example 20.1 Weibull's Distribution

An example of wind speed time sequence generated using a Weibull's distribution is depicted in Figure 20.2. Wind data are $v_{wn} = 16$ m/s, $\Delta t = 0.1$ s, $T_w = 0.5$ s, $c_w = 20$ and $k_w = 2$.

20.1.2 Composite Wind Speed Model

This subsection describes a composite wind model similar to what proposed in [343] and [9]. This model considers that the wind speed is composed of four parts:

1. Average and initial wind speed v_w^a.
2. Ramp component of the wind speed v_w^r.
3. Gust component of the wind speed v_w^g.
4. Wind speed turbulence v_w^t.

The resulting wind speed $\breve{v}_w$ is:

$$\breve{v}_w(t) = v_w^a + v_w^r(t) + v_w^g(t) + v_w^t(t) \tag{20.5}$$

where all components are time-dependent except for the average speed v_w^a.

Wind Ramp Component

The wind ramp component is defined by an amplitude A_w^r and starting and ending times, t_s^r and t_e^r respectively:

$$\begin{aligned} t < t_s^r &: \ v_w^r(t) = 0 \\ t_s^r \leq t \leq t_e^r &: \ v_w^r(t) = A_w^r \left(\frac{t - t_s^r}{t_e^r - t_s^r} \right) \\ t > t_e^r &: \ v_w^r(t) = A_w^r \end{aligned} \tag{20.6}$$

Wind Gust Component

The wind gust component is defined by an amplitude A_w^g and starting and ending times, t_s^g and t_e^g respectively:

$$\begin{aligned} t < t_s^g &: \ v_w^g(t) = 0 \\ t_s^g \leq t \leq t_e^g &: \ v_w^g(t) = \frac{A_w^g}{2} \left(1 - \cos\left(2\pi \frac{t - t_s^g}{t_e^g - t_s^g} \right) \right) \\ t > t_e^g &: \ v_w^g(t) = A_w^g \end{aligned} \tag{20.7}$$

Wind Turbulence Component

The wind turbulence component is described by a power spectral density, as follows:

$$S_w^t = \frac{\frac{1}{(\ln(h_w/z_0))^2} \ell v_w^a}{\left(1 + 1.5 \frac{\ell f}{v_w^a}\right)^{\frac{5}{3}}} \tag{20.8}$$

where f is the frequency, h_w the wind turbine tower height, z_0 is the roughness length and ℓ is the turbulence length scale:

$$\begin{aligned} h_w < 30 &: \ \ell = 20h \\ h_w \geq 30 &: \ \ell = 600 \end{aligned} \tag{20.9}$$

Table 20.2 depicts roughness values z_0 for a variety of ground surfaces.

Table 20.2 Roughness length z_0 for a variety of ground surfaces [231, 281]

Ground surface	Roughness length z_0 [m]
Open sea, sand	$10^{-4} \div 10^{-3}$
Snow surface	$10^{-3} \div 5 \cdot 10^{-3}$
Mown grass, steppe	$10^{-3} \div 10^{-2}$
Long grass, rocky ground	$0.04 \div 0.1$
Forests, cities, hilly areas	$1 \div 5$

The spectral density is then converted in a time domain cosine series as illustrated in [283]:

$$v_w^t(t) = \sum_{i=1}^{n_{\text{har}}} \sqrt{S_w^t(f_i)\Delta f} \cos(2\pi f_i t + \phi_i + \Delta\phi) \tag{20.10}$$

where f_i and ϕ_i are the frequency and the initial phase of the ith frequency component, being ϕ_i random phases ($\phi_i \in [0, 2\pi)$). The frequency step Δf should be $\Delta f \in (0.1, 0.3)$ Hz. Finally $\Delta\phi$ is a small random phase angle introduced to avoid periodicity of the turbulence signal.

Example 20.2 Composite Wind Model

An example of wind speed time series generated using a composite model is depicted in Figure 20.3. Wind data are $v_{wn} = 16$ m/s, $\Delta t = 0.1$ s, $T_w = 0.5$ s, $t_s^r = 5$ s, $t_e^r = 15$ s, $v_w^r = 0.5$ m/s, $t_s^g = 5$ s, $t_e^g = 15$ s and $v_w^g = 0.2$ m/s, $h_w = 50$ m, $z_0 = 0.01$ m, $\Delta f = 0.1$ Hz and $n = 50$.

20.1.3 Mexican Hat Wavelet Model

The Mexican hat wavelet is a deterministic wind gust that has been standardized by IEC [139]. The Mexican hat wavelet is the normalized second derivative of the Gauss' distribution function, i.e., up to scale normalization, the second Hermite's function, as follows:

$$\breve{v}_w(t) = v_w^a + (v_w^g - v_w^a)\left(1 - \frac{(t - t_0)^2}{\sigma^2}\right)e^{-\frac{(t-t_0)^2}{2\sigma^2}} \tag{20.11}$$

where t_0 is the centering time of the gust, σ is the gust shape factor, v_w^g is the peak wind speed value and v_w^a the average speed value.

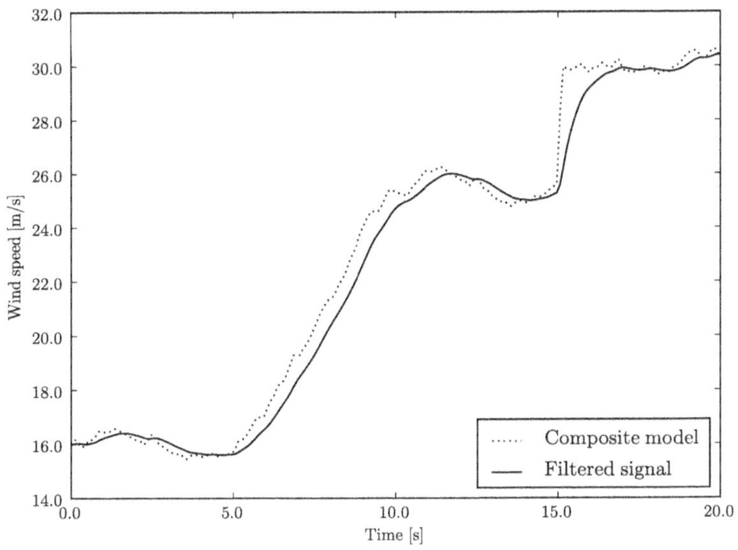

Fig. 20.3 Composite model of the wind speed

Example 20.3 Mexican Hat Wavelet Wind Model

Figure 20.4 shows an example of wind gust modeled through a Mexican hat wavelet with $v_{wn} = 16$ m/s, $v_w^a = 12$ m/s, $\Delta t = 0.1$ s, $T_w = 0.5$, $t_0 = 10$ s, $\sigma = 1$, and $v_w^g = 25$ m/s.

20.2 Wind Turbines

This section describes the three most common wind turbine types: the non-controlled speed wind turbine with squirrel-cage induction generator, the controlled speed wind turbine with doubly-fed (wound rotor) asynchronous generator and the direct-drive synchronous generator [4]. Figure 20.5 depicts three wind turbines types, while Table 20.3 illustrates a few recent wind turbines data as documented in [91].

The squirrel-cage induction generator type is the oldest one. It has the advantages of being relatively cheap and electrically efficient. However, due to the lack of speed regulation, it is not aerodynamically efficient. This kind of wind generators is also noisy, requires a gearbox and the shaft suffers mechanical stresses (i.e., oscillations due the shadow effect and blade torsional modes). It has to be noted that the speed control is theoretically possible for this machines. However, this would require refurbishing existing turbines.

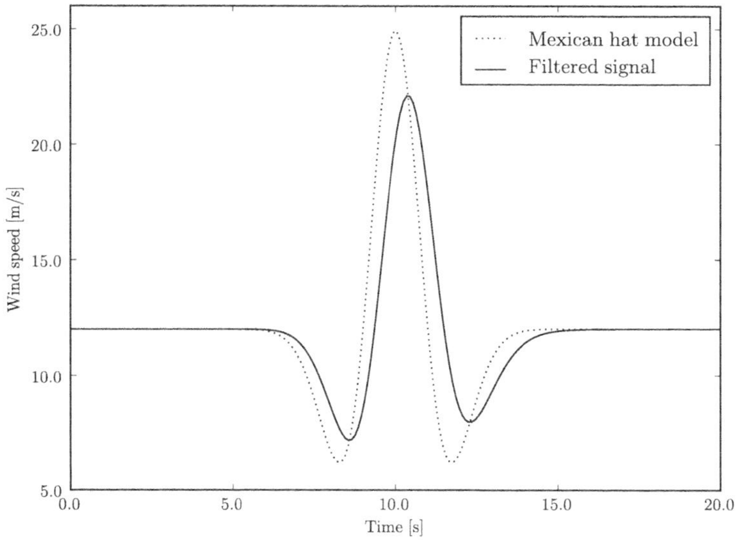

Fig. 20.4 Mexican hat model of the wind speed

The doubly-fed asynchronous generator and direct-drive synchronous generator are aerodynamically more efficient than the squirrel-cage induction generator type thanks to the speed control. However, the electrical efficiency is lower. The need of two VSC converters with a back-to-back connection increases the cost but provides the ability of controlling the voltage and the power output. In the case of the synchronous generator, the converters are relatively more expensive than for the asynchronous generator because have to stand the entire generator power output. This fact has limited the diffusion the the direct-drive type. However, VSC converter cost is rapidly decreasing, hence the increasing interest in synchronous generators for wind power applications. Finally, no gearbox is required for the synchronous generator since the VSC converters fully decouple the machine from the grid.

20.2.1 Single Machine and Aggregate Models

The wind turbine and generators models that are described in following subsections are adequate for a single machine as well as for a wind park composed of several generators (i.e., aggregate wind turbine model). The rules for a correct definition of the wind turbine data are:

1. The nominal power S_n is the total power in MVA of the single or aggregate wind turbine. If the wind turbine is an aggregate equivalent model of a

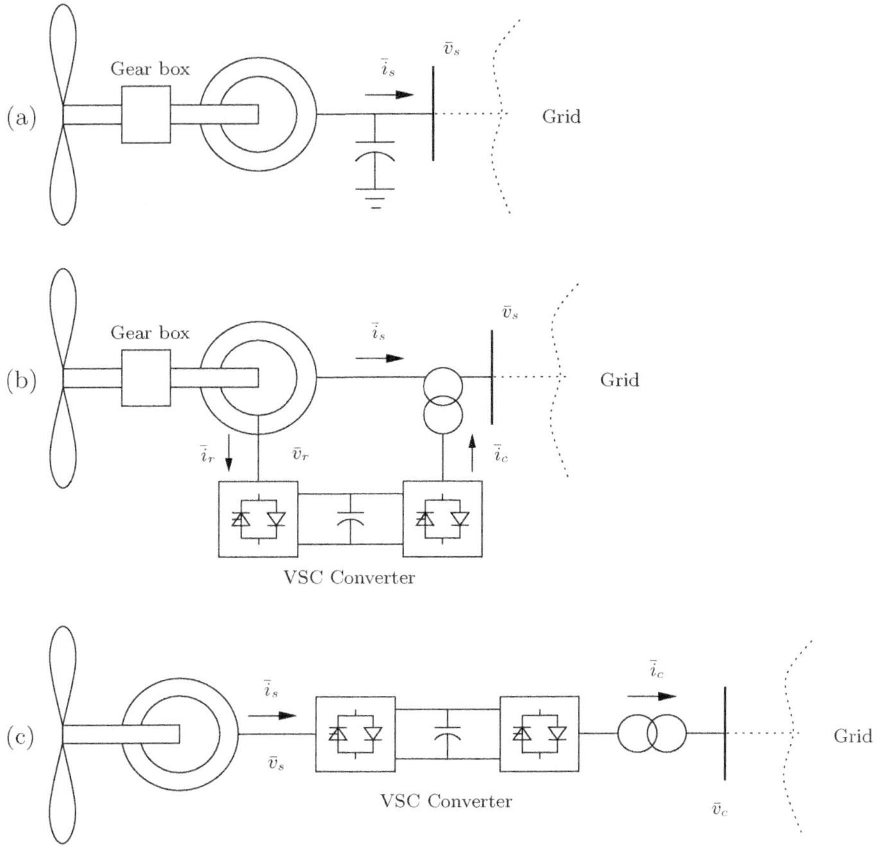

Fig. 20.5 Wind turbine types. (a) Non-controlled speed wind turbine with squirrel-cage induction generator; (b) Controlled speed wind turbine with doubly-fed asynchronous generator; (c) Controlled speed wind turbine with direct-drive synchronous generator

Table 20.3 Recent wind turbines [91]

Type	Power [MW]	Diam. [m]	Height [m]	Control	Speed [rpm]
Bonus	2	86	80	GD/TS/PS	17
NEC NM 1500/72	1.5	72	98	GD/TS/PS	17.3
Nordex N-80	2.5	80	80	GD/VS/PC	19
Vestas V-80	2	80	78	GD/VS/PC	19
Enercon e-66	1.5	66	85	GD/VS/PC	22

GD gearbox drive DD direct drive VS variable speed
TS two speed PC pitch control PS shift pitch by stall

wind park, then the number of generators $n_{\mathrm{gen}} > 1$. The capacity of each generator is about $[0.5, 5]$ MVA (being 1 and 2 MVA the most common capacities), thus $n_{\mathrm{gen}} \in [0.2S_n, 2S_n]$. An inconsistent value of n_{gen} can lead to an inadequate initialization of the wind speed.
2. Machine impedances and inertiæ are equivalent weighted pu values of the machines that compose the aggregate wind park.

20.2.2 Wind Turbine Initialization

Wind turbines are initialized after power flow analysis and a static generator is needed to impose the desired voltage and active power at the wind turbine bus. Once the power flow solution is available, v_0, θ_0, p_0 and q_0 at the generation bus are used for initializing the state and input variables, the latter being the wind speed v_{w0}, which is used as the average wind speed v_w^a for the wind speed models. Due to the nonlinearity of generator, converter and turbine models, the initialization requires the implementation of a Newton's method. For the interested reader, further insights on wind turbine initialization are given in [285] and [133].

The following subsections are organized taking into account implementation issues. The models that are common to all wind turbine are described first. These are the turbine model and the shaft model. Then, each specific wind generator type is described. With this aim, electrical machine and VSC converter models are described together.

20.2.3 Turbine Model

The mechanical model of the wind turbine is independent from the generator configuration. Thus, the mechanical equations of the turbine can be implemented as a separate class and then imported in the wind generator model.

Following items describe two models, namely fixed-blade and variable pitch angle position blade turbines. Table 20.4 defines all parameters used below.

Table 20.4 Turbine mechanical parameters

Variable	Description	Unit
K_p	Pitch control gain	rad/pu
n_{blade}	Number of blades	int
n_{gen}	Number of machines that compose the wind park	int
n_{pole}	Number of poles	int
S_n	Power rating	MVA
R	Rotor radius	m
T_p	Pitch control time constant	s
η_{GB}	Gear box ratio	-
ρ	Air density	kg/m^3

Fixed-Blade Turbine Model

This model assumes fixed turbine blades and is adequate for wind generators without speed regulation. The mechanical power p_w extracted from the wind is a function of the wind speed v_w and the turbine rotor speed ω_t and can be approximated as follows:

$$p_w = \frac{n_{\text{gen}}\rho}{2S_n} c_p(\lambda) A_r v_w^3 \tag{20.12}$$

where c_p is the performance coefficient or power coefficient, λ the tip speed ratio and $A_r = \pi R^2$ the area swept by the rotor. The speed tip ratio λ is the ratio between the blade tip speed v_{bt} and the wind upstream the rotor v_w:

$$\lambda = \frac{v_{bt}}{v_w} = \eta_{GB} \frac{2R\omega_t}{n_{\text{pole}} v_w} \tag{20.13}$$

Finally, the $c_p(\lambda)$ curve is approximated as follows:

$$c_p = 0.44\left(\frac{125}{\lambda_i} - 6.94\right) e^{-\frac{16.5}{\lambda_i}} \tag{20.14}$$

with

$$\lambda_i = \frac{1}{\frac{1}{\lambda} + 0.002} \tag{20.15}$$

Turbine Model with Pitch Angle Control

In this model, turbine blades can rotate in order to reduce the rotor speed in case of super-synchronous conditions. The angular position θ_p of the blades is called *pitch angle*. This turbine model is adequate for wind generators with speed control.

The mechanical power p_w extracted from the wind is a function of the wind speed v_w, the rotor speed ω_m and the pitch angle θ_p. The mechanical power p_w can be approximated as:

$$p_w = \frac{n_{\text{gen}}\rho}{2S_n} c_p(\lambda, \theta_p) A_r v_w^3 \tag{20.16}$$

where parameters and variables are the same as in (20.12) and the speed tip ratio λ is defined in (20.13). The $c_p(\lambda, \theta_p)$ curve is approximated as follows [126]:

$$c_p = 0.22\left(\frac{116}{\lambda_i} - 0.4\theta_p - 5\right) e^{-\frac{12.5}{\lambda_i}} \tag{20.17}$$

with

$$\frac{1}{\lambda_i} = \frac{1}{\lambda + 0.08\theta_p} - \frac{0.035}{\theta_p^3 + 1} \tag{20.18}$$

Alternative equations are given in [284], as follows:

$$c_p = 0.73\left(\frac{151}{\lambda_i} - 0.58\theta_p - 0.002\theta^{2.14} - 13.2\right)e^{-\frac{18.4}{\lambda_i}} \tag{20.19}$$

with

$$\frac{1}{\lambda_i} = \frac{1}{\lambda - 0.02\theta_p} - \frac{0.003}{\theta_p^3 + 1} \tag{20.20}$$

As discussed above, the pitch angle θ_p is controlled to avoid super-synchronous speeds. The control diagram is shown in Figure 20.6 and described by the differential equation:

$$\dot{\theta}_p = (K_p\phi(\omega_m - \omega^{\rm ref}) - \theta_p)/T_p \tag{20.21}$$

where ϕ is a function which allows varying the pitch angle set point only when the difference $(\omega_m - \omega^{\rm ref})$ exceeds a predefined value $\pm\Delta\omega$. Since the pitch control works only for super-synchronous speeds, an anti-windup limiter locks the pitch angle to $\theta_p = 0$ for sub-synchronous speeds.

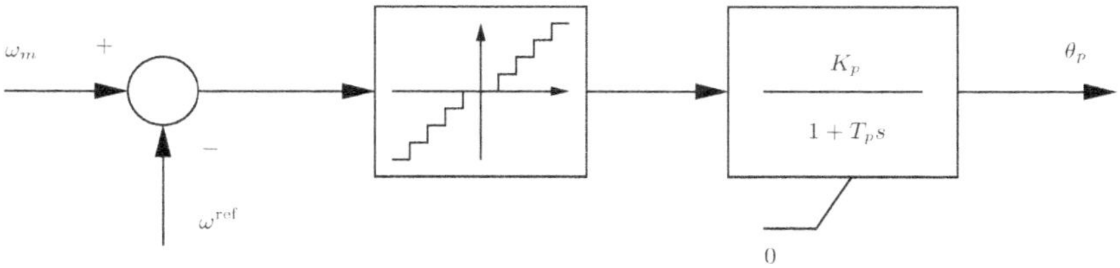

Fig. 20.6 Pitch angle control diagram

The speed control is aimed to maximize the power production of the wind turbine. Figure 20.7 shows the dependence of the mechanical power p_w produced by the wind turbine on the wind speed v_w and the turbine rotor speed ω_t. The solid line is the maximum mechanical power locus for each wind and rotor speeds. This curve is used for defining, for each value of the rotor speed, the optimal mechanical power p_w^* that the turbine has to produce. Figure 20.8 shows a possible implementation of the (p_w^*, ω_t) characteristic. For super-synchronous speeds, the reference power is fixed to 1 pu to avoid overloading the generator. For $\omega_t < 0.5$ pu, the reference mechanical power is set to zero.

The detailed (p_w^*, ω_t) characteristic is:

$$p_w^*(\omega_t) = \begin{cases} 0 \quad \text{if} \quad \omega_t < 0.5 \\ p_w(\omega_t^*) \ \text{that satisfies} \ \dfrac{dp_w(\omega_t^*, v_w, \theta_p)}{d\omega_t} = 0 \quad \text{if} \quad 0.5 \leq \omega_t \leq 1 \\ 1 \quad \text{if} \quad \omega_m > 1 \end{cases} \tag{20.22}$$

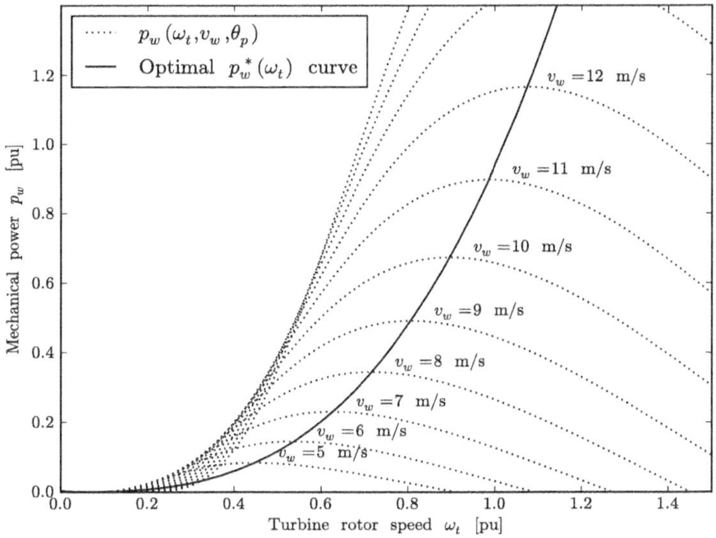

Fig. 20.7 Speed-power characteristic of the wind turbine. The pitch angle is assumed $\theta_p = 0$ to plot the $p_w(\omega_t, v_w, \theta_p)$ curve

The simplified (p_w^*, ω_t) characteristic is:

$$p_w^*(\omega_t) = \begin{cases} 0 \quad \text{if} \quad \omega_t < 0.5 \\ 2\omega_t - 1 \quad \text{if} \quad 0.5 \leq \omega_t \leq 1 \\ 1 \quad \text{if} \quad \omega_t > 1 \end{cases} \tag{20.23}$$

20.2.4 Dynamic Shaft

The shaft of wind generators can be modelled as two masses connected by a spring. The masses represent the turbine shaft and the generator shaft, while the spring models the shaft stiffness. The resulting model is similar to the one described in Section 15.1.10 of Chapter 15.

Mechanical differential equations are:

$$\begin{aligned} \dot{\omega}_t &= (\tau_t - K_s \delta_{tm})/(2H_t) \\ \dot{\omega}_m &= (K_s \delta_{tm} - \tau_e)/(2H_m) \\ \dot{\delta}_{tm} &= \Omega_b(\omega_t - \omega_m) \end{aligned} \tag{20.24}$$

where Ω_b is the system rated frequency in rad/s, ω_t is the wind turbine angular speed, ω_m is the generator rotor speed, δ_{tm} is the relative angle

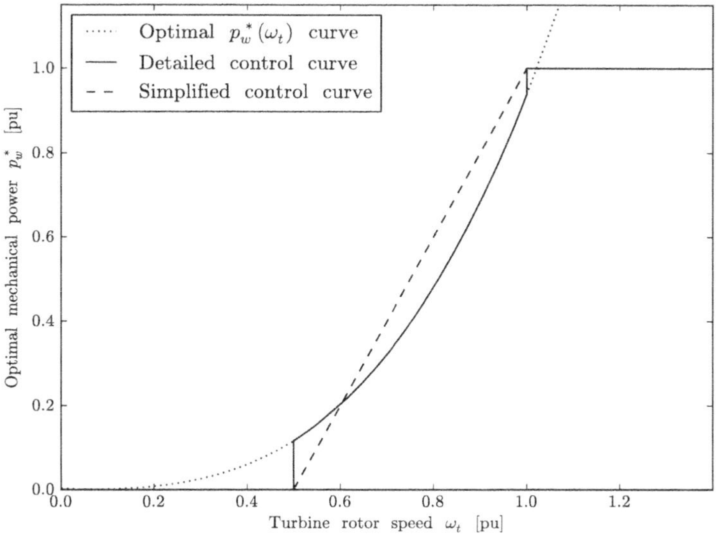

Fig. 20.8 Optimal and implemented control speed-power characteristics

displacement of the two shafts, τ_e is the electrical torque and τ_t is the mechanical torque:

$$\tau_t = \frac{p_w}{\omega_t} \tag{20.25}$$

and all other parameters are defined in Table 20.5.

A periodic torque pulsation can be added to τ_t to simulate the tower shadow effect. The shadow-effect frequency depends on the rotor speed ω_t, the gear box ratio η_{GB}, and the number of blades $n_{\rm blade}$, as follows:

$$\tilde{\tau}_t = \tau_t\Big(1 + \alpha_s \sin\Big(\eta_{GB}\frac{\Omega_b \omega_t}{n_{\rm blade}} t\Big)\Big) \tag{20.26}$$

where the torque pulsation amplitude can be fixed equal to $\alpha_s = 0.025$ [7]. The expression (20.26) substitutes the mechanical torque in the first equation of (20.24).

Shaft oscillations cannot be removed in the squirrel-cage induction generator type. For other wind turbine types that include VSC devices, the converter controls can effectively damp shadow effect modes and shaft oscillations [187]. If the control is efficient enough, the shaft can be considered rigid, i.e., $\omega_t = \omega_m$. Hence:

$$\dot{\omega}_t = (\tau_t - \tau_e)/(2H_t + 2H_m) \tag{20.27}$$

Table 20.5 Wind turbine shaft parameters

Variable	Description	Unit
H_m	Machine rotor inertia constant	MWs/MVA
H_t	Wind turbine inertia constant	MWs/MVA
K_s	Shaft stiffness	pu
α_s	Shadow effect factor	-

20.2.5 Non-Controlled Speed Wind Turbine

The simplified electrical circuit used for the squirrel-cage induction generator is the same as the one for the single-cage induction motor, shown in Figure 15.14 of Chapter 15. The only difference with respect to the induction motor is that the currents are positive if injected into the network. The equations are formulated in terms of the real (d-) and imaginary (q-) axes, with respect to the network reference angle. Table 20.6 summarizes and defines all parameters of the squirrel-cage induction machine.

Table 20.6 Squirrel-cage induction machine parameters

Variable	Description	Unit
r_r	Rotor resistance	pu
r_s	Stator resistance	pu
x_r	Rotor reactance	pu
x_s	Stator reactance	pu
x_μ	Magnetizing reactance	pu

Network Interface

In a synchronously rotating reference frame, the link between the network and the stator machine voltages is:

$$\begin{aligned} v_d &= -v_h \sin\theta_h \\ v_q &= v_h \cos\theta_h \end{aligned} \tag{20.28}$$

and the active and reactive power productions are:

$$\begin{aligned} p_h &= v_d i_d + v_q i_q \\ q_h &= v_q i_d - v_d i_q + b_c(v_d^2 + v_q^2) \end{aligned} \tag{20.29}$$

where b_c is the fixed capacitor conductance which is determined at the initialization step to impose the required bus voltage level.

Machine Electro-Magnetic Equations

The differential equations in terms of the voltage behind the stator resistance r_S are:

$$\begin{aligned} e'_d - v_d &= r_s i_d - x' i_q \\ e'_q - v_q &= r_s i_q + x' i_d \end{aligned} \tag{20.30}$$

whereas the link between voltages, currents and state variables is as follows:

$$\begin{aligned} \dot{e}'_d &= \Omega_b(1 - \omega_m)e'_q - (e'_d - (x_0 - x')i_q)/T'_0 \\ \dot{e}'_q &= -\Omega_b(1 - \omega_m)e'_d - (e'_q + (x_0 - x')i_d)/T'_0 \end{aligned} \tag{20.31}$$

where ω_m is the rotor angular speed, and x_0, x' and T'_0 can be obtained from generator parameters:

$$\begin{aligned} x_0 &= x_s + x_\mu \\ x' &= x_s + \frac{x_r x_\mu}{x_r + x_\mu} \\ T'_0 &= \frac{x_r + x_\mu}{\Omega_b r_R} \end{aligned} \tag{20.32}$$

Turbine and Machine Mechanical Equations

The mechanical DAE system is (20.12) and (20.24)-(20.26), where the electrical torque τ_e is defined as:

$$\tau_e = e'_d i_d + e'_q i_q \tag{20.33}$$

20.2.6 Doubly-Fed Asynchronous Generator

The model of the doubly-fed asynchronous generator is assumed steady-state, as the stator and rotor flux dynamics are fast with respect to grid dynamics. As a result of these assumptions, one has the DAE system described below. Table 20.7 summarizes and defines all parameters required by wind turbine with doubly-fed asynchronous generator.

Network Interface

Stator voltages depends on the bus voltage $\bar{v}_h$:

$$\begin{aligned} v_{s,d} &= -v_h \sin\theta_h \\ v_{s,q} &= v_h \cos\theta_h \end{aligned} \tag{20.34}$$

Table 20.7 Doubly-fed asynchronous generator parameters

Variable	Description	Unit
K_V	Voltage control gain	pu/pu/s
$p^{\max}$	Maximum active power	pu
$p^{\min}$	Minimum active power	pu
$q^{\max}$	Maximum reactive power	pu
$q^{\min}$	Minimum reactive power	pu
r_r	Rotor resistance	pu
r_s	Stator resistance	pu
T_ϵ	Power control time constant	s
x_r	Rotor reactance	pu
x_s	Stator reactance	pu
x_μ	Magnetizing reactance	pu

The generator active and reactive power productions depend on the stator and converter currents $i_{s,d}+ji_{s,q}$ and $i_{c,d}+ji_{c,q}$, respectively, and stator and converter voltages $v_{s,d}+jv_{s,q}$ and $v_{c,d}+jv_{c,q}$, respectively, as follows:

$$\begin{aligned} p_h &= v_{s,d}i_{s,d} + v_{s,q}i_{s,q} + v_{c,d}i_{c,d} + v_{c,q}i_{c,q} \\ q_h &= v_{s,q}i_{s,d} - v_{s,d}i_{s,q} + v_{c,q}i_{c,d} - v_{c,d}i_{c,q} \end{aligned} \tag{20.35}$$

The expressions above can be rewritten as a function of stator and rotor currents $i_{s,d}+ji_{s,q}$ and $i_{r,d}+ji_{r,q}$, respectively, and stator and rotor voltages $v_{s,d}+jv_{s,q}$ and $v_{r,d}+jv_{r,q}$, respectively. In fact, the converter powers on the grid side are:

$$\begin{aligned} p_c &= v_{c,d}i_{c,d} + v_{c,q}i_{c,q} \\ q_c &= v_{c,q}i_{c,d} - v_{c,d}i_{c,q} \end{aligned} \tag{20.36}$$

whereas, on the rotor side:

$$\begin{aligned} p_r &= v_{r,d}i_{r,d} + v_{r,q}i_{r,q} \\ q_r &= v_{r,q}i_{r,d} - v_{r,d}i_{r,q} \end{aligned} \tag{20.37}$$

Assuming a loss-less converter model, the active power of the converter coincides with the rotor active power, thus $p_c = p_r$. The reactive power injected into the grid can be approximated neglecting stator resistance and assuming that the d-axis coincides with the maximum of the stator flux. Therefore, the powers injected in the grid are:

$$\begin{aligned} p_h &= v_{s,d}i_{s,d} + v_{s,q}i_{s,q} + v_{r,d}i_{r,d} + v_{r,q}i_{r,q} \\ q_h &= -\frac{x_\mu v_h i_{r,d}}{x_s + x_\mu} - \frac{v_h^2}{x_\mu} \end{aligned} \tag{20.38}$$

Machine Electro-Magnetic Equations

The machine stator and rotor voltages are a function of stator and rotor currents and the rotor speed ω_m:

$$\begin{aligned}
v_{s,d} &= -r_s i_{s,d} + ((x_s + x_\mu) i_{s,q} + x_\mu i_{r,q}) \\
v_{s,q} &= -r_s i_{s,q} - ((x_s + x_\mu) i_{s,d} + x_\mu i_{r,d}) \\
v_{r,d} &= -r_r i_{r,d} + (1 - \omega_m)((x_s + x_\mu) i_{r,q} + x_\mu i_{s,q}) \\
v_{r,q} &= -r_r i_{r,q} - (1 - \omega_m)((x_s + x_\mu) i_{r,d} + x_\mu i_{s,d})
\end{aligned} \tag{20.39}$$

whereas the links between stator fluxes and generator currents are:

$$\begin{aligned}
\psi_{s,d} &= -((x_s + x_\mu) i_{s,d} + x_\mu i_{r,d}) \\
\psi_{s,q} &= -((x_s + x_\mu) i_{s,q} + x_\mu i_{r,q})
\end{aligned} \tag{20.40}$$

Turbine and Machine Mechanical Equations

The generator motion equation is modeled as a single shaft, i.e., (20.27), as it is assumed that the converter controls are able to filter shaft dynamics. For the same reason, no tower shadow effect is considered in this model. In (20.27), the electrical torque is:

$$\tau_e = \psi_{s,d} i_{s,q} - \psi_{s,q} i_{s,d} \tag{20.41}$$

Substituting the stator flux equations (20.40) in (20.41) leads to:

$$\tau_e = x_\mu (i_{r,q} i_{s,d} - i_{r,d} i_{s,q}) \tag{20.42}$$

In [286], the following approximation of the electrical torque τ_e is proposed:

$$\tau_e \approx -\frac{x_\mu v_h i_{r,q}}{\Omega_b (x_s + x_\mu)} \tag{20.43}$$

where Ω_b is the system rated frequency in rad/s.

Mechanical equations are completed by the mechanical torque τ_t equation (20.25) and the turbine model (20.16) and (20.17)-(20.18) or (20.19)-(20.20) and the pitch angle control (20.21).

VSC Regulators

Since VSC dynamics are quite fast with respect to the electro-mechanical transients, the converter can be modeled as an ideal current source, where $i_{r,q}$ and $i_{r,d}$ are state variables and are used for controlling the rotor speed and the bus voltage, respectively. VSC control diagrams are depicted in Figures 20.9 and 20.10. The DAE system of the converter currents is:

$$\dot{i}_{r,q} = \left(-\frac{x_s + x_\mu}{x_\mu v} p_w^*(\omega_m)/\omega_m - i_{r,q}\right)\frac{1}{T_\epsilon} \qquad (20.44)$$
$$\dot{i}_{r,d} = K_V(v_h - v^{\mathrm{ref}}) - v_h/x_\mu - i_{r,d}$$

where v^{ref} is the reference voltage computed at the initialization step and $p_w^*(\omega_m)$ is the power-speed characteristic (20.22) or (20.23).

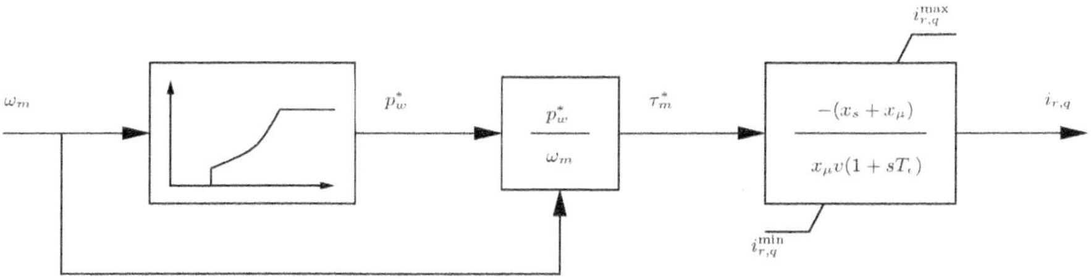

Fig. 20.9 Rotor speed control diagram

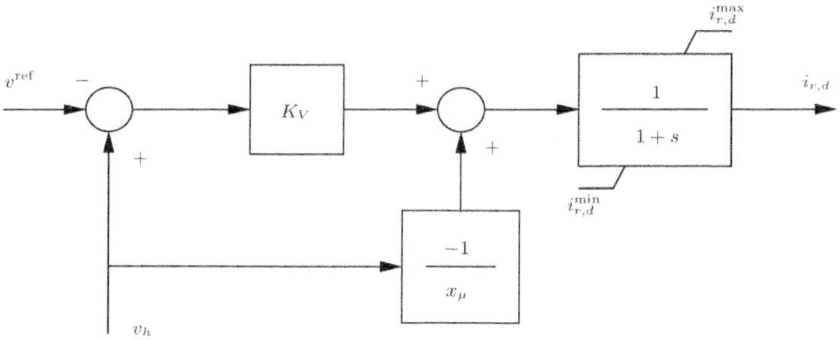

Fig. 20.10 Voltage control diagram of the doubly-fed asynchronous generator

Hard Limits

Both the speed and voltage controls undergo anti-windup limiters to avoid converter over-currents. Rotor current limits are computed based on active and reactive limits, and assuming bus voltage $v_h \approx 1$ one has:

$$i_{r,q}^{\max} \approx -\frac{x_s + x_\mu}{x_\mu} p^{\min} \qquad (20.45)$$
$$i_{r,q}^{\min} \approx -\frac{x_s + x_\mu}{x_\mu} p^{\max}$$
$$i_{r,d}^{\max} \approx -\frac{x_s + x_\mu}{x_\mu} q^{\min} - \frac{x_s + x_\mu}{x_\mu^2}$$
$$i_{r,d}^{\min} \approx -\frac{x_s + x_\mu}{x_\mu} q^{\max} - \frac{x_s + x_\mu}{x_\mu^2}$$

20.2.7 Direct-Drive Synchronous Generator

The model of the direct-drive synchronous generator is assumed steady-state, as the stator and rotor flux dynamics are fast with respect to grid dynamics. Furthermore, the converter decouples the generator from the grid. As a result of these assumptions, one has the DAE system described below. Table 20.8 summarizes and defines all parameters required by wind turbine with direct-drive synchronous generator.

Table 20.8 Direct-drive synchronous generator parameters

Variable	Description	Unit
K_{dc}	Gain of the bus voltage control	pu/pu
K_{ds}	Gain of the generator reactive power control	pu/pu
K_{qc}	Gain of the active power control	pu/pu
$i^{\max}$	Maximum current	pu
r_s	Stator resistance	pu
T_{dc}	Time constant of the bus voltage control	s
T_{ds}	Time constant of the generator reactive power control	s
T_{qc}	Time constant of the active power control	s
T_{qs}	Time constant of the speed control	s
x_d	d-axis reactance	pu
x_q	q-axis reactance	pu
ψ_p	Permanent field flux	pu

Network Interface

The active and reactive powers injected into the grid depend on the grid side current $i_{c,d} + ji_{c,q}$ and voltage $v_{c,d} + jv_{c,q}$ of the converter:

$$\begin{aligned} p_h = p_c &= v_{c,d}i_{c,d} + v_{c,q}i_{c,q} \\ q_h = q_c &= v_{c,q}i_{c,d} - v_{c,d}i_{c,q} \end{aligned} \tag{20.46}$$

where the converter voltages are functions of the grid voltage magnitude and phase, as follows:

$$\begin{aligned} v_{c,d} &= -v_h \sin\theta_h \\ v_{c,q} &= v_h \cos\theta_h \end{aligned} \tag{20.47}$$

Machine Electro-Magnetic Equations

Assuming a permanent magnet synchronous generator, machine equations are

$$\begin{aligned} v_{s,d} &= -r_s i_{s,d} + \omega_m x_q i_{s,q} \\ v_{s,q} &= -r_s i_{s,q} - \omega_m (x_d i_{s,d} - \psi_p) \end{aligned} \tag{20.48}$$

where the permanent field flux ψ_p represents the rotor circuit. The active and reactive power produced by the generator are as follows:

$$\begin{aligned} p_s &= v_{s,d} i_{s,d} + v_{s,q} i_{s,q} \\ q_s &= v_{s,q} i_{s,d} - v_{s,d} i_{s,q} \end{aligned} \tag{20.49}$$

The link between stator fluxes and generator currents:

$$\begin{aligned} \psi_{s,d} &= -x_d i_{s,d} + \psi_p \\ \psi_{s,q} &= -x_q i_{s,q} \end{aligned} \tag{20.50}$$

Turbine and Machine Mechanical Equations

The generator motion equation is modeled as a single shaft, i.e., (20.27), as it is assumed that the converter controls are able to filter shaft dynamics. For the same reason, no tower shadow effect is considered in this model. In (20.27), the electrical torque is:

$$\tau_e = \psi_{s,d} i_{s,q} - \psi_{s,q} i_{s,d} \tag{20.51}$$

Substituting (20.50) in (20.51) leads to:

$$\tau_e = (\psi_{s,d} + (x_q - x_d) i_{s,d}) i_{s,q} \tag{20.52}$$

Mechanical equations are completed by the mechanical torque τ_t equation (20.25) and the turbine model (20.16) and (20.17)-(20.18) or (20.19)-(20.20) and the pitch angle control (20.21).

VSC Regulators

The back-to-back VSC that connect the generator to the grid allow controlling four quantities. Controllable quantities are the converter currents $i_{s,d}$, $i_{s,q}$, $i_{c,d}$ and $i_{c,q}$. Typical controlled quantities are the active power injected into the grid, the grid side voltage, the dc voltage of the capacitor in the dc back-to-back connection, and the reactive power on the generator side. Several combinations have been proposed [3, 4, 119, 351]. A possible control scheme is as follows.

The currents on the generator side control the rotor speed and the generator reactive power:

$$\dot{i}_{s,q} = \frac{1}{T_{qs}}\left(\frac{p_w^*(\omega_m)}{\omega_m(\psi_p - x_d i_{s,d})} - i_{s,q}\right) \tag{20.53}$$
$$\dot{i}_{s,d} = (K_{ds}(q_{s0} - q_s) - i_{s,d})/T_{ds}$$

where $p_w^*(\omega_m)$ is the power-speed characteristic (20.22) or (20.23) and q_{s0} is the reactive power determined at the initialization step.

The currents on the grid side control the active power and bus voltage:

$$\dot{i}_{c,q} = (K_{qc}(p_s - p_c) - i_{c,q})/T_{qc} \tag{20.54}$$
$$\dot{i}_{c,d} = (K_{dc}(v^{\mathrm{ref}} - v_h) - i_{c,d})/T_{dc}$$

The first of the previous equations indirectly models the dynamic of the dc system of the back-to-back connection. The steady-state error $p_s - p_c$ is a model of VSC and capacitor losses. A pure integrator can be used for modelling a loss-less back-to-back VSC:

$$\dot{i}_{c,q} = K_{qc}(p_s - p_c)/T_{qc} \tag{20.55}$$

If the dynamics of the VSC dc connection are fast, one can simply impose:

$$0 = p_s - p_c \tag{20.56}$$

Limits

All currents in (20.53) and (20.54) undergo anti-windup limiters. The limits have to be carefully calculated to avoid overloading. On the dc side, one has:

$$\sqrt{i_{c,d}^2 + i_{c,q}^2} \le i^{\max} \tag{20.57}$$

where $i^{\max}$ is the VSC maximum current. Since (20.57) contains two variables, additional conditions are required to define $i_{c,d}^{\max}$ and $i_{c,q}^{\max}$. For example:

$$i_{c,q}^{\max} = i^{\max} \tag{20.58}$$
$$i_{c,d}^{\max} = \sqrt{(i^{\max})^2 - i_{c,q}^2}$$

Thus, one of the limits, e.g., $i_{c,d}^{\max}$ is a function of the other current. Then, for the lower limits:

$$i_{c,q}^{\min} = -i^{\max} \tag{20.59}$$
$$i_{c,d}^{\min} = -i_{c,d}^{\max}$$

Similar expressions can be defined for the limits of generator stator currents $i_{s,d}$ and $i_{s,q}$.

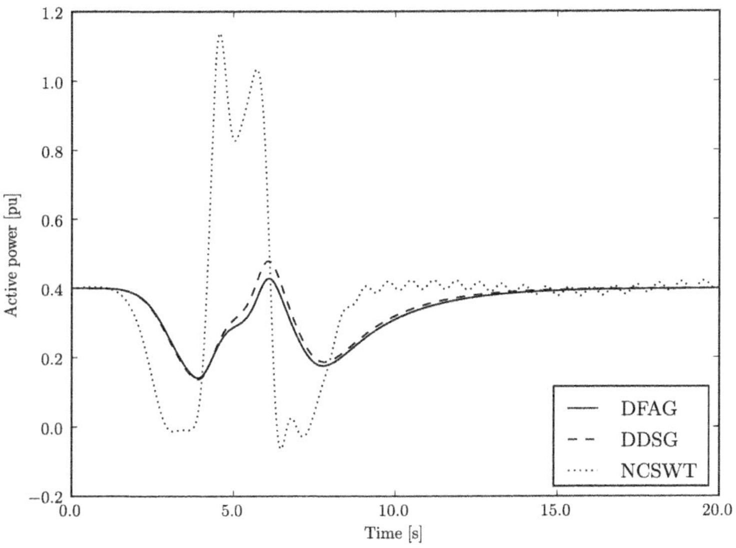

Fig. 20.11 Comparison of transient behavior of different wind turbine types: doubly-fed asynchronous generator (DFAG), and direct-drive synchronous machine (DDSG), and non-controlled speed wind turbine (NCSWT)

Example 20.4 Comparison of Wind Turbine Transient Behaviors

Figure 20.11 shows a comparison of the transient behavior of the three wind turbine types described in this section. In particular, the plot shows the power injected by the wind park at bus 2 of the IEEE 14-bus system.[1] The disturbance is a Mexican hat wavelet centered at $t = 5$ s, with a peak of 25 m/s. The initial power of the wind turbine is 0.4 pu and the wind park is composed of 40 machines (hence, the capacity of each machine is 1 MW). Mechanical data are the same for all wind turbine types. Machine and control data are provided in Appendix D.

The transient behaviors of the controlled speed wind turbines are quite similar: the controls of both the doubly-fed asynchronous generator and the direct-drive synchronous generator are able to slightly smooth the wind peak. On the other hand, the non-controlled wind turbine with squirrel-cage induction machine shows high oscillations that follows the wind peak and leads to undamped oscillations due to the shadow effect.

[1] The synchronous machine at bus 2 is substituted for a wind turbine.

Part IV
Spare Material and Concluding Remarks

Chapter 21
Data Formats

This chapter provides a taxonomy of existing data formats for power power system analysis. These include most commonly used formats of free and proprietary software packages as well as the IEC common information model. The chapter is completed by a discussion about the desirable features of a data format for power system analysis.

21.1 Data Format Taxonomy

The number of existing formats for power system analysis is huge. In general, each software application has is own specific data format. However, among all existing formats, few basic characteristics can be identified. The taxonomy can be made based on different features, as follows.

1. The way data are stored, organized and structured.
2. The kind of data and analysis supported.
3. The number of files that compose the full system data set.
4. The way default values and data manipulation is handled.

Each feature is described in the following subsections. Table 21.1 shows a synoptic scheme of the features of a variety of formats for power system analysis. In Table 21.1 as well as in the whole chapter, only formats available as plain ASCII files are available. Most commercial software applications store data as binary files, with the clear intention of making impossible to use those data by other applications. Such binary formats are intentionally ignored.

21.1.1 Data Organization and Structures

The way data are organized and structured affects the aspect of the resulting data file. Old data formats uses a fixed position and fixed order format, while the modern trend is to use mark-up languages. The following items describes some examples.

F. Milano: Power System Modelling and Scripting, Power Systems, pp. 459–474.
springerlink.com

Table 21.1 Features of a variety of data formats for power system analysis

Format Name	Data Position	Data Order	Dynamic Data	Market Data	Short Circuit Data	Graphic Data	Custom Data	Number of Files	Default Values	Alter Command
CEPEL	Fixed	Fixed	No	No	No	No	No	Unique	No	No
CYME	Fixed	Fixed	Yes	No	Yes	No	No	Multiple	Yes	No
DigSilent	Free	Free	Yes	No	Yes	Yes	No	Unique	Prototypes	No
EPRI/BPA	Fixed	Fixed	Yes	No	Yes	No	No	Unique	Yes	No
Eurostag	Fixed	Fixed	Yes	No	Yes	No	No	Multiple	Yes	No
FlowDemo	Free	Fixed	No	No	No	Yes	No	Unique	No	No
GE-PSLF	Free	Fixed	No	No	No	No	No	Unique	Yes	No
IEEE CDF	Fixed	Fixed	No	No	No	No	No	Unique	No	No
INPTC1	Fixed	Fixed	No	No	No	No	No	Multiple	No	Yes
Matpower	Free	Free	No	Yes	No	No	No	Any	No	No
Neplan	Free	Free	Yes	No	Yes	Yes	Yes	Multiple	Yes	No
PowerWorld	Free	Free	No	Yes	Yes	Yes	No	Unique	Yes	No
PSAT	Free	Free	Yes	Yes	No	Yes	Yes	Any	No	Yes
PSS/E	Free	Fixed	Yes	No	Yes	No	Yes	Unique	Yes	No
PST	Free	Free	Yes	No	Yes	No	Yes	Any	No	Yes
Simpow	Free	Free	Yes	No	Yes	Yes	Yes	Any	Yes	Yes
UCTE	Fixed	Fixed	No	No	No	No	No	Unique	Yes	No

- *Fixed position, fixed order.* This is the oldest method of storing data and date back to the paper cards that were used in the seventies for loading and saving the information used by computers. Any data has is starting and ending columns and is assigned a maximum number of digits. Also the order in which data are listed is fixed. For example, in the IEEE CDF the bus card comes before the branch card, and the voltage magnitude must occupy columns 28 to 33 of the bus card [350]. The formats created by WSCC-EPRI [90], Eurostag [92], UCTE [318], and ENEL (INPTC1) [87] are other examples of this kind of format.
- *Free position, fixed order.* The position of the data is free, however, the order in which the data is listed in the file is fixed. For example, in the PSS/E format, the load data must follow the bus data, and the generator data must follow the load data [245]. In the bus data, the bus name must follow the bus identification number, etc. However, there is no restriction on the token position, and data can be separated by commas or by spaces. Other examples of this kind of formats are those of GE-PSLF [88] and FlowDemo.net [83].
- *Free position, free order.* The position and the order of the data are free. Typically, an *ad-hoc* parser is needed to read this kind of data files, which makes quite difficult to create an import/export utility. Some examples of such formats are the formats used by Simpow [302], DigSilent [75], PowerWorld [246] and InterPSS [359]. A special example of this format are also all data files written in Matlab (see for example, PSAT [194], PST [59], and Matpower [363]).

  In this case, the data file is parsed by the Matlab interpreter, which makes extremely flexible the information that can be stored in the file, but also hard to export to other platforms if not using the Matlab interpreter.
- *Mark-up languages* Mark-up languages allows the maximum freedom for organizing data. The idea of mark-up languages is simple but powerful: each data is introduced by a pre-defined syntax which allows clearly identifying the data itself. Thus the position and the order of the data is not relevant. For example a widely used mark-up language is XML [342]. Another example is the Resource Description Framework (RDF) that is used for CIM data bases.

21.1.2 Kind of Supported Data

This feature affects the contents of the data file. Clearly, the more data kinds the format can support, the more complete and general the format is. A format that pretends to be application independent should provide as many kind of data as possible. For example, typical data kinds are as follows.

- *Static Data.* (e.g., power flow data).
- *Dynamic Data.* (e.g., synchronous machine and regulator parameters).
- *Market Data.* (e.g., generator and load bids).

- *Short Circuit Analysis Data.* (e.g., negative and zero sequence of generators and transformers).
- *Graphical Data.* (e.g., network scheme, geographical information system, etc.).
- *Other Data.* (e.g., FACTS data, user defined component data, etc.).

21.1.3 Number of Files

Having multiple files for defining a network can be an useful feature if working with several scenarios for the same network. For example, one can use a single power flow data and work on several dynamic scenarios. This feature allows reducing the amount of data stored on the computer. However, nowadays, this is a not so critical issue taking into account that a 15000 bus system takes about 10 MB of disk space while modern hard disks contain hundreds of GB. Furthermore, the "modification command" described in the next subsection is a better option than the multiple-file feature for multi-scenario studies. The following items describes some examples.

- *Single file.* Most of the data formats requires a single file for defining the whole network. This is typical of most formats.
- *Multiple fixed number of files.* Some formats uses different files for different information. For example the power flow data is in some case separated from the dynamic data (e.g., CYME [68] and Eurostag [92] formats).
- *Any number of files.* The Simpow format [302] provides the possibility of including any number of files. Nested file inclusion is also allowed. This feature is useful in case of large networks, where the amount of data is cumbersome. The user can be interested in modifying only a small part of the network and it is thus easier to work on a small file that is then included in the main data file.

21.1.4 Default Values, Prototypes and Data Manipulation

Accepting default data is both a feature of the data format and of the application that reads the data format. The application that reads the data file must assign a known default value to all parameters that are not specifically defined in the file. Thus, it is necessary that the documentation of the data format clearly specifies default values.

Similar to default value definition is data prototyping, which consists in referring certain device data to a common device prototype. Any device instance inherits the data of the prototypes. This technique is similar to creating a class (prototype) and then instantiating several times that class.[1] The advantage is that a change in the prototype data automatically update all the data referring to that prototype.

Finally, being able to modify data "on the fly" is an useful feature for scripting applications. For example, if one wants to solve the power flow for different load levels, it could be convenient to have a compact command included in the data file that modifies the load powers without the need of rewriting all load data.

The following items describes some examples.

- *Default Values.* Most formats, especially fixed position formats, does not support default values and force writing all data of a component, even if those data are not known or could be easily deduced by default.
- *Device Prototyping.* This feature is provided by the DigSilent format [75]. In case of big distribution networks where most of the transformers and protections are the same, data prototyping can save space. The Simpow format also provides a sort of device prototyping feature in the definition of synchronous machine regulators [302]. Synchronous machine data include the code of a certain AVRs and/or turbine governor prototype. This technique allows reducing the data file size if most of AVRs and turbine governors have same data.
- *Modification Command.* The Simpow format provides the powerful ALTER command for modifying the *base case* data. Matlab-based formats implicitly include modification commands, since any Matlab function and matrix manipulation can be included in the data file.

21.2 Canonical Model

From the previous section, it is clear that if one wants to use different software tools, some data import/export utility is needed. Software companies such as PowerWorld and most open-source tools, such as UWPFLOW, InterPSS and PSAT, develop and provide adapters to import data. However, the direction is mainly one way, i.e., importing data in different formats into simulation applications, but generally not the other way round, i.e., exporting data to different data formats. This makes very difficult (if not impossible) to exchange power system simulation study case in a robust and reliable way.

Another strong restriction to the diffusion and creation of data format adapters is the fact that the documentation of most commercial data formats is not freely available and, in some cases, is intentionally not complete.

Even if one can access the full and detailed documentation of any formats in use in the power system community, there is another important issue that

[1] The difference between "classes" and "prototypes" is that a class defines a abstract type that in general does not contains any specific data, whereas a prototype is a template that defines data.

should be solved. If there are n data formats and m software applications, the possible filtering/converting routes that connect each format to each application are $n \times m$. This fact is illustrated in Figure 21.1. Actually, the number of routes is even higher than $n \times m$, because some formats come in several versions and require a specific filter for each version (e.g., the family of PSS/E v. 2x and 3x formats).

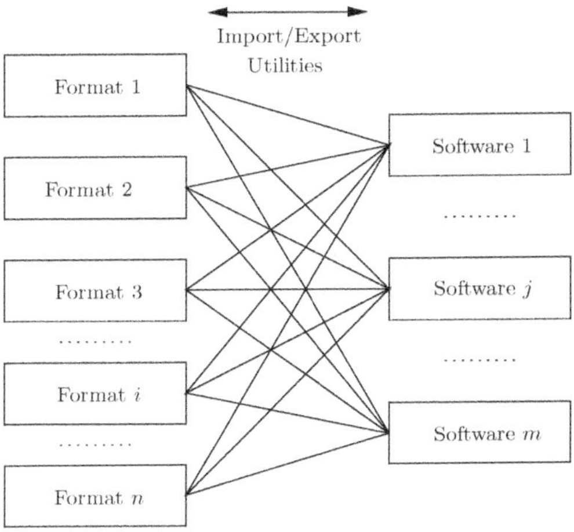

Fig. 21.1 Current state of data exchange structure

A more efficient solution is to use a *canonical model* [201]. In the information model theory, a canonical model indicates a common format into which any data file written in other formats can be converted and that can be read by any software application. In other words, a canonical model is a "super-set" of all other data formats. Using the canonical model, the possible filtering/converting routes for n formats and m applications are only $n + m$, as illustrated in Figure 21.2.

The critical assumption is that the canonical model is a standard for power system analysis. Without this assumption the canonical model would be just another available format. A tentative to create such canonical model actually exists and is described in the following section.

21.3 Common Information Model

The Common Information Model (CIM) is defined by international IEC standards, e.g., [137, 138]. Reference [188] provides the following good definition of CIM:

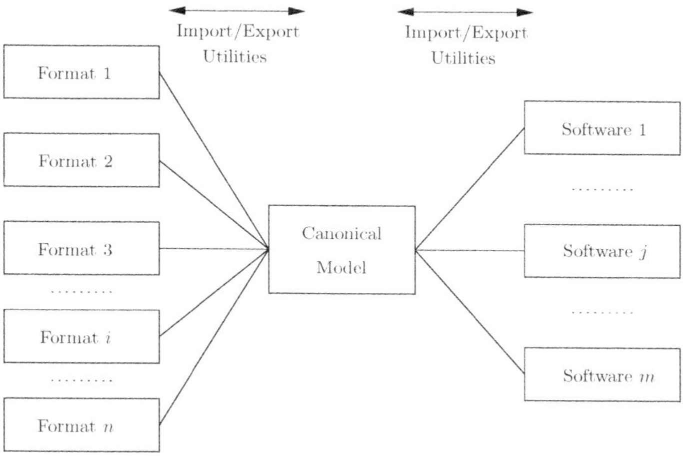

Fig. 21.2 Proposed data exchange structure

"The IEC standard 61970-301 [138] is a semantic model that describes the components of a power system at an electrical level and the relationships between each component. The IEC 61968-11 [137] extends this model to cover the other aspects of power system software data exchange such as asset tracking, work scheduling and customer billing. These two standards, 61970-301 and 61968-11 are collectively known as the Common Information Model (CIM) for power systems and currently have two primary uses: to facilitate the exchange of power system network data between companies; and to allow the exchange of data between applications within a company."

In its purpose, CIM is designed to cover all aspects of power systems and uses RDF for organizing data. There are several ways to implement a CIM data file. A common structure consists in dividing the power system elements into two sections, namely topological data and physical resources. The topological data organizes the system into islands and each island into topological nodes that contain *terminals*. The attributes of each topological object define its connectivity to other terminals. Resources can be further subdivided into two groups, operational resources and equipments. Operational resources are *containers* such as control areas, bays, substations, voltage levels, etc, while equipments are the physical devices, such as loads, that belong to some container. Terminals provide the links between topological data and physical resources, as illustrated in Figure 21.3. Topological nodes and equipments belong to containers. Then, equipments can be connected to a node through one or more terminals.

The CIM scheme is sufficiently general to be able to describe any aspect of power systems. And actually this is the goal of the task forces that are defining the IEC standards that describes the CIM format. Furthermore, being based

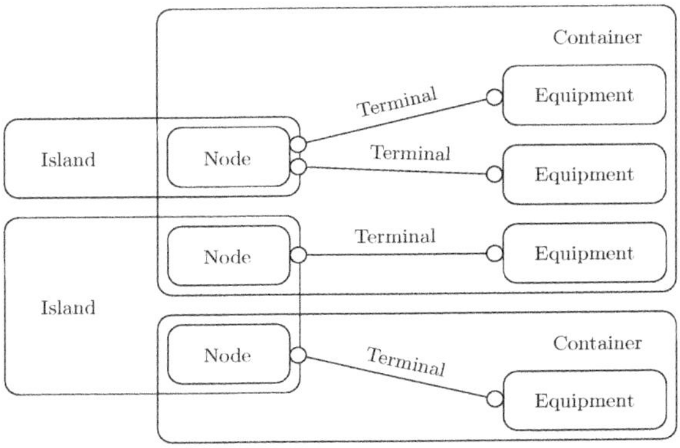

Fig. 21.3 Structure of a possible CIM implementation

on international standards, CIM is gaining more and more popularity among utilities and practitioners. However, there are some intrinsic weaknesses that make CIM quite inadequate for being systematically adopted for research- and education-oriented power system analysis.

1. The CIM structure is rather complex and requires a time to be understood and efficiently used. This is against the well-known "kiss" rule.
2. Due to the internal division into topological and physical elements, most devices are defined twice. This is against the Occam's razor principle that forbids unnecessarily multiplying objects.[2]
3. CIM is based on RDF, which is similar but much more difficult to process than the XML schema. In any case, RDF is designed to be processed by machines, since it is nearly impossible for a human to read a CIM document describing a real power system. Furthermore, RDF has the drawback of being particularly lengthy. The use of meaningful tags and the intrinsic CIM redundancy lead to files of abnormal size. Generally, a CIM file is 500 to 1000 times bigger than data files written in some common power system format. Thus, a power system that could be described by a 4 MB files requires 4 GB if written in the CIM format. CIM supporters claim that disk space is not an issue anymore. However, the time required for reading a 4 GB file cannot be neglected if the operation has to be repeated hundreds or thousands of times (e.g., for some Montecarlo simulation).
4. A byproduct of the issues above is that writing, reading and maintaining CIM data bases are not straightforward tasks. The RDF scheme underlying the CIM format is quite difficult to implement. There are some early stage implementations of CIM version based on XML schemes, but only proprietary and commercial solutions are currently available for handling complete CIM data bases. Clearly, there is nothing wrong in that there

[2] The Occam's razor states that *entia non sunt multiplicanda præter necessitatem.*

are commercial interests behind the definition of CIM. However, a monopolistic proprietary approach is certainly to be avoided if a format claims to be an international standard.

5. CIM has been designed to cover all aspects of power systems. In the idea of its creators, CIM can be actually able to describe any aspect of reality. Due to the Gödel's theorem, if a system is complete is inconsistent, and if it is consistent is incomplete. If the goal of CIM is to be complete, then CIM is inexorably destined to be inconsistent.
6. The name itself, i.e., common information model, indicates the inadequacy of CIM for describing data for power system analysis. CIM is designed to be a mere "information" data base. However, as discussed in the following section, a data format cannot be independent from the mathematical models used for describing power system devices.

21.4 Consistent Data Schemes

Any data format is implicitly based on a certain set of mathematical models of power system devices. Part III has discussed a variety of devices, but the list is well far to be complete and more devices will be invented in the future. Furthermore, each device has a variety of possible models, depending on the detail level and on the analysis that has to be solved. According to this premise, the definition of a complete data format is simply impossible. But, fortunately, it is not desirable. Rather than aiming to classify any power system device in a closed structure, a consistent data format should simply provide a structure (or a *scheme*) able to describe as many devices as possible. In other words, it is not relevant how many devices a data format includes, but how versatile and well-designed is the scheme on which the format is built. In this regard, the approach of CIM is correct, since it provides a general scheme for describing power systems.[3]

A scheme is an abstract object and should not be based on a specific language. A common issue of most data formats is to confuse syntax rules with the data scheme. Mark-up languages avoid this confusion and should be preferred. However, the scheme should not depend on a specific mark-up language.

Based on the features described in Section 21.1, other desired properties are the possibility of spreading data in multiple files, data prototyping and data manipulating commands. Least but not last, a scheme should be as readable as possible so that it is not necessary to use dedicated software to write a simple data file for didactic use. This feature is irrelevant for industrial applications because nobody actually writes by hand a data file composed of thousands of buses. However, designing a data format only thinking in

[3] The success of the XML language is also due to the fact that XML provides the means for designing as complex as desired schemes.

industrial applications (e.g., CIM) and neglecting didactic issues implies a training cost when it comes to employ just-graduated engineers.

Nevertheless, it would be very useful if a data format for power system analysis aligns with the concepts that are included in the CIM for power system data (i.e., standards IEC 61970, 61968, 61850, 60870-6 (TASE.2), ISO9506 (MMS), etc.) In fact, it is not desirable to have corporate systems, SCADA, and substation protection and control working on a common information model, and the analysis software moving to a completely different direction.

Finally, according to the discussion given in Subsection 22.2.4 of Chapter 22, it is desirable that a data format integrates device parameters and geographical information so that one does not have to maintain two separate databases that almost never agree.

Example 21.1 Data Format Example

This example provides a simple yet versatile data format.

The proposed scheme is as follows. A network is composed of a set of devices. A device can be topological (e.g., buses and areas) or physical (e.g., generators and loads). Each device is defined by a list of parameters. The device identification code (called `idx` in the following example) allows connecting devices together. For example, to indicate that a load is connected to a certain bus, it suffices to assign the bus identification code to the load parameter `bus`.

As discussed above, the syntax is not really important, especially if using mark-up languages. However, to maintain as readable as possible the file, the following rules are used:

1. The name of the device starts in the first column.
2. Device data can span multiple lines.
3. Lines following the first one must be indented.
4. Data follows in `Property_Name = Value` pairs, separated by commas.
5. Data in form of strings must be delimited by `""`.
6. Data in form of arrays must be delimited by brackets `[]`.
7. Each element of the array must be separated by semicolons.
8. Float data support simple operations (sum, multiplication, etc.).
9. Comments starts with a `#` in the first column

For example this is how the IEEE 14-bus system looks like in this format:

```
Bus, Vn = 69.0, idx = 1, name = "Bus 1"
Bus, Vn = 69.0, idx = 2, name = "Bus 2"
Bus, Vn = 69.0, idx = 3, name = "Bus 3"
Bus, Vn = 69.0, idx = 4, name = "Bus 4"
Bus, Vn = 69.0, idx = 5, name = "Bus 5"
Bus, Vn = 13.8, idx = 6, name = "Bus 6"
Bus, Vn = 13.8, idx = 7, name = "Bus 7"
```

```
Bus, Vn = 18.0, idx = 8, name = "Bus 8"
Bus, Vn = 13.8, idx = 9, name = "Bus 9"
Bus, Vn = 13.8, idx = 10, name = "Bus 10"
Bus, Vn = 13.8, idx = 11, name = "Bus 11"
Bus, Vn = 13.8, idx = 12, name = "Bus 12"
Bus, Vn = 13.8, idx = 13, name = "Bus 13"
Bus, Vn = 13.8, idx = 14, name = "Bus 14"

Area, idx = 1, name = "14-Bus"

Region, Ptol = 9.9999, idx = 1, name = "IEEE 14 Bus",
        slack = 1.0

Line, Vn = 69.0, Vn2 = 69.0, b = 0.0528, bus1 = 1, bus2 = 2,
      idx = "Line_1", name = "Line 1", r = 0.01938, x = 0.05917
Line, Vn = 69.0, Vn2 = 69.0, b = 0.0492, bus1 = 1, bus2 = 5,
      idx = "Line_2", name = "Line 2", r = 0.05403, x = 0.22304
Line, Vn = 69.0, Vn2 = 69.0, b = 0.0438, bus1 = 2, bus2 = 3,
      idx = "Line_3", name = "Line 3", r = 0.04699, x = 0.19797
Line, Vn = 69.0, Vn2 = 69.0, b = 0.0374, bus1 = 2, bus2 = 4,
      idx = "Line_4", name = "Line 4", r = 0.05811, x = 0.17632
Line, Vn = 69.0, Vn2 = 69.0, b = 0.034, bus1 = 2, bus2 = 5,
      idx = "Line_5", name = "Line 5", r = 0.05695, x = 0.17388
Line, Vn = 69.0, Vn2 = 69.0, b = 0.0346, bus1 = 3, bus2 = 4,
      idx = "Line_6", name = "Line 6", r = 0.06701, x = 0.17103
Line, Vn = 69.0, Vn2 = 69.0, b = 0.0128, bus1 = 4, bus2 = 5,
      idx = "Line_7", name = "Line 7", r = 0.01335, x = 0.04211
Line, Vn = 69.0, Vn2 = 13.8, bus1 = 4, bus2 = 7, idx = "Line_8",
      name = "Line 8", tap = 0.978, trasf = True, x = 0.20912
Line, Vn = 69.0, Vn2 = 13.8, bus1 = 4, bus2 = 9, idx = "Line_9",
      name = "Line 9", tap = 0.969, trasf = True, x = 0.55618
Line, Vn = 69.0, Vn2 = 13.8, bus1 = 5, bus2 = 6, idx = "Line_10",
      name = "Line 10", tap = 0.932, trasf = True, x = 0.25202
Line, Vn = 13.8, Vn2 = 13.8, bus1 = 6, bus2 = 11, idx = "Line_11",
      name = "Line 11", r = 0.09498, x = 0.19890
Line, Vn = 13.8, Vn2 = 13.8, bus1 = 6, bus2 = 12, idx = "Line_12",
      name = "Line 12", r = 0.12291, x = 0.25581
Line, Vn = 13.8, Vn2 = 13.8, bus1 = 6, bus2 = 13, idx = "Line_13",
      name = "Line 13", r = 0.06615, x = 0.13027
Line, Vn = 13.8, Vn2 = 18.0, bus1 = 7, bus2 = 8, idx = "Line_14",
      name = "Line 14", trasf = True, x = 0.17615
Line, Vn = 13.8, Vn2 = 13.8, bus1 = 7, bus2 = 9, idx = "Line_15",
      name = "Line 15", x = 0.11001
Line, Vn = 13.8, Vn2 = 13.8, bus1 = 9, bus2 = 10, idx = "Line_16",
      name = "Line 16", r = 0.03181, x = 0.08450
Line, Vn = 13.8, Vn2 = 13.8, bus1 = 9, bus2 = 14, idx = "Line_17",
      name = "Line 17", r = 0.12711, x = 0.27038
Line, Vn = 13.8, Vn2 = 13.8, bus1 = 10, bus2 = 11, idx = "Line_18",
      name = "Line 18", r = 0.08205, x = 0.19207
Line, Vn = 13.8, Vn2 = 13.8, bus1 = 12, bus2 = 13, idx = "Line_19",
      name = "Line 19", r = 0.22092, x = 0.19988
Line, Vn = 13.8, Vn2 = 13.8, bus1 = 13, bus2 = 14, idx = "Line_20",
      name = "Line 20", r = 0.17093, x = 0.34802
```

```
PQ, Vn = 69.0, bus = 2, idx = "PQ load_1", name = "PQ Bus 2",
    p = 0.217, q = 0.127
PQ, Vn = 69.0, bus = 3, idx = "PQ load_2", name = "PQ Bus 3",
    p = 0.942, q = 0.19
PQ, Vn = 69.0, bus = 4, idx = "PQ load_3", name = "PQ Bus 4",
    p = 0.478, q = -0.039
PQ, Vn = 69.0, bus = 5, idx = "PQ load_4", name = "PQ Bus 5",
    p = 0.076, q = 0.016
PQ, Vn = 13.8, bus = 6, idx = "PQ load_5", name = "PQ Bus 6",
    p = 0.112, q = 0.075
PQ, Vn = 13.8, bus = 9, idx = "PQ load_6", name = "PQ Bus 9",
    p = 0.295, q = 0.166
PQ, Vn = 13.8, bus = 10, idx = "PQ load_7", name = "PQ Bus 10",
    p = 0.09, q = 0.058
PQ, Vn = 13.8, bus = 11, idx = "PQ load_8", name = "PQ Bus 11",
    p = 0.035, q = 0.018
PQ, Vn = 13.8, bus = 12, idx = "PQ load_9", name = "PQ Bus 12",
    p = 0.061, q = 0.016
PQ, Vn = 13.8, bus = 13, idx = "PQ load_10", name = "PQ Bus 13",
    p = 0.135, q = 0.058
PQ, Vn = 13.8, bus = 14, idx = "PQ load_11", name = "PQ Bus 14",
    p = 0.149, q = 0.05

Breaker, Vn = 69.0, bus = 2, fn = 60.0, idx = 1, line = "Line_4",
         name = "Breaker 1", t1 = 1.0, u1 = 1

PV, Vn = 69.0, bus = 2, busr = 2, idx = 2, name = "PV Bus 2",
    pg = 0.4, pmax = 1.0, pmin = 0, qmax = 0.5, qmin = -0.4,
    v0 = 1.045
PV, Vn = 69.0, bus = 3, busr = 3, idx = 3, name = "PV Bus 3",
    pmax = 1.0, pmin = 0, qmax = 0.4, v0 = 1.01
PV, Vn = 13.8, bus = 6, busr = 6, idx = 6, name = "PV Bus 6",
    pmax = 1.0, pmin = 0, qmax = 0.24, qmin = -0.06, v0 = 1.07
PV, Vn = 18.0, bus = 8, busr = 8, idx = 8, name = "PV Bus 8",
    pmax = 1.0, pmin = 0, qmax = 0.24, qmin = -0.06, v0 = 1.09

Shunt, Vn = 13.8, b = 0.19, bus = 9, idx = "Shunt_1",
       name = "Shunt Bus 9"

SW, Vn = 69.0, bus = 1, busr = 1, idx = 1, name = "SW Bus 1",
    pg = 2.324, pmax = 999.9, pmin = -999.9, qmax = 9.9,
    qmin = -9.9, v0 = 1.06

Syn5a, D = 2.0, M = 2*5.143, Sn = 615.0, Td10 = 7.4, Td20 = 0.03,
       Tq20 = 0.033, Vn = 69.0, bus = 1, fn = 60.0, gen = 1,
       idx = 1, name = "Syn 1", xd = 0.8979, xd1 = 0.2995, xd2 = 0.23,
       xl = 0.2396, xq = 0.646, xq1 = 0.646, xq2 = 0.4
Syn6a, D = 2.0, M = 2*6.54, Sn = 60.0, Td10 = 6.1, Td20 = 0.04,
       Tq10 = 0.3, Tq20 = 0.099, Vn = 69.0, bus = 2, fn = 60.0,
       gen = 2, idx = 2, name = "Syn 2", ra = 0.0031, xd = 1.05,
       xd1 = 0.185, xd2 = 0.13, xq = 0.98, xq1 = 0.36, xq2 = 0.13
Syn6a, D = 2.0, M = 2*6.54, Sn = 60.0, Td10 = 6.1, Td20 = 0.04,
       Tq10 = 0.3, Tq20 = 0.099, Vn = 69.0, bus = 3, fn = 60.0,
       gen = 3, idx = 3, name = "Syn 3", ra = 0.0031, xd = 1.05,
```

```
       xd1 = 0.185, xd2 = 0.13, xq = 0.98, xq1 = 0.36, xq2 = 0.13
Syn6a, D = 2.0, M = 2*5.06, Sn = 25.0, Td10 = 4.75, Td20 = 0.06,
       Tq10 = 1.5, Tq20 = 0.21, Vn = 13.8, bus = 6, fn = 60.0,
       gen = 6, idx = 4, name = "Syn 4", ra = 0.0041, xd = 1.25,
       xd1 = 0.232, xd2 = 0.12, xl = 0.134, xq = 1.22, xq1 = 0.715,
       xq2 = 0.12
Syn6a, D = 2.0, M = 2*5.06, Sn = 25.0, Td10 = 4.75, Td20 = 0.06,
       Tq10 = 1.5, Tq20 = 0.21, Vn = 18.0, bus = 8, fn = 60.0,
       gen = 8, idx = 5, name = "Syn 5", ra = 0.0041, xd = 1.25,
       xd1 = 0.232, xd2 = 0.12, xl = 0.134, xq = 1.22, xq1 = 0.715,
       xq2 = 0.12

Avr1, Ka = 200.0, Kf = 0.0012, Ta = 0.02, Te = 0.19, Tf = 1.0,
      bus = 1, idx = 1, name = "AVR 1", syn = 1, vmax = 9.99,
      vmin = 0.0
Avr1, Ka = 20.0, Kf = 0.001, Ta = 0.02, Te = 1.98, Tf = 1.0,
      bus = 2, idx = 2, name = "AVR 2", syn = 2, vmax = 2.05,
      vmin = 0.0
Avr1, Ka = 20.0, Kf = 0.001, Ta = 0.02, Te = 1.98, Tf = 1.0,
      bus = 3, idx = 3, name = "AVR 3", syn = 3, vmax = 1.7,
      vmin = 0.0
Avr1, Ka = 20.0, Kf = 0.001, Ta = 0.02, Te = 0.7, Tf = 1.0,
      bus = 6, idx = 4, name = "AVR 4", syn = 4, vmax = 2.2,
      vmin = 1.0
Avr1, Ka = 20.0, Kf = 0.001, Ta = 0.02, Te = 0.7, Tf = 1.0,
      bus = 8, idx = 5, name = "AVR 5", syn = 5, vmax = 2.2,
      vmin = 1.0
```

The format can be completed by the `INCLUDE` and `ALTER` commands, that allows including external files and modifying data previously defined, respectively.

For example, the syntax of the `INCLUDE` command is:

```
INCLUDE, include_file
```

where `include_path` is the full or relative path to the external file. The syntax of the `ALTER` command is as follows:

```
ALTER, device_name, action, filter, property, value
```

where

`device_name` is the system name of any device previously defined.
`actions` are: `MUL`, `DIV`, `SUM`, `SUB`, `POW`, `REP`.
`filter` is the reg-exp to be used for selecting the device (based on the name).
`property` is any numerical property of the device.
`value` the numerical value to be used by `action`. The value must be a float.

For example, to multiply by 1.2 all loads and PV generator powers, one has:

```
ALTER, PQ, MUL, *, p, 1.2
ALTER, PQ, MUL, *, q, 1.2
ALTER, PV, MUL, *, pg, 1.2
```

Script 21.1 Data Parser

The following Python code implements a parser for the data format described in the previous section. A parser consists in a procedure that reads a file and extracts data and/or information based on a given syntax. Regular expressions (imported through the standard library `re`) allows efficiently parsing the data.[4] Finally, the command `INCLUDE` is supported through a recursive call to the function `read`.

```
import re
import system

def alter(data):

    device = data[0]
    action = data[1]
    if data[2] == '*': data[2] = '.*'
    regex = re.compile(data[2])
    prop = data[3]
    value = float(data[4])

    if action == 'MUL':
        for item in xrange(system.__dict__[device].n):
            if regex.search(system.__dict__[device].name[item]):
                system.__dict__[device].__dict__[prop][item] *= value
    elif action == 'REP':
        for item in xrange(system.__dict__[device].n):
            if regex.search(system.__dict__[device].name[item]):
                system.__dict__[device].__dict__[prop][item] = value
    elif action == 'DIV':
        if not value:
            return
        for item in xrange(system.__dict__[device].n):
            if regex.search(system.__dict__[device].name[item]):
                system.__dict__[device].__dict__[prop][item] /= value
    elif action == 'SUM':
        for item in xrange(system.__dict__[device].n):
            if regex.search(system.__dict__[device].name[item]):
                system.__dict__[device].__dict__[prop][item] += value
    elif action == 'SUB':
        for item in xrange(system.__dict__[device].n):
            if regex.search(system.__dict__[device].name[item]):
                system.__dict__[device].__dict__[prop][item] -= value
    elif action == 'POW':
        for item in xrange(system.__dict__[device].n):
            if regex.search(system.__dict__[device].name[item]):
                system.__dict__[device].__dict__[prop][item] **= value
```

[4] Regular expressions are a fascinating world apart in computer programming. The regular expression parser is an extremely powerful tool for manipulating text and data and is provided as a standard library by most programming languages, including Perl, Python and Java. The interested reader can find a complete description of regular expressions in [102].

```
    else:
        print 'ALTER action <%s> is not defined' % action


def read(fid, header = True):

    # useful regular expressions and constants
    sep = re.compile(r'\s*,\s*')
    comment = re.compile(r'^#\s*')
    equal = re.compile(r'\s*=\s*')
    math = re.compile(r'[*/+-]')
    double  = re.compile(r'[+-]? *(?:\d+(?:\.\d*)?|\.\d+)(?:[eE][+-]?\d+)?')

    # parse data
    while 1:

        line = fid.readline()
        if not line: break
        line = line.replace('\n', '')
        line = line.strip()
        if not line: continue
        if comment.search(line): continue

        # span multiple line
        while line.endswith(',') or line.endswith(';'):
            newline = fid.readline()
            line = line.replace('\n', '')
            if not newline: break
            newline = newline.strip()
            if not newline: continue
            if comment.search(newline): continue
            line += ' ' + newline

        data = sep.split(line)
        device = data.pop(0)
        device = device.strip()

        if device == 'ALTER':
            alter(data)
            continue

        if device == 'INCLUDE':
            print 'Parsing include file <%s>.' % data[0]
            newfid = open(data[0], 'rt')
            read(newfid, header = False) # recursive call
            newfid.close()
            print 'Parsing of include file <%s> completed.' % data[0]
            continue

        kwargs =

        for item in data:
            pair = equal.split(item)
            key = pair[0].strip()
```

```
            value = pair[1].strip()
            if value.startswith('"'):
                value = value[1:-1]
            elif value.startswith('['):
                array = value[1:-1].split(';')
                if math.search(value): # execute simple operations
                    value = map(lambda x: eval(x), array)
                else:
                    value = map(lambda x: float(x), array)
            elif double.search(value):
                if math.search(value): # execute simple operations
                    value = eval(value)
                else:
                    value = float(value)
            elif value == 'True':
                value = True
            elif value == 'False':
                value == False
            else:
                value = int(value)
            kwargs[key] = value

        index = kwargs.pop('idx', None)
        namex = kwargs.pop('name', None)

        try:
            system.__dict__[device].add(idx=index, name=namex, **kwargs)
        except KeyError:
            print 'Device <%s> is will be skipped' % device

    return True
```

Chapter 22
Visualization Matters

This chapter discusses visualization matters related to power system analysis. In particular, the aspects covered in the chapter are the adequacy of graphical user interfaces versus the command line usage (Section 22.1) and available approaches for displaying results (Section 22.2). The latter describes standard two-dimensional plots (Subsection 22.2.1), temperature maps (Subsection 22.2.2), three-dimensional plots (Subsection 22.2.3), and the integration of graphical information systems into power system analysis software packages (Subsection 22.2.4).

22.1 Graphical Interface vs. Command Line Approach

The current trend of proprietary operating systems and software applications is to mask the functioning of the software behind a Graphical User Interface (GUI). There is no doubt that intuitive GUIs simplify the learning process of the application, at least at the first approach. Furthermore, GUIs are certainly useful for didactic purposes. On the other hand, the weakness of any GUI is that the only allowed operations are those embedded in the GUI itself. Thus, GUIs reduce user freedom. This is a real issue for most scientific (i.e., number-crunching) applications.[1]

Opposite to GUIs is the command-line approach. The advantages of using a pure command-line approach for scientific applications are as follows.

1. The command-line approach allows efficient use on remote servers.
2. Batch programming is possible only with the command-line approach (see Script 22.1).
3. The size of the application can be very reduced if GUIs are taken apart.
4. GUIs are often operating system dependent. Thus, using a command line approach eases portability.

[1] It is important not to confuse GUIs with CAD/CAM applications and result visualization. The latter is discussed in the following section.

F. Milano: Power System Modelling and Scripting, Power Systems, pp. 475–488.
springerlink.com

5. The command-line approach can provide a finer tuning than GUI-based ones.
6. Due to the lack of a user-friendly interface, the user is forced to carefully study the command-line options that, in turn, leads to a better understanding of the application functioning.

GUIs are compatible with the command-line approach but only if the GUI is maintained strictly separated from the main application that performs the operations. GUIs built on top of an existing command-line application are often called *skins*. The advantage of separating the skin from the kernel application is twofold: (i) the user can chose if using the GUI or not, and (ii) several skins can be built for the same application.[2] In conclusion, the command line approach results in more flexibility and more freedom for the user.

Unfortunately, few power system applications have adopted the efficient command-line approach, e.g., [42, 363], and very few have provided a command-line version and a separate skin [195]. This is probably due to the fact that most power system software packages are proprietary applications and works only on Microsoft® operating systems.

Script 22.1 Batch Script for Power Flow Analysis

Batch scripting is a powerful technique that consists in executing a series of operations in a non-interactive way. All commands are stored in a file (batch script) and then the file executed. Batch scripting can be considered a sort of meta-programming.

Python provides an easy way to take advantage of batch scripting. In fact any Python procedure can be distributed as a library rather than as stand-alone applications.[3] Since Python libraries can be imported by any Python script, the same Python language can be used for creating batch files. This technique is illustrated in the script below, which produces the results presented in Example 4.4 of Chapter 4.

The script works as follows. For the sake of example, assume that the power flow analysis tool described in Script 3.2 of Chapter 3 is available as a Python library called `powerlib`. Thus, importing this library provides all classes and routines described in Parts II and III. For example, settings can be adjusted using the attributes of the `powerlib.system.Settings` class (see Script 3.2 for further details on the architecture of the `system` structure).

In the example below, the main data file is `ieee14.dm` that, apart from the power flow data, contains the statement:

```
INCLUDE, load.txt
```

[2] This explains why, for example, there exist several window managers for Linux systems, while Windows system have an unique user interface.

[3] The `setuptools` library allows creating custom libraries in few easy steps.

This command imposes that the parser reads a file called `load.txt` (see Example 21.1 and Script 21.1 of Chapter 21 for further details on `INCLUDE` statement). Then the script calls repeatedly the solution of the power flow analysis changing the load level in the file `load.txt`. The load level is modified through the `ALTER` statement (see again Example 21.1 for details). Finally results are plotted and saved to a file using the Matplotlib library.

```
import powerlib
import numpy
from cvxopt.base import matrix, div
from matplotlib import pyplot

def run():

    load = numpy.linspace(0.6, 2.6, num=50, endpoint=True)
    a = []
    b = []
    powerlib.system.Settings.pv2pq = False
    powerlib.system.Settings.pq2z = False

    powerlib.system.Settings.pfsolver = 'NR'
    for m in load:
        powerlib.system.Settings.iteration = 0
        fid = open('load.txt', 'wt')
        fid.write('ALTER, PQ, MUL, *, p, %.5f\n' % m)
        fid.write('ALTER, PQ, MUL, *, q, %.5f\n' % m)
        fid.close()
        powerlib.run('ieee14.dm', path='tests/powerlib')
        vec = powerlib.system.DAE.y[powerlib.system.Bus.a]
        a.append(vec)

    powerlib.system.Settings.pfsolver = 'DC'
    for m in load:
        powerlib.system.Settings.iteration = 0
        fid = open('load.txt', 'wt')
        fid.write('ALTER, PQ, MUL, *, p, %.5f\n' % m)
        fid.write('ALTER, PQ, MUL, *, q, %.5f\n' % m)
        fid.close()
        powerlib.run('ieee14.dm', path='tests')
        vec = powerlib.system.DAE.y[powerlib.system.Bus.a]
        b.append(vec)

    n = load.size
    c = matrix(0, (n, 1), 'd')
    d = matrix(0, (n, 1), 'd')
    m = matrix(load)

    for i, ai, bi in zip(range(n), a, b):
        am = matrix(ai)
        bm = matrix(bi)
        ad = matrix(ai)
        ad[0] = 1.0
        er = 100*div(abs(am - bm), abs(ad))
```

```
        d[i] = max(er)
        c[i] = sum(er)/(n - 1)

    fig = pyplot.figure()
    pyplot.hold(True)
    pyplot.plot(m, c, 'k', label='mean error')
    pyplot.plot(m, d, 'k:', label='max error')
    pyplot.legend(loc='upper left')
    pyplot.xlabel('load')
    pyplot.ylabel('error')
    pyplot.xlim([min(m), max(m)])
    pyplot.savefig('dcerror.eps', format='eps')
    pyplot.show()
```

22.2 Result Visualization

The importance of an intuitive and fully informative visualization of power system results has been recognized and formalized in early nineties. In [180], the authors specify three guidelines for setting up a good graphical representation of physical phenomena: (i) natural encoding of information; (ii) task specific graphics; and (iii) no gratuitous graphics. In [225, 346], two-dimensional contour plots are proposed for the visualization of voltage bus levels with inclusion of the topological information of the network. Standard two-dimensional plots, temperature maps and three-dimensional plots and geographical information systems comply with the three guidelines mentioned above and are described in the following subsections.

22.2.1 Standard Two-Dimensional Plots

Standard two-dimensional (2D) plots are extensively used throughout the whole book. Moreover, the features of this visualization technique are very well-known. This subsection only describes an efficient approach for manipulating simulations results.

One of the most important strengths of UNIX systems is that every manipulable object is a file. Using this concept, UNIX system has proved to be the most adequate for scientific applications. A standard UNIX application can be launched from the command line and accepts files for both input data and settings and then provides the output as a set of files. For example, the LaTeX compiler accepts as input .tex files and provides as outputs a .dvi file. Then, the .dvi file can be viewed as is or translated into postscript or pdf formats by other applications.

Following the UNIX teaching, the outputs of a power system analysis tool, such as time domain simulations or continuation power flow analysis, can be conveniently exported as plain ASCII files. For example, one file can report variable values in columns and another file can specify to which variable each column corresponds. Then the two files can be parsed and results plotted according to user settings.

This approach provides several advantages with respect to a all-in-one approach which is the standard of most power system applications. In particular, the main advantage is the possibility of keeping the application that performs the simulation separated from the application that plots results. This provides more freedom, similarly to the command-line/skin approach described in previous Section 22.1.

Script 22.2 Parser for Simulations Results

The following script provides an example of parser for simulation results. The parser reads one or more files, stores the specified values in lists and finally calls a `plot` function that actually creates the 2D plot. The syntax for calling the script for the command line is as follows:

```
parse_results [-options] datafile1 listfile1 x y1 [y2 y3 ...]
              [datafile2 listfile2 x y1 [y2 y3 ...]] ...
```

where `x` is the number the variable used for the x-axis and `y1`, `y2`, etc., are the numbers of the variables used for the y-axis.

```
import os
import re
import sys
from optparse import OptionParser

def parse_results():

    parser = OptionParser()
    parser.add_option('-p', '--path', dest='path',
                      default=cwd, help='Path of the data file')
    parser.add_option('-l', '--legend', dest='legend',
                      action='store_true', default=False,
                      help='Add legend to the plot')
    parser.add_option('-g', '--grid', dest='grid',
                      action='store_true', default=False,
                      help='Add grid to the plot')
    parser.add_option('-s', '--style', dest='style', default='colors',
                      help=plotstyle)
    parser.add_option('-y', '--ylabel', dest='ylabel',
                      default=None, help='Label for the y axis')
    parser.add_option('-k', '--position', dest='position',
                      default='lower right', help=legendhelp)

    options, args = parser.parse_args(sys.argv[1:])

    # set up arguments
    if len(args) == 0:
        print '* Error: At least one data file must be defined!'
        sys.exit(1)
```

```
interval = re.compile(r'\d+:\d+')
stepint = re.compile(r'\d+:\d+:\d+')
xname = ''
ylegend = []
xout = []
yout = []
nval = 0

# parse argument list
while len(args):

    # data file name
    datafile = args.pop(0)
    listfile = args.pop(0)

    # output file
    outfile, ext = os.path.splitext(options.output)
    outfile += '.eps'

    # populate index vector
    xyidx = []
    yname = []
    while len(args):
        if stepint.search(args[0]):
            values = re.findall(r'\d+', args.pop(0))
            for item in range(int(values[0]), int(values[2]) + 1, \
                              int(values[1])):
                xyidx.append(item)
        elif interval.search(args[0]):
            values = re.findall(r'\d+', args.pop(0))
            for item in range(int(values[0]), int(values[1]) + 1):
                xyidx.append(item)
        elif not args[0].isdigit():
            break
        else:
            xyidx.append(int(args.pop(0)))

    if len(xyidx) == 0:
        xyidx = [0, 1]
    elif len(xyidx) == 1:
        xyidx.append(1)

    # scan list file
    fid = open(listfile, 'r')
    for line in fid:
        data = line.split(',')
        if int(data[0].strip()) == xyidx[0]:
            xname = data[1].strip()
        if int(data[0].strip()) in xyidx[1:]:
            yname.append(data[1].strip())

    # order variable names according to command line order
    yidx = xyidx[1:]
    ysorted = sorted(yidx)
```

```
        mapy = map(yidx.index, ysorted)
        yname_temp = yname[:]
        for idx, item in enumerate(mapy):
            yname[idx] = yname_temp[item]

        fid.close()

        # scan data file
        fid = open(datafile, 'r')
        xx = [[] for x in range(len(xyidx) - 1)]
        yy = [[] for y in range(len(xyidx) - 1)]

        """ Note: xx = [[]]*(len(xyidx) - 1) does not work since each
        element is a copy of the others and updating one element will
        update all the remaining ones."""

        xidx = xyidx[0]
        ref = False
        if not options.reference is None:
            ref = True
            rdx = int(options.reference)

        for line in fid:
            data = line.split()
            for item, yidx in enumerate(xyidx[1:]):
                xx[item].append(float(data[xidx]))
                yy[item].append(float(data[yidx]))

        fid.close()

        xout += xx
        yout += yy
        ylegend += yname

    # call the plot function
    plot(outfile, xout, yout, xname, ylegend,
         style    = options.style,
         legend   = options.legend,
         names    = options.names,
         ylabel   = options.ylabel,
         grid     = options.grid,
         position = options.position)

if __name__ == '__main__':

    parse_results()
```

The standard library `OptionParser` provides the user with Unix-like command line options. The function `plot` imports the Matplotlib library for creating the figure. A possible implementation of the function `plot` is as follows.

```
from matplotlib import pyplot

def plot(output, x, y, xname, yname, style='colors',
         legend=True, grid=False,
         ylabel=None, position='lower right'):

    colors = ['b', 'r', 'g', 'c', 'm', 'y']
    black = ['k-', 'k--', 'k:', 'k-.', 'k,']

    fig = pyplot.figure()
    pyplot.hold(True)

    if style in ('black', 'b'):
        s = black
    else: # default is "colors" or "c"
        s = colors

    for item in xrange(len(y)):
        pyplot.plot(x[item], y[item], \
                    s[item % len(s)], label=yname[item])

    if legend:
        if position.isdigit():
            position = int(position)
        pyplot.legend(loc=position)
    pyplot.xlabel(xname)
    if not ylabel is None:
        pyplot.ylabel(ylabel)
    pyplot.grid(grid)

    # plot figure and save to file
    pyplot.savefig(output, format='eps')
    pyplot.show()
```

All plots provided in this book were obtained using the Python functions described in this example.

22.2.2 Temperature Maps

Contour and temperature maps have proved to be an effective solution for visualizing results of a variety of power system analyses. In [157, 158, 222, 223, 224, 303], 2D maps are used for visualizing a variety of results, such as power flows in transmission lines, locational marginal prices, available transfer capability, contingency analysis, etc. All the previous references are based on a proprietary software package [247] but, nowadays, scripting languages are mature enough for producing high quality temperature maps [197, 295].

From the implementation viewpoint, producing 2D maps is just a matter of assigning topological data (i.e., coordinates) to technical data (e.g., bus voltage magnitudes). This can be obtained using a one-line diagram editor (e.g., PSAT leans on the Simulink editor [195]) or using graphical information systems (see Subsection 22.2.4).

Example 22.1 Temperature Map of the IEEE 14-Bus System

Figure 22.1 shows the voltage temperature map for the power flow solution of the IEEE 14-bus system. The procedure for obtaining the map is as follows.

1. Topological data have to be assigned to buses, lines and other devices that compose the system.
2. Once the power flow analysis is solved, bus coordinates and voltage magnitude values are collected as a set of triplets.
3. A map is created using the functions `meshgrid` and `griddata` from the libraries NumPy and Matplotlib, respectively. The function `meshgrid` creates a coordinate matrix of uniformly spaced points while `griddata` fits a surface of the form $z = f(x, y)$ to the data in the non-uniformly spaced vectors of bus coordinates (x, y) and voltage magnitude values z.
4. The surface computed in the previous point is plotted using any of the 2D functions provided by the Matplotlib library. For example, Figure 22.1 is obtained using the function `imshow`.

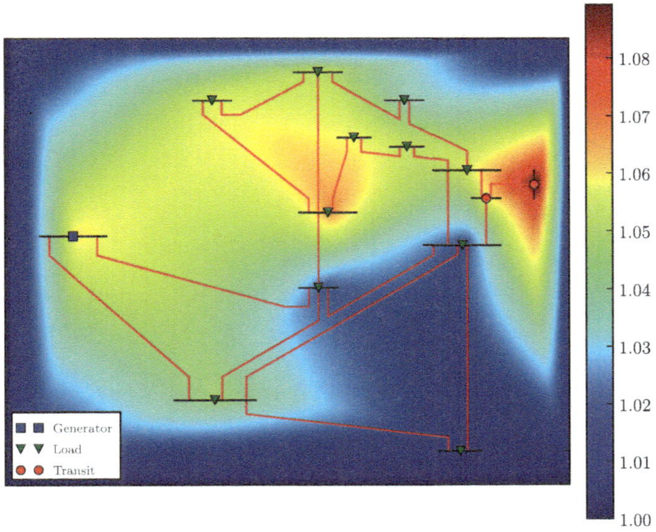

Fig. 22.1 Voltage temperature map for the IEEE 14-bus system. Values in the legend are in pu

22.2.3 Three-Dimensional Plots

Three-dimensional (3D) visualization has been little exploited for power system analysis, although in [348], the advantages of the 3D visualization are discussed and recognized. In [208], rotor speeds of a multi-machine system are displayed in a kind of 3D plot, however the topological information is missing. Reference [93] proposes a variety of 3D visualizations and animations of traveling waves in transmission lines. Finally, [197] proposes 3D animations for visualizing the voltage collapse and undamped oscillation phenomena.

The main issue with 3D plots is how to create a smooth surface that qualitatively captures the "shape" of a set of data (e.g., bus voltage magnitude values). An effective solution to this problem is to compute the *convex hull* of a set of points, which "is considered one of the most elementary interesting problem in computational geometry, just as minimum spanning tree is the most elementary interesting problem in graph algorithms" [282]. Using mathematical terms, the convex hull for a set of points $\mathcal{X}$ in a real vector space $\boldsymbol{V}$ is the minimal convex set containing $\mathcal{X}$ [248].

The idea of convex hull can be easily visualized in two dimensions, i.e., for data sets that lie in the plane. In this case, the convex hull can be thought as an elastic band stretched open to encompass the given object. If released, the elastic band assumes the shape of the convex hull (see Figure 22.2).

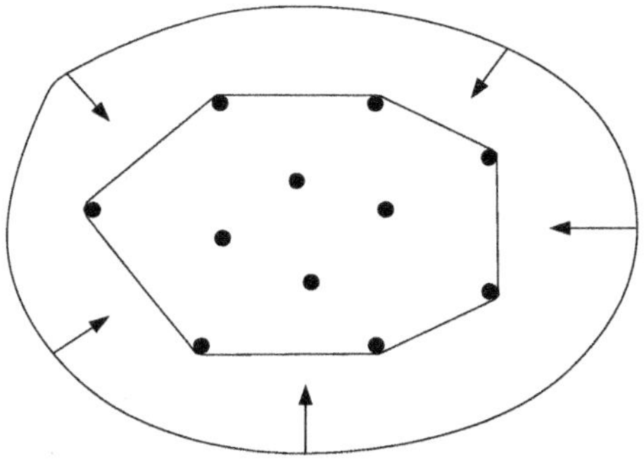

Fig. 22.2 2D representation of the convex hull [282]

The convex hull of a set $\mathcal{X}$ in a real vector space $\boldsymbol{V}$ certainly exists since $\mathcal{X}$ is contained at least in $\boldsymbol{V}$, which is a convex set. Furthermore, any intersection containing $\mathcal{X}$ is also a convex set containing $\mathcal{X}$. This fact, is useful for a mathematical definition of the convex hull. In particular, the Carathéodory's theorem states that the convex hull of $\mathcal{X}$ is the union of all simplexes with at most $n+1$ vertices from $\mathcal{X}$.

A convex hull can be defined for any set composed of points in a vector space. The dimension of the data set can be any. However, the convex hull of finite sets of points in a two or three dimensions are the cases of most practical importance.

The determination of the convex hull is an important problem of computational geometry. Several algorithms with various computational burdens have been proposed for a finite set of points [67, 73]. The complexity of the corresponding algorithms is usually estimated in terms of n, of the number of input points, and of the number of points on the convex hull. An open source implementation of algorithms related to the convex hull problem are provided by the `qhull` project, which is a general dimension code for computing convex hulls, Delaunay's triangulations, and Voronoi's diagrams [249].

The problem of finding convex hulls has several practical applications. Fields where the convex hull is widely used include, for example, pattern recognition, image processing, statistics and GIS. Furthermore, several important geometrical problems are based on the determination of the convex hull. For example, just think of the determination of the diameter given a set of points describing a circle is based on the convex hull. In fact, any diameter will always connect to points laying on the convex hull (e.g., the circumference) of the circle.

Example 22.2 3D Visualization of the IEEE 14-Bus System

Figure 22.3 shows a 3D voltage temperature map for the power flow solution of the IEEE 14-bus system. The procedure for obtaining the map is as follows. The 3D plot is obtained using the Mayavi suite for Python [253]. The procedure is practically the same as that described in Example 22.1 except for the last step for which the function `mesh` of the Mayavi library was used.

An important aspect that is difficult to "feel" from Figure 22.3 is the fact that 3D maps are interactive, i.e., the user can rotate, zoom and manipulate the map. In a 3D map, "peaks" are generally easy to see, while "valleys" can be hidden. However, rotating the 3D map allows viewing the map from all perspectives and creates in the user the impression of "flying" over the power system. Since one can see the map from any point of view, there is actually no part of the map that remains hidden.

22.2.4 Geographic Information System

Geographical Information Systems (GIS) are used for *visualizing*, *digitizing* and *analyzing* data by linking geographic locations to information. Geospatial data is used for creating maps, assigning data (or extracting "features", which is the name given to individual geometrical objects) and performing spatial analysis.

Network operators use GIS for their infrastructure and utilities management and network construction planing. As a part of the network information system, the geographical data of the network is associated with the database of the utility. An interface to enterprise resource planning (ERP) software

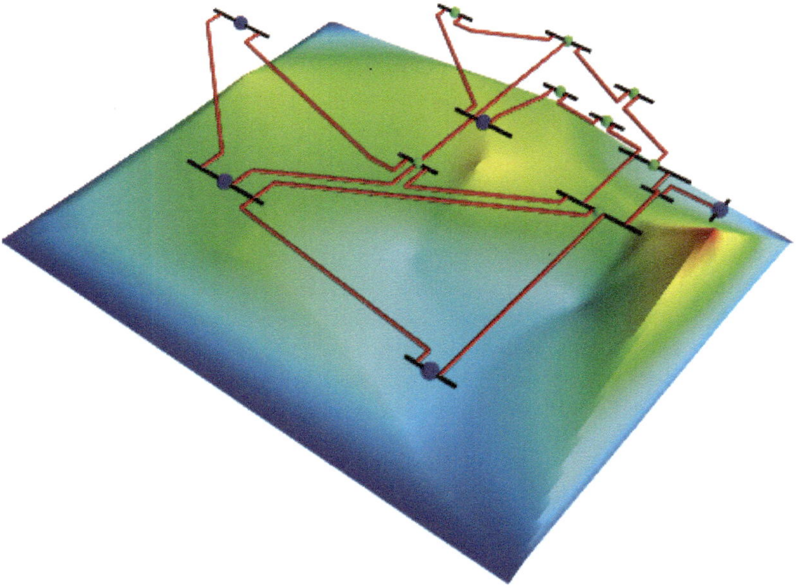

Fig. 22.3 Voltage level 3D visualization for the IEEE 14-bus system

may exist for the organization of the resources. Furthermore, network planners often need the GIS data for simulating future planned network assets. In most cases, these two systems are separated and operate independently. The only link is an identification or location reference in a database table.

In summary, a GIS-based power system information platform can provide:

- Geographical, topological and schematic representation.
- Graphical representation of simulation analysis results.
- Cable layout and detailed local area network plan.
- Search and query functions.
- Live information about the network status.

Recently, the major GIS software houses have being trying to interface power system simulation software. This interface allows only one common and consistent database. For example, a project of EDF Energy and GE Energy for a network planning system comprises of the Smallworld GIS application with an embedded network analysis engine [70]. Other examples are Smallworld and PTI/PSSEngines. All these attempts to combine GIS and simulation tools are proprietary solutions, thus showing all the issues associated with "closed" systems that have to be avoided.

In recent years, several open source GIS applications have reached the required maturity for being used in power system analysis [101, 214, 216, 217, 242, 250, 306, 319]. Open source geospatial libraries are the base of many open-source desktop GIS applications. Since these libraries are distributed as extensible open-source code, one can freely use, access and extend the functionality of these libraries by means of plug-ins or script interfaces (Python, Java, etc.).

The quickest way to incorporate a GIS system in a power system application is to create the geographical map using an external open source program and then import GIS data into the power system application. In this way, the programming effort is reduced to the code necessary for importing the data. Clearly, a necessary condition for implementing the bridge between the GIS application (e.g., OpenJump [217]) and Python is that the GIS application allows creating user-defined data. This is possible only if the GIS software packages is "open".

As discussed in Script 9.2 of Chapter 9, any device is referenced using unique identification codes (ids). These ids must be assigned to their geometrical representation. For example, the simplest geometrical representation of topological buses is a point type primitive; transmission lines can be represented using polylines, etc.

Example 22.3 Italian System Temperature Map

OpenJump allows exporting topological data in GML format, which is a special XML scheme particularly suited for defining GIS [105]. The GML/XML format can be easily exchanged and parsed by other applications. The next step is to parse the XML file containing the GIS data and assign the topological information to each electrical device. Parsing XML data is rather simple in Python [153].

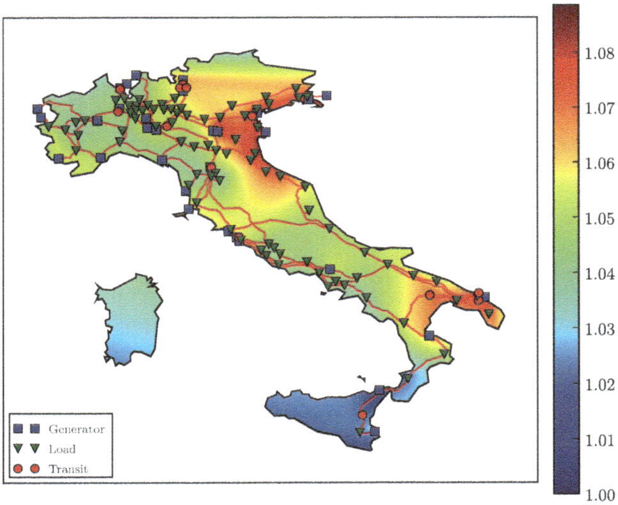

Fig. 22.4 Bus voltage magnitude map for the Italian HV transmission system. Values in the legend are in pu

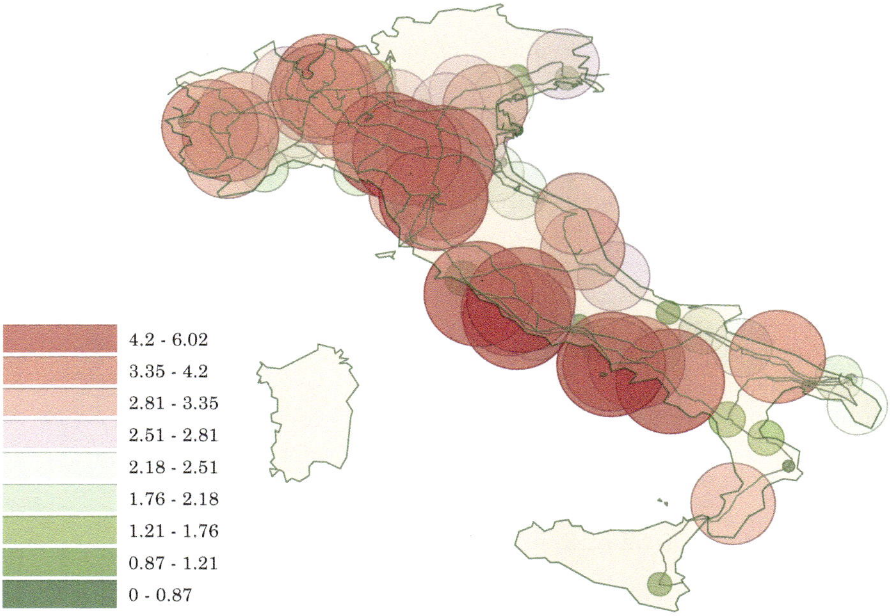

Fig. 22.5 Load active power visualization for the Italian grid obtained using the JML-OSGIS tools. Values in the legend are in pu

Figure 22.4 depicts the Italian 400 kV transmission grid. The system includes 32 generators and 82 loads. The mainland system has been drawn in OpenJump and then parsed using Python. In particular, Figure 22.4 shows a temperature map of bus voltage magnitude levels of a power flow solution. In the maps, generators are represented as blue squares while loads are indicated by green triangles.

The procedure for generating the topological scheme is as follows. The first step is to acquire and generate the geographical information as a data set for the GIS platform. The picture of the geographical layout of the Italian Grid is georeferenced with the help of QuantumGIS to generate the spatial representation of the buses and lines. Then, the bus connections of the lines and the length are analyzed. Finally a GML file is generated. This file holds the geospatial features and attributes of the Italian HV grid. The map of Figure 22.4 was obtained using the Matplotlib library, which allows automatically clipping the map using the border paths.

An entire family of maps can be also generated the other way round, e.g., by importing power flow results into the JML file once the power flow analysis is solved. Figure 22.5 shows an example of the possibilities of the JML format and GIS visualization tools [295].[4]

[4] The picture is courtesy of Mr. Matthias Stifter, arsenal research, Vienna.

Chapter 23
Challenges of Scripting for Power System Education

Most power system software packages are commercial proprietary products that require a generally costly license. This fact is implicitly accepted as normal in the power system community. However, there can be a reliable and costless alternative. This alternative is provided by Free and Open Source Software (FOSS). This chapter shows that FOSS is a valid platform to distribute educational and research-oriented tools for power system analysis as it has proved to be in several other scientific fields.

23.1 Concepts and Definitions

In Chapter 3, scripting is presented as an effective approach for producing *open* software projects, as opposed to system programming that lead to *closed* projects. In this section, the concepts "open" and "close" are formalized using commonly accepted definitions.

Software can be divided into three main categories based on the development method: proprietary software, open source software and free software. Free and open source software merges together the common characteristics of free software and open source software. This section introduces concepts and definitions of each type of software development.

23.1.1 Proprietary Software

Proprietary software refers to software that has restrictions for its use, modification and, more importantly, restrictions on copying, distributing, and publishing unmodified or modified versions of it. The restrictions are placed by the proprietors of the software and are detailed in the software license. In the U.S., copyright laws provide severe penalties for unlawful distribution of copyrighted material. Reverse engineering of the software could also violate the U.S. Copyright Law or the Digital Millennium Copyright Act [321].

F. Milano: Power System Modelling and Scripting, Power Systems, pp. 489–493.
springerlink.com

Proprietary software is also referred as commercial, non-open, or non-free software. Sometimes these terms are considered derogatory, which is actually unintended, due to the fact that terms like *freedom* and *openness* are more appealing than their opposites. In [170], the term *closed software* has been coined to refer to all this type of software and avoid the derogatory misunderstanding.

23.1.2 Open Source Software

Open Source Software (OSS) is one that complies with the Open Source Definition (OSD) [63], published by the Open Source Initiative (OSI) [215]. Formed in 1998, OSI is a non-profit corporation created by a group of programmers to promote the adoption of OSS licenses. The OSD gives the criteria to which software wishing to adopt an OSS license must comply with. These criteria are that users must be free to use the software for any purpose, make copies and distribute the software without paying to the issuer of the license, to create derived works and to distribute them without paying royalties, to view and use the source code and to use the open source software in combination with other software *including proprietary software.*

In summary, open source software permits anyone, anywhere and for any purpose to copy, modify, and distribute the software for free or for a fee. Therefore, anyone has to have full access to the source code.

23.1.3 Free Software

Free Software (FS) is software that can be used, studied, and modified without restriction. The term *free software* was coined by Richard Stallman who founded the Free Software Foundation, its formally defined by the Free Software Definition [100]. One of the most important instances of the definition is that software can be copied and distributed in modified or unmodified form without restrictions. Restrictions may be used only for ensuring that future recipients of the software are guaranteed to copy, study, modify and distribute the software (this is the main difference with respect to OSS). Moreover, the source code of the software must be made available, and it may be accompanied by a software license. The license should state that the copyright holder permits these acts, or alternatively the software can be released into public domain so that the rights mentioned above automatically hold. The Free Software Foundation maintains a list of Free Software Licenses [103], being the most common the GNU General Public License (GPL) [293].

It is customary to explain the FS concept by saying that the idea is not that the software should be free "as in beer", or available at no charge, but that it should be free "as in speech", so that the software can be reviewed and modified.

23.1.4 Free Open Source Software

Free Open Source Software (FOSS), also known as Free/Libre Open Source Software (FLOSS), is a merger of the FS and OS concepts focusing on the characteristics for software development and distribution without addressing subtle differences between the two of them. To this extent, Free Software and Open Source Software are near synonyms. Interested readers can find details on the philosophical background and differences between FS and OSS in [58, 167, 254, 292].

23.2 Education-Oriented FOSS

Software for educational purposes should be user-friendly, easy to use and reliable. In particular, software for power system education should contain an user interface that allows drawing one-line diagrams, displays results and plots time domain simulations. Most proprietary software for power system analysis presents these features (see for example PSS/E [244]). However, proprietary software has two main drawbacks: it needs a costly license and it is generally difficult (if not impossible) to modify models and/or algorithms provided with the software. The first drawback limits the diffusion of commercial software in developing countries, while the second issue imposes a severe limitation to the software development by Ph.D. students and researchers.

Opposite to proprietary software, free and open source software provides the user with the freedom of reading, copying, and modifying the source code. It is also possible with FOSS to redistribute the modified code, with the only condition that the resulting program must also be distributed as free and open source software [293]. Despite initial skepticism shown by commercial software houses, a huge number of FOSS projects have been developed and improved thanks to the cooperation of thousands of users. Some FOSS projects have also obtained worldwide success (see for example the Linux, Perl and LaTeX experiences).

If applied to the power system academic community, the FOSS approach would allow deploying tools that are suitable for education and research, and at the same time creating a community of learners [292]. From the educational viewpoint, FOSS projects have the drawback of being barely understood by an undergraduate student. Even for Ph.D. students, to be familiar with the details of large C++ or Java projects requires a lot of time, which should be better dedicated to their research topics.

23.2.1 Pedagogical Issues

The current generation of students is accustomed to sophisticated software suites that aid academic work via the computer. However, the most serious drawback of proprietary software for educational purposes is to reduce the freedom of the students and to not develop their skill of analyzing results. On the other hand, open source software, while not limiting the user freedom, tends to be less complete and intuitive than proprietary one. Thus, a compromise is needed. Open source software should have a reasonably user-friendly interface and should be written in a simple and high-level programming language (e.g., Octave or Python).

An education-oriented FOSS should merge the positive features of educational software and open source philosophy. Educational free and open source software should have a reasonably user-friendly interface and should be written in a simple and high-level programming language (e.g., Python). Educational software should develop the learning process and the curiosity of the student. In summary, the use of an open source package for education should be preferred for the following reasons.

1. The mind of the student should be opened. The student should not become accustomed to a program that gives all the answers.
2. The learning process should develop the curiosity of the student. Only if the code is open can the student explore all software features.
3. The students should understand that knowledge should be free and available to everyone [292].

23.2.2 Failure of FOSS for Power System Analysis

One of the most interesting phenomena of open source software is that users feel involved in the development of the project. Instead of complaining about missing features or bugs, users often contribute suggestions, bug fixes, and even new code [292]. However, this cooperative attitude is not the case of open source projects for power system analysis. The miracle of a rapid growth and community-based development that is typical of most open source projects (for example the Linux case) does not happen in the power system community. This situation is a result of four main issues.

1. The typical users of an open source power system software package are students attending the last year of their undergraduate courses or at the beginning of their Master or Ph.D. program. However, these students are typically not yet experienced enough to write code by themselves.
2. Students attending the last years of their Master or Ph.D. courses are in principle the ideal candidates for contributing with new code. However, in this case, the students are more likely developing their own software tools and uses the open software only as a benchmark or as a store from which getting ideas.
3. Researchers seldom use open source power system software package or contribute new code. The conservativeness (i.e., closed view) of the scientific world is unfortunately a common practice. Furthermore, a surprisingly high percentage of researchers is not aware of the advantages of the "open source" way of thinking.

4. The users of an open source power system software package are a very reduced subset of the total number of university students and researchers. Thus, an open source project aimed to power system analysis cannot be compared with other open source packages, such as Apache or LaTeX, which are used by a broad range of people.

Despite several attempts [199, 200, 326, 327], FOSS projects for power system analysis are still in a very early stage. A generational change is required. The next generation of electrical engineering students should learn to question commonly accepted assumptions and simplifications, should understand the deep, intriguing mechanisms that link modelling and scripting, and should not acritically accept results provided by an opaque proprietary software application. This is a currently open challenge.

Part V
Appendices

Appendix A
Python Libraries

This appendix provides a quick reference for the Python libraries used in the examples provided in the book. In particular, Section A.1 describes the CVXOPT library, Section A.2 describes the NumPy library, and Section A.3 describes the Matplotlib library. Following sections only concern functions and methods that are used in the book. The interested reader can find the complete documentation on the websites of these software packages (see Appendix E).

A.1 CVXOPT

CVXOPT is a free software package for convex optimization based on the Python programming language. It can be used with the interactive Python interpreter, on the command line by executing Python scripts, or integrated in other software via Python extension modules. Its main purpose is to make the development of software for convex optimization applications straightforward by building on Python's extensive standard library and on the strengths of Python as a high-level programming language. Material presented in this section was obtained by the CVXOPT on-line documentation, available at `http://abel.ee.ucla.edu/cvxopt`.

A.1.1 *cvxopt.base*

This module defines a Python type matrix for storing and manipulating dense matrices, a Python type spmatrix for storing and manipulating sparse matrices, routines for generating sparse dense matrices, and routines for sparse matrix-vector and matrix-matrix multiplication.

Dense Matrices

`matrix(x[, size[, tc]])` creates dense matrices.

F. Milano: Power System Modelling and Scripting, Power Systems, pp. 497–509.
springerlink.com

`x` can be a number, a sequence of numbers, a dense or sparse matrix, a one- or two-dimensional NumPy array, or a list of lists of matrices and numbers.

`size` is a tuple of length two with the matrix dimensions. The number of rows and/or the number of columns can be zero.

`tc` stands for type code. The possible values are `'i'`, `'d'` and `'z'`, for integer, double float and complex matrices, respectively.

Examples:

```
>>> from cvxopt.base import matrix
>>> A = matrix([1., 2., 3., 4., 5., 6.], (2,3))
>>> print A
[ 1.00e+00 3.00e+00 5.00e+00]
[ 2.00e+00 4.00e+00 6.00e+00]
>>> A1 = matrix([1, 2], (2,1))
>>> B1 = matrix([6, 7, 8, 9, 10, 11], (2,3))
>>> B2 = matrix([12, 13, 14, 15, 16, 17], (2,3))
>>> B3 = matrix([18, 19, 20], (1,3))
>>> C = matrix([[A1, 3.0, 4.0, 5.0], [B1, B2, B3]])
>>> print C
[ 1.00e+00 6.00e+00 8.00e+00 1.00e+01]
[ 2.00e+00 7.00e+00 9.00e+00 1.10e+01]
[ 3.00e+00 1.20e+01 1.40e+01 1.60e+01]
[ 4.00e+00 1.30e+01 1.50e+01 1.70e+01]
[ 5.00e+00 1.80e+01 1.90e+01 2.00e+01]
```

Sparse Matrices

`spmatrix(x, I, J[, size[, tc]])` constructs a sparse matrix from a triplet description.

`I` and `J` are sequences of integers (lists, tuples, array arrays, xrange objects, ...) or integer matrices (matrix objects with type code `'i'`), containing the row and column indexes of the nonzero entries. The lengths of `I` and `J` must be equal. If `I` and `J` are matrices, they are treated as lists of indexes stored in column-major order, i.e., as lists `list(I)`, respectively, `list(J)`.

`size` is a tuple of non-negative integers with the row and column dimensions of the matrix. The size argument is only needed when creating a matrix with a zero last row or last column. If the size is not specified, it is determined from `I` and `J`: the default value for `size[0]` is `max(I)+1` if `I` is nonempty and zero otherwise. The default value for `size[1]` is `max(J)+1` if `J` is nonempty and zero otherwise.

`tc` is the type code, `'d'` or `'z'`, for double and complex matrices, respectively. Integer sparse matrices are not implemented.

`x` can be a number, a sequence of numbers, or a dense matrix. This argument specifies the numerical values of nonzero entries.

Example:

```
>>> from cvxopt.base import spmatrix
>>> A = spmatrix([2,-1,2,-2,1,4,3], [1,2,0,2,3,2,0],
                 [0,0,1,1,2,3,4])
>>> print A
[    0       2.00e+00     0           0        3.00e+00]
[ 2.00e+00     0          0           0           0    ]
[-1.00e+00 -2.00e+00      0        4.00e+00       0    ]
[    0         0       1.00e+00       0           0    ]
```

`sparse(x[, tc])` constructs a sparse matrix from a block-matrix description.

`tc` is the type code, `'d'` or `'z'`, for double and complex matrices, respectively.

`x` can be a matrix, spmatrix, or a list of lists of matrices (matrix or spmatrix objects) and numbers (Python integer, float or complex).

Example:

```
>>> from cvxopt.base import matrix, spmatrix, sparse
>>> A = matrix([[1, 2, 0], [2, 1, 2], [0, 2, 1]])
>>> B = spmatrix([], [], [], (3,3))
>>> C = spmatrix([3, 4, 5], [0, 1, 2], [0, 1, 2])
>>> D = sparse([[A, B], [B, C]])
>>> print D
[ 1.00e+00 2.00e+00     0        0          0          0    ]
[ 2.00e+00 1.00e+00 2.00e+00     0          0          0    ]
[    0     2.00e+00 1.00e+00     0          0          0    ]
[    0         0        0     3.00e+00      0          0    ]
[    0         0        0        0       4.00e+00      0    ]
[    0         0        0        0          0       5.00e+00]
```

`spdiag(x)` constructs a block-diagonal sparse matrix from a list of matrices.

`x` is a matrix with a single row or column, or a list of square dense or sparse matrices or scalars. If `x` is matrix, a sparse diagonal matrix is returned with the entries of `x` on its diagonal. If `x` is list, a sparse block-diagonal matrix is returned with the element in the list as its diagonal blocks.

Example:

```
>>> from cvxopt.base import matrix, spmatrix, spdiag
>>> A = 3.0
>>> B = matrix([[1,-2],[-2,1]])
>>> C = spmatrix([1,1,1,1,1],[0,1,2,0,0,],[0,0,0,1,2])
>>> D = spdiag([A, B, C])
>>> print D
[ 3.00e+00     0            0          0          0          0    ]
[    0      1.00e+00   -2.00e+00       0          0          0    ]
```

```
[     0  -2.00e+00  1.00e+00      0         0         0    ]
[     0      0          0    1.00e+00  1.00e+00  1.00e+00]
[     0      0          0    1.00e+00      0         0    ]
[     0      0          0    1.00e+00      0         0    ]
```

Matrix Attributes, Methods and Operations

Dense matrix attributes and methods. The attribute and methods of dense matrices that are used in the book are as follows:

`size` A tuple with the dimensions of the matrix. The size of the matrix can be changed by altering this attribute, as long as the number of elements in the matrix remains unchanged.

`typecode` A char, either `'i'`, `'d'`, or `'z'`, for integer, real and complex matrices, respectively. It is a read-only attribute.

`trans()` Returns the transpose of the matrix as a new matrix. The notation `A.T` is an alias of `A.trans()`.

`ctrans()` Returns the conjugate transpose of the matrix as a new matrix. The notation `A.H` is an alias of `A.ctrans()`.

`real()` For complex matrices, returns the real part as a real matrix. For integer and real matrices, returns a copy of the matrix.

`imag()` For complex matrices, returns the imaginary part as a real matrix. For integer and real matrices, returns an integer or real zero matrix.

Sparse matrix attributes and methods. The attribute and methods of sparse matrices that are used in the book are as follows:

`V` A single-column dense matrix containing the numerical values of the nonzero entries in column-major order. Making an assignment to the attribute is an efficient way of changing the values of the sparse matrix, without changing the sparsity pattern. When the attribute `V` is read, a copy of `V` is returned, as a new dense matrix. This implies, for example, that an indexed assignment `A.V[I] = B` does not work. Instead, the attribute `V` has to be read and returned as a new matrix; then the elements of this new matrix are modified.

`I` A single-column integer matrix with the row indexes of the entries in `V`. It is a read-only attribute.

`J` A single-column integer matrix with the column indexes of the entries in `V`. It is a read-only attribute. size `A` tuple with the dimensions of the matrix. The size of the matrix can be changed by altering this attribute, as long as the number of elements in the matrix remains unchanged.

`trans()` Returns the transpose of a sparse matrix as a new sparse matrix. The notation `A.T` is an alias of `A.trans()`.

`ctrans()` Returns the complex conjugate transpose of a sparse matrix as a new sparse matrix. The notation `A.H` is an alias of `A.ctrans()`.

Arithmetic operations. The following table lists the arithmetic operations defined for dense matrices. In this table `A` and `B` are dense matrices with compatible dimensions, `c` is a scalar (a Python number or a dense 1 by 1 matrix), and `d` is a Python number.

`+A, -A` Unary plus/minus.
`A+B, A+c, c+A` Addition.
`A-B, A-c, c-A` Subtraction.
`A*B` Matrix multiplication.
`c*A, A*c, A/c` Scalar multiplication and division.
`A%c` Remainder after division.
`A**d` Element-wise exponentiation.

The following in-place operations are also defined, but only if they do not change the type of the matrix `A`:

`A+=B, A+=c` In-place addition.
`A-=B, A-=c` In-place subtraction.
`A*=c, A/=c` In-place scalar multiplication and division.
`A%=c` In-place remainder.

Indexing and slicing. There are four indexing methods, as follows.

1. The index can be a single integer. This returns a number. For example, `A[0]` is the first element of `A`.
2. The index can be an integer matrix. This returns a column matrix. For example, the command `A[matrix([0,1,2,3])]` returns the 4 by 1 matrix consisting of the first four elements of `A`. The size of the index matrix is ignored: `A[matrix([0,1,2,3], (2,2))]` returns the same 4 by 1 matrix.
3. The index can be a list of integers. This returns a column matrix. For example, `A[[0,1,2,3]]` is the 4 by 1 matrix consisting of elements 0, 1, 2, 3 of `A`.
4. The index can be a Python slice. This returns a matrix with one column (possibly 0 by 1, or 1 by 1). For example, `A[::2]` is the column matrix defined by taking every other element of `A`, stored in column-major order. `A[0:0]` is a matrix with size $(0, 1)$.

Matrix operations. The following operations are supported by both dense and sparse matrices.

`sqrt(x)` The element-wise square root of `x`. The result is returned as a real matrix if `x` is an integer or real matrix and as a complex matrix if `x` is a complex matrix. This method raises an exception when `x` is an integer or real matrix with negative elements.

`sin(x)` The sine function applied element wise to `x`. The result is returned as a real matrix if `x` is an integer or real matrix and as a complex matrix otherwise.

`cos(x)` The cosine function applied element wise to `x`. The result is returned as a real matrix if `x` is an integer or real matrix and as a complex matrix otherwise.

`exp(x)` The exponential function applied element wise to `x`. The result is returned as a real matrix if `x` is an integer or real matrix and as a complex matrix otherwise.

`log(x)` The natural logarithm applied element wise to `x`. The result is returned as a real matrix if `x` is an integer or real matrix and as a complex matrix otherwise. This method raises an exception when `x` is an integer or real matrix with non-negative elements, or a complex matrix with zero elements.

`mul(x, y)` The element-wise product of `x` and `y`. The two matrices must have the same size and type.

`div(x, y)` The element-wise division of `x` by `y`. The two matrices must have the same size and type.

A.1.2 cvxopt.blas

The `cvxopt.blas` module provides an interface to the double-precision real and complex Basic Linear Algebra Subprograms (BLAS). The names and calling sequences of the Python functions in the interface closely match the corresponding FORTRAN BLAS routines (described in the references below) and their functionality is exactly the same.

`dotu(x, y)` returns $\boldsymbol{x}^T\boldsymbol{y}$ where $\boldsymbol{x}$ and $\boldsymbol{y}$ are matrices of the same type (`'d'` or `'z'`).

A.1.3 cvxopt.lapack

The module cvxopt.lapack includes functions for solving dense sets of linear equations, for the corresponding matrix factorizations (LU, Cholesky, $\boldsymbol{LDL}^T$), for solving least-squares and least-norm problems, for QR factorization, for symmetric eigenvalue problems and for singular value decomposition.

`gees(...)` Schur's factorization of a real of complex matrix.
The syntax is as follows:

```
sdim = gees(A, w=None, V=None, select=None, n=A.size[0],
            ldA=max(1,A.size[0]), ldV=max(1,Vs.size[0]),
            offsetA=0, offsetw=0, offsetV=0)
```

Purpose:

Computes the real Schur's form $\boldsymbol{A} = \boldsymbol{V}\boldsymbol{S}\boldsymbol{V}^T$ or the complex Schur's form $\boldsymbol{A} = \boldsymbol{V}\boldsymbol{S}\boldsymbol{V}^H$, the eigenvalues, and, optionally, the matrix of Schur's vectors of an $n \times n$ matrix $\boldsymbol{A}$. The real Schur's form is computed if $\boldsymbol{A}$ is real, and the complex Schur's form is computed if $\boldsymbol{A}$ is complex. On exit, $\boldsymbol{A}$ is replaced with $\boldsymbol{S}$. If the argument `w` is provided, the eigenvalues are returned in `w`. If `V` is provided, the Schur's vectors are computed and returned in `V`. The argument `select` can be used for obtaining an ordered Schur's factorization. It must be a Python function that can be called as $f(s)$ with s complex, and returns 0 or 1. The eigenvalues s for which $f(s)$ is 1 are placed first in the Schur's factorization. For the real case, eigenvalues s for which $f(s)$ or $f(\text{conj}(s))$ is 1, are placed first. If `select` is provided, `gees()` returns the number of eigenvalues that satisfy the selection criterion. Otherwise, it returns 0.

Arguments:

`A` 'd' or 'z' matrix.
`w` 'z' matrix of length at least `n`.
`V` 'd' or 'z' matrix. It must have the same type as `A`.
`select` Python function that takes a complex number as argument and returns `True` or `False`.
`n` integer. If `n` is negative, the default value is used.
`ldA` non-negative integer. `ldA >= max(1,n)`. If `ldA` is zero, the default value is used.
`ldV` non-negative integer. `ldV >= 1` and `ldV >= n` if `V` is present. If `ldV` is zero, the default value is used (with `V.size[0]` replaced by 0 if `V` is `None`).
`offsetA` non-negative integer.
`offsetW` non-negative integer.
`offsets` non-negative integer.
`sdim` number of eigenvalues that satisfy the selection criterion specified by select.

A.1.4 cvxopt.umfpack

The module `cvxopt.umfpack` includes four functions for solving sparse non-symmetric sets of linear equations. These functions call routines from the UMFPACK library, with all control options set to the default values described in the UMFPACK user guide.

`linsolve(A, B[, trans='N'])` solves a sparse set of linear equations:

`trans = 'N'`: $\boldsymbol{A}\boldsymbol{X} = \boldsymbol{B}$
`trans = 'T'`: $\boldsymbol{A}^T\boldsymbol{X} = \boldsymbol{B}$
`trans = 'C'`: $\boldsymbol{A}^H\boldsymbol{X} = \boldsymbol{B}$

where A is a sparse matrix and B is a dense matrix of the same type ('d' or 'z') as A. On exit, B contains the solution. It raises an ArithmeticError exception if the coefficient matrix is singular.

Example:

```
>>> from cvxopt.base import spmatrix, matrix
>>> from cvxopt import umfpack
>>> V = [2,3, 3,-1,4, 4,-3,1,2, 2, 6,1]
>>> I = [0,1, 0, 2,4, 1, 2,3,4, 2, 1,4]
>>> J = [0,0, 1, 1,1, 2, 2,2,2, 3, 4,4]
>>> A = spmatrix(V,I,J)
>>> B = matrix(1.0, (5,1))
>>> umfpack.linsolve(A,B)
>>> print B
[ 5.79e-01]
[-5.26e-02]
[ 1.00e+00]
[ 1.97e+00]
[-7.89e-01]
```

The function umfpack.linsolve() is equivalent to the following three functions called in sequence.

symbolic(A) reorders the columns of A to reduce fill-in and performs a symbolic LU factorization. A is a sparse, possibly rectangular, matrix. It returns the symbolic factorization as an opaque C object that can be passed on to umfpack.numeric().

numeric(A, F) performs a numeric LU factorization of a sparse, possibly rectangular, matrix A. The argument F is the symbolic factorization computed by umfpack.symbolic() applied to the matrix A, or another sparse matrix with the same sparsity pattern, dimensions, and type. The numeric factorization is returned as an opaque C object that that can be passed on to umfpack.solve(). It raises an ArithmeticError if the matrix is singular.

solve(A, F, B[, trans='N']) solves a set of linear equations:

trans = 'N': $\boldsymbol{AX} = \boldsymbol{B}$
trans = 'T': $\boldsymbol{A}^T\boldsymbol{X} = \boldsymbol{B}$
trans = 'C': $\boldsymbol{A}^H\boldsymbol{X} = \boldsymbol{B}$

where A is a sparse matrix and B is a dense matrix of the same type as A. The argument F is a numeric factorization computed by umfpack.numeric(). On exit, B is overwritten by the solution.

The functions symbolic, numeric and solve are useful for solving several sets of linear equations with the same coefficient matrix and different right-hand sides, or with coefficient matrices that share the same sparsity pattern. The symbolic factorization depends only on the sparsity pattern of the matrix, and not on the numerical values of the nonzero coefficients. The numerical factorization on the other hand depends on the sparsity pattern of the matrix and on its the numerical values. The following is an example of the use of the functions `symbolic`, `numeric` and `solve`.

```
>>> from cvxopt.base import spmatrix, matrix
>>> from cvxopt import umfpack
>>> VA = [2,3, 3,-1,4, 4,-3,1,2, 2, 6,1]
>>> VB = [4,3, 3,-1,4, 4,-3,1,2, 2, 6,2]
>>> I = [0,1, 0, 2,4, 1, 2,3,4, 2, 1,4]
>>> J = [0,0, 1, 1,1, 2, 2,2,2, 3, 4,4]
>>> A = spmatrix(VA, I, J)
>>> B = spmatrix(VB, I, J)
>>> x = matrix(1.0, (5,1))
>>> Fs = umfpack.symbolic(A)
>>> FA = umfpack.numeric(A, Fs)
>>> FB = umfpack.numeric(B, Fs)
>>> umfpack.solve(A, FA, x)
>>> umfpack.solve(B, FB, x)
>>> umfpack.solve(A, FA, x, trans='T')
>>> print x
[ 5.81e-01]
[-2.37e-01]
[ 1.63e+00]
[ 8.07e+00]
[-1.31e-01]
```

A.2 NumPy

NumPy provides types and packages for scientific computing [334]. Basic types provided by the NumPy suite are n-dimensional arrays and universal functions. Relevant packages are basic linear algebra (`linalg`), discrete Fourier transforms (`dft`), random number generators (`random`) and automatic wrapping of FORTRAN code (`f2py`).

`linspace(start, stop, num=50, endpoint=True, retstep=False)`
returns `num` evenly spaced samples, calculated over the interval `[start, stop]`. The end point of the interval can optionally be excluded.

Parameters:

`start` the starting value of the sequence.
`stop` the end value of the sequence, unless `endpoint` is set to `False`. In that case, the sequence consists of all but the last of `num + 1` evenly spaced samples, so that `stop` is excluded. Note that the step size changes when `endpoint` is `False`.
`num` integer, number of samples to generate. The default value is 50.
`endpoint` if `True`, `stop` is the last sample. Otherwise, it is not included. The default value is `True`.
`retstep` if `True`, return `(samples, step)`, where `step` is the spacing between samples.

Returns:

`samples` there are `num` equally spaced samples in the closed interval `[start, stop]` or the half-open interval `[start, stop)` (depending on whether `endpoint` is `True` or `False`).

`step` (only if `retstep` is `True`) size of spacing between samples.

Examples:

```
>>> numpy.linspace(2.0, 3.0, num=5)
array([ 2.  ,  2.25,  2.5 ,  2.75,  3.  ])
>>> numpy.linspace(2.0, 3.0, num=5, endpoint=False)
array([ 2. ,  2.2,  2.4,  2.6,  2.8])
>>> numpy.linspace(2.0, 3.0, num=5, retstep=True)
(array([ 2.  ,  2.25,  2.5 ,  2.75,  3.  ]), 0.25)
```

`linalg.eigvals(A)` computes the eigenvalues of a general matrix.

Parameters:

`A` a complex or real square matrix whose eigenvalues and eigenvectors have to be computed.

Returns:

`w` the eigenvalues. Each eigenvalue is repeated according to its multiplicity. Eigenvalues are not ordered and can be complex.

`linalg.eig(A)` computes eigenvalues and right eigenvectors of an array.

Parameters:

`A` a complex or real square matrix.

Returns:

`w` the eigenvalues, each repeated according to its multiplicity. The eigenvalues are not necessarily ordered, nor are they necessarily real for real matrices.

`v` the normalized eigenvector corresponding to the eigenvalue `w[i]` is the column `v[:,i]`.

`meshgrid(x, y)` returns coordinate matrices from two coordinate vectors.

Parameters:

x, y two one-dimensional arrays representing the x and y coordinates of a grid.

Returns:

`X`, `Y` for vectors `x`, `y` with lengths `Nx=len(x)` and `Ny=len(y)`, return `X`, `Y` where `X` and `Y` are `(Ny, Nx)` shaped arrays with the elements of `x` and `y` repeated to fill the matrix along the first dimension for `x`, the second for `y`.

A.3 Matplotlib

Matplotlib is a library for making 2D plots of arrays in Python. The syntax is similar to that of Matlab graphic commands. However, Matplotlib is independent of Matlab, and can be used in a object oriented way. Matplotlib is based on NumPy.

`axhline(y=0, xmin=0, xmax=1, **kwargs)` draws a horizontal line at `y` from `xmin` to `xmax`. The line always spans the horizontal extent of the axes.

Optional arguments:

`color` any matplotlib color, e.g., `'k'` for black.
`linestyle` any line style expression, e.g., `'-'`, `'--'`, `'-.'`, `':'`.

`axvline(x=0, ymin=0, ymax=1, **kwargs)` draws a vertical line at `x` from `ymin` to `ymax`. The line always spans the vertical extent of the axes. Optional arguments are same as for `axhline`.

`figure(num=None)` creates a new figure and returns an instance of the class `matplotlib.figure.Figure`. If `num = None`, the figure number is incremented and a new figure is created. The returned figure objects have a number attribute holding this number. If `num` must be an integer. If `figure(num)` already exists, the figure becomes active and a handle to that figure is returned. If `figure(num)` does not exist, a new figure is created.

`grid(self, b=None, **kwargs)` sets the axis grids on or off; `b` is a Boolean. If `b` is `None` and `len(kwargs)==0`, the grid state is toggled. If `kwargs` are supplied, `b` is set to `True`. `kwargs` are used for setting the grid line properties.

`hold(b=None)` sets the hold state. If `b` is `None` (default), the hold state is toggled, otherwise the hold state is set to `b`. Examples:

```
hold()       # toggle hold
hold(True)   # hold is on
hold(False)  # hold is off
```

If `hold` is `True`, subsequent plot commands are added to the current axes. If `hold` is `False`, the current axes and figure are cleared on the next plot command.

`legend(*args, **kwargs)` places a legend on the current axes at location `loc`. Labels are a sequence of strings and `loc` can be a string or an integer that specifies the legend location. Available locations are `'upper right'`, `'upper left'`, `'lower left'`, `'lower right'`, etc.

`plot(*args, **kwargs)` plots lines and/or markers to the current axes. `args` is a variable length argument, allowing for multiple `x`, `y` pairs with an optional format string. For example, each of the following statement is allowed.

```
plot(x, y)          # plot x and y using default settings
plot(x, y, 'bo')    # plot x and y using blue circle markers
plot(y)             # plot y using x as index array 0..N-1
plot(y, 'r+')       # ditto, but with red plusses
```

If `x` and/or `y` is two-dimensional, then the corresponding columns is plotted.

An arbitrary number of `x`, `y`, `fmt` groups can be specified. The returned value is a list of the lines that have been added.

The following format string characters are accepted to control the line style or marker:

'-'	solid line style.
'--'	dashed line style.
'-.'	dash-dot line style.
':'	dotted line style.
'.'	point marker.
'o'	circle marker.
'v'	triangle-down marker.
'^'	triangle-up marker.
'<'	triangle-left marker.
'>'	triangle-right marker.
'*'	star marker.
'x'	x marker.

The following color abbreviations are supported:

'b'	blue.
'g'	green.
'r'	red.
'c'	cyan.
'm'	magenta.
'y'	yellow.
'k'	black.
'w'	white.

The `kwargs` can be used for setting line properties. One can use this to set a line label (for auto legends), line width, anitialising, marker face color, etc. Some examples are as follows.

```
plot([1,2,3], [1,2,3], 'go-', label='line 1', linewidth=2)
plot([1,2,3], [1,4,9], 'rs',  label='line 2')
axis([0, 4, 0, 10])
legend()
```

`savefig(fname, **kwargs)` saves the current figure. The output formats depend on the availability of the back-end in use.

Arguments:

`fname` a string containing a path to a file name, or a Python file-like object. If `format` is `None` and `fname` is a string, the output format is deduced from the extension of the file name.

`format` one of the file extensions supported by the active back-end. Most back-ends support png, pdf, ps, eps and svg formats.

`show()` shows all the figures. This function should conclude the script.

`xlabel(s)` sets the x-axis label of the current axis to string `s`.

`xlim(*args, **kwargs)` sets or gets the x-limits of the current axes. The new x-axis limits are returned as a length 2 tuple. Examples:

```
xmin, xmax = xlim()    # return the current xlim
xlim( (xmin, xmax) )  # set the xlim to xmin, xmax
xlim( xmin, xmax )    # set the xlim to xmin, xmax
```

`ylabel(s)` sets the y-axis label of the current axis to string `s`.

`ylim(*args, **kwargs)` sets or gets the y-limits of the current axes. The new y-axis limits are returned as a two-element tuple. Examples:

```
ymin, ymax = ylim()    # return the current ylim
ylim( (ymin, ymax) )  # set the ylim to ymin, ymax
ylim( ymin, ymax )    # set the ylim to ymin, ymax
```

Appendix B
System Classes

This appendix provides a quick reference of the classes `DAE`, `Settings`, `CPF` and `OPF`. These classes are used in the scripts provided in the book.

B.1 System Properties and Settings

`DAE` Differential and algebraic equations, functions and Jacobians matrices.

- `Ac` Complete DAE Jacobian matrix for numerical integration.
- `f` Differential equations $\boldsymbol{f}$.
- `factorize` If `True`, it forces the factorization of the system Jacobian matrix.
- `Fx` Jacobian matrix of differential equations $\boldsymbol{f}_{\boldsymbol{x}}$.
- `Fy` Jacobian matrix of differential equations $\boldsymbol{f}_{\boldsymbol{y}}$.
- `h` Inequality constraints $\boldsymbol{h}$.
- `g` Algebraic equations $\boldsymbol{g}$.
- `Gmu` Jacobian matrix of algebraic equations $\boldsymbol{g}_{\mu}$.
- `Gx` Jacobian matrix of algebraic equations $\boldsymbol{g}_{\boldsymbol{x}}$.
- `Gy` Jacobian matrix of algebraic equations $\boldsymbol{g}_{\boldsymbol{y}}$.
- `Gz` Jacobian matrix of algebraic equations $\boldsymbol{g}_{\boldsymbol{z}}$.
- `Hes` Hessian matrix multiplied by dual variables.
- `Hz` Jacobian matrix of inequality constraints $\boldsymbol{h}_{\boldsymbol{z}}$.
- `kg` Variable for distributing losses among generators.
- `mu` Continuation parameter or loading level μ.
- `ng` Number of equality constraints.
- `nh` Number of inequality constraints.
- `nx` Number of state variables.
- `ny` Number of algebraic variables.
- `nz` Number of algebraic variables used in OPF analysis.
- `obj` Objective function φ.
- `Oz` Jacobian matrix of the objective function $\varphi_{\boldsymbol{z}}$.
- `pi` Dual variables $\boldsymbol{\pi}$ of inequality constraints.

F. Milano: Power System Modelling and Scripting, Power Systems, pp. 511–513.
springerlink.com

- `q` Vector $\boldsymbol{q}$ used in numerical integration.
- `rho` Dual variables $\boldsymbol{\rho}$ of equality constraints.
- `s` Slack variables $\boldsymbol{s}$ for OPF analysis.
- `t` Current simulation time.
- `x` State variables $\boldsymbol{x}$.
- `y` Algebraic variables $\boldsymbol{y}$.
- `z` Algebraic variables $\boldsymbol{z}$ used in OPF analysis.

`Settings` General settings and parameters for power flow computations and numerical integration.

- `deltat` Time step for time domain integrations in seconds.
- `deltatmax` Maximum time step in seconds.
- `deltatmin` Minimum time step in seconds.
- `disturbance` If `True`, it forces the call of an external disturbance function during numerical integration.
- `dynmit` Maximum number of iterations for each step of the numerical integration.
- `error` Maximum equation mismatch.
- `fixt` If `True`, it forces a fixed time step for numerical integration.
- `freq` System frequency rating in Hz.
- `tol` Mismatch tolerance for static analyses (default 10^{-5}).
- `method` Numerical integration method (e.g., trapezoidal method, backward Euler, etc.).
- `mva` System power rating in MVA (default 100 MVA).
- `pf_max_iter` Maximum number of iteration of the power flow solver.
- `pfsolver` Power flow solver.
  - `'DC'` Dc power flow.
  - `'NR'` Newton's method.
  - `'XB'` XB variation of fast decoupled power flow.
  - `'BX'` BX variation of fast decoupled power flow.
  - `'RK'` Runge-Kutta-based continuous Newton's method.
  - `'IW'` Iwamoto's method.
  - `'SR'` Simple robust method.
- `rad` System frequency rating in rad/s.
- `t0` Initial simulation time in seconds.
- `tf` Final simulation time in seconds.
- `tstep` Fixed step length Δt in seconds for numerical integration.

`CPF` Continuation power flow settings.

- `method` Method for corrector step.
  - `'perpendicular intersection'` Perpendicular intersection.
  - `'local parametrization'` Local parametrization.

`OPF` Optimal power flow settings.

`eps1` Error tolerance of the power flow equations.
`eps2` Error tolerance of the objective function.
`eps_mu` Error tolerance of the barrier parameter $\hat{\mu}$.
`gamma` Safety factor γ.
`max_iter` Maximum number of iterations of the NLP solver.
`method` Method used for computing the variable directions and increments.

`'Newton'`	Newton's directions.
`'Mehrotra'`	Mehrotra's predictor-corrector.

`sigma` Centering parameter σ.

Appendix C
Control Diagrams

This appendix describes the representation of transfer functions through control diagrams and defines the notation used in the book. The control diagram approach is particularly suited for illustrating regulators but, in principle, can be used for any DAE system. Modelling and implementation of windup and anti-windup limits are also discussed.

C.1 Representation of Basic Functions

A control diagram is an oriented graph where arcs (*arrows*) indicate signal flows and nodes (*blocks*) indicate transfer functions (e.g., gain and integrator) or operations (e.g., sum and product). For example, Figure C.1 shows the detailed and compact notation of a lag transfer function:

$$\dot{x} = \frac{1}{T}(Ky - x) \tag{C.1}$$

In Figure C.1, s represents the differential operator, i.e., $s = \frac{d}{dt}$.

Another example is shown in Figure C.2 that shows three possible representations of a lead-lag transfer function. The "parallel" model is described by:

$$\begin{aligned} \dot{x} &= \left(\left(1 - \frac{T_1}{T_2}\right) y - x\right) \frac{1}{T_2} \\ w &= \frac{T_1}{T_2} y + x \end{aligned} \tag{C.2}$$

whereas the "series" model leads to the DAE system:

$$\begin{aligned} \dot{x}' &= \frac{1}{T_2}(y - x') \\ w &= T_1 \dot{x}' + x' = \frac{T_1}{T_2}(y - x') + x' \end{aligned} \tag{C.3}$$

F. Milano: Power System Modelling and Scripting, Power Systems, pp. 515–521.
springerlink.com

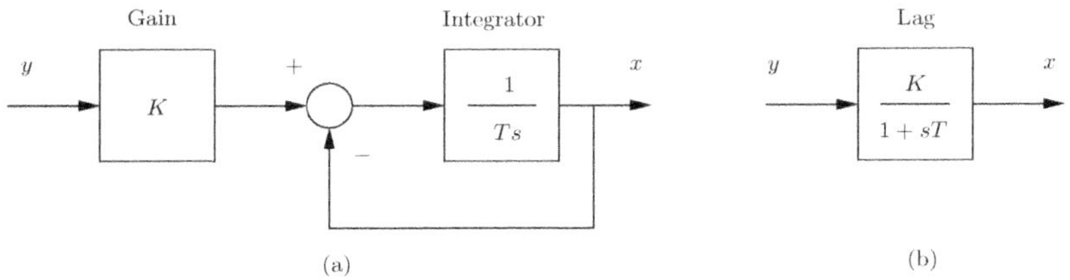

Fig. C.1 Lag diagram: (a) detailed diagram, and (b) compact diagram

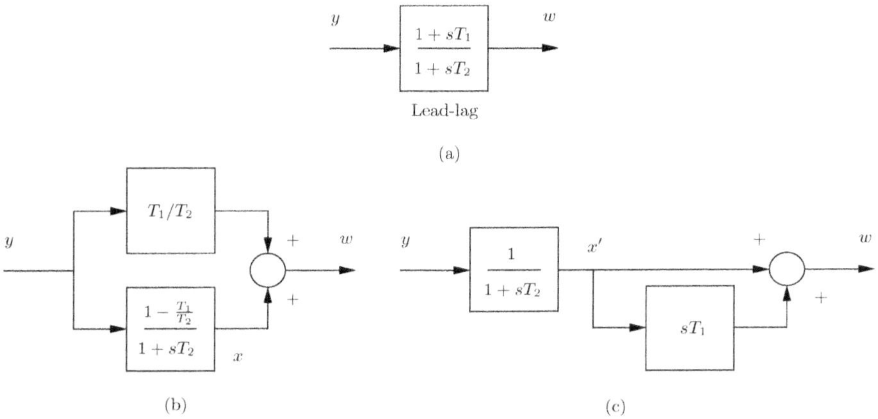

Fig. C.2 Lead-lag diagram: (a) compact notation, (b) "parallel" model, and (c) "series" model

The state variable x in (C.2) takes different values than x' in (C.3). Clearly, the output w is the same for the two models. Multi-pole and multi-zero transfer functions can be decomposed by both the "series" and the "parallel" approaches, but the "series" approach is generally more intuitive an easier to apply (see for example the transfer function of Figure 16.17 in Chapter 16).

C.2 Hard Limits

An important issue related to control systems is the way hard limits are handled. Hard limits consist in locking the output of a variable when it reaches a limit (or *saturates*). The simplest manner for modelling a hard limit is the *windup* model shown in Figure C.3.a. Mathematically, one has:

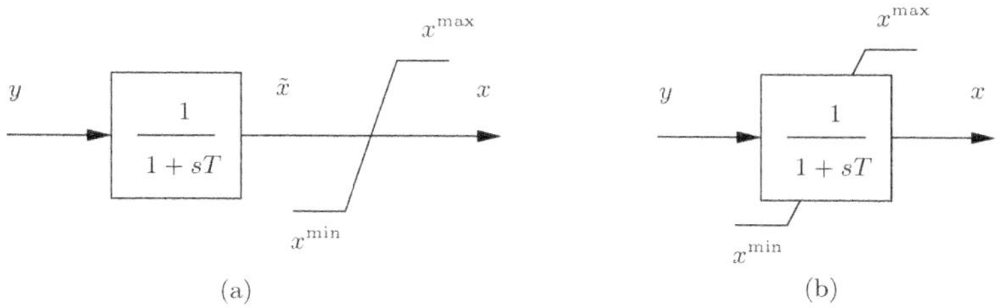

Fig. C.3 Hard limit diagrams: (a) windup model, (b) anti-windup model

$$x = \begin{cases} x^{\max}, & \text{if } \tilde{x} \geq x^{\max} \\ \tilde{x}, & \text{if } x^{\min} < \tilde{x} < x^{\max} \\ x^{\min}, & \text{if } \tilde{x} \leq x^{\min} \end{cases} \tag{C.4}$$

The windup limiter is also the simplest saturation model. Other saturation models are described in Subsection 15.1.8 of Chapter 15.

The windup limiter introduces a delay, also known as *windup effect*. Figure C.4 illustrates the windup effect of the system of Figure C.3.a assuming $y(t) = \sin(t)$, $T = 1$ s, $x^{\max} = 0.5$ and $x^{\min} = -0.5$. When the state variable $\tilde{x} > x^{max}$ at $t = 1.34$ s, the output x is locked. Afterwards, x remains locked until $\tilde{x} < x^{\max}$, which happens at $t = 3.18$ s.

The windup limiter is not adequate for modelling the behavior of the majority of real controllers, for which x is decreased when $\dot{x} < 0$. Figure C.4 also shows the output of the anti-windup limiter that, as expected, removes the windup effect, i.e., x starts decreasing at $t = 2.62$ s, which is the instant for which $\dot{x}$ becomes negative. Mathematically, the anti-windup limiter imposes conditions on both the state variable x and its time derivative $\dot{x}$, as follows:

$$\begin{aligned} \text{if } x \geq x^{\max} \text{ and } \dot{x} \geq 0 &\Rightarrow x = x^{\max} \text{ and } \dot{x} = 0 \\ \text{if } x \leq x^{\min} \text{ and } \dot{x} \leq 0 &\Rightarrow x = x^{\min} \text{ and } \dot{x} = 0 \\ \text{otherwise} &\Rightarrow \dot{x} = (y - x)/T \end{aligned} \tag{C.5}$$

The amount of the delay introduced by windup limiters depends on how fast is the response of the system. The higher is the time constant T, the higher is the windup effect, i.e., the delay. Hence, for low time constants (e.g., $T < 0.1$ s), the output of windup and anti-windup limiters is practically the same. Moreover, the windup effect does not exist for algebraic variables, since these can be considered state variables with an infinitely fast dynamic (i.e., $T \rightarrow 0$).

While it is relatively easy to implement an anti-windup limiter for a pure integrator or a lag transfer function, the model of an anti-windup limiter for lead-lags or PI (i.e., proportional and integral) controllers are not so

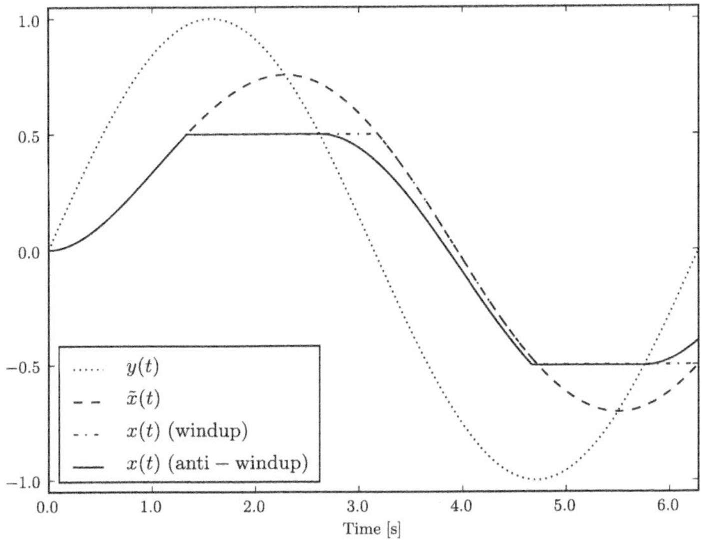

Fig. C.4 Transient response of windup and anti-windup limiters

straightforward.[1] The issue consists in that the output w of lead-lags or PI controllers is an explicit function of the input y, whose first time derivative, in some cases, cannot be determined. Figure C.5 shows some common solutions for implementing an anti-windup limiter of a PI controller. Figure C.5.e is the most precise model, which is also the most complex.

Script C.1 Implementation of Windup and Anti-Windup Limiters

The object of the scripts below is to define a vector, say `self.z`, whose elements are 1 if the associated state variable is not saturated and belongs to an on-line element; and 0 otherwise. Thus, `self.z` is a copy of the status array `self.u` if no state variable is saturated.

The following code implements the windup limiter. The input arguments are the DAE system (`dae`) and the names of the state variable (`x`), of the maximum and minimum value (`xmax` and `xmin`, respectively) and of the saturation status (`z`).

[1] Difficulties arises whenever the transfer function has the same number of poles and zeros.

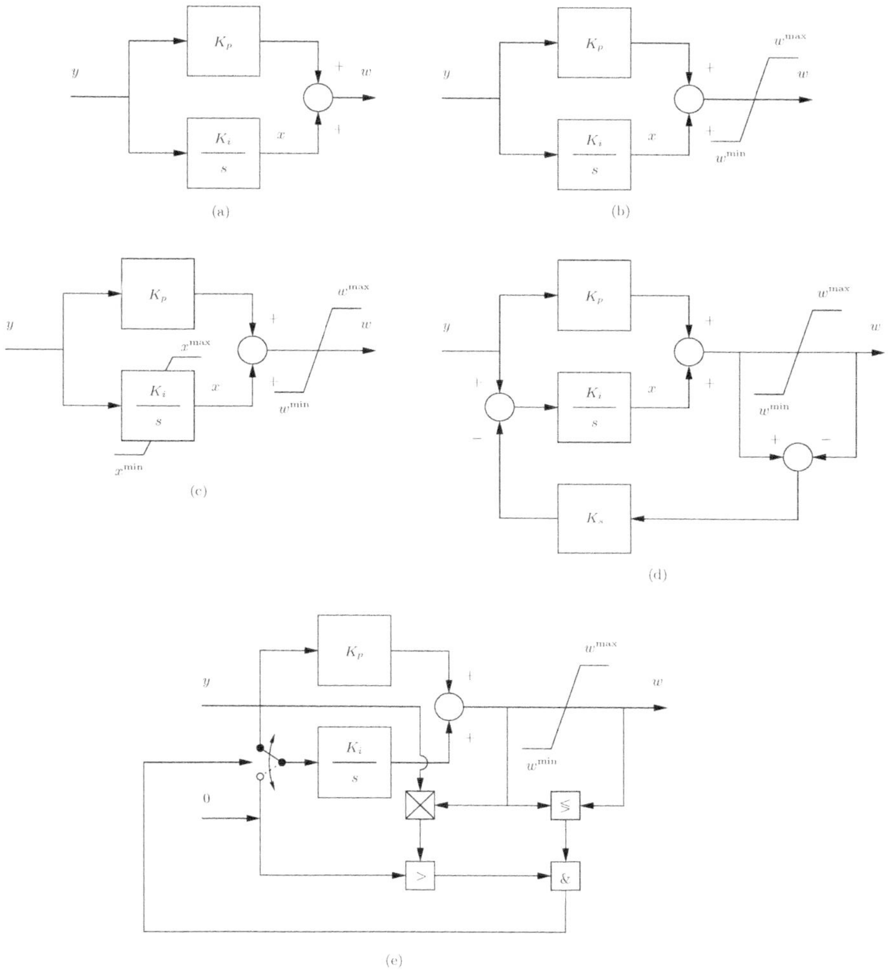

Fig. C.5 PI controller: (a) basic PI scheme without hard limits, (b) windup limiter, (c) limited integrator, (d) tracking anti-windup, and (e) integrator clamping

```
def windup(self, dae, x, xmax, xmin, z):

    sumz = sum(self.__dict__[z])
    self.__dict__[z] = matrix(self.u)
    zmax = matrix(self.u)
    zmin = matrix(self.u)
    for idx, item in enumerate(self.__dict__[x]):
        if dae.x[item] >= self.__dict__[xmax][idx]:
            dae.f[item] = 0
            dae.x[item] = self.__dict__[xmax][idx]
```

```
                self.__dict__[z][idx] = 0
                zmax[idx] = 0
            elif dae.x[item] <= self.__dict__[xmin][idx]:
                dae.f[item] = 0
                dae.x[item] = self.__dict__[xmin][idx]
                self.__dict__[z][idx] = 0
                zmin[idx] = 0
    if sumz != sum(zmax) or sumz != sum(zmin):
        dae.factorize = True
```

A similar method can be used for limiting algebraic variables. The last line of the previous code is needed to impose a re-factorization of the system Jacobian matrix. If a state variable reaches a limit, its value is constant. Thus, if a variable has a position k in the system Jacobian matrix $\boldsymbol{A}_C$, then one has to set to zero the entire k^{th} row and k^{th} column. Only the diagonal element (k,k) of the matrix $\boldsymbol{A}_C$ has to be non-zero to avoid singularity. Since $\boldsymbol{A}_C = [[\boldsymbol{f}_{\boldsymbol{x}}, \boldsymbol{f}_{\boldsymbol{y}}]; [\boldsymbol{g}_x, \boldsymbol{g}_{\boldsymbol{y}}]]$, the saturation of the state variable with index k corresponds to set to zero the k-th row of $\boldsymbol{f}_{\boldsymbol{x}}$ and $\boldsymbol{f}_{\boldsymbol{y}}$ and the k-th column of $\boldsymbol{f}_{\boldsymbol{x}}$ and $\boldsymbol{g}_{\boldsymbol{x}}$ and to assign a non-zero value to the element (k,k) of $\boldsymbol{f}_{\boldsymbol{x}}$. The following code is an efficient way to do that (it is assumed that the function is a method of the class `system.DAE`).

```
def set_Ac(self, idx):

    H = spmatrix(1.0, idx, idx, (self.ny, self.ny))
    I = spdiag([1.0]*self.ny) - H
    self.Gy = I * (self.Gy * I) + H
    self.Fy = self.Fy * I
    self.Gx = I * self.Gx
```

The following code does the same as the previous one, but it is less efficient (about 5 times slower):

```
def set_Ac(self, idx):

    self.Gy[idx, :] = 0
    self.Gy[:, idx] = 0
    self.Gy += spmatrix(1.0, idx, idx, (self.ny, self.ny), 'd')
    self.Fy[:, idx] = 0.0
    self.Gx[idx, :] = 0.0
    # needed to maintain the minimum sparsity level
    self.Gy = sparse(self.Gy)
    self.Fy = sparse(self.Fy)
    self.Gx = sparse(self.Gx)
```

The difference in the efficiency of the previous functions is typical of scripting languages. In fact, matrix initialization, indexing and operations are obtained through interfaces to external libraries. Since the bottleneck are the calls to the external library, the less the number of calls, the faster the resulting code.

The following method implements an anti-windup limiter. The main difference with the previous method is the check on the sign of the differential equation, i.e., the state variable time derivative.

```
def anti_windup(self, dae, x, xmax, xmin, z):

    sumz = sum(self.__dict__[z])
    self.__dict__[z] = matrix(self.u)
    zmax = matrix(self.u)
    zmin = matrix(self.u)
    for idx, item in enumerate(self.__dict__[x]):
        if dae.x[item] >= self.__dict__[xmax][idx] and \
                dae.f[item] > 0:
            dae.f[item] = 0
            self.__dict__[z][idx] = 0
            zmax[idx] = 0
            dae.x[item] = self.__dict__[xmax][idx]
        elif dae.x[item] <= self.__dict__[xmin][idx] and \
                dae.f[item] < 0:
            dae.f[item] = 0
            self.__dict__[z][idx] = 0
            zmin[idx] = 0
            dae.x[item] = self.__dict__[xmin][idx]
    if sumz != sum(zmax) or sumz != sum(zmin):
        dae.factorize = True
```

Appendix D
IEEE 14-Bus System Data

This appendix provides the data of the IEEE 14-bus test systems used in most examples of this book. The topological scheme of this system in depicted in Figure 2.4 of Chapter 2.

D.1 Common Data

System bases are:

1. Power base: 100 MVA.
2. Frequency base: 60 Hz.
3. Voltage bases: 69 kV for buses 1 to 5, 13.8 kV for buses 6, 7 and 9 to 14, and 18 kV for bus 8.

Unless otherwise indicated, nominal ratings of devices are equal to system bases. Parameters are in pu with respect to device nominal ratings. For power flow analysis, bus 1 is the slack bus and the reference angle. For optimal power flow analysis, $v^{\max} = 1.06$ pu and $v^{\min} = 0.94$ pu are used as voltage security limits. Following sections provides the data of all devices used in the examples based on the IEEE 14-bus system. Parameter symbols are defined in the chapters where the devices are presented.

D.2 Static Data

This section provides the static data used throughout this book, unless explicitly specified. Bus, PQ load and shunt data are depicted in Table D.1. Static generator data are shown in Table D.2 whereas transmission line and transformer data are depicted in Table D.3.

D.3 Market Data

Table D.4 shows the generator bid data used in the examples of Chapter 6.

F. Milano: Power System Modelling and Scripting, Power Systems, pp. 523–528.
springerlink.com

Table D.1 Bus, PQ load and shunt data

Bus #	Voltage Rating kV	p_L pu	q_L pu	b_{sh} pu
1	69.0	-	-	-
2	69.0	0.217	0.127	-
3	69.0	0.942	0.19	-
4	69.0	0.478	-0.039	-
5	69.0	0.076	0.016	-
6	13.8	0.112	0.075	-
7	13.8	-	-	-
8	18.0	-	-	-
9	13.8	0.295	0.166	0.19
10	13.8	0.09	0.058	-
11	13.8	0.035	0.018	-
12	13.8	0.061	0.016	-
13	13.8	0.135	0.058	-
14	13.8	0.149	0.05	-

Table D.2 Static generator data

Bus #	Type	p_G pu	v_G pu	$q_G^{\max}$ pu	$q_G^{\min}$ pu
1	Slack	2.324	1.06	9.9	-9.9
2	PV	0.4	1.045	0.5	-0.4
3	PV	0	1.01	0.4	0
6	PV	0	1.07	0.24	-0.06
8	PV	0	1.09	0.24	-0.06

D.4 Dynamic Data

Tables D.5 and D.6 show the synchronous machine and automatic voltage regulator data used throughout this book, unless otherwise indicated. Dynamic shaft data used in Example 15.5 are shown in Table D.7. Turbine governor data used in Example 16.1 are depicted in Table D.8. Power System stabilizer data used in Example 16.3 are shown in Table D.9.

D.5 FACTS Data

SVC Type I and Type II data used in Example 19.1 are depicted in Tables D.10 and D.11.

The data of the detailed STATCOM device used in Example 19.2 are as follows.

Table D.3 Transmission line and transformer data

From bus #	To Bus #	Type	r_{hk} pu	x_{hk} pu	$b_h = b_k$ pu	m pu/pu
1	2	Line	0.01938	0.05917	0.0528	-
1	5	Line	0.05403	0.22304	0.0492	-
2	3	Line	0.04699	0.19797	0.0438	-
2	4	Line	0.05811	0.17632	0.0374	-
2	5	Line	0.05695	0.17388	0.034	-
3	4	Line	0.06701	0.17103	0.0346	-
4	5	Line	0.01335	0.04211	0.0128	-
4	7	Transf.	0	0.20912	0	0.978
4	9	Transf.	0	0.55618	0	0.969
5	6	Transf.	0	0.25202	0	0.932
6	11	Line	0.09498	0.19890	0	-
6	12	Line	0.12291	0.25581	0	-
6	13	Line	0.06615	0.13027	0	-
7	8	Transf.	0	0.17615	0	1.0
7	9	Line	0	0.11001	0	-
9	10	Line	0.03181	0.08450	0	-
9	14	Line	0.12711	0.27038	0	-
10	11	Line	0.08205	0.19207	0	-
12	13	Line	0.22092	0.19988	0	-
13	14	Line	0.17093	0.34802	0	-

Table D.4 Generator bid data

Bus #	C_{P1} €/MWh	C_{P2} €/MW^{2}h	$p_G^{\max}$ pu	$p_G^{\min}$ pu
1	20	0.04303	2.0	0
2	20	0.25	1.4	0
3	40	0.01	1.0	0
6	40	0.01	1.0	0
8	40	0.01	1.0	0

Dc nodes: The dc networks is composed of two dc nodes with $V_{\mathrm{dc},n} = 10$ kV. One of the node is the ground ($v = 0$).
Parallel RC: $G = 0.0017$ S, $C = 0.0432$ F.
VSC device: $r_T = 0.02$ pu, $x_T = 0.1$ pu, $i^{\max} = 1.2$ pu.
STATCOM regulators: $v_{\mathrm{dc}}^{\mathrm{ref}} = 1.06$ pu, $v_{\mathrm{ac}}^{\mathrm{ref}} = 1.0563$ pu, $T_1 = T_2 = T_{\mathrm{ac}} = T_{\mathrm{dc}} = 0.01$ s, $K = 100$ pu/pu, $K_D = 0$, $K_I = 5$ rad/pu/s, $K_P = 10$ rad/pu, $K_{\mathrm{ac}} = K_{\mathrm{dc}} = 1$ pu/pu, $m^{\max} = 3$, $m^{\min} = 0.5$, $\alpha^{\max} = \pi$ rad, $\alpha^{\min} = -\pi$ rad.

Table D.5 Synchronous machine data

Param.	Unit	Mach. 1	Mach. 2	Mach. 3	Mach. 4	Mach. 5
Bus	#	1	2	3	6	8
Model		5.a	6.a	6.a	6.a	6.a
S_n	MVA	615	60	60	25	25
x_ℓ	pu	0.2396	0	0	0.134	0.134
r_a	pu	0	0.0031	0.0031	0.0041	0.0041
x_d	pu	0.8979	1.05	1.05	1.25	1.25
x'_d	pu	0.2995	0.185	0.185	0.232	0.232
x''_d	pu	0.23	0.13	0.13	0.12	0.12
T'_{d0}	s	7.4	6.1	6.1	4.75	4.75
T''_{d0}	s	0.03	0.04	0.04	0.06	0.06
x_q	pu	0.646	0.98	0.98	1.22	1.22
x'_q	pu	0.646	0.36	0.36	0.715	0.715
x''_q	pu	0.4	0.13	0.13	0.12	0.12
T'_{q0}	s	0	0.3	0.3	1.5	1.5
T''_{q0}	s	0.033	0.099	0.099	0.21	0.21
D	pu	2	2	2	2	2
H	MWs/MVA	5.148	6.54	6.54	5.06	5.06

Table D.6 Automatic voltage regulator data

Param.	Unit	AVR 1	AVR 2	AVR 3	AVR 4	AVR 5
Machine	#	1	2	3	4	5
Model		1	1	1	1	1
$v_r^{\max}$	pu	9.9	2.05	1.7	2.2	2.2
$v_r^{\min}$	pu	0	0	0	1.0	1.0
K_a	pu/pu	200	20	20	20	20
T_a	s	0.02	0.02	0.02	0.02	0.02
K_f	s pu/pu	0.0012	0.001	0.001	0.001	0.001
T_f	s	1.0	1.0	1.0	1.0	1.0
K_e	pu	1.0	1.0	1.0	1.0	1.0
T_e	s	0.19	1.98	1.98	0.7	0.7
T_r	s	0.001	0.001	0.001	0.001	0.001
A_e	-	0.0006	0.0006	0.0006	0.0006	0.0006
B_e	1/pu	0.9	0.9	0.9	0.9	0.9

D.6 Wind Turbine Data

Wind turbine data used in Example 20.4 of Chapter 20 are given below.

Mechanical data: $n_{\rm gen} = 40$, $n_{\rm pole} = 4$, $n_{\rm blade} = 3$, $\eta_{\rm GB} = 1/89$, $H_t = 1.5$ MWs/MVA, $H_m = 0.5$ MWs/MVA, $K_s = 1$ pu, $\rho = 1.225$ kg/m^3, $R = 35$ m, $T_p = 3$ s, and $K_p = 10$ rad/pu.

Table D.7 Dynamic shaft data

H_{HP} $\frac{\mathrm{MWs}}{\mathrm{MVA}}$	H_{IP} $\frac{\mathrm{MWs}}{\mathrm{MVA}}$	H_{LP} $\frac{\mathrm{MWs}}{\mathrm{MVA}}$	H_{EX} $\frac{\mathrm{MWs}}{\mathrm{MVA}}$	D_{HP} pu	D_{IP} pu	D_{LP} pu	D_{EX} pu
0.3348	0.7306	0.8154	0.0452	0.518	0.2240	0.224	0.145
D_{12} pu	D_{23} pu	D_{34} pu	D_{45} pu	K_{12} pu	K_{23} pu	K_{34} pu	K_{45} pu
0.0518	0.0224	0.0224	0.0145	33.07	28.59	44.68	21.984

Table D.8 Turbine governor data

Machine #	R pu	T_s s	T_c s	T_3 s	T_4 s	T_5 s	p^{max} pu	p^{min} pu
1	0.02	0.1	0.45	0	0	50.0	1.2	0.3
2	0.02	0.1	0.45	0	0	50.0	1.2	0.3

Table D.9 PSS data

AVR #	K_w pu/pu	T_w s	T_1 s	T_2 s	T_3 s	T_4 s	v_s^{max} pu	v_s^{min} pu
1	5.0	10.0	0.28	0.02	0.28	0.02	0.1	-0.1

Table D.10 SVC Type I data

Bus #	v^{ref} pu	K rad/pu	K_D -	K_M pu/pu	T_1 s
9	1.0563	10.0	0	1.0	0.01
T_2 s	T_m s	x_C pu	x_L pu	α^{max} rad	α^{min} rad
0.01	0.01	0.1	0.1	3.14	-3.14

Table D.11 SVC Type II data

Bus #	v^{ref} pu	K_r pu/pu	T_r s	b^{max} pu	b^{min} pu
9	1.0563	10.0	0.01	5.0	5.0

Non-controlled speed wind turbine data: $r_r = 0.01$ pu, $r_s = 0.01$ pu, $x_\mu = 3.0$ pu, $x_r = 0.08$ pu, and $x_s = 0.1$ pu.

Doubly-fed asynchronous generator data: $K_V = 10$ pu/pu/s, $T_\epsilon = 0.01$ s, $r_r = 0.01$ pu, $r_s = 0.01$ pu, $x_\mu = 3.0$ pu, $x_r = 0.08$ pu, and $x_s = 0.1$ pu.

Direct-drive synchronous machine data: $K_{dc} = K_{ds} = 1.0$ pu/pu, $K_{qc} = 200$ pu/pu, $T_{dc} = T_{ds} = T_{qc} = T_{qs} = 0.01$ s, $\psi_p = 1.2$ pu, $r_s = 0.01$, pu, $x_d = 1.0$, and $x_q = 0.8$ pu.

Appendix E
Software Packages and Links

This appendix describes the software packages used throughout the book, discusses the requirements for working with these software tools and provides the links of such tools. Useful links of webpages related to power system analysis are also provided in Section E.2.

E.1 Software Packages Used in the Book

All scripts and simulations of this book are carried out using a Unix-based system. All script examples implicitly assume that the user is familiar with the command lines of a Unix shell. However, it is not an issue if the user is not, since shell commands are reduced to the minimum and are as much as possible self-explicative.

The basic applications that have to be installed on the system are: Python 2.5 or Python 2.6 with a few external packages that are not provided with the main Python distribution. These are: NumPy, Matplotlib and CVXOPT. The latest versions of these modules should work fine for the examples provided in this book. However, the specific versions used in the book are NumPy 1.3, CVXOPT 1.1.2 and Matplotlib 0.99.

At the time of writing the book, the latest Python version is the 3.1, however, this is the first intentionally backward-incompatible version of Python, and most third-party modules (included all used in this book) are not currently providing a version for Python 3.1. Fortunately, even though incompatible with previous versions, changes introduced in Python 3.1 do not alter the essence of the Python syntax. Only a limited number of functions have been modified and the changes are mostly immaterial for the purposes of the book.

For those that cannot avoid using Windows, it is recommended to install Python 2.5 and IPython, which is the Python idle and reproduce a kind of Unix terminal. Likely, also installing Python using Cygwin should work. Users of Mac OS X 10.4 or newer can install XCode and MacPorts in order to have a full working Unix environment.

F. Milano: Power System Modelling and Scripting, Power Systems, pp. 529–530.
springerlink.com

For the sake of completeness, the web pages of the software packages discussed above are:

Python	`www.python.org`
NumPy	`numpy.scipy.org`
CVXOPT	`abel.ee.ucla.edu/cvxopt`
Matplotlib	`matplotlib.sourgeforge.net`
IPython	`ipython.scipy.org/moin`
Cywin	`www.cygwin.com`
Xcode	`developer.apple.com/TOOLS/Xcode`
MacPorts	`www.macports.org`

Unfortunately, since most of these packages are free and open source, their web pages can move or, worse, disappear. The maintenance of some packages could be also discontinued in the future. However, one of the advantages of free and open source projects is that even if the original maintainer drops his project, the latest stable version will be always available and, in most cases, other developers will take care of that project. Thus, using Python and other open source projects ensures long-lasting software applications.

E.2 Links related to Power System Analysis

Other useful links related to power system analysis are:

- IEEE Task Force on Open Source Software for Power Systems, available at:

  `ewh.ieee.org/cmte/psace/CAMS_taskforce/index.htm`

- IEEE PES PEEC Digital Educational Resources, available at:

  `www.ece.mtu.edu/faculty/ljbohman/peec/Dig_Rsor.htm`

- IEEE Power Systems Test Case Archive, available at:

  `www.ee.washington.edu/research/pstca/`

- Power Systems Dynamic Test Cases Archive, available at:

  `psdyn.ece.wisc.edu/IEEE_benchmarks/index.htm`

References

[1] Acevedo, S., Linares, L.R., Martí, J.R., Fujimoto, Y.: Efficient HVDC Converter Model for Real Time Transient Simulation. IEEE Transactions on Power Systems 14(1), 166–171 (1999)

[2] Acha, E., Fuerte-Esquivel, C.R., Ambriz-Pérez, H., Angeles-Camacho, C.: FACTS - Modelling and Simulation in Power Networks. John Wiley & Sons, New York (2004)

[3] Achilles, S., Pöller, M.: Direct Drive Synchronous Machine Models for Stability Assessment of Wind Farms. In: Proceedings of the 4th International Workshop on Large-Scale Integration of Wind Power and Transmission Networks for Offshore Wind Farms (October 2003)

[4] Ackermann, T.: Wind Power in Power Systems. John Wiley & Sons, Chichester (2005)

[5] Ajjarapu, V., Christy, C.: The Continuation Power Flow: a Tool for Steady State Voltage Stability Analysis. IEEE Transactions on Power Systems 7(1), 416–423 (1992)

[6] Ajjarapu, V., Lee, B.: Bifurcation Theory and its Application to Nonlinear Dynamical Phenomena in an Electrical Power System. IEEE Transactions on Power Systems 7(1), 424–431 (1992)

[7] Akhmatov, V., Knudsen, H., Nielsen, A.H.: Advanced Simulation of Windmills in the Electric Power Supply. International Journal of Electrical Power and Energy Systems 22(6), 421–434 (2000)

[8] Alsac, O., Bright, J., Prais, M., Stott, B.: Further Developments in LP-based Optimal Power Flow. IEEE Transactions on Power Systems 5(3), 697–711 (1990)

[9] Anderson, P.M., Bose, A.: Stability Simulation of Wind Turbine Systems. IEEE Transactions on Power Apparatus and Systems 102(12), 3791–3795 (1983)

[10] Anderson, P.M., Fouad, A.A.: Power System Control and Stability. Wiley-IEEE Press, New York (2002)

[11] Arabi, S., Kundur, P.: Stability Modelling of Storage Devices in FACTS Applications. In: Proceedings of the IEEE PES Summer Meeting, Edmonton, Alberta (July 2001)

[12] Arabi, S., Kundur, P., Sawada, J.H.: Appropriate HVDC Transmission Simulation Models for Various Power System Stability Studies. IEEE Transactions on Power Systems 13(4), 1292–1297 (1998)

[13] Arabi, S., Rogers, G.J., Wong, D.Y., Kundur, P., Lauby, M.G.: Small Signal stability Program Analysis of SVC and HVDC in AC Power Systems. IEEE Transactions on Power Systems 6(3), 1147–1153 (1991)

[14] Arrillaga, J., Arnold, C.P.: Computer Analysis Power Systems. John Wiley & Sons, New York (1990)

[15] Arrillaga, J., Arnold, C.P., Camacho, J.R., Sankar, S.: AC-DC Load Flow with Unit-Connected Generator-Converter Infeeds. IEEE Transactions on Power Systems 8(2), 701–706 (1993)

[16] Arrillaga, J., Smith, B.: AC-DC Power System Analysis. power and Energy Series. The Institution of Electrical Engineers, London (1998)

[17] Arrillaga, J., Watson, N.R.: Computer Modelling of Electrical Power Systems, 2nd edn. John Wiley & Sons, York (2001)

[18] Ascher, D., Martelli, A., Ravenscroft, A.: Python Cookbook, 2nd edn. O'Reilly, Sebastopol (2006)

[19] Astic, J.Y., Binhain, A., Jerosolimski, M.: The Mixed Adams-BDF Variable Step Size Algorithm to Simulate Transient and Long Term Phenomena in Power Systems. IEEE Transactions on Power Systems 9(2), 929–935 (1994)

[20] Athay, T., Podmore, R., Virmani, S.: A Practical Method for the Direct Analysis of Transient Stability. IEEE Transactions on Power Apparatus and Systems 98(2), 573–584 (1979)

[21] Ávalos, R.J., Cañizares, C.A., Milano, F., Conejo, A.J.: Equivalency of Continuation and Optimization Methods to Determine Saddle-node and Limit-induced Bifurcations in Power Systems. IEEE Transactions on Circuits and Systems - I: Fundamental Theory and Applications 56(1), 210–223 (2009)

[22] Ayasun, S., Nwankpa, C.O., Kwatny, H.G.: Computation of Singular and Singularity Induced Bifurcation Points of Differential-Algebraic Power System Model. IEEE Transactions on Circuits and Systems - I: Fundamental Theory and Applications 51(8), 1525–1538 (2004)

[23] Barcelo, W.R., Lemmon, W.W.: Standardized Sensitivity Coefficients for Power System Networks. IEEE Transactions on Power Systems 3(4), 1591–1599 (1988)

[24] Bazaraa, M.S., Sherali, H.O., Shetty, C.M.: Nonlinear Programming: Theory and Algorithms, 2nd edn. John Wiley & Sons, New York (1993)

[25] Berg, G.L.: Power System Load Representation. Proceedings of the IEEE 120(3), 344–348 (1973)

[26] Bhattacharya, S., Dommel, H.W.: A New Commutation Margin Control Representation for Digital Simulation of HVDC System Transient. IEEE Transactions on Power Systems 3(3), 1127–1132 (1988)

[27] Bijwe, P.R., Kelapure, S.M.: Nondivergent Fast Power Flow Methods. IEEE Transactions on Power Systems 18(2), 633–638 (2003)

[28] Billington, R., Aborehaid, S., Fotuhi-Firuzabad, M.: Well-Being Analysis for HVDC Transmission Systems. IEEE Transactions on Power Systems 12(2), 913–918 (1997)

[29] Braz, L.M.C., Castro, C.A., Murari, C.A.F.: A Critical Evaluation of Step Size Optimization Based Load Flow Methods. IEEE Transactions on Power Systems 15(1), 202–207 (2000)

[30] Brenan, K.E., Campbell, S.L., Petzold, L.: Numerical Solution of Initial-Value Problems in Differential-Algebraic Equations. SIAM, Philadelphia (1995)
[31] Brooke, A., Kendrick, D., Meeraus, A., Raman, R., Rosenthal, R.E.: GAMS, a User's Guide, GAMS Development Corporation, 1217 Potomac Street, NW, Washington, DC 20007, USA (December 1998), http://www.gams.com
[32] Brueck, D., Tanner, S.: Python 2.1 Bible. Hungry Minds, Inc., New York (2006)
[33] Buresh, M.: Photovoltaic Energy Systems, Design and Installation. McGraw-Hill, New York (1983)
[34] Butcher, J.C.: Numerical Methods for Ordinary Differential Equations. John Wiley & Sons, New York (2003)
[35] Cañizares, C.A.: On Bifurcation Voltage Collapse and Load Modeling. IEEE Transactions on Power Systems 10(1), 512–522 (1995)
[36] Cañizares, C.A.: Applications of Optimization to Voltage Collapse Analysis. In: Proceedings of the IEEE PES Summer Meeting, San Diego, USA (July 1998)
[37] Cañizares, C.A.: Calculating Optimal System Parameters to Maximize the Distance to Saddle-Node Bifurcations. IEEE Transactions on Circuits and Systems - I: Fundamental Theory and Applications 45(3), 225–237 (1998)
[38] Cañizares, C.A.: Modeling of TCR and VSI Based FACTS Controllers. ENEC, Milan, Italy, Tech. Rep. (December 1999)
[39] Cañizares, C.A.: Voltage Stability Assessment: Concepts, Practices and Tools. IEEE/PES Power System Stability Subcommittee, Final Document, Tech. Rep. (August 2002), http://www.power.uwaterloo.ca
[40] Cañizares, C.A., Alvarado, F.L.: Point of Collapse Methods and Continuation Methods for Large AC/DC Systems. IEEE Transactions on Power Systems 8(1), 1–8 (1993)
[41] Cañizares, C.A., Alvarado, F.L., DeMarco, C.L., Dobson, I., Long, W.F.: Point of Collapse Methods applied to AC/DC Power Systems. IEEE Transactions on Power Systems 7(2), 673–683 (1992)
[42] Cañizares, C.A., Alvarado, F.L., Zhang, S.: UWPFLOW Program, university of Waterloo (2006), http://www.power.uwaterloo.ca
[43] Cañizares, C.A., Chen, H., Rosehart, W.: Pricing System Security in Electricity Markets. In: Proceedings of the Bulk Power systems Dynamics and Control V, Onomichi, Japan (September 2001)
[44] Cañizares, C.A., Hranilovic, S.: Transcritical and Hopf Bifurcation in AC/DC Systems. In: Proceedings of the Bulk Power System Voltage Phenomena III - Seminar, Davos, Switzerland (August 1994)
[45] Cañizares, C.A., Mithulananthan, N., Milano, F., Reeve, J.: Linear Performance Indices to Predict Oscillatory Stability Problems in Power Systems. IEEE Transactions on Power Systems 19(2), 1104–1114 (2004)
[46] Cañizares, C.A., Rosehart, W., Berizzi, A., Bovo, C.: Comparison of Voltage Security Constrained Optimal Power flow Techniques. In: Proceedings of the IEEE PES Summer Meeting, Vancouver, BC, Canada (July 2001)
[47] Cañizares, C.A., Rosehart, W., Quintana, V.: Costs of Voltage Security in Electricity Markets. In: Proceedings of the IEEE PES Summer Meeting, Seattle, WA, USA (July 2000)
[48] Carpentier, J.: Contribution á l'Étude du Dispatching Économique. Bulletin de la Société Française des Electriciens 3(6), 431–447 (1962)

[49] Carpentier, J.: Differential Injection Method, a General Method for Secure and Optimal Load Flows. In: Proceedings of the Power Industry Computer Application (PICA), 255–262 (1973)
[50] Carpentier, J.: Optimal Power Flows. International Journal of Electrical Power and Energy Systems 1(1), 3–15 (1979)
[51] Castillo, E., Conejo, A.J., Pedregal, P., García, R., Alguacil, N.: Building and Solving Mathematical Programming Models in Engineering and Science. John Wiley & Sons, New York (2001)
[52] Chapman, S.J.: Electric Machinery and Power System Fundamentals. McGraw Hill, New York (2002)
[53] Chen, A.H.L., Nwankpa, C.O., Kawatny, H.G., Ming Yu, X.: Voltage Stability Toolbox: An Introduction and Implementation. In: Proceedings of the North American Power Symposium (NAPS), MIT, Cambridge (November 1996)
[54] Chen, Y., Shen, C.: A Jacobian-Free Newton-GMRES(m) Method with Adaptive Preconditioner and its Application for Power Flow Calculations. IEEE Transactions on Power Systems 21(3), 1096–1103 (2006)
[55] Chiang, H.D., Dobson, I., Thomas, R.J.: On Voltage Collapse in Electric Power Systems. IEEE Transactions on Power Systems 5(2), 601–611 (1990)
[56] Chiang, H.D., Flueck, A.J., Shah, K.S., Balu, N.: CPFLOW: A Practical Tool for Tracing Power System Steady-State Stationary Behavior due to Load and Generation Variations. IEEE Transactions on Power Systems 10(2), 623–634 (1995)
[57] Chiang, H.D., Wu, F.F., Varaiya, P.P.: Foundations of the Direct Methods for Power System Transient Stability Analysis. IEEE Transactions on Circuits and Systems - I: Fundamental Theory and Applications 34(1), 160–173 (1987)
[58] Chopra, S., Dexter, S.D.: Decoding Liberation: The Promise of Free and Open Source Software. Routledge Taylor & Francis Group, New York (2008)
[59] Chow, J.: Power System Toolbox (2002), `http://www.eagle.ca/~cherry`
[60] Chow, J.H., Cheung, K.W.: A Toolbox for Power System Dynamics and Control Engineering Education and Research. IEEE Transactions on Power Systems 7(4), 1559–1564 (1992)
[61] Chun, L., Qirong, J., Xiaorong, X., Zhonghong, W.: Rule-based Control for STATCOM to Increase Power System Stability. In: Proceedings of the PowerCon, (August 1998)
[62] Chun, W.J.: Core Python Programming, 1st edn. Prentice Hall, Inc., Upper Saddle River (2000)
[63] Coar, K.: Open Source Definition (2007), `http://www.opensource.org/docs/osd`
[64] Cole, S.: MatDyn. Katholieke Universiteit Leuven, Belgium, `http://www.esat.kuleuven.be/electa/teaching/matdyn`
[65] Conejo, A.J., Arroyo, J.M.: Optimal Response of a Thermal Unit to an Electricity Spot Market. IEEE Transactions on Power Systems 15, 1098–1104 (2000)
[66] Conejo, A.J., Arroyo, J.M.: Multiperiod Auction for a Pool-based electricity market. IEEE Transactions on Power Systems 17(4), 1225–1231 (2002)
[67] Cormen, T.H., Leiserson, C.E., Rivest, R.L., Stein, C.: Introduction to Algorithms. MIT Press and McGraw-Hill (2001)
[68] CYME International Inc., CYMSTAB - User's Guide and Reference Manual, Burlington, MA (July 1994)

[69] Dahlquist, G.G.: A Special Stability Problem for Linear Multistep Methods. BIT Numerical Mathematics 3(1), 27–43 (1963)

[70] D'Albertanson, B., Hawkins, D.: An Integrated MV Distributed Generation Connection Planning Tool. In: Proceedings of the IET-CIRED SmartGrids for Distribution (June 2008)

[71] Das, D., Kothari, D.P., Kalam, A.: Simple and Efficient Method for Load Flow Solution of Radial Distribution Networks. International Journal of Electrical Power and Energy Systems 17(5), 335–346 (1995)

[72] Davidenko, D.F.: On a New Method of Numerical Solution of Systems of Nonlinear Equations. Mathematical Reviews 14, 906 (1953)

[73] de Berg, M., van Kreveld, M., Overmars, M., Schwarzkopf, O.: Computational Geometry, Algorithms and Applications. Springer, Heidelberg (2000)

[74] Demmel, J.W.: Applied Numerical Linear Algebra. SIAM, Philadelphia (1997)

[75] Power System Engineering and Software, DigSilent, `http://www.digsilent.de`

[76] Dobson, I.: Observations on the Geometry of Saddle Node Bifurcation and Voltage Collapse in Electrical Power Systems. IEEE Transactions on Circuits and Systems - I: Fundamental Theory and Applications 39(3), 240–243 (1992)

[77] Dobson, I., Alvarado, F., DeMarco, C.L.: Sensitivity of Hopf Bifurcation to Power System Parameters. IEEE Decision and Control 3, 2928–2933 (1992)

[78] Dommel, H.W.: Digital Computer Solution of Electromagnetic Transients in Single and Multiphase Networks. IEEE Transactions on Power Apparatus and Systems 88(4), 388–398 (1969)

[79] Dommel, H.W.: Nonlinear and Time-Varying Elements in Digital Simulation of Electromagnetic Transients. IEEE Transactions on Power Apparatus and Systems 90(6), 2561–2567 (1971)

[80] Dommel, H.W., Tinney, W.F.: Optimal Power Flow Solutions. IEEE Transactions on Power Apparatus and Systems 87(10), 1866–1876 (1968)

[81] Drud, A.S.: GAMS/CONOPT, ARKI Consulting and Development, Bagsvaerdvej 246A, DK-2880 Bagsvaerd, Denmark (1996), `http://www.gams.com/`

[82] Dular, P., Kuo-Peng, P.: Three-Dimensional Modeling of Both Inductive and Capacitive Effects in Massive Inductors. IEEE Transactions on Magnetics 42(4), 743–746 (2006)

[83] FlowDemo.net, EEH - Power Systems Laboratory, Zürich, `http://flowdemo.net`

[84] El-Hawary, M.E.: Electrical Energy Systems. CRC Press, Boca Raton (2000)

[85] El-Samahy, I., Bhattacharya, K., Cañizares, C., Anjos, M.F., Pan, J.: A Procurement Market Model for Reactive Power Services Considering System Security. IEEE Transactions on Power Systems 23(1), 137–149 (2008)

[86] OpenDSS, Electric Power Research Institute, `http://sourceforge.net/projects/electricdss`

[87] ENEL, `http://www.enel.com`

[88] Energy, G.: GE-PSLF Load Flow Data Export/Import File for PSLF Version 15.1, General Electric International, Inc., 1 River Road, Schenectady, NY 12345, USA (June 2005)

[89] Energy Development and Power Generating Committee of the Power Engineering Society, "IEEE Recommended Practice for Excitation System Models for Power System Stability Studies," IEEE Std 421.5-1992, New York, Tech. Rep. (March 1992)
[90] Extended Transient-Midterm Stability Package: User's Manual for the Power Flow Program, EPRI, ePRI computer code manual EL-2002-CCM (January 1987)
[91] European Wind Energy Association, Wind Force 12-A Blueprint to Archive 12% of the World's Electricity from Wind Power by 2020, EWEA, Tech. Rep., 56 pages (2001)
[92] Eurostag, `http://www.eurostag.be`
[93] Evrenosoğlu, C.Y., Abur, A., Akleman, E.: Three Dimensional Visualization and Animation of Traveling Waves in Power Systems. Electric Power Systems Research 77(7), 876–883 (2007)
[94] Feng, Z., Xu, W.: Fast Computation of Post-contingency System Margins for Voltage Stability Assessments of Large-scale Power Systems. IEE Proceedings on Generation, Transmission and Distribution 147(2), 76–80 (1990)
[95] Fiacco, A.V., McCormick, G.P.: Nonlinear Programming: Sequential Unconstrained Minimization. John Wiley & Sons, New York (1968)
[96] Fletcher, R.: Practical Methods of Optimization. John Wiley & Sons, New York (1987)
[97] Flueck, A.J., Chiang, H.D.: Solving the Nonlinear Power Flow Equations with an Inexact Newton Method Using GMRES. IEEE Transactions on Power Systems 13(2), 267–273 (1998)
[98] Fouad, A.A., Vittal, V.: Power System Transient Stability Analysis Using the Transient Energy Function Method. Prentice Hall, Upper Saddle River (1992)
[99] Fourer, R., Gay, D.M., Kernighan, B.W.: AMPL: A Modeling Language for Mathematical Programming, 2nd edn. Duxbury Press/Brooks/Cole Publishing Company, Boston (2002)
[100] Free Software Foundation (FSF), `http://www.fsf.org/`
[101] FreeGIS, `http://www.freegis.org`
[102] Friedl, J.E.F.: Mastering Regular Expressions, 2nd edn. O'Reilly, Sebastopol (2002)
[103] FSF Free Software Licensing and Compliance Lab, `http://www.fsf.org/licensing/`
[104] Gao, B., Morison, G.K., Kundur, P.: Voltage Stability Evaluation using Modal Analysis. IEEE Transactions on Power Systems 7(4), 1529–1542 (1992)
[105] Geography Markup Language, `http://www.opengeospatial.org/standards/gml`
[106] Gill, P.E., Murray, W., Wright, M.H.: Practical Optimization. Academic Press, London (1981)
[107] Gisin, B.S., Obessis, M.V., Mitsche, J.V.: Practical Methods for Transfer Limit Analysis in the Power Industry Deregulated Environment. In: Proceedings of the Power Industry Computer Application (PICA), pp. 261–266 (1999)
[108] Goderya, F., Metwally, A.A., Mansour, O.: Fast Detection and Identification of Islands in Power Systems. IEEE Transactions on Power Apparatus and Systems 99(1), 217–221 (1980)

[109] Gole, A.M., Sood, V.K.: A Static Compensator Model for use with Electromagnetic Transients Simulation Programs. IEEE Transactions on Power Systems 5(3), 1389–1407 (1990)

[110] Gómez-Expósito, A., Conejo, A.J., Cañizares, C.: Electric Energy Systems - Analysis and Operation. CRC Press, Boca Raton (2008)

[111] Granville, S.: Optimal Reactive Dispatch through Interior Point Methods. IEEE Transactions on Power Systems 9(1), 136–146 (1994)

[112] Granville, S., Mello, J.C.O., Melo, A.C.G.: Application of Interior Point Methods to Power Flow Unsolvability. IEEE Transactions on Power Systems 11(4), 1096–1103 (1996)

[113] Grijalva, S., Sauer, P.W.: A Necessary Condition for Power Flow Power Jacobian Singularity Based on Branch Complex Flows. IEEE Transactions on Circuits and Systems - I: Fundamental Theory and Applications 52(7), 1406–1413 (2005)

[114] Gross, C.A.: Power System Analysis, 2nd edn. John Wiley & Sons, Chichester (1986)

[115] Gu, W., Milano, F., Jang, P., Tang, G.: Hopf Bifurcations Induced by SVC Controllers: A Didactic Example. Electric Power Systems Research 77(3-4), 234–240 (2007)

[116] Guoyu, X., Galiana, F.D., Low, S.: Decoupled Economic Dispatch using the Participation Factors Load Flow. IEEE Transactions on Power Apparatus and Systems 104(6), 1377–1384 (1985)

[117] Hairer, E., Nørsett, S.: Solving Ordinary Differential Equations I: Nonstiff Problems, 2nd edn., Berlin, Germany. Springer Series in Computational Mathematics, vol. 8 (2000)

[118] Haley, P.H., Ayres, M.: Super Decoupled Loadflow with Distributed Slack Bus. IEEE Transactions on Power Apparatus and Systems 104(1), 104–113 (1985)

[119] Hansen, A.D., Michalke, G.: Modelling and Control of Variable speed Multipole Permanent Magnet Synchronous Generator Wind Turbine. Wind Energy 11(5), 537–554 (2008)

[120] Happ, H.H.: Optimal Power Dispatch: A Comprehensive Survey. IEEE Transactions on Power Apparatus and Systems PAS-96(3), 841–854 (1977)

[121] Haque, M.H.: A General Load Flow Method for Distribution Systems. Electric Power Systems Research 54(1), 47–54 (2000)

[122] Haque, M.H.: Improvement of First Swing Stability Limit by Utilizing Full Benefit of Shunt FACTS Devices. IEEE Transactions on Power Systems 19(4), 1894–1902 (2004)

[123] Hassan, I.D., Bucci, R.M., Swe, K.T.: 400 MW SMES Power Conditioning System Development and Simulation. IEEE Transactions on Power Electronics 8(3), 237–249 (1993)

[124] Hatziadoniu, C.J., Lobo, A.A., Pourboghrat, F., Daneshdoost, M.: A Simplified Dynamic Model of Grid-Connected Fuel-Cell Generators. IEEE Transactions on Power Systems 17(2), 467–473 (2002)

[125] Haug, E.J., Arora, J.S.: Applied Optimal Design. John Wiley & Sons, New York (1979)

[126] Heier, S.: Grid Integration of Wind Energy Conversion Systems. John Wiley & Sons, England (1998)

[127] Hetzler, S.M.: A Continuous Version of Newton's Method. The College Mathematical Journal 28(5), 348–351 (1997)
[128] Hill, D.J.: Nonlinear Dynamic Load Models with Recovery for Voltage Stability Studies. IEEE Transactions on Power Systems 8(1), 166–176 (1993)
[129] Hingorani, G., Gyugyi, L.: Understanding FACTS: Concepts and Technology of Flexible AC Transmission Systems. IEEE Press, Los Alamitos (1999)
[130] Hirsch, P.: Extended Transient-Midterm Stability Program (ETMSP) Ver. 3.1: User's Manual, EPRI, TR-102004-V2R1 (May 1994)
[131] Hiskens, I.A.: Power System Modeling for Inverse Problems. IEEE Transactions on Circuits and Systems - I: Regular Papers 51(3), 539–551 (2004)
[132] Hoffman, K., Kunze, R.: Linear Algebra, 2nd edn. Prentice-Hall, Englewood Cliffs (1971)
[133] Holdsworth, L., Wu, X.G., Ekanayake, J.B., Jenkins, N.: Direct Solution Method for Initialising Doubly-fed Induction Wind Turbines in Power System Dynamic Models. IEE Proceedings on Generation, Transmission and Distribution 150(3), 334–342 (2003)
[134] Huneault, M., Galiana, F.D.: A Survey of the Optimal Power Flow Literature. IEEE Transactions on Power Systems 6(2), 762–770 (1991)
[135] Iba, K., Suzuki, H., Egawa, M., Watanabe, T.: A Method for Finding a Pair of Multiple Load Flow Solutions in Bulk Power Systems. IEEE Transactions on Power Systems 5(2), 582–591 (1990)
[136] Iba, K., Suzuki, H., Egawa, M., Watanabe, T.: Calculation of Critical Loading Condition with Nose Curve using Homotopy Continuation Method. IEEE Transactions on Power Systems 6(2), 584–593 (1991)
[137] IEC, IEC 61968 Application Integration at Electric Utilities - System Interfaces for Distribution Management - Part 11: Common Information Model (CIM), draft
[138] IEC, IEC 61970 Energy Management System Application Program Interface (EMS-API) - Part 301: Common Information Model (CIM) Base, edition 1.0 (November 2003)
[139] IEC 61400-1, Wind Turbines - Part 1: Design Requirements, International Electrotechnical Commission, Geneva, Switzerland, Tech. Rep. (August 2005)
[140] IEEE Commettee Report, Reader's Guide to SSR. IEEE Transactions on Power Apparatus and Systems 7(2), 150–157 (1992)
[141] IEEE Power System Engineering Committee Report, Terms, Definitions & Symbols for Subsynchronous Oscillations. IEEE Transactions on Power Apparatus and Systems 104(6), 1326–1334 (1985)
[142] IEEE Subsynchronous Resonance Task Force, First Benchmark Model for Computer Simulation of Subsynchronous Resonance. IEEE Transactions on Power Apparatus and Systems 96(5), 1565–1572 (1977)
[143] IEEE Task Force on Excitation Limiters, Recommended Models for Overexcitation Limiting Devices. IEEE Transactions on Energy Conversion 10(4), 706–713 (1995)
[144] IEEE Task Force on Excitation Limiters, Underexcitation Limiter Models for Power System Stability Studies. IEEE Transactions on Energy Conversion 10(3), 524–531 (1995)
[145] IEEE Working Group on Computer Modelling of Excitation Systems, Excitation System Models for Power System Stability Studies. IEEE Transactions on Power Apparatus and Systems 100(2), 494–509 (1981)

[146] IEEE/CIGRE Joint Task Force on stability Terms and Definitions, Definition and Classification of Power System Stability. IEEE Transactions on Power Systems 19(2), 1387–1401 (2004)
[147] Ilić, M., Zaborszky, J.: Dynamic and Control of Large Electric Power Systems. Wiley-Interscience Publication, New York (2000)
[148] Irisarri, G.D., Wang, X., Tong, J., Mokhtari, S.: Maximum Loadability of Power Systems using Interior Point Nonlinear Optimization Method. IEEE Transactions on Power Systems 12(1), 162–172 (1997)
[149] Iwamoto, S., Tamura, Y.: A Fast Load Flow Method Retaining Nonlinearity. IEEE Transactions on Power Apparatus and Systems PAS-97(5), 1586–1599 (1978)
[150] Iwamoto, S., Tamura, Y.: A Load Flow Calculation Method for Ill-Conditioned Power Systems. IEEE Transactions on Power Apparatus and Systems 100(4), 1736–1743 (1981)
[151] Jalili-Marandi, V., Dinavahi, V., Strunz, K., Martinez, J.A., Ramirez, A.: Interfacing Techniques for Transient Stability and Electromagnetic Transient Programs. IEEE Transactions on Power Delivery 24(4), 2385–2395 (2009)
[152] Jimma, K., Tomac, A., Liu, C.C., Vu, K.T.: A Study of Dynamic Load Models for Voltage Collapse Analysis. In: Proceedings of the Bulk Power System Voltage Phenomena II - Voltage Stability and Security, August 1991, pp. 423–429 (1991)
[153] Jones, C.A., Drake Jr., F.L.: Python & XML. O'Reilly, Sebastopol (2001)
[154] Kang, Y., Lavers, J.D.: Transient Analysis of Electric Power Systems: Reformulation and Theoretical Basis. IEEE Transactions on Power Systems 11(2), 754–760 (1996)
[155] Karlsson, D., Hill, D.J.: Modelling and Identification of Nonlinear Dynamic Loads in Power Systems. IEEE Transactions on Power Systems 9(1), 157–166 (1994)
[156] Kezunovic, M., Abur, G.H.A., Bose, A., Tomsovic, K.: The Role of Digital Modeling and Simulation in Power Engineering Education. IEEE Transactions on Power Systems (1), 64–72 (2004)
[157] Klump, R., Wu, W., Dooley, G.: Displaying Aggregate Data, Interrelated Quantities, and Data Trends in Electric Power Systems. In: Proceedings of the 36th Hawaii International Conference on System Sciences, Hawaii (2003)
[158] Klump, R.P., Weber, J.D.: Real-Time Data Retrieval and New Visualization Techniques for the Energy Industry. In: Proceedings of the 35th Hawaii International Conference on System Sciences, Hawaii (2002)
[159] Knyazkin, V., Söder, L., Cañizares, C.: Control Challenges of Fuel Cell-Driven Distributed Generation. In: Proceedings of the IEEE PES General Meeting, Toronto (July 2003)
[160] Krause, P.C., Wasynczuk, O., Sudhoff, S.D.: Analysis of Electric Machinery and Drive Systems, 2nd edn. John Wiley & Sons, New York (2002)
[161] Kumkratug, P., Haque, M.: Versatile Model of a Unified Power Flow Controller a Simple Power System. IEE Proceedings on Generation, Transmission and Distribution 150, 155–161 (2003)
[162] Kumkratug, P., Haque, M.H.: Improvement of Stability Region and Damping of a Power System by Using SSSC. In: Proceedings of the IEEE PES General Meeting, Denver, CO, vol. 3 (2003)

[163] Kundur, P.: Power System Stability and Control. McGraw-Hill, New York (1994)
[164] Langtangen, H.P.: Python Scripting for Computational Science, 3rd edn. Springer, Heidelberg (2002)
[165] Lapidus, L., Seinfeld, J.H.: Numerical Solution of Ordinary Differential Equations. Academic Press, New York (1971)
[166] Larsson, M.: ObjectStab – An Educational Tool for Power System Stability Studies. IEEE Transactions on Power Systems 19(1), 56–63 (2004)
[167] Laurent, A.M.S.: Understanding Open Source & Free Software Licensing. O'Reilly Media, Sebastopol (2004)
[168] Lee, K., Mohammadi, S., Bhattacharya, P.K., Katehi, L.P.B.: Compact Models Based on Transmission-Line Concept for Integrated Capacitors and Inductors. IEEE Transactions on Microwave Theory and Techniques 54(12), 4141–4148 (2006)
[169] Lee, K.Y., El-Sharkawi, A.: Modern Heuristic Optimization Techniques. IEEE Press Series on Power Engineering. John Wiley & Sons, Hoboken (2008)
[170] Lessing, L.: The Future of Ideas: The Fate of the Commons in a Connected World. Vintage (2002)
[171] Li, H., Tesfatsion, L.: The AMES Wholesale Power Market Test Bed: A Computational Laboratory for Research, Teaching, and Training. In: Proceedings of the IEEE PES General Meeting, Calgary, Alberta (July 2009)
[172] Lincoln, R.W.: Pylon. University of Strathclide, UK, `http://pylon.eee.strath.ac.uk/`
[173] Liu, C.W., Thorp, J.S., Lu, J., Thomas, R.J., Chiang, H.D.: Detection of Transient Chaotic Swings in Power Systems using Real-Time Phasor Measurements. IEEE Transactions on Power Systems 9(3), 1285–1292 (1994)
[174] Liu, S., Dougal, R.A.: Dynamic Multi-physics Model for Solar Array. IEEE Transactions on Energy Conversion 17(3), 285–294 (2002)
[175] Luenberger, D.G.: Linear and Nonlinear Programming, 2nd edn. Addison-Wesley Publishing Company, Reading (1984)
[176] Luo, G.X., Semlyen, A.: Efficient Load Flow for Large Weakly Meshed Networks. IEEE Transactions on Power Systems 5(4), 1309–1316 (1990)
[177] Luque, A., Hegedus, S.: Handbook of Photovoltaic Science and Engineering. John Wiley & Sons, Chichester (2003)
[178] Lyapunov, A.M.: Stability of Motion. Academic Press, New York (1966)
[179] Machowski, J., Bialek, J.W., Bumby, J.R.: Power System Dynamics and Stability. John Wiley & Sons, New York (1998)
[180] Mahadev, P.M., Christie, R.D.: Envisioning Power System Data: Concepts and a Prototype System State Representation. IEEE Transactions on Power Systems 8(3), 1084–1090 (1993)
[181] Mahseredjian, J., Alvarado, F.: Creating an Electromagnetic Transient Program in MATLAB: MatEMTP. IEEE Transactions on Power Delivery 12(1), 380–388 (1997)
[182] Malek-Zavarei, M., Jamshidi, M.: Time-Delay Systems Analysis, Optimization and Applications. Elsevier Science Publishers B. V., Amsterdam (1987)
[183] Manitoba HVDC Research Center, "PSCAD/EMTDC," Manitoba Hydro, Canada, `https://pscad.com`
[184] Marconato, R.: Electric Power Systems, vol. 2. CEI, Italian Electrotechnical Committee, Milano, Italy (2002)

[185] Marcus, M., Minc, H.: Introduction to Linear Algebra. Dover, New York (1988)
[186] Martelli, A.: Python in a Nutshell, 2nd edn. O'Reilly, Sebastopol (2006)
[187] Mauricio, J.M., León, A.E., Gómez-Expósito, A., Solsona, J.A.: An Electrical Approach to Mechanical Effort Reduction in Wind Energy Conversion Systems. IEEE Transactions on Energy Conversion 23(4), 1108–1611 (2008)
[188] McMorran, A.W.: An Introduction to IEC 61970-301 & 61968-11: The Common Information Model (January 2007), `http://cimphony.org/cimphony/cim-intro.pdf`
[189] Medanić, J., Ilić-Spong, M., Christensen, J.: Discrete Models of Slow Voltage Dynamics for Under Load Tap-Changing Transformer Coordination. IEEE Transactions on Power Systems 2(4), 873–880 (1987)
[190] Mehrotra, S.: On the Implementation of a Primal-dual Interior Point Method. SIAM Journal on Optimization 2(3), 575–601 (1992)
[191] Mekhamer, S.F., Soliman, S.A., Moustafa, M.A., El-Hawary, M.E.: Load Flow Solution of Radial Distribution Feeders: A New Contribution. International Journal of Electrical Power and Energy Systems 24(9), 701–707 (2002)
[192] Meng, Z.J., So, P.L.: A Current Injection UPFC Model for Enhancing Power System. In: Proceedings of the IEEE PES Winter Meeting, 2nd edn., pp. 1544–1549 (2000)
[193] Mihalic, R., Gabrijel, U.: A Structure-Preserving Energy Function for a Static Series Synchronous Compensator. IEEE Transactions on Power Systems 19(3), 1501–1507 (2004)
[194] Milano, F.: PSAT, Matlab-based Power System Analysis Toolbox (2002), `http://www.uclm.es/area/gsee/Web/Federico`
[195] Milano, F.: An Open Source Power System Analysis Toolbox. IEEE Transactions on Power Systems 20(3), 1199–1206 (2005)
[196] Milano, F.: Continuous Newton's Method for Power Flow Analysis. IEEE Transactions on Power Systems 24(1), 50–57 (2009)
[197] Milano, F.: Three-Dimensional Visualization and Animation for Power Systems Analysis. Electric Power Systems Research 79(12), 1638–1647 (2009)
[198] Milano, F., Cañizares, C.A., Invernizzi, M.: Multi-objective Optimization for Pricing System Security in Electricity Markets. IEEE Transactions on Power Systems 18(2), 596–604 (2003)
[199] Milano, F., Vanfretti, L.: State of the Art and Future of OSS for Power Systems. In: Proceedings of the IEEE PES General Meeting, Calgary, Canada (July 2009)
[200] Milano, F., Vanfretti, L., Morataya, J.C.: An Open Source Power System Virtual Laboratory: The PSAT Case and Experience. IEEE Transactions on Education 51(1), 17–23 (2008)
[201] Milano, F., Zhou, M., Hou, G.: Open Model for Exchanging Power System Data. In: Proceedings of the IEEE PES General Meeting, Calgary, Canada (July 2009)
[202] Milias-Argitis, J., Zacharias, T., Hatzadoniu, C., Galanos, G.D.: Transient Simulation of Integrated AC/DC Systems, Part I: Converter Modeling and Simulation. IEEE Transactions on Power Systems 3(1), 166–172 (1988)
[203] Milias-Argitis, J., Zacharias, T., Hatzadoniu, C., Galanos, G.D.: Transient Simulation of Integrated AC/DC Systems, Part II: System Modeling and Simulation. IEEE Transactions on Power Systems 3(1), 173–179 (1988)

[204] Mínguez, R., Milano, F., Zárate-Miñano, R., Conejo, A.J.: Optimal Network Placement of SVC Devices. IEEE Transactions on Power Systems 22(4), 1851–1860 (2007)
[205] Mitani, Y., Tsuji, K.: Bifurcations Associated with Sub-Synchronous Resonance. IEEE Transactions on Power Systems 10(4), 1471–1478 (1995)
[206] Mitani, Y., Tsuji, K., Varghese, M., Wu, F., Varaiya, P.: Bifurcation Associated with Sub-Synchronous Resonance. IEEE Transactions on Power Systems 13(1), 139–144 (1998)
[207] Mithulananthan, N., Cañizares, C.A., Reeve, J.: Indices to Detect Hopf Bifurcation in Power Systems. In: Proceedings of the North American Power Symposium (NAPS), vol. 23, pp. 15–18–15–23. University of Waterloo, Waterloo (2000)
[208] Mithulananthan, N., Cañizares, C.A., Reeve, J., Rogers, G.J.: Comparison of PSS, SVC and STATCOM Controllers for Damping Power System Oscillations. IEEE Transactions on Power Systems 18(2), 786–792 (2003)
[209] Monmoh, J.A., Guo, S.X., Ogbuobiri, E.C., Adapa, R.: The Quadratic Interior Point Method solving Power System Optimization Problems. IEEE Transactions on Power Systems 9(3), 1327–1336 (1994)
[210] Monticelli, A., Garcia, A.: Modeling Zero Impedance Branches in Power System State Estimation. IEEE Transactions on Power Systems 6(4), 1561–1570 (1991)
[211] Morison, G.K., Gao, B., Kundur, P.: Voltage Stability Analysis using Static and Dynamic Approaches. IEEE Transactions on Power Systems 8(3), 1159–1171 (1993)
[212] Motto, A.L., Galiana, F.D., Conejo, A.J., Arroyo, J.M.: Network-constrained Multiperiod Auction for a Pool-based Electricity Market. IEEE Transactions on Power Systems 17(3), 646–653 (2002)
[213] Noroozian, M., Angquist, L., Ghandhari, M., Andersson, G.: Improving Power System Dynamics by Series-Connected FACTS Devices. IEEE Transactions on Power Delivery 12, 1635–1641 (1997)
[214] Open Geospatial Consortium, `http://www.opengeospatial.org`
[215] Open Source Initiative (OSI), `http://www.opensource.org/`
[216] OpenGIS® Standards and Specifications, `http://www.opengeospatial.org/standards`
[217] The Free, Java-based and Open Source Geographic Information System for the World, OpenJump, `http://openjump.org`
[218] Ortega, J.M., Rheinbolt, W.C.: Iterative Solutions of Nonlinear Equations in Several Variables. Academic, New York (1969)
[219] Overbye, T.J.: A Power Flow Measure for Unsolvable Cases. IEEE Transactions on Power Systems 9(3), 1359–1365 (1994)
[220] Overbye, T.J.: Computation of a Practical Method to Restore Power Flow Solvability. IEEE Transactions on Power Systems 10(1), 280–287 (1995)
[221] Overbye, T.J., Klump, R.P.: Effective Calculation of Power System Low-Voltage Solutions. IEEE Transactions on Power Systems 11(1), 75–82 (1996)
[222] Overbye, T.J., Weber, J.D.: Visualizing Power System Data. In: Proceedings of the 33th Hawaii International Conference on System Sciences, Hawaii (2000)
[223] Overbye, T.J., Weber, J.D.: Visualizing the Electric Grid. IEEE Spectrum 38(2), 52–58 (2001)

[224] Overbye, T.J., Weber, J.D., Patten, K.J.: Analysis and Visualization of Market Power in Electric Power Systems. In: Proceedings of the 32nd Hawaii International Conference on System Sciences, Hawaii (1999)
[225] Overbye, T.J., Wiegmann, D.A., Rich, A.M., Sun, Y.: Human Factors Aspects of Power System Voltage Contour Visualizations. IEEE Transactions on Power Systems 18(1), 76–82 (2003)
[226] Padullés, J., Ault, G.W., McDonald, J.R.: An Integrated SOFC Plant Dynamic Model for Power Systems Simulation. International Journal of Power Sources 86, 495–500 (2000)
[227] Pai, M.A.: Computer Techniques in Power System Analysis. Tata McGraw-Hill Publishing Company, New Delhi (1979)
[228] Pai, M.A.: Power System Stability. North-Holland Company, New York (1981)
[229] Pai, M.A.: Energy Function Analysis for Power System Stability. Kluwer Academic Publishers, Boston (1989)
[230] Pal, M.K.: Voltage Stability Conditions considering Load Characteristics. IEEE Transactions on Power Systems 7(1), 243–249 (1992)
[231] Panofsky, H.A., Dutton, J.A.: Atmospheric Turbulence; Models and Methods for Engineering Applications. John Wiley & Sons, New York (1984)
[232] Papic, I., Zunko, P., Povh, D.: Basic Control of Unified Power Flow Controller. IEEE Transactions on Power Systems 12(4), 1734–1739 (1997)
[233] Parashar, M., Thorp, J.S., Seyler, C.E.: Continuum Modeling of Electromechanical Dynamics in Large-Scale Power Systems. IEEE Transactions on Circuits and Systems - I: Fundamental Theory and Applications 51(9), 1848–1858 (2004)
[234] Park, R.H.: Two-reaction Theory of Synchronous Machines. Generalized Method of Analysis - Part I. AIEE Transactions 48, 716–727 (1929)
[235] Patel, M.R.: Wind and Solar Power Systems. CRC Press, Boca Raton (1999)
[236] Pavella, M., Ernst, D., Ruiz-Vega, D.: Transient Stability of Power systems: A Unified Approach to Assessment and Control. Kluwer Academic Publisher, Boston (2000)
[237] Pereira, L., Undrill, J., Kosterev, D., Patterson, S.: A New Thermal Governor Modeling Approach in the WECC. IEEE Transactions on Power Systems 18(2), 819–829 (2003)
[238] Petersen, N.M., Scott Meyer, W.: Automatic Adjustment of Transformer and Phase-Shifter Taps in the Newton Power Flow. IEEE Transactions on Power Apparatus and Systems 90(1), 103–108 (1971)
[239] Pierce, B.: Types and Programming Languages. MIT Press, Cambridge (2002)
[240] Pilotto, L.A.S., Roitman, M., Alves, J.E.R.: Digital Control HVDC Converters. IEEE Transactions on Power Systems 4(2), 704–711 (1989)
[241] Popović, D., Hiskens, I.A., Hill, D.J.: Investigations of Load-Tap Changer Interaction. International Journal of Electrical Power and Energy Systems 18(2), 81–97 (1996)
[242] PostGIS, `http://postgis.refractions.net`
[243] PSS/E Program Application Guide, Power Technologies, Inc. (December 1996)
[244] Online Documentation PSS/E 30, Power Technologies, Inc. (2004)
[245] Siemens PTI, Power Technologies International, `http://www.pti-us.com`

[246] PowerWorld Corporation, Auxiliary File Format for Simulator 11.0, Powerwold Corporation (April 2005), http://www.powerworld.com
[247] PowerWorld Corporation, POWERWORLD Simulator Version 11.0: Interactive Power System Simulator Analysis and Visualization, Powerwold Corporation (April 2005), http://www.powerworld.com
[248] Preparata, F.P., Hong, S.J.: Convex Hulls of Finite Sets of Points in Two and Three Dimensions. Commun. ACM 20(2), 87–93 (1977)
[249] Qhull, http://www.qhull.org
[250] QGIS - user friendly Open Source Geographic Information System, Quantum GIS, http://www.qgis.org/
[251] Quintana, V.H., Torres, G.L., Medina-Palomo, J.: Interior-Point Methods and their Applications to Power Systems: A Classification of Publications and Software Codes. IEEE Transactions on Power Systems 15(1), 170–176 (2000)
[252] Rahim, A.H.M.A., Al-Baiyat, S.A., Al-Maghrabi, H.M.: Robust damping controller design for a static compensator. IEE Proceedings on Generation, Transmission and Distribution 149(4), 491–496 (2002)
[253] Ramachandran, P., Varoquaux, G.: Mayavi User Guide Release 3.2.1 (July 2009), https://svn.enthought.com/
[254] Raymond, E.S.: The Cathedral and the Bazaar. Thyrsus Enterprises (2000), http://www.tuxedo.org/~esr/
[255] Reid, G.F., Hasdorff, L.: Economic Dispatch using Quadratic Programming. IEEE Transactions on Power Apparatus and Systems 92(6), 2015–2023 (1973)
[256] Riaza, R.: Singularity-Induced Bifurcations in Lumped Circuits. IEEE Transactions on Circuits and Systems - I: Fundamental Theory and Applications 52(7), 1442–1450 (2005)
[257] Ribbens-Pavella, M., Evans, F.J.: Direct Methods for Studying Dynamics of Large-Scale electric Power Systems - A Survey. Automatica 32(1), 1–21 (1985)
[258] Richard, J.P.: Time-Delay Systems: An Overview of some Recent Advances and Problems. Automatica 39, 1667–1694 (2003)
[259] Rosehart, W., Cañizares, C.A., Quintana, V.H.: Optimal Power Flow Incorporating Voltage Collapse Constraints. In: Proceedings of the IEEE PES Summer Meeting, Edmonton, Alberta (July 1999)
[260] Rosehart, W.D., Cañizares, C.A., Quintana, V.: Multi-Objective Optimal Power Flows to Evaluate Voltage Security Costs in Power Networks. IEEE Transactions on Power Systems 18(2), 578–587 (2003)
[261] Rostamkolai, N., Wegner, C.A., Piwko, R.J., Elahi, H., Eitzmann, M.A., Garzi, G., Tietz, P.: Control Design of Santo Tomé Bak-to-Back HVDC Link. IEEE Transactions on Power Systems 8(3), 1250–1256 (1993)
[262] Ruiz-Vega, D., Asiaín Olivares, T.I., Olguín Salinas, D.: An Approach to the Initialization of Dynamic Induction Motor Models. IEEE Transactions on Power Systems 17(3), 747–751 (2002)
[263] Saccomanno, F.: Electric Power Systems - Analysis and Control. John Wiley & Sons, New York (2003)
[264] Salameh, Z.M., Casacca, M.A., Lynch, W.A.: A Mathematical Model for Lead-Acid Batteries. IEEE Transactions on Energy Conversions 7(1), 93–97 (1992)

[265] Sasson, A.M.: Combined Use of the Powell and Fletcher-Powell Nonlinear Programming Methods for Optimal Load Flows. IEEE Transactions on Power Apparatus and Systems 88(10), 1530–1537 (1969)

[266] Sasson, A.M.: Optimal Load Flow Solution using the Hessian Matrix. IEEE Transactions on Power Apparatus and Systems 92(1), 31–41 (1973)

[267] Sasson, A.M., Merrill, H.M.: Some Applications of Optimization Techniques to Power System Problems. Proceedings of the IEEE 62(7), 959–972 (1974)

[268] Sato, M., Yamaji, K., Sekita, M.: Development of a Hybrid Margin Angle Controller for HVDC Continuous Operation. IEEE Transactions on Power Systems 11(4), 1792–1798 (1996)

[269] Sauer, P.W., Pai, M.A.: Power System Dynamics and Stability. Prentice Hall, Upper Saddle River (1998)

[270] Schaffer, M.D., Tylavsky, D.J.: A Nondiverging Polar-Form Newton-Based Power Flow. IEEE Transactions on Industry Applications 24(5), 870–877 (1988)

[271] Schoder, K., Feliachi, A., Hasanović, A.: PAT: A Power Analysis Toolbox for Matlab/Simulink. IEEE Transactions on Power Systems 18(1), 42–47 (2003)

[272] Schultz, R.P.: Synchronous Machine Modeling. In: Symposium on Adequacy and Philosophy of Modeling: System Dynamic Performance, San Francisco, CA (July 1972)

[273] Sedghisigarchi, K., Feliachi, A.: Dynamic and Transient Analysis of Power Distribution Systems With Fuel Cells, Part I: Fuel-Cell Dynamic Model. IEEE Transactions on Power Systems 19(2), 423–428 (2004)

[274] Semlyen, A.: Analysis of Disturbance Propagation in Power Systems Based on a Homogeneous Dynamic Model. IEEE Transactions on Power Apparatus and Systems 93(2), 676–684 (1974)

[275] Semlyen, A.: Fundamental Concepts of a Krylov Subspace Power Flow Methodology. IEEE Transactions on Power Systems 11(3), 1528–1537 (1996)

[276] Seydel, R.: Practical Bifurcation and Stability Analysis: From Equilibrium to Chaos, 2nd edn. Springer, New York (1994)

[277] Sheblé, G.B.: Computational Auction Mechanism for Restructured Power Industry Operation. Kluwer Academic Publishers, Boston (1998)

[278] Shepherd, C.M.: Design of Primary and Secondary Cells. Journal of The Electrochemical Society 112(3), 252–257 (1965)

[279] Shirmohammadi, D., Hong, H.W., Semlyen, A., Luo, G.X.: A Compensation-based Power Flow Method for Weakly Meshed Distribution and Transmission Systems. IEEE Transactions on Power Systems 3(2), 753–762 (1988)

[280] Shome, T., Gole, A.M., Brandt, D.P., Hamlin, R.J.: Adjusting Converter Control for Paralled DC Converters Using a Digital Transient Simulation Program. IEEE Transactions on Power Systems 5(1), 12–19 (1990)

[281] Simiu, E., Scanlan, R.H.: Wind Effects on Structures; an Introduction to Wind Engineering. John Wiley & Sons, New York (1986)

[282] Skiena, S.S.: The Algorithm Design Manual. Springer, New York (1998)

[283] Slootweg, J.G.: Wind Power: Modelling and Impact on Power System Dynamics. Ph.D. dissertation, Delft University of Technology, Delft, Netherlands (2003)

[284] Slootweg, J.G., de Haan, S.W.H., Polinder, H., Kling, W.L.: General Model for Representing Variable Speed Wind Turbines in Power System Dynamics Simulations. IEEE Transactions on Power Systems 18(1), 144–151 (2003)

[285] Slootweg, J.G., Polinder, H., Kling, W.L.: Initialization of Wind Turbine Models in Power System Dynamics Simulations. In: Proceedings of the IEEE PowerTech Conference, Porto, Portugal (September 2001)
[286] Slootweg, J.G., Polinder, H., Kling, W.L.: Representing Wind Turbine Electrical Generating Systems in Fundamental Frequency Simulations. IEEE Transactions on Power Systems 18(4), 516–524 (2003)
[287] Soman, S.A., Khaparde, S.A., Pandit, S.: Computational Methods for Large Sparse Power Systems Analysis: An Object Oriented Approach. Kluwer Academic Publisher, Boston (2002)
[288] Song, Y.H., Johns, A.T.: Flexible AC Transmission System (FACTS). The Institute of Electrical Engineers, London (1999)
[289] Souza, A.C.Z., Cañizares, C.A., Quintana, V.H.: New Techniques to Speed up Voltage Collapse Computations using Tangent Vectors. IEEE Transactions on Power Systems 12(3), 1380–1387 (1997)
[290] Squires, R.B.: Economic Dispatch of Generation Directly From Power System Voltages and Admittances. AIEE Transactions 79(3), 1235–1244 (1961)
[291] Stagg, G.W., El-Abiad, A.H.: Computer Methods in Power System Analysis. McGraw-Hill, New York (1968)
[292] Stallman, R.M.: Free Software, Free Society: Selected Essays of Richard M. Stallman. Free Software Foundation, Boston (2002)
[293] Stallman, R.M.: GNU General Public License.plus 0.5em minus 0.4emFree Software Foundation (2007), `http://www.gnu.org/copyleft/gpl.html`
[294] Stevenson, W.D.: Elements of Power System Analysis. McGraw-Hill, New York (1975)
[295] Stifter, M., Milano, F.: An Example of Integrating Open Source Modelling Frameworks: The Integration of GIS in PSAT. In: Proceedings of the IEEE PES General Meeting, Calgary, Canada (July 2009)
[296] Stott, B.: Effective Starting Process for Newton-Raphson Load Flows. In: Proceedings of the Institute of Electrical Engineering, vol. 118(8), 983–987 (August 1971)
[297] Stott, B.: Review of load-Flow Calculation Methods. Proceedings of the IEEE 62(7), 916–929 (1974)
[298] Stott, B.: Power System Dynamic Response Calculations. Proceedings of the IEEE 67(2), 219–241 (1979)
[299] Stott, B., Alsac, O.: Fast Decoupled Load Flow. IEEE Transactions on Power Apparatus and Systems PAS-93(3), 859–869 (1974)
[300] Stott, B., Alsac, O., Monticelli, A.J.: Security Analysis and Optimization. Proceedings of the IEEE 75(12), 1623–1644 (1987)
[301] Stott, B., Marino, J.L., Alsac, O.: Review of Linear Programming Applied to Power System Rescheduling. In: Proceedings of the Power Industry Computer Application (PICA), 142–154 (1979)
[302] ABB Simpow, STRI, `http://www.stri.se`
[303] Sun, Y., Overbye, T.J.: Visualizations for Power System Contingency Analysis Data. IEEE Transactions on Power Systems 19(4), 1859–1866 (2004)
[304] Tate, J.E., Overbye, T.J.: A Comparison of the Optimal Multiplier in Polar and Rectangular Coordinates. IEEE Transactions on Power Systems 20(4), 1667–1674 (2005)
[305] Taylor, C.W., Lefebvre, S.: HVDC Controls for System Dynamic Performance. IEEE Transactions on Power Systems 6(2), 743–752 (1991)

[306] The Open Source Geospatial Foundation, http://www.osgeo.org
[307] Thomas, R.J., Barnard, R.D., Meisel, J.: The Generation of Quasi Steady-State Load Flow Trajectories and Multiple Singular Point Solutions. In: Proceedings of the IEEE PES Winter Meeting, New York, NY (1971)
[308] Thorp, J.S., Seyler, C.E., Phadke, A.G.: Electromechanical Wave Propagation in Large Electric Power Systems. IEEE Transactions on Circuits and Systems - I: Fundamental Theory and Applications 45(6), 614–622 (1998)
[309] Tinney, W.F., Enns, M.K.: Controlling and Optimizing Power Systems. IEEE Spectrum 11, 56–60 (1974)
[310] Tinney, W.F., Hart, C.E.: Power Flow Solution by Newton's Method. IEEE Transactions on Power Apparatus and Systems PAS-86, 1449–1460 (1967)
[311] Torres, G.L., Quintana, V.H.: An Interior Point Method for Nonlinear Optimal Power Flow using Voltage Rectangular Coordinates. IEEE Transactions on Power Systems 13(4), 1211–1218 (1998)
[312] Torres, G.L., Quintana, V.H.: Introduction to Interior-Point Methods. In: Proceedings of the Power Industry Computer Application (PICA), Santa Clara, CA (May 1999)
[313] Trefethen, L.N.: Numerical Linear Algebra. SIAM, Philadelphia (1997)
[314] Tremblay, O., Dessaint, L.A., Dekkiche, A.I.: A Generic Battery Model for the Dynamic Simulation of Hybrid Electric Vehicles. In: Proceedings of the IEEE Vehicle Power and Propulsion Conference, Arlington, TX, USA, September 2007, pp. 284–289 (2007)
[315] Tripathy, S.C., Durga Prasad, G., Malik, O.P., Hope, G.S.: Load-Flow Solutions for Ill-Conditioned Power Systems by a Newton-like Method. IEEE Transactions on Power Apparatus and Systems 101(10), 3648–3657 (1982)
[316] Tripathy, S.C., Rao, N.D.: A-Stable Numerical Integration Method for Transmission System Transients. IEEE Transactions on Power Apparatus and Systems 96(4), 1399–1407 (1977)
[317] Tylavsky, D.J., Crouch, P., Jarriel, L.F., Adapa, R.: Improved Power Flow Robustness for Personal Computers. IEEE Transactions on Industry Applications 28(5), 1102–1108 (1992)
[318] UCTE, Approximate Model of European Interconnected System, http://www.see.ed.ac.uk/~jbialek/Europe_load_flow/
[319] User-friendly Desktop Internet GIS, uDig, http://udig.refractions.net
[320] Unnewehr, L.E., Nasar, S.A.: Electric vehicle technology. Wiley, New York (1982)
[321] U.S. Congress. Senate, Digital Millennium Copyright Act, http://thomas.loc.gov/
[322] Uzunovic, E., Cañizares, C.A., Reeve, J.: Fundamental Frequency Model of Static Synchronous Compensator. In: Proceedings of the North American Power Symposium (NAPS), Laramie, Wyoming, 49–54 (October 1997)
[323] Uzunovic, E., Cañizares, C.A., Reeve, J.: Fundamental Frequency Model of Unified Power Flow Controller. In: Proceedings of the North American Power Symposium (NAPS), Cleveland, Ohio, 294–299 (October 1998)
[324] van Amerongen, R.: A General-Purpose Version of the Fast Decoupled Loadflow. IEEE Transactions on Power Systems 4(2), 760–770 (1989)
[325] Van Cutsem, T., Vournas, C.: Voltage Stability of Electric Power Systems. Springer Science, New York (1998)

[326] Vanfretti, L., Milano, F.: Application of the PSAT, an Open Source Software, for Educational and Research Purposes. In: Proceedings of the IEEE PES General Meeting, Tampa, USA (June 2007)

[327] Vanfretti, L., Milano, F.: The Experience of PSAT (Power System Analysis Toolbox) as a Free and Open Source Software for Power System Education and Research. International Journal of Electrical Engineering Education 43(3) (July 2009) (Accepted for future publication), http://www.uclm.es/area/gsee/Federico

[328] Varaiya, P.P., Wu, F.F., Chen, R.L.: Direct Methods for Transient Stability Analysis of Power Systems: Recent Results. Proceedings of the IEEE 73(5), 266–276 (1985)

[329] Venikov, V.A.: Theory of Similarity and Simulation with Applications to Problems in Electrical Power Engineering. Macdonald Techincal & Scientific, London (1969)

[330] Venikov, V.A.: Transient Processes in Electrical Power Systems. Mir Publishers, Moscow (1977)

[331] Venkataraman, S., Khammash, M.H., Vittal, V.: Analysis and Synthesis of HVDC Controls for Robust Stability of Power Systems. IEEE Transactions on Power Systems 10(4), 1933–1938 (1995)

[332] Venkatasubramanian, V., Schättler, H., Zaborszky, J.: Dynamics of Large Constrained Nonlinear Systems-A Taxonomy Theory. Proceedings of the IEEE 83(11), 1530–1560 (1995)

[333] Verboomen, J., Van Hertem, D., Schavemaker, P., Kling, W., Belmans, R.: Phase shifting transformers: Principles and applications. In: International Conference on Future Power Systems, Amsterdam, Netherlands (November 2005), http://www.esat.kuleuven.be/electa/publications/fulltexts/pub_1502.pdf

[334] Virtanen, P., Gouillart, E.: Numpy and Scipy Documentation. Scipy.org, http://docs.scipy.org/doc

[335] Vittal, V.: Electric Energy Systems - Analysis and Operation. IEEE Power & Energy Magazine 7(5), 75–76 (2009) (book review)

[336] Vournas, C.D., Nikolaidis, V.C., Tassoulis, A.A.: Postmortem Analysis and Data Validation in the Wake of the 2004 Athens Blackout. IEEE Transactions on Power Systems 21(3), 1331–1339 (2004)

[337] Vournas, C.D., Pai, M.A., Sauer, P.W.: The Effect of Automatic Voltage Regulation on the Bifurcation Evolution in Power Systems. IEEE Transactions on Power Systems 11(4), 37–43 (1996)

[338] Vournas, C.D., Potamianakis, E.G., Moors, C., Van Cutsem, T.: An Educational Simulation Tool for Power System Control and Stability. IEEE Transactions on Power Systems 19(1), 48–55 (2004)

[339] Vournas, C.D., Sakellaridis, N.G.: Region of Attraction in a Power System with Discrete LTCs. IEEE Transactions on Circuits and Systems - I: Fundamental Theory and Applications 53(7), 1610–1618 (2006)

[340] Vournas, C.D.: Scientific Coordinator, Software Development for Voltage Stability Analysis, National Technical University of Athens, Greece, project 96 SYN 95

[341] Vu, K.T., Liu, C.-C., Taylor, C.W., Jimma, K.M.: Voltage instability: Mechanisms and Control Strategy. Proceedings of the IEEE 83(11), 1442–1455 (1995)

[342] Extensible Markup Language, W3.ORG, http://www.w3.org/XML
[343] Wasynczuk, O., Man, D.T., Sullivan, J.P.: Dynamic Behavior of a Class of Wind Turbine Generators during Random Wind Fluctuations. IEEE Transactions on Power Apparatus and Systems 100(6), 2837–2845 (1981)
[344] Watson, L.T., Billups, S.C., Morgan, A.P.: HOMPACK: A Suite of Codes for Globally Convergent Homotopy Algorithms. ACM Transactions on Mathematical Software 13, 281–310 (1987)
[345] Watson, N., Arrillaga, J.: Power Systems Electromagnetic Transients Simulation. The Institution of Engineering and Technology, London (2003)
[346] Weber, J.D., Overbye, T.J.: Voltage Contours for Power System Visualization. IEEE Transactions on Power Systems 15(1), 404–409 (2000)
[347] Weidlich, A.: Engineering Interrelated Electricity Markets - An Agent-Based Computational Approach. Physica-Verlag, Heidelberg (2008)
[348] Wiegmann, D.A., Overbye, T.J., Hoppe, S.M., Essemberg, G.R., Sun, Y.: Human Factors Aspects of Three-Dimensional Visualization of Power System Information. In: Proceedings of the IEEE PES General Meeting, Montreal (July 2006)
[349] Working Group H-7 of the Relaying Channels Subcommittee of the IEEE Power System Relaying Committee, Synchronized Sampling and Phasor Measurements for Relaying and Control. IEEE Transactions on Power Delivery 9(1), 442–452 (1994)
[350] Working Group on a Common Format for Exchange of Solved Load Flow Data, Common Format for the Exchange of Solved Load Flow Data. IEEE Transactions on Power Apparatus and Systems 92(6), 1916–1925 (1973)
[351] Wu, F., Zhang, X., Ju, P.: Small Signal Stability Analysis and Control of the Wind Turbine with the Direct-Drive Permanent Magnet Generator Integrated to the Grid. Electric Power Systems Research 79(12), 1661–1667 (2009)
[352] Wu, Y., Debs, A.S., Marsten, R.E.: A Direct Nonlinear Predictor-Corrector Primal-Dual Interior Point Algorithm for Optimal Power Flow. IEEE Transactions on Power Systems 9(3), 876–883 (1994)
[353] Xie, K., Song, Y.-H., Stonham, J., Yu, E., Liu, G.: Decomposition Model and Interior Point Methods for Optimal Spot Pricing of Electricity in Deregulation Environments. IEEE Transactions on Power Systems 15(1), 39–50 (2000)
[354] Xu, W., Mansour, Y.: Voltage Stability Analysis Using Generic Dynamic Load Models. IEEE Transactions on Power Systems 9(1), 479–493 (1994)
[355] Yao-Nan-Yu: Electric Power System Dynamic. Academic Press, New York (1983)
[356] Zárate-Miñano, R., Conejo, A.J., Milano, F.: OPF-based Security Redispatching Including FACTS Devices. IET Generation, Transmission & Distribution 2(6), 821–833 (2008)
[357] Zhang, F., Cheng, C.S.: A Modified Newton Method for Radial Distribution System Power Flow Analysis. IEEE Transactions on Power Systems 12(1), 389–397 (1997)
[358] Zhong, J., Bhattacharya, K.: Toward a Competitive Market for Reactive Power. IEEE Transactions on Power Systems 17(4), 1206–1215 (2002)
[359] Zhou, M.: InterPSS, http://www.interpss.org
[360] Zhou, M., Zhou, S.: Internet, Open-source and Power System Simulation. In: Proceedings of the IEEE PES General Meeting, Montreal, Quebec (June 2007)

[361] Zhu, W., Mohler, R., Spee, R., Mittelstadt, W., Maratukulam, D.: Hopf Bifurcations in a SMIB Power System with SSR. IEEE Transactions on Power Systems 11(3), 1579–1584 (1996)

[362] Zhu, Y., Tomsovic, K.: Development of Models for Analyzing the Load-Following Performance of Microturbines and Fuel Cells. Electric Power Systems Research 62(1), 1–11 (2002)

[363] Zimmerman, R.D., Murrillo-Sánchez, C.E.: MatPower: A Matlab Power System Simulation Package. User's Manual, Power System Engineering Research Center, Cornell University (2007), version 3.2, http://www.pserc.cornell.edu/matpower/matpower.html

[364] Zobian, A., Ilić, M.D.: Unbundling of Transmission and Ancillary Services. Part I: Technical Issues. IEEE Transactions on Power Systems 12(2), 539–548 (1997)

Index

D

E

F

G

GPSR Compliance

The European Union's (EU) General Product Safety Regulation (GPSR) is a set of rules that requires consumer products to be safe and our obligations to ensure this.

If you have any concerns about our products, you can contact us on ProductSafety@springernature.com

In case Publisher is established outside the EU, the EU authorized representative is:

Springer Nature Customer Service Center GmbH
Europaplatz 3
69115 Heidelberg, Germany

Batch number: 10755461

Printed by Printforce, the Netherlands